Springer Series in Computational Physics

W0106808

Editors:
W. Beiglböck,   H. Cabannes,
H. B. Keller,   J. Killeen,   S. Orszag

Demetri P. Telionis

Unsteady Viscous Flows

With 132 Illustrations

Springer-Verlag
New York   Heidelberg   Berlin

Demetri P. Telionis

Department of Engineering Science and Mechanics
Virginia Polytechnic Institute
  and State University
Blacksburg, Virginia 24061
USA

Editors:

Wolf B

Institut für Angewandte Mathematik
Universität Heidelberg
Im Neuenheimer Feld 5
D-6900 Heidelberg 1
Federal Republic of Germany

Stephen A. Orszag

Department of Mathematics
Massachusetts Institute of Technology
Cambridge, Massachusetts 02139
USA

John Killeen

Lawrence Livermore Laboratory
P. O. Box 808
Livermore, California 94551
USA

Henri Cabannes

Mécanique Théoretique
Université Pierre et Marie Curie
Tour 66, 4, Place Jussieu
F-75005 Paris
France

H. B. Keller

Applied Mathematics 101-50
Firestone Laboratory
California Institute of Technology
Pasadena, California 91125
USA

Library of Congress Cataloging in Publication Data
Telionis, Demetri.
  Unsteady viscous flows.
  (Springer series in computational physics)
  Includes bibliographical references and index.
  1. Viscous flow. I. Title. II. Series.
QA929.T44   531'.0533   81-816AACR2

9 8 7 6 5 4 3 2 1

ISBN-13: 978-3-642-88569-3   e-ISBN-13: 978-3-642-88567-9
DOI: 10.1007/978-3-642-88567-9

This book is dedicated to
Professor W. R. Sears

Preface

Most of the fundamental concepts of unsteady viscous flows have been known since the early part of the century. However, the past decade has seen an unprecedented number of publications in this area. In this monograph I try to connect materials of earlier contributions and synthesize them into a comprehensive entity. One of the main purposes of a monograph, in my opinion, is to fit together in a comprehensive way scattered contributions that provide fragmented information to the readers. The collection of such contributions should be presented in a unified way; continuity of thought and logical sequence of the presentation of ideas and methods are essential. The reader should be able to follow through without having to resort to other references, something that is unavoidable in the case of a research paper or even a review paper.

Many of the solutions discussed in the literature address specific practical problems. In fact, in the process of collecting information, I discovered independent lines of investigations, dealing with the same physical problem, but inspired by different practical applications. For example, I found that two groups of investigators have been studying independently the response of a viscous layer to a harmonic external disturbance. One group is concerned with mass transport and the transport of sediment over the bottom of the ocean, and the other is interested in the aerodynamics of lifting surfaces in harmonically changing environments. Interest is concentrated here on the behavior of viscous fluids and, in particular, their response to time-dependent external conditions. This monograph does not cover any specific discipline or application. The approach is computational, and I like to think of this term as encompassing any method producing numerical results. Even a closed-formed solution may qualify as a computational method, since today, evaluating such a solution at N points may take more computer time and space than solving the problem of finite differences on an N-point grid. Descriptions of the physical significance of the results are also included, while possible practical applications and relevant engineering problems are only mentioned.

An important tool in computational physics is perturbation theory. This theory gives valuable insight into the actual behavior of the mathematical models, information that is vital for the construction of efficient algorithms.

Due to the complexity of the mathematical problems under consideration, very few exact solutions exist. Perturbation methods can be used to generate approximate numerical solutions, but most importantly, they are called upon to complement the purely numerical calculations in regions where particular features of the solution make it difficult or impossible to generate numerical results. Perturbation methods thus provide unambiguous and well-accepted test cases against which numerical results can be compared and, quite often, provide the necessary initial numerical values to start a purely numerical calculation. Numerical analysis and perturbation methods are therefore inseparable.

Working on review articles and notes of a graduate course I have been teaching at VPI and SU, I found that the process was extremely helpful in understanding and organizing the material in my mind. I soon felt the need and the challenge to expand my notes into the form of a monograph which I, and perhaps others, can use as a text in a related graduate course. I gained valuable experience on how to present the material by discussing various topics with graduate students. The diversity of their backgrounds helped me identify some of the potential difficulties of my audiences. This work is therefore addressed to the novices in the area of unsteady viscous flows who have a reasonable understanding of classical fluid mechanics and, in particular, boundary-layer theory. In writing this monograph I had in mind the second-year graduate student, or a young scientist in industry, who is about to begin work on unsteady viscous fluid mechanics. He will find here the fundamental concepts adequately explained. Advanced topics are also discussed, but details can be obtained from the original publications given in the references.

A natural separation of the material could have been based on particular methods of computation. This would be well justified since this monograph is a volume in the Springer Series in Computational Physics. Although the emphasis is certainly on computational aspects, these are viewed as tools rather than the ultimate goal. To identify the physical characteristics of the problem, the chapters are instead devoted to different types of unsteadiness as, for example, transient or oscillatory flows, or different types of flows as, for example, attached laminar flows, turbulent boundary layers, or separated flows. Each chapter begins with an introductory section that describes briefly the physical problem and defines the mathematical boundary conditions. The first few sections are devoted to the basic principles and are written with the true novice in mind. Some of the examples worked out are classical and are included in all popular texts on boundary-layer theory and viscous flows. The experienced reader may find the level of the first few sections in each chapter elementary, in comparison with the ones that follow. I find that detailed expanded descriptions are very helpful to students attending a course in this or related areas. In subsequent sections more advanced topics are discussed and references are given to very recent contributions. Computer listings are also included, with short introductions and a brief guide for the user.

The material included in this monograph of course carries the strong bias of the author. The title is deceiving and overly ambitious. A more appropriate title, more faithful to the material actually included, would be "Unsteady laminar and turbulent, attached and separated boundary-layer flows with transient and periodic disturbances excluding structural interactions, acoustic phenomena, compressibility effects, etc." The introduction provides a more detailed account of the topics discussed. Topics that perhaps could have been included but were left out due to bias of the author are also enumerated.

Working on a monograph like this volume, the author is asked to describe mostly the work of others. This is a great responsibility that proved to be a very agonizing and frustrating experience. I found out very soon that it is impossible to do justice to all the important contributions. Some of the discussions are inevitably brief, and the emphasis is on whatever I considered important. I believe that some of the authors whose works are described here may not agree with my interpretation. One of them, who has seen a chapter of my manuscript, found my discussion acceptable, but politely expressed his disapproval by saying that this would not be "exactly" the way he would have done it. Moreover, some important developments are mentioned only in passing and others are not even referenced. I do not doubt that I have unwittingly overlooked others. Because this monograph was planned to be brief, citing all the contributions in the area would make it look like a reference book. Besides, an effort to include all the existing literature would make the accidental omissions even more embarrassing to the author. The choices I made of course simply reflect my personal taste, the sequence with which I was exposed to the work of others, and the areas that I have worked in and, therefore, gained better understanding.

My most frustrating experience has been the realization of inadequacy and incompleteness that grows as the work progresses. This feeling is reinforced by the ever-increasing number of research findings that continue to appear. As time passes, it becomes more and more difficult to let the manuscript go to the printer. At the end, one realizes that a manuscript is never finished. It is just "abandoned."

My interest in unsteady aerodynamics was stimulated at Cornell University by Professor W. R. Sears. I consider this man to have played the most important role in my graduate education and in the shaping of my subsequent technical interests and professional attitudes and ethics. To him I dedicate this monograph. Another professor of mine, Professor N. Athanassiadis of the National Technical University of Athens has also greatly influenced me. I am very grateful to both.

My work in the area was supported by grants from the National Science Foundation, the Air Force Office of Scientific Research, and the Army

Research Office, and was monitored by Dr. George Lea, the late Paul Thurston, Captain Westy Smith, and Dr. Robert Singleton. Their interest in my work, their helpful suggestions, and their support are gratefully acknowledged.

Graduate students have read chapters of my manuscripts and made suggestions, and some have helped me check computer programs. Among them, I am particularly indebted to Maria Romaniuk, Thomas Mezaris, and Takis Konstantinopoulos. I am also very thankful for the excellent typing job of Marlene Taylor, Vanessa McCoy, and June Harrison. The largest part of the typing has been the work of Peggy Epperly, whom I truly admire for her excellent work and her ability to decipher my scribblings.

I was impressed with the warm interest and meaningful suggestions of Professor Beiglböck, the Editor of this Springer Series. I am thankful to him and the reviewers for their constructive criticism. I also feel indebted to the staff of Springer–New York.

In the last stages of the work some colleagues have looked at some of my chapters and made useful suggestions. Among these I would like to mention Drs. W. J. McCroskey and L. W. Carr, and Professors J. C. Williams, S.-F. Shen, and W. R. Sears. However, I feel I should single out Drs. U. B. Mehta and L. van Dommelen who have done a very thorough job reviewing the first few chapters and the last chapter, respectively. Finally, Miss H. L. Reed has scrutinized the galleys, catching an embarrassing number of errors. I thank them all deeply.

Contents

Introduction

Unsteady flows are those whose properties depend on time if referenced with respect to an Eulerian frame. The peculiar distinction between steady and unsteady motions in fluid mechanics has no counterpart in solid-mechanics problems. In solid mechanics, steady motion would be the trivial case of motion with no acceleration. In fluid mechanics, a very large class of problems which involves extensive acceleration regions is classified as steady flows, if in the Eulerian system of reference all properties are independent of time.

There are two large groups of problems in unsteady viscous flows. In the first, we study the response of viscous flows to dynamic disturbances introduced extraneously. In internal fluid mechanics, this could be a periodic change of the driving pressure gradient or a dynamic change of the shape of the containing wall. In external fluid mechanics, unsteadiness may be introduced by changing the shape or the orientation of rigid bodies or by disturbing the oncoming stream. Mathematically, all these dynamic changes appear in the boundary conditions. The appropriate mathematical models then can be used to predict the response of the flow to the unsteady external disturbances.

In the second group of problems, we study unsteady fields that are self-generated and self-sustained. The solid boundaries of such flows as well as the conditions of the oncoming flow are time-independent, and yet, unsteadiness sets in by itself. Such phenomena have been predicted with some success by analytical methods. A straightforward approach to the problem has been possible via the solution of the full Navier–Stokes equations. Typical problems of the second group are the instability and transition of boundary layers, free shear layers and jets, the shedding of vortices, the development of unsteady wakes, the unsteady interaction of a shock wave and a boundary layer and of course all problems involving turbulence. What seems to be unexpected and most exciting is the fact that quite often the natural unsteadiness is well organized. For example, the wakes of blunt bodies contain coherent large scale vortices in a relatively organized manner. The frequency of shedding vortices over circular cylinders is proportional to the oncoming velocity for a wide range of the Reynolds number. The study of the spontaneous generation of unsteadiness has been termed *stability theory* and the sustained unsteadiness has been known to fluid mechanicists as *turbulence*.

This monograph is devoted exclusively to the first class of problems, namely, the response of flows to extraneously introduced transient or periodic disturbances. The response of turbulence to external disturbances is also discussed. However, the emphasis is on the organized part of the motion.

Unsteady phenomena are worth studying only if they depart substantially from the quasi-steady state. It is important to note here that potential flows are always, to some extent, quasi-steady; their kinematic properties at each instant satisfy the steady state governing equations. In terms of the potential function, ϕ, the governing equation for steady or unsteady incompressible flow is simply $\nabla^2 \phi = 0$. The flow can be identified as unsteady only if the pressure field is examined. The instantaneous velocity field and the instantaneous streamlines at time t_0 are identical to those of a steady flow with boundary conditions the instantaneous conditions of the unsteady flow at t_0.

It is remarkable that in most cases the departure from the quasi-steady flows is due to viscous phenomena. The distinction between viscous and inviscid flows sprang from the mathematical models and their approximations. No flow in real life can be truly inviscid. However, the mathematical models and the experiments designed to prove their practical value, help the investigators to analyze the competing effects. In this sense it is possible to identify viscosity as the triggering effect or the actual cause of certain phenomena. At the beginning of this century and following the pioneering contribution of Prandtl (1904), it was recognized that even though viscosity is for all practical purposes extremely small, its effect essentially controls basic properties of the flow, as for example the size and shape of the wake. In unsteady flow again, viscosity plays a very significant role. It is due to shearing effects and their subsequent development, that natural instabilities grow into small or large scale turbulence. For externally driven unsteady flows, the viscous effects seem to control the response of the entire flow field. Unsteady separation, for example, responds with some inertia to an abrupt increase in the adverse pressure gradient. This leads the flow over an airfoil into a domain of attached flow, even at large angles of attack. Its lift may thus increase considerably, beyond its quasi-steady maximum value. The situation is depicted schematically in Fig. 1. Sketches *a* and *b* correspond to steady flow over an airfoil at angles of attack α_1 and α_2 respectively. The angle α_2 is large enough to bring the airfoil into full stall, whereby the entire suction surface is covered with separated flow. However, if the change from α_1 to α_2 is achieved dynamically, the flow remains attached for a while as shown in sketch *c*, leading to a much larger value of lift. This condition of course cannot be sustained for long; the airfoil eventually stalls, or if the angle of attack decreases again, its lift enters into a hystersis loop.

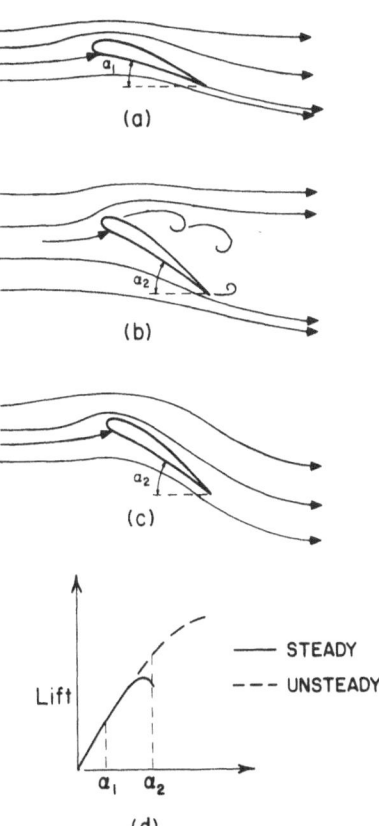

Fig. 1 Schematic of flow about an airfoil. a and b, steady flow; c, unsteady flow; d, plot of lift versus angle of attack, α.

Examples of unsteady flows are many. In fact, there is no actual flow situation, natural or artificial, that does not involve some unsteadiness. It is emphasized that here the term "unsteadiness" implies the response of the flow to extraneously introduced changes and not the natural unsteadiness we described before. For a long time, the flows in all engineering applications were arbitrarily assumed to be steady. For example, the lifting characteristics of an airfoil or the drag characteristics of a blunt body, were problems attacked both analytically and experimentally as steady problems. It is well known, however, that in practice such devices encounter smooth or sudden changes in their aerodynamic environment. In many other engineering applications, unsteadiness is an integral part of the problem. The helicopter rotor, the cascades of blades of turbomachinery, and the ship propeller normally operate in an unsteady aerodynamic environment.

For a hovering helicopter rotor, it is only the wake of the preceeding blades that may influence the environment of a rotating blade. For an advancing disk, however, the situation is drastically different. A blade's rotational motion is superimposed on the steady oncoming free stream. The

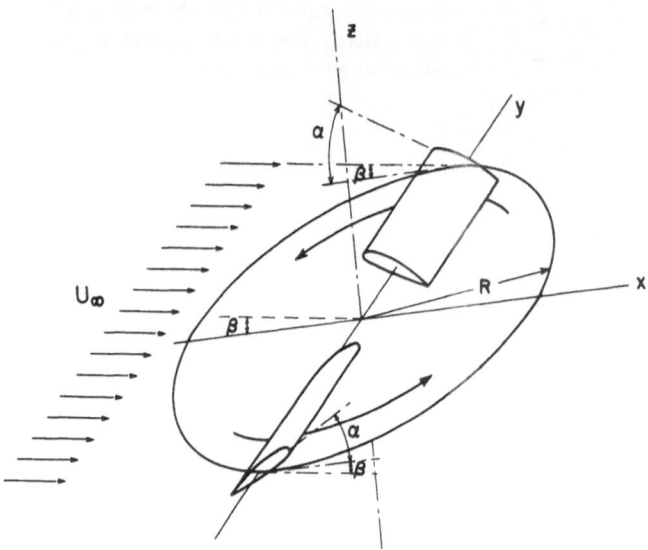

Fig. 2 An idealized helicopter rotor consisting of two blades at a fixed and uniform angle of attack, α.

apparent outer flow for a blade contains a periodic disturbance in both magnitude and direction. The frequency of the disturbance equals the frequency of the rotor. A forward component of the thrust is obtained by tilting the lifting disk forward. This introduces an extra complexity to the problem. The situation is shown in Fig. 2. The frame of references x, y, z is attached to the lifting disk which is inclined with respect to the oncoming stream, U_∞, by an angle β. The blades are given an angle of attack, α, with respect to the disk plane, the plane x, y. The blades rotate with an angular velocity ω which generates tip speeds ωR, considerably larger than U_∞. For an advancing blade, the effective angle of attack is decreased to $\alpha - \beta$, whereas for a retreating blade the effective angle of attack becomes $\alpha + \beta$. A blade therefore encounters an oncoming stream which fluctuates in magnitude and in two directions. Due to these periodic changes a blade may periodically enter domains of dynamic stall.

Another typical example of truly unsteady phenomena is the flow through turbomachinery cascades. A cascade of blades operates in the wake of the previous cascade as shown schematically in Fig. 3. This is periodically contaminated by low speed turbulent regions that emanate at the trailing tips of the blades of the previous cascade. A small change in the angle of attack is also induced. The frequency of this periodicity depends on the relative speed of the two cascades. The downstream row of blades is thus forced to cut through an oncoming flow, which contains spatially periodic changes in the magnitude and direction of the flow, as well as in the turbulence level. The frequency of the periodic disturbances in this case is the relative angular

Fig. 3 Schematic representation showing the flow downstream of one cascade which becomes the oncoming stream for the following cascade.

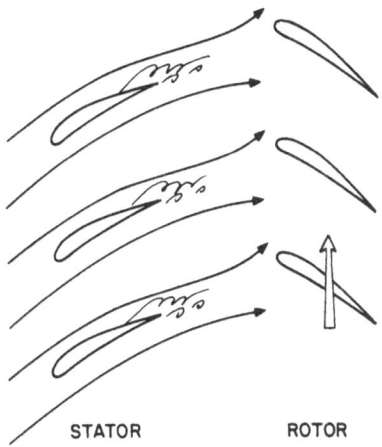

STATOR   ROTOR

velocity of the two cascades multiplied by the number of blades. Such periodic disturbances often lead to rotating stall. Due to some inhomogeneity in the flow, separation over one or two blades, say blades B and C in Fig. 4, is advanced towards the leading edge, and their wakes are expanded to cover their entire suction side. The blades are stalled and the passages between them are practically blocked. This forces the oncoming flow to alter its direction in its effort to bypass the blocked region. The flow is turned downward on blade A, more aligned with its shape, but upward on blade D. In other words, as shown in Fig. 4 for $t = t_2$ the angle of attack decreases on blade A but increases on blade D for the instant $t = t_1$. As a result, blade D soon stalls but the conditions lead to gradual reattachment of the flow over blade B ($t = t_2$ in Fig. 4). The events are repeated and eventually the blades D and E stall ($t = t_3$ in Fig. 4). The contaminated area thus rotates bringing momentarily into stall different sections of the cascade. The phenomenon is known as "rotating stall" and is due to unsteady separation and wake. Other types of unsteady phenomena are also encountered in the aerodynamics of turbomachinery. Of great practical interest is also the situation whereby the oncoming flow encounters the engine axis at an angle. The mouth of the intake then "shadows" a portion of the fan (the first stage of blades). A nonuniformity of this type generates periodic disturbances with a frequency equal to the frequency of rotation.

Another class of important engineering problems involves the study of unsteady forces on bluff bodies as for example buildings and structures in the atmosphere, or piles and cables submerged in rivers or the sea. A classical problem which finds immediate application in engineering designs is that of a circular cylinder. Unsteadiness in this case forces the shedding frequency to shift and to "lock on" the forcing frequency. This is essentially an interaction between the natural frequency of the hydrodynamics and the frequency imposed externally. Such unsteady flow problems are necessary

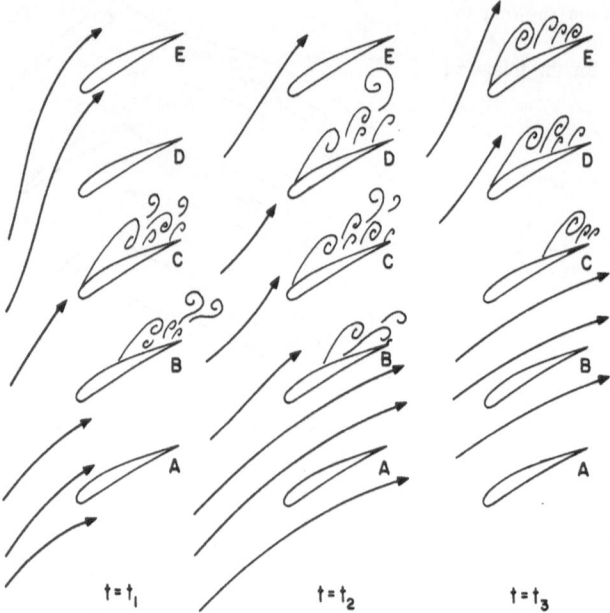

Fig. 4 A cascade in rotating stall. The three sketches represent the same cascade of blades at three different time intervals; $t_1 < t_2 < t_3$.

input for the study of the dynamic deformation characteristics of structures. Flutter is the spontaneous vibration of a structural element exposed to a fluid stream; it is the result of interaction between the aerodynamics and the mechanical vibrations of the element and is a very significant area of study in aircraft and missile design. Flutter is also of great concern in the design and operation of many other engineering devices. Elements like the blades of gas or liquid turbines, blades of propellers, hydrofoils, transmission lines, the tubes of heat exchangers, and even suspension bridges are subject to flutter and a related phenomenon, buffeting.

Unsteady viscous flow phenomena are very common in geophysical flows. For example, the periodic disturbances of ocean surface waves generate secondary viscous flows on the bottom of the ocean which give rise to sand ripples. Many examples of significant unsteady viscous effects can be found in biological flows as well. The pulsating flow of blood in blood vessels is a typical example: the problem of separation in expanding sections of the arteries or in regions of bifurcation has been of great interest.

Many other problems of fluid mechanics involve some type of unsteadiness but are essentially independent of viscous effects. Typical examples are the theory of sound propagation and the propagation of various types of waves.

The first unsteady flow problem to stimulate human interest was the flow of air about the wings of a flying bird. Flying fascinated man for thousands of years, but the first to look carefully into the mechanisms of the process was Leonardo da Vinci. Time-dependent viscous flows were studied in the 19th century by Stokes and Rayleigh long before Prandt (1904) laid the foundations of boundary-layer theory at the turn of the century. However, the need to consider unsteady viscous flows as an entity, as a separate chapter of fluid mechanics, arose in the sixties. Classical volumes edited by Dryden and Von Kármán (1960), Moore (1964), and Rosenhead (1964) contain articles on the topic by Stewartson, Rott, and Stuart respectively. The significance of unsteady viscous flow problems was recognized by the International Union of Theoretical and Applied Mechanics (I.U.T.A.M.), which in 1970 organized an international symposium on unsteady boundary layers (Eichelbrenner, 1972). A number of national and international meetings on unsteady aerodynamics followed as shown in Table 1 which is taken from McCroskey (1977), edited and supplemented with a few recent conferences. Review articles in the area appeared, most of which have been keynote presentations at conferences listed in Table 1. Chronologically listed, review articles concerned exclusively or partly with unsteady viscous flows are due to Stuart (1972), Riley (1975), Telionis (1975), Wirz (1975), McCroskey (1977), Telionis (1977, 1979) and Shen (1978).

In this monograph emphasis is given to the principles of the phenomena involved. There is very little discussion of applications to particular engineering problems of the type described earlier in this section. The reader may trace back contributions concerned with practical engineering problems in the proceedings of the conferences listed in Table 1.

Some types of problems that would qualify as unsteady viscous flow problems according to the definition adopted in this introduction have not been included in this monograph. Flutter is a classical example. Flutter is a discipline in itself and excellent detailed contributions have been written about it. Another example is the formation of wakes. Wakes are inherently unsteady. There is evidence that the unsteady motion in the wake is organized in the form of large scale eddies, at least for medium Reynolds numbers. Moreover, experimental evidence indicates that unsteadiness tends to organize the wake vortices even at higher Reynolds numbers. However, very little has been done to study analytically the motion in the wake, and, to the knowledge of the author, no attempt has been made to model the response of wakes to external disturbances. This topic is therefore discussed only briefly in the chapter on separation. If compressibility is introduced, the problem becomes even more complex. For example, a classical phenomenon of great importance is the unsteady interaction of a shock wave and a boundary layer. Compressibility effects have not been addressed at all in this monograph. In sum, the reasons for excluding a specific topic are either the lack of an analytical literature, adequate for synthesis into a concise body, or the existence of literature which extensively covers areas traditionally considered to be self-contained. Examples of the two types of topics are wakes and flutter, respectively.

Table 1 Meetings on unsteady aerodynamics

RECENT RESEARCH OF UNSTEADY BOUNDARY LAYERS
  IUTAM International Symposium,
  Quebec, Canada, June 1971
FLUID DYNAMICS OF UNSTEADY, THREE-DIMENSIONAL AND
SEPARATED FLOWS
  Project SQUID Workshop
  Atlanta, Georgia, June 1971
UNSTEADY FLOWS IN JET ENGINES
  Project SQUID Workshop
  UAR Laboratories, Connecticut, July 1974
UNSTEADY AERODYNAMICS
  Sponsored by the Air Force Office of Scientific Research and the
  University of Arizona
  Tucson, Arizona, January 1975
UNSTEADY PHENOMENA IN TURBOMACHINERY
  AGARD PEP Panel Meeting
  Monterey, California, September 1975
UNSTEADY AERODYNAMICS
  AGARD FDP Panel Symposium
  Ottawa, Canada, September 1977
UNSTEADY TURBULENT BOUNDARY LAYERS AND SHEAR FLOWS
  EUROVISC Workshops
  Toulouse, France, January 1977
  Liverpool, England, April 1979
UNSTEADY FLUID DYNAMICS
  Sponsored by the ASME Fluid Mechanics Committee and presented at
  the 1978 ASME Winter Annual Meeting
  San Francisco, California, December 1978
SPECIAL COURSE ON UNSTEADY AERODYNAMICS
  AGARD Special Course
  Rhode-St-Genèse, Belgium, March 1980
UNSTEADY SEPARATION AND REVERSED FLOW IN EXTERNAL FLOW
DYNAMICS
  EUROMECH 155,
  Marseille, France, October 1980
UNSTEADY TURBULENT SHEAR FLOWS
  IUTAM International Symposium
  Toulouse, France, May 1981

It is hoped that after studying the material of this book, the reader will become able to apply the same principles and techniques to other practical problems not specifically mentioned here, or even envisioned by the author.

REFERENCES

Dryden, H. L., and Kármán, Th. von., eds., 1960. *Advances in Applied Mechanics*, Academic Press, New York.
Eichelbrenner, E. A., ed., 1972. *Recent Research of Unsteady Boundary Layers*, Laval University Press, Quebec, Canada.

McCroskey, W. J., 1977. *J. Fluids Eng.* **99**, 8–38.

Moore, F. K., ed., 1964. *Theory of Laminar Flows*, Princeton University Press.

Prandtl, L., 1904. "Uber Flussigkeitsbewegung bei sehr kleiner Reibung," III, Int. Congress of Mathematics, Heidelberg, 484–491.

Riley, N., 1975. *SIAM Review*, **17**, 274–297.

Rosenhead, L., ed., 1964. *Laminar Boundary Layers*, Oxford University Press.

Shen, S. F., 1978. In *Advances in Applied Mechanics*, ed., **18**, 177–220.

Stuart, J. T., 1972. In *Recent Research of Unsteady Boundary Layers*, ed. Eichelbrenner, E. Q., Laval University Press, Quebec, Canada **2**, 1–59.

Telionis, D. P., 1975. In *Unsteady Aerodynamics*, ed. Kinney, R. B., **1**, 155–190.

Telionis, D. P., 1977. In AGARD Symposium on Unsteady Aerodynamics, Ottawa.

Telionis, D. P., 1979. *J. Fluids Eng.*, **101**, 29–43.

Wirz, H. J., 1975. In *Progress in Numerical Fluid Dynamics*, *Lecture Notes in Physics*, ed. Wirz, H. J., **41**, 442–476, Springer, New York.

CHAPTER 1

Basic Concepts

1.1 Introduction

This chapter could also be termed "general introduction" since it contains material of basic, introductory character, which is well documented and relatively familiar to fluid dynamicists. Elementary concepts of methods of computation are also included, which should be familiar to the reader. The material is therefore presented parsimoniously with references to other sources for more details. Most of the sections in this chapter contain basic ideas and equations that will serve as the starting point for the derivations described in the following chapters.

Simple concepts of numerical analysis are introduced in Chapter 2 where the character of the differential equations is discussed. Special attention is directed toward the parabolic features of the mathematical models, and recent contributions on the topic are described in detail. In particular, numerical schemes of solution are proposed that allow integration to proceed through regions of partially reversed flow.

Throughout this monograph, it will be convenient to use the same symbols to denote dimensional and dimensionless quantities. Since most of the equations are expressed in terms of dimensionless quantities, we denote such quantities with plain symbols. Unless otherwise stated, the same symbols with an asterisk denote a dimensional quantity. For example, if U and L are a characteristic constant velocity and a typical length of the problem, the dimensional and dimensionless velocity and space coordinates are defined by the equations

$$u = \frac{u^*}{U}, \qquad x = \frac{x^*}{L} . \tag{1.1.1}$$

This notation will hold mainly for dependent and independent variables. Moreover, if all the quantities that appear in an equation are dimensional and there is no danger of confusion, plain symbols will be used to denote dimensional quantities as well. Such equations will be identified by an asterisk before the equation number. For example, the equation of heat conduction in its dimensional form is

$$\frac{\partial T^*}{\partial t^*} = \alpha \frac{\partial^2 T^*}{\partial y^{*2}} , \tag{1.1.2}$$

where T^* is the temperature, t^* and y^* are time and a space coordinate, and α is the thermal diffusivity. Dimensionless quantities can be defined, if typical temperature, time, and length are available, as T_0, t_0, and L, respectively.

$$T = \frac{T^*}{T_0}, \qquad t = \frac{t^*}{t_0}, \qquad y = \frac{y^*}{L}. \tag{1.1.3}$$

The dimensionless form of Eq. (1.1.2) then becomes

$$\frac{\partial T}{\partial t} = A \frac{\partial^2 T}{\partial y^2}, \tag{1.1.4}$$

where A is a dimensionless parameter equal to $\alpha t_0 / L^2$. If all the symbols in the equation are dimensional, then the symbols without asterisks may also be used to denote the dimensional quantities as described before, provided an asterisk precedes the number of the equation,

$$\frac{\partial T}{\partial t} = \alpha \frac{\partial^2 T}{\partial y^2}. \tag{*(1.1.5)}$$

No asterisk will be necessary for dimensional parameters like viscosity μ, conductivity k, etc., or for typical lengths, velocities, etc., as, for example, L, U, etc.

1.2 The Governing Equations

The motion of a fluid is governed by the conservation laws of mass, momentum, and energy (Schlichting, 1968; White, 1974):

$$\frac{\partial \rho}{\partial t} + u_j \frac{\partial \rho}{\partial x_j} + \rho \frac{\partial u_j}{\partial x_j} = 0, \tag{*(1.2.1)}$$

$$\rho \frac{\partial u_i}{\partial t} + \rho u_j \frac{\partial u_i}{\partial x_j} = \rho g_i - \frac{\partial p}{\partial x_i} + \frac{\partial}{\partial x_j} \left[\mu \left(\frac{\partial u_i}{\partial x_j} + \frac{\partial u_j}{\partial x_i} + \delta_{ij} \lambda \frac{\partial u_l}{\partial x_l} \right) \right], \tag{*(1.2.2)}$$

$$\rho \frac{\partial h}{\partial t} + \rho u_j \frac{\partial h}{\partial x_j} = \frac{\partial p}{\partial t} + u_j \frac{\partial p}{\partial x_j} + \frac{\partial}{\partial x_j} \left(k \frac{\partial T}{\partial x_j} \right) + \Phi, \tag{*(1.2.3)}$$

where u_i, p, ρ, T, and h are the velocity vector, pressure, density, temperature, and enthalpy, respectively; g_i is the acceleration of gravity; μ and λ are the first and second coefficients of viscosity, respectively; k is the coefficient of heat conduction; t and x_j are the independent variables of time and space, respectively; and Φ is the dissipation function. Indicial notation will be used whenever vector equations appear and the summation convention will be invoked, unless otherwise stated.

To close the system of equations, the thermodynamic and the transport properties will have to be related by state equations. In external aerodynamics, which is mainly the area of interest here, the flow properties are expected to approach their undisturbed value at infinity. On the solid boundaries, the velocity vector should match the skin velocity, unless there is blowing or suction, and the temperature or its gradient should be specified.

If the density and viscosity are constants, then the system of the governing equations is simplified considerably,

$$\frac{\partial u_i}{\partial x_i} = 0, \qquad\qquad\qquad *(1.2.4)$$

$$\rho\left(\frac{\partial u_i}{\partial t} + u_j \frac{\partial u_i}{\partial x_j}\right) = \rho g_i - \frac{\partial p}{\partial x_i} + \mu \frac{\partial^2 u_i}{\partial x_j \partial x_j}, \qquad *(1.2.5)$$

$$\rho c_v\left(\frac{\partial T}{\partial t} + u_j \frac{\partial T}{\partial x_j}\right) = k \frac{\partial^2 T}{\partial x_j \partial x_j} + \Phi, \qquad *(1.2.6)$$

where c_v is the specific heat at constant volume. The energy equation then uncouples and the velocity field can be found independently of the thermal field.

Many viscous flow problems are formulated in terms of the vorticity $\Omega_i = \text{curl}_i(u_j)$. Mathematically, such formulations have the advantage of eliminating pressure altogether. Physically, vorticity is directly related to viscous effects, since its generation is an entirely viscous phenomenon, and its convection and diffusion can be used to describe all viscous flow phenomena. The vorticity equation in its general form reads (White, 1974)

$$\frac{\partial \Omega_i}{\partial t} + u_j \frac{\partial \Omega_i}{\partial x_j} = \Omega_j \frac{\partial u_i}{\partial x_j} + \nu \frac{\partial^2 \Omega_i}{\partial x_j \partial x_j}. \qquad *(1.2.7)$$

For two-dimensional flows it is convenient to express the governing equations in terms of the stream function $\psi(x_i, t)$, which is defined by

$$u_1 = \frac{\partial \psi}{\partial x_2}, \qquad u_2 = -\frac{\partial \psi}{\partial x_1}. \qquad *(1.2.8)$$

The equation of continuity is then identically satisfied. For two-dimensional flow the first two components of vorticity vanish identically and the third is directly related to the stream function

$$\Omega_3 = -\nabla^2 \psi. \qquad *(1.2.9)$$

A straightforward substitution of Eqs. (1.2.8) and (1.2.9) in (1.2.7) yields

$$\frac{\partial}{\partial t}(\nabla^2 \psi) + \frac{\partial \psi}{\partial x_2}\frac{\partial}{\partial x_1}(\nabla^2 \psi) - \frac{\partial \psi}{\partial x_1}\frac{\partial}{\partial x_2}(\nabla^2 \psi) = \nu \nabla^4 \psi. \qquad *(1.2.10)$$

In this monograph we shall be mostly concerned with constant property flows. We shall therefore seek solutions to (1.2.4)–(1.2.10) in their general form or in their well-known approximate form, the boundary-layer equations.

Let U, L, and T_0 be some characteristic values of the velocity, length scale, and temperature of the problem. Dimensionless quantities are then defined according to the formulas:

$$x_i = \frac{x_i^*}{L}, \qquad t = \frac{t^* U}{L}, \tag{1.2.11}$$

$$u_i = \frac{u_i^*}{U}, \qquad p = \frac{p^*}{\rho U^2}, \qquad T = \frac{T^*}{T_0}. \tag{1.2.12}$$

Expressing the dependent and independent variables of (1.2.4)–(1.2.6) in terms of the above quantities and dropping the buoyancy effect, one can arrive at the following system of equations:

$$\frac{\partial u_i}{\partial x_i} = 0, \tag{1.2.13}$$

$$\frac{\partial u_i}{\partial t} + u_j \frac{\partial u_i}{\partial x_j} = -\frac{\partial p}{\partial x_i} + \frac{1}{Re} \frac{\partial^2 u_i}{\partial x_j \partial x_j}, \tag{1.2.14}$$

$$\frac{\partial T}{\partial t} + u_j \frac{\partial T}{\partial x_j} = \frac{1}{RePr} \frac{\partial^2 T}{\partial x_j \partial x_j} + \frac{Ec}{Re} \Phi, \tag{1.2.15}$$

The dimensionless numbers that appear in these equations are

the Reynolds number, $Re = UL/\nu$,
the Prandtl number, $Pr = \mu c_v/k$,
the Eckert number, $Ec = U^2/c_v T_0$.

For large Reynolds numbers, Eqs. (1.2.13)–(1.2.15) can be approximated by their boundary-layer form. The process involves an asymptotic expansion whereby the outer part of the flow can be approximated by the Euler equations and the inner part by the boundary-layer equations. Both sets of equations govern the behavior of a first term in asymptotic expansions. This has been described in detail in many modern texts. The boundary-layer form of (1.2.13)–(1.2.15), sometimes referred to in literature as the Prandtl equations, is written here in an expanded form for three-dimensional flow:

$$\frac{\partial u_1}{\partial x_1} + \frac{\partial u_2}{\partial x_2} + \frac{\partial u_3}{\partial x_3} = 0, \tag{1.2.16}$$

$$\frac{\partial u_1}{\partial t} + u_1 \frac{\partial u_1}{\partial x_1} + u_2 \frac{\partial u_1}{\partial x_2} + u_3 \frac{\partial u_1}{\partial x_3} = -\frac{\partial p}{\partial x_1} + \frac{1}{Re} \frac{\partial^2 u_1}{\partial x_2^2}, \tag{1.2.17}$$

$$\frac{\partial u_3}{\partial t} + u_1 \frac{\partial u_3}{\partial x_1} + u_2 \frac{\partial u_3}{\partial x_2} + u_3 \frac{\partial u_3}{\partial x_3} = -\frac{\partial p}{\partial x_3} + \frac{1}{Re} \frac{\partial^2 u_3}{\partial x_2^2}, \tag{1.2.18}$$

$$\frac{\partial T}{\partial t} + u_1 \frac{\partial T}{\partial x_1} + u_2 \frac{\partial T}{\partial x_2} + u_3 \frac{\partial T}{\partial x_3} = \frac{1}{RePr} \frac{\partial^2 T}{\partial x_2^2} + \frac{Ec}{Re} \Phi. \tag{1.2.19}$$

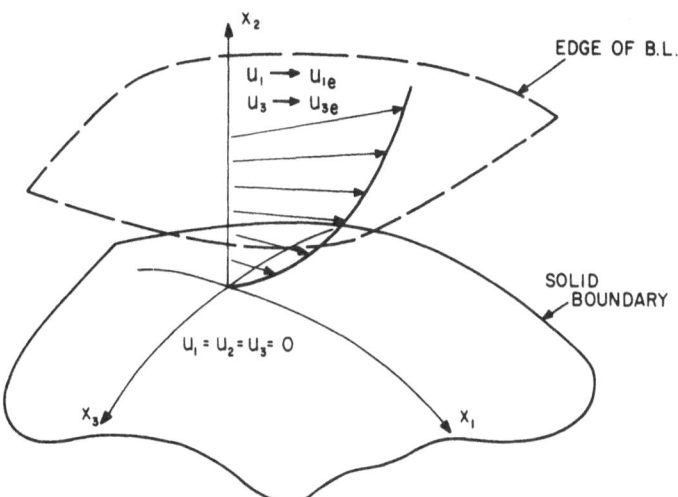

Fig. 1.1 Definition of the coordinate system.

In these equations x_1, x_3 and u_1, u_3 are the coordinates and velocity components, respectively, parallel to the wall, (see Fig. 1.1), while x_2 and u_2 are the coordinate and velocity components, respectively, perpendicular to the wall. The last two quantities are usually stretched by the factor $\sqrt{Re}$. The process of stretching is a necessary step in the method of inner and outer expansions, and the reader may find lucid descriptions of all the steps involved in recent monographs (Van Dyke, 1964; Cole, 1968; Nayfeh, 1973). The concept of stretching will be used in one of the chapters that follow when higher-order corrections to oscillatory flow are considered. Moreover, a brief description is included in Section 1.4, where the method of inner and outer expansions is outlined for a model equation.

The velocity components satisfy the no-penetration and no-slip condition at the wall, while the temperature matches with the wall temperature or meets a specific condition on the rate of heat transfer:

$$u_1 = u_{1w}, \qquad u_2 = u_{2w}, \qquad u_3 = u_{3w}, \quad \text{at } x_2 = 0, \tag{1.2.20}$$

$$T = T_w \quad \text{or} \quad \left.\frac{\partial T}{\partial x_2}\right|_w = \text{const} \quad \text{at } x_2 = 0, \tag{1.2.21}$$

where the subscript w denotes evaluation at the wall. At the outer edge of the layer, the boundary-layer flow merges into the outer flow,

$$u_1 \rightarrow U_{1e}, \quad u_3 \rightarrow U_{3e}, \qquad\qquad \text{as } x_2 \rightarrow \infty, \tag{1.2.22}$$

$$T \rightarrow T_e, \qquad\qquad\qquad\qquad\qquad \text{as } x_2 \rightarrow \infty. \tag{1.2.23}$$

Initial profiles are specified at $x_1 = 0$, $x_2 = 0$ for all times

$$u_1 = u_{11}(t, x_2, x_3), \qquad u_2 = u_{21}(t, x_2, x_3) \quad \text{at } x_1 = 0, \tag{1.2.24}$$

$$u_1 = u_{12}(t, x_1, x_3), \qquad u_2 = u_{22}(t, x_1, x_3) \quad \text{at } x_2 = 0, \tag{1.2.25}$$

as well as throughout the three-dimensional space for $t = 0$

$$u_1 = u_{10}(x_1, x_2, x_3), \qquad u_2 = u_{20}(x_1, x_2, x_3),$$
$$u_3 = u_{30}(x_1, x_2, x_3) \quad \text{for } t = 0. \tag{1.2.26}$$

Many of the practical problems are usually framed in terms of the traditional notation, whereby x_1, x_2, x_3 and u_1, u_2, u_3 are replaced by x, y, z and u, v, w, respectively. The boundary-layer equations (1.2.16)–(1.2.19) in their dimensional form then become

$$\frac{\partial u}{\partial x} + \frac{\partial v}{\partial y} + \frac{\partial w}{\partial z} = 0, \qquad\qquad *(1.2.27)$$

$$\frac{\partial u}{\partial t} + u\frac{\partial u}{\partial x} + v\frac{\partial u}{\partial y} + w\frac{\partial u}{\partial z} = -\frac{1}{\rho}\frac{\partial p}{\partial x} + \nu\frac{\partial^2 u}{\partial y^2}, \qquad *(1.2.28)$$

$$\frac{\partial w}{\partial t} + u\frac{\partial w}{\partial x} + v\frac{\partial w}{\partial y} + w\frac{\partial w}{\partial z} = -\frac{1}{\rho}\frac{\partial p}{\partial z} + \nu\frac{\partial^2 w}{\partial y^2}, \qquad *(1.2.29)$$

$$\frac{\partial T}{\partial t} + u\frac{\partial T}{\partial x} + v\frac{\partial T}{\partial y} + w\frac{\partial T}{\partial z} = \alpha\frac{\partial^2 T}{\partial y^2} + \Phi. \qquad\qquad *(1.2.30)$$

1.3 Characteristics

The theory of characteristics is an important prerequisite for the understanding of basic principles in numerical analysis. It is helpful in identifying the zones of influence and dependence. It therefore provides a necessary constraint in the construction of stable algorithms for the numerical calculation of partial differential equations. The theory of characteristics has been adequately described in many classical texts. We provide here only an outline of the results in order to introduce the appropriate terminology and nomenclature. Moreover, it appears more convenient in the later chapters to refer to equations numbered in this monograph than to other publications.

Consider the system of N-differential equations for the N-independent variables u_j; $j = 1, 2, \ldots, N$

$$\sum_{j, k_0, k_1, \ldots, k_n} a_{ij}^{(k_0 \cdots k_n)} \frac{\partial^{n_j} u_j}{\partial x_0^{k_0} \partial x_1^{k_1} \ldots \partial x_n^{k_n}} + f_i = 0. \tag{1.3.1}$$

In the above equations, $a_{ij}^{(k_0 \cdots k_n)}$ and f_i are functions of lower-order derivatives of the dependent variables. The highest-order derivative for the jth variable, that is, the derivative of order n_j appears explicitly in, and is the dominant part of, the differential equation. Note that n_j is not necessarily the same for all variables; however, the highest-order derivative may be a partial derivative involving two, three, or all of the independent variables x_0, $x_1, \ldots, x_n$. All the dependent variables but not necessarily all the independent variables appear in (1.3.1). The superscripts on the coefficient a_{ij} denote the order of the derivatives with respect to the variables $x_0, \ldots, x_n$, which, if summed up, represent all the highest derivatives of the jth variable in the ith equation. The surface

$$\phi(x_1, x_2, \ldots, x_n) = 0 \tag{1.3.2}$$

is a characteristic surface for the system (1.3.1), if at every point of this surface the following determinant vanishes (Petrovskii, 1967; Courant and Hilbert, 1962):

$$\left| \sum_{k_0 + \cdots + k_n = n_j} a_{ij}^{(k_0 \cdots k_n)} \left(\frac{\partial \phi}{\partial x_0} \right)^{k_0} \left(\frac{\partial \phi}{\partial x_1} \right)^{k_1} \cdots \left(\frac{\partial \phi}{\partial x_n} \right)^{k_n} \right| = 0. \tag{1.3.3}$$

The equation

$$\left| \sum_{k_0 + \cdots + k_n = n_j} a_{ij}^{(k_0 \cdots k_n)} \alpha_0^{k_0} \alpha_1^{k_1} \cdots \alpha_n^{k_n} \right| = 0 \tag{1.3.4}$$

with α_i constant parameters such that $\sum_{i=0}^{n} \alpha_i^2 \neq 0$ is called the characteristic equation for the system (1.3.1), while the tangent hyperplane P to the characteristic surface at the point $(\tilde{x}_0, \ldots, \tilde{x}_n)$

$$\sum_{k=0}^{n} \alpha_k (x_k - \tilde{x}_k) = 0 \tag{1.3.5}$$

defines the characteristic direction at this point. Without loss of generality, we may normalize the quantities α_i so that

$$\sum_{k=0}^{n} \alpha_k^2 = 1. \tag{1.3.6}$$

Then α_k can be interpreted as the directional cosine between the normal to the characteristic plane e and the x_k axis as shown schematically in Fig. 1.2.

If all the functions u_j and their derivatives up to order n_{j-1} are given on a

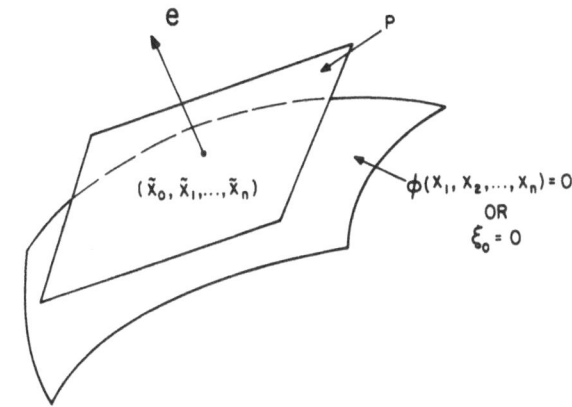

Fig. 1.2 The characteristic surface $\phi = 0$ [Eq. (1.3.4)], the characteristic plane P [Eq. (1.3.5)] at the point $\tilde{x}_0, \tilde{x}_1, \ldots, \tilde{x}_n$ and the normal to both, $e(\alpha_1, \alpha_2, \ldots, \alpha_k)$.

sufficiently smooth surface S, then the system (1.3.1) can be solved for the highest-order derivative in the direction perpendicular to S, provided the surface S is nowhere a characteristic surface, that is, the surface S is nowhere tangent to a characteristic plane.

One usually defines a new set of coordinates $(\xi_0, \xi_1, \ldots, \xi_n)$, such that the data surface is the surface $\xi_0 = 0$. Kovalevsky's theorem then states that the generalized Cauchy problem always has a unique solution in some neighborhood of the surface S, if this surface is not tangent at any point to any characteristic plane and if, moreover, the following conditions are met: (a) the coefficients of the system (1.3.1) are analytic functions of $x_1, x_2, \ldots, x_n$; (b) the functions $x_i = X_i(\xi_0, \xi_1, \ldots, \xi_n)$ that define the transformation to the new coordinate system are analytic; and (c) the initial conditions defined on the data surface are analytic.

If the function $\phi(x_1, x_2, \ldots, x_n)$ is an analytic function of its arguments, and the surface $\phi = 0$ does not contain any singular points, that is, points at which the first-order derivatives of ϕ vanish, then it is possible again to define an analytic system of coordinate lines, such that $\xi_0 = 0$ represents the surface $\phi = 0$.

If the surface S has a characteristic direction at some point, that is, if (1.3.3) is met at a point, then we cannot always give arbitrary values to the functions u_i and their derivatives on the data surface and still be sure that the generalized Cauchy problem has a solution.

Consider for example a second-order quasi-linear differential equation of the form

$$a_{11} \frac{\partial^2 u}{\partial x_1^2} + 2a_{12} \frac{\partial^2 u}{\partial x_1 \partial x_2} + a_{22} \frac{\partial^2 u}{\partial x_2^2} + f(u, x_1, x_2) = 0, \tag{1.3.7}$$

where the coefficients a_{ij} may depend on x_i and u. Following our notation, we should denote the three coefficients a_{11}, a_{12}, a_{22} by the symbols $a^{(11)}, a^{(12)}, a^{(22)}$, respectively. Note that no subscripts i and j are necessary since we only have one equation and one dependent variable. The determinant in (1.3.3) is reduced to one element and the equation of the characteristic surface is given by

$$a_{11} \left(\frac{\partial \phi}{\partial x_1} \right)^2 + 2a_{12} \left(\frac{\partial \phi}{\partial x_1} \right) \left(\frac{\partial \phi}{\partial x_2} \right) + a_{22} \left(\frac{\partial \phi}{\partial x_2} \right)^2 = 0. \tag{1.3.8}$$

In terms of the slope of the characteristic

$$\frac{dx_2}{dx_1} = - \frac{\partial \phi / \partial x_1}{\partial \phi / \partial x_2}, \quad \text{on } \phi = 0, \tag{1.3.9}$$

Eq. (1.3.8) becomes

$$a_{11} \left(\frac{dx_2}{dx_1} \right)^2 - 2a_{12} \left(\frac{dx_2}{dx_1} \right) + a_{22} = 0. \tag{1.3.10}$$

Second-order differential equations are traditionally classified according to the discriminant of Eq. (1.3.10):

(a) $a_{12}^2 - a_{11}a_{22} < 0$, elliptic.

No real roots of the algebraic equation (1.3.10), no real characteristic surfaces exist:

(b) $a_{12}^2 - a_{11}a_{22} > 0$, hyperbolic.

Two real directions of characteristic surfaces exist at each point:

(c) $a_{12}^2 - a_{11}a_{22} = 0$, parabolic.

The characteristic equation (1.3.10) has one double root. The two solutions of (1.3.10) and the two characteristic surfaces coincide.

A typical example of elliptic equations is Laplace's equation:

$$\sum_{i=1}^{n} \frac{\partial^2 u}{\partial x_i^2} = 0, \tag{1.3.11}$$

with a characteristic equation

$$\sum_{i=1}^{n} \left(\frac{\partial \phi}{\partial x_i} \right)^2 = 0. \tag{1.3.12}$$

An example of a higher-order elliptic equation is

$$\sum_{i,k=1}^{n} \frac{\partial^4 u}{\partial x_i^2 \partial x_k^2} = 0, \tag{1.3.13}$$

with a characteristic equation

$$\left(\sum_{i=1}^{n} \left(\frac{\partial \phi}{\partial x_i} \right)^2 \right)^2 = 0. \tag{1.3.14}$$

In fluid mechanics, inviscid, incompressible flow is governed by Laplace's equation, which is simply an expression of irrotationality and incompressibility. With x_i the physical space variables, we may interpret its mathematical properties described in this section as follows. Any surface may serve as an initial surface S, since no characteristic direction exists anywhere in the domain of integration. Any initial data therefore influence both sides of the initial surface. In the traditional terminology we say that disturbances propagate in all directions of space instantly or, alternatively, the properties of the flow at a point depend on all other points in the domain of integration and therefore on all the points of the boundary. As a result the problem is properly posed, only if boundary conditions are given on a closed contour that contains the domain of integration. The idea of disturbance propagation enters the picture naturally in compressible flow where disturbances propagate with the speed of sound. In incompressible flow, the form of the governing equations is equivalent to the assumption of an infinite speed of sound and, therefore, instant propagation of all disturbances to all points in space.

A typical example of a parabolic equation is the heat equation

$$\frac{\partial u}{\partial t} = \sum_{i=1}^{n} \frac{\partial^2 u}{\partial x_i^2} , \tag{1.3.15}$$

with a characteristic equation given by

$$\sum_{i=1}^{n} \left(\frac{\partial \phi}{\partial x_i} \right)^2 = 0. \tag{1.3.16}$$

Here the characteristic form is degenerate since derivatives with respect to t do not appear at all. This implies that disturbances propagate instantly in all space directions. However, propagation in the direction of t is possible only in the positive t direction, as we shall demonstrate in a later section, and has a finite rate. As a result, posing the problem correctly requires boundary conditions on every point of a domain defined on the space x_i but only an initial condition at a specific value of t. Traditionally the variable t represents time. As a result whenever other independent variables appear to play the role of parabolic variables, they are often termed "timelike" variables.

Finally the most common example of a hyperbolic equation is the wave equation

$$\frac{\partial^2 u}{\partial x_1^2} - \frac{\partial^2 u}{\partial x_2^2} = 0, \tag{1.3.17}$$

with a characteristic equation

$$\left(\frac{\partial \phi}{\partial x_1} \right)^2 - \left(\frac{\partial \phi}{\partial x_2} \right)^2 = 0. \tag{1.3.18}$$

This equation has two characteristic surfaces, more familiar in literature as the Mach surfaces, inclined with respect to the x_1 axis by angles of 45°. Disturbances are now confined within the Mach surfaces.

1.4 Subcharacteristics

Perturbation methods have recently proved to be a powerful tool for solving differential equations (van Dyke, 1964; Cole, 1968; Nayfeh, 1973). According to one of the most common methods, differential equations can be simplified by dropping terms of small magnitude. This may be possible, for example, if a small parameter ε multiplies certain terms in an equation, provided the coefficients of ε do not grow out of bounds. Such simplifications may change considerably the character of a differential equation, especially if the small parameter ε multiplies one or more of the highest derivatives. This is indeed the case with the Navier–Stokes equations and deserves a careful examination. A typical model example will be described here briefly (Cole, 1968). This is a simplified version of mathematical models of steady or unsteady aerodynamics, which often contain regions of sharp variations commonly known as boundary layers. This model demonstrates the significance of the concept of subcharacteristics and the necessity of analyzing separately boundary-layer regions through appropriate stretching of variables.

Consider the differential equation

$$\varepsilon \left[a_{11} \frac{\partial^2 u}{\partial x_1^2} + 2a_{12} \frac{\partial^2 u}{\partial x_1 \partial x_2} + a_{22} \frac{\partial^2 u}{\partial x_2^2} \right] = a_1 \frac{\partial u}{\partial x_1} + a_2 \frac{\partial u}{\partial x_2} , \tag{1.4.1}$$

with ε a small constant parameter and a_{ij}, a_i constant coefficients. Let us assume that

$$a_{12}^2 - a_{11} a_{22} < 0. \tag{1.4.2}$$

Equation (1.4.1) is then elliptic and the problem is posed properly if boundary conditions on u or its derivatives are imposed all around a closed contour. No real characteristics exist. However, if $\varepsilon \to 0$, then the equation can be approximated by

$$a_1 \frac{\partial u}{\partial x_1} + a_2 \frac{\partial u}{\partial x_2} = 0. \tag{1.4.3}$$

The characteristic surfaces of (1.4.3) are given by

$$a_1 \frac{\partial \phi}{\partial x_1} + a_2 \frac{\partial \phi}{\partial x_2} = 0, \tag{1.4.4}$$

and they are lines with slope

$$\frac{dx_2}{dx_1} = \frac{a_2}{a_1} . \tag{1.4.5}$$

Cole defines the characteristics of the simplified version of the equation as the "subcharacteristics" of the original equation. The term subcharacteristic is perhaps unfortunate since the prefix "sub" is traditionally used in mathematics to imply "part of," i.e., a subset of a set. A more appropriate term here would be "limiting characteristics." Moreover, we should emphasize here that the characteristics of a set of differential equations are invariants of the problem, whereas the subcharacteristics depend on the particular approximation employed. However, the characteristics as well as the subcharacteristics have a very significant physical meaning as we shall describe in this and the following section. Here the subcharacteristics of (1.4.1) are the lines

$$\xi_1 = a_2 x_1 - a_1 x_2 = \text{const.} \tag{1.4.6}$$

Figure 1.3 shows schematically the domain of integration of Eq. (1.4.1), the boundary and the subcharacteristics $\xi_1 = $ constant. A new set of orthogo-

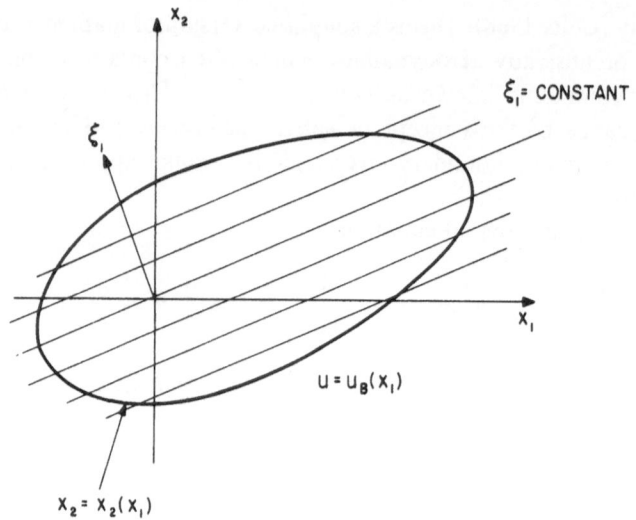

Fig. 1.3 The domain of integration in the (x_1, x_2) plane. The value of the function is prescribed along the boundary $x_2 = x_2(x_1)$; $u = u_B(x_1)$.

nal coordinates is defined:

$$\xi_1 = a_2 x_1 - a_1 x_2, \qquad \xi_2 = a_1 x_1 + a_2 x_2, \tag{1.4.7}$$

and the original equation is expressed in terms of ξ_1 and ξ_2:

$$\varepsilon \left[A_{11} \frac{\partial^2 u}{\partial \xi_1^2} + 2A_{12} \frac{\partial^2 u}{\partial \xi_1 \partial \xi_2} + A_{22} \frac{\partial^2 u}{\partial \xi_2^2} \right] = \frac{\partial u}{\partial \xi_2} \tag{1.4.8}$$

with

$$A_{11} = \left(a_1^2 + a_2^2 \right)^{-1} \left(a_{11} a_2^2 - 2 a_{12} a_1 a_2 + a_{22} a_1^2 \right), \tag{1.4.9}$$

$$A_{12} = \left(a_1^2 + a_2^2 \right)^{-1} \left[a_{11} a_1 a_2 + a_{12} \left(a_2^2 - a_1^2 \right) - a_{22} a_1 a_2 \right], \tag{1.4.10}$$

$$A_{22} = \left(a_1^2 + a_2^2 \right)^{-1} \left(a_{11} a_1^2 + 2 a_{12} a_1 a_2 + a_{22} a_2^2 \right). \tag{1.4.11}$$

Figure 1.4 shows the domain of integration in the (ξ_1, ξ_2) plane. Points A and B are the points of the minimum and maximum ξ_1 of the boundary, respectively. The function u is prescribed in the upper and lower parts of the boundary $u = u_U(\xi_1)$ on $\xi_2 = \xi_{2U}(\xi_1)$, and $u = u_L(\xi_1)$ on $\xi_2 = \xi_{2L}(\xi_1)$, respectively. The equation is still elliptic and therefore we know that

$$A_{12}^2 - A_{11} A_{22} < 0. \tag{1.4.12}$$

However if $\varepsilon \to 0$, we would expect that an approximate form of (1.4.8) is

$$\frac{\partial u}{\partial \xi_2} = 0, \tag{1.4.13}$$

Fig. 1.4 The domain of integration
in the (ξ_1, ξ_2) plane.

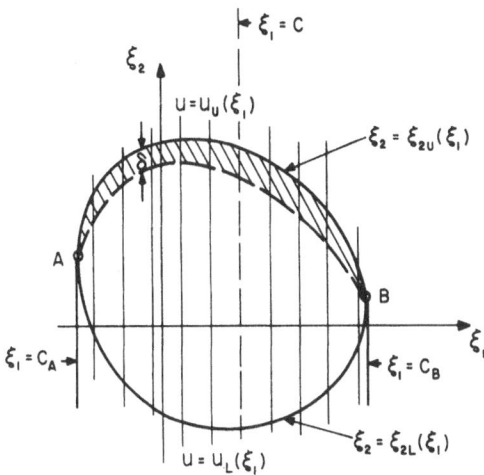

with a solution

$$u = u(\xi_1). \tag{1.4.14}$$

In the process of simplifying our equation, we have reduced it from a second-order equation to a first-order equation. We are therefore unable to meet the boundary conditions on both the upper and the lower parts of the boundary. Moreover, our approximation is valid only if the function $u(\xi_1)$ and its derivatives are smooth throughout the domain of integration. These two statements actually lead to the existence of a boundary layer. In many real-life problems, this situation is indeed true. A property behaves nicely throughout the domain of variation including one of the two sides of the boundary. Let us assume that this is the case here and that we can meet the boundary condition at $\xi_2 = \xi_{2L}$. Our solution then becomes

$$u_0 = u(\xi_1) = u_L(\xi_1). \tag{1.4.15}$$

This will be called the outer solution.

In the neighborhood of the other boundary and within a thin region, which is shown shaded in Fig. 1.4, our function must change sharply with ξ_2 in order to meet the boundary condition $u_U(\xi_1)$. A representative cross section of the solution at $\xi_1 = C$ is shown schematically in Fig. 1.5. The outer solution has the value $u_L(C)$ at $\xi_2 = \xi_2(C)$ and, therefore, violates the boundary condition $u = u_U(C)$. Within the boundary layer, that is, the layer of thickness $\delta(\varepsilon, \xi_1)$, the function $u(\xi)$ changes drastically with ξ_2 and our approximation is invalid.

To investigate the singular behavior of our solution within this boundary, we stretch the coordinate ξ_2 by introducing a new variable

$$\tilde{\xi}_2 = \frac{\xi_2 - \xi_{2U}(\xi_1)}{\delta(\varepsilon)}, \tag{1.4.16}$$

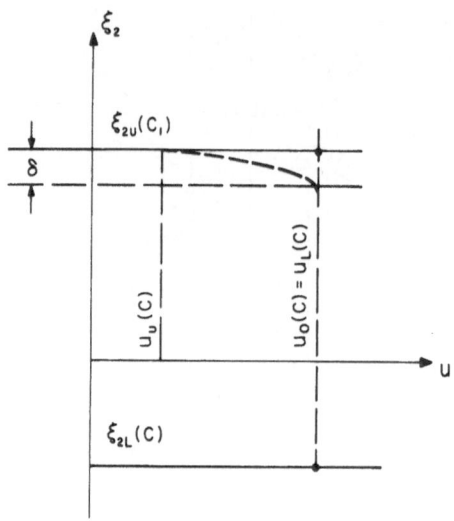

and expect that some terms on the left-hand side of (1.4.8) will balance the
term $\partial u/\partial \xi_2$ on the right-hand side if expressed in terms of $\tilde{\xi}_2$-derivatives.
This is indeed the case if $\delta = O(\varepsilon)$. The symbol $O(\ )$ stands for "order of"
and implies here that δ is of the same order of magnitude as ε. If all terms of
higher order are neglected, our equation may be further approximated by

$$\left[A_{11} \left(\frac{d\xi_{2U}}{d\xi_1} \right)^2 - 2A_{12} \frac{d\xi_{2U}}{d\xi_1} + A_{22} \right] \frac{\partial^2 u}{\partial \tilde{\xi}_2^2} = \frac{\partial u}{\partial \tilde{\xi}_2}. \tag{1.4.17}$$

At this level of approximation, the boundary-layer equation is an ordinary
differential equation. The correction necessary to match the outer solution
$u_0 = u_L(\xi_1)$ with the inner boundary condition $u = u_U(\xi_1)$, that is, the dashed
line of Fig. 1.5, is the solution to Eq. (1.4.17). This is a second-order
differential equation and its solution must satisfy the conditions

$$u(\tilde{\xi}_2; \xi_1) = u_U(\xi_1), \quad \text{as } \tilde{\xi}_2 \to 0, \tag{1.4.18}$$

and in the limit of $\varepsilon \to 0$,

$$u(\tilde{\xi}_2; \xi_1) \to u_0(\xi_1), \quad \text{as } \tilde{\xi}_2 \to \infty. \tag{1.4.19}$$

Note that the independent variable ξ_1 enters in this problem only as a
parameter.

The situation is drastically changed if $d\xi_{2U}/d\xi_1 \to \infty$, that is, if the
contour of the boundary is a subcharacteristic. This is of great interest to us
here, because in fact the contours of solid boundaries are subcharacteristics
of the Navier–Stokes equations. This case is shown schematically in Fig. 1.6
where the "upper" boundary is parallel to the ξ_2 axis beyond point B and,
therefore, coincides with a subcharacteristic. The upper part of the boundary
is simply $\xi_1 = \xi_{1U} = C_B$, and our solution is expected to meet on this line the

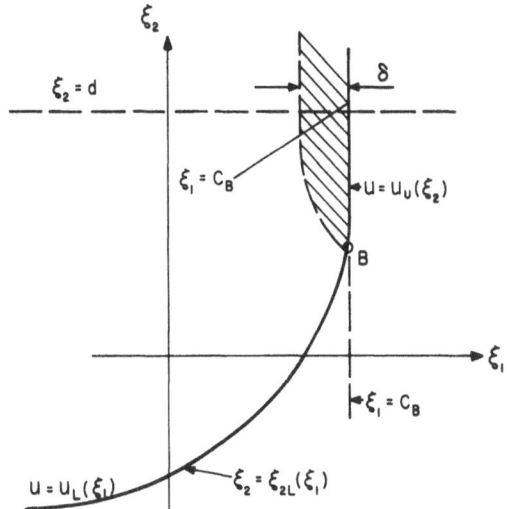

Fig. 1.6 The domain of integration in the (ξ_1, ξ_2) plane when a portion of the boundary coincides with a subcharacteristic.

boundary condition $u = u_U(\xi_2)$. Clearly this is not possible since our outer solution (1.4.15) takes the constant value $u_L(C_B)$ for all points on $\xi_1 = C_B$. We now anticipate that a boundary layer exists again in the neighborhood of $\xi_2 = \xi_{2U}$. In other words, we expect that our outer solution is approximately valid throughout the domain of integration except in a thin layer of thickness δ next to the line $\xi_1 = C_B$. In the boundary layer the function u will change sharply now with ξ_1 in order to meet the boundary condition $u = u_U$. A typical cross section is shown schematically in Fig. 1.7 for $\xi_2 = d$. To bring about the character of the solution within the boundary layer, we need to introduce now a stretched coordinate in the ξ_1 direction:

$$\tilde{\xi}_1 = \frac{\xi_1 - C_B}{\delta} . \tag{1.4.20}$$

The dominant term on the left-hand side of Eq. (1.4.8) is the term $\partial^2 u / \partial \xi_1^2$

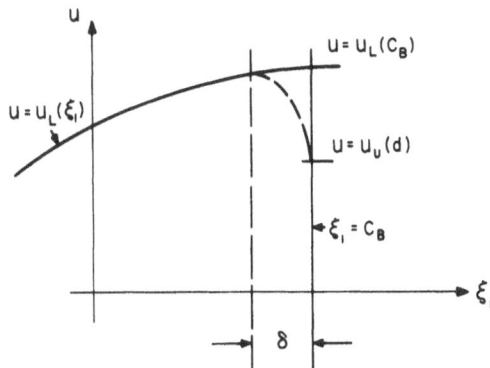

Fig. 1.7 Cross section of the plane (ξ_1, ξ_2) of Fig. 1.6 at $\xi_2 = d$.

and if expressed in terms of $\tilde{\xi}_1$ derivatives, Eq. (1.4.8) becomes

$$\varepsilon \left[\frac{A_{11}}{\delta^2} \frac{\partial^2 u}{\partial \tilde{\xi}_1^2} + O\left(\frac{1}{\delta}\right) \right] = \frac{\partial u}{\partial \xi_2} . \tag{1.4.21}$$

The right-hand side can be balanced by the left-hand side only if

$$\delta = O(\sqrt{\varepsilon}), \tag{1.4.22}$$

and thus, (1.4.21) can be approximated within an error of $O(\sqrt{\varepsilon})$ by

$$A_{11} \frac{\partial^2 u}{\partial \tilde{\xi}_1^2} = \frac{\partial u}{\partial \xi_2} . \tag{1.4.23}$$

Notice that this equation is parabolic, and the effect of the boundary condition at $\xi_1 = C_B$ is penetrating the outer solution according to the diffusion process as dictated by the heat equation, that is, (1.4.23). The proper boundary conditions in the limit of $\varepsilon \to 0$ are

$$u(\tilde{\xi}_1, \xi_2) = u_U(\xi_2), \quad \text{for } \tilde{\xi}_1 = 0, \tag{1.4.24}$$

$$u(\tilde{\xi}_1, \xi_2) \to u_L(C_B), \quad \text{for } \tilde{\xi}_1 \to \infty. \tag{1.4.25}$$

It should be emphasized that now the coordinate ξ_2 is a timelike coordinate and therefore an initial condition is necessary at $\xi_2 = \xi_{2L}(C_B)$, that is, the smallest value of ξ_2.

The most common boundary layer in fluid mechanics is the viscous layer in the immediate vicinity of solid boundaries. In fact the term has been originally introduced to denote this boundary layer and was given a more general meaning only in the past decade or two with the development of perturbation theories. The steps described above can be followed one by one to derive the Euler and Prandtl equations from the Navier–Stokes equations. This, of course, has been included in classical publications as referenced by Schlichting (1968) and White (1974), and it is beyond the scope of the present monograph. We only emphasize in the following section the significance of these concepts in the numerical computations of steady and unsteady viscous flows. Moreover, we shall encounter in the following chapters examples of boundary layers that emerge naturally in the time domain, typical examples of the properties of unsteady flows.

1.5 Navier–Stokes versus Euler and Prandtl

The study of the characteristic and subcharacteristic surfaces of a certain problem determines the main structure of the flow, indicates the regions of influence and dependence, and guides the numerical analyst in the development of stable numerical schemes. The theory presented in the earlier sections will now be employed in the study of the unsteady Navier–Stokes equations and their approximate forms. Wang (1971) has studied the characteristics and subcharacteristics for the steady Navier–Stokes equations and their approximate form, the Euler and Prandtl equations. In this section we present an analysis that includes the unsteady effects and the energy equation. Wang (1975) later investigated the characteristic and subcharacteristic structure of the unsteady boundary-layer equations discussing numerical calculations that already appeared in literature. A brief description of these ideas and appropriate references are also included in this section.

Consider the system of the five equations (1.2.13)–(1.2.15). To avoid the complexities of some nonlinearities, let us assume here that dissipation is negligible. The characteristic surfaces in the four dimensional space (x_i, t), $i = 1, 2, 3$ are then given by (1.3.3). In our particular case, with columns corresponding to the five dependent variables u_1, u_2, u_3, T, and p, and rows corresponding to the five equations, this determinant becomes

$$\begin{vmatrix} \dfrac{\partial \phi}{\partial x_1} & \dfrac{\partial \phi}{\partial x_2} & \dfrac{\partial \phi}{\partial x_3} & 0 & 0 \\[2ex] Q & 0 & 0 & 0 & \dfrac{\partial \phi}{\partial x_1} \\[2ex] 0 & Q & 0 & 0 & \dfrac{\partial \phi}{\partial x_2} \\[2ex] 0 & 0 & Q & 0 & \dfrac{\partial \phi}{\partial x_3} \\[2ex] 0 & 0 & 0 & \dfrac{Q}{Pr} & 0 \end{vmatrix} = 0, \tag{1.5.1}$$

where

$$Q = -\frac{1}{Re}\left[\left(\frac{\partial \phi}{\partial x_1}\right)^2 + \left(\frac{\partial \phi}{\partial x_2}\right)^2 + \left(\frac{\partial \phi}{\partial x_3}\right)^2 \right]. \tag{1.5.2}$$

Equation (1.5.1) can be written as

$$\frac{1}{Pr}\left[\left(\frac{\partial \phi}{\partial x_1}\right)^2 + \left(\frac{\partial \phi}{\partial x_2}\right)^2 + \left(\frac{\partial \phi}{\partial x_3}\right)^2 \right] Q^3 = 0. \tag{1.5.3}$$

It should be noticed that derivatives with respect to time do not appear. The system of equations is therefore parabolic and integration can proceed by marching in the positive direction of the parabolic variable, time t. For steady flows, time is eliminated from the problem, but (1.5.1)–(1.5.3) do not change. Then, there is no real solution of (1.5.3) and the system of Eqs. (1.2.13)–(1.2.15) is elliptic.

If the parameter Re is very large, then the partial differential Eqs. (1.2.13)–(1.2.15) fall in the category of problems considered in the previous section. Elimination of the highest derivative reduces the order of the differential equation. Dropping all the terms that multiply the quantity $1/Re$, one readily arrives at the Euler equations:

$$\frac{\partial u_i}{\partial x_i} = 0, \tag{1.5.4}$$

$$\frac{\partial u_i}{\partial t} + u_j \frac{\partial u_i}{\partial x_j} + \frac{\partial p}{\partial x_i} = 0, \tag{1.5.5}$$

$$\frac{\partial T}{\partial t} + u_j \frac{\partial T}{\partial x_j} = 0. \tag{1.5.6}$$

If no sharp changes occur in the domain of integration to render our approximation invalid, the above equations represent a good approximation for our problem, and their solution is expected to be accurate within an error of order $1/Re$. This is the outer part of our asymptotic solution. The characteristic determinant of this system is now (Wang, 1971)

$$\begin{vmatrix} \dfrac{\partial \phi}{\partial x_1} & \dfrac{\partial \phi}{\partial x_2} & \dfrac{\partial \phi}{\partial x_3} & 0 & 0 \\[2ex] S & 0 & 0 & 0 & \dfrac{\partial \phi}{\partial x_1} \\[2ex] 0 & S & 0 & 0 & \dfrac{\partial \phi}{\partial x_2} \\[2ex] 0 & 0 & S & 0 & \dfrac{\partial \phi}{\partial x_3} \\[2ex] 0 & 0 & 0 & S & 0 \end{vmatrix} = 0, \tag{1.5.7}$$

where

$$S = \frac{\partial \phi}{\partial t} + u_1 \frac{\partial \phi}{\partial x_1} + u_2 \frac{\partial \phi}{\partial x_2} + u_3 \frac{\partial \phi}{\partial x_3}. \tag{1.5.8}$$

This results in the characteristic equation

$$\left[\left(\frac{\partial \phi}{\partial x_1} \right)^2 + \left(\frac{\partial \phi}{\partial x_2} \right)^2 + \left(\frac{\partial \phi}{\partial x_3} \right)^2 \right] S^3 = 0. \tag{1.5.9}$$

It was pointed out by Wang (1971) that the second factors in Eqs. (1.5.3) and (1.5.9) have their origin in the pressure terms and the incompressibility of the flow. They are present in both characteristic equations and they are elliptic in character. However the factors Q and S are entirely different. The term Q stems from the diffusion terms of the original Navier–Stokes equations and it is clearly an elliptic element, whereas the term S in Eq. (1.5.9) is nothing but the particle derivative. Wang notes that for steady flow, the streamlines are subcharacteristics but the system of the Euler equations is elliptic because the elliptic character of the equations dominates as we shall argue in the next section.

Elimination of the diffusion terms in Eqs. (1.2.13)–(1.2.15) results in reduction of the order of equations from two to one. One of the boundary conditions should therefore be dropped. It is a well-known fact that the Euler equation cannot meet the proper boundary conditions and the no-slip condition on the skin of the body is violated. In the spirit of the method of inner and outer expansions briefly outlined in the previous section, we expect that there exists a boundary layer in the immediate neighborhood of the solid boundary. In other words, we expect that in a thin layer next to the solid boundary, the solution to the original problem changes sharply with distance normal to the wall in order to meet the proper boundary condition. To capture this behavior, an inner approximation to the Navier–Stokes equations is necessary, and this is nothing but the classical equations of Prandtl, the boundary-layer equations (1.2.16)–(1.2.19). To order $1/Re$, these equations describe the flow in the narrow layer next to the solid boundary. One of the results of the boundary-layer approximation is that to order Re^{-1} pressure does not vary across the boundary layer. Bernoulli's equation holds at the edge of the boundary layer and therefore pressure is determined by the equations

$$-\frac{\partial p(x_1, x_2, x_3)}{\partial x_1} = -\frac{\partial p(x_1, \infty, x_3)}{\partial x_1} = \frac{\partial U_{1e}}{\partial t} + U_{1e}\frac{\partial U_{1e}}{\partial x_1} + U_{3e}\frac{\partial U_{1e}}{\partial x_3},$$

$$(1.5.10)$$

$$-\frac{\partial p(x_1, x_2, x_3)}{\partial x_3} = -\frac{\partial p(x_1, \infty, x_3)}{\partial x_3} = \frac{\partial U_{3e}}{\partial t} + U_{1e}\frac{\partial U_{3e}}{\partial x_1} + U_{3e}\frac{\partial U_{3e}}{\partial x_3}.$$

$$(1.5.11)$$

The unknown functions in Eqs. (1.2.16)–(1.2.18) are u_i and T. In a coordinate system aligned with the contour of the body, such that the x_2 coordinate is perpendicular to it, and with $\Phi = 0$ the characteristic determinant becomes

$$\begin{vmatrix} \dfrac{\partial \phi}{\partial x_1} & \dfrac{\partial \phi}{\partial x_2} & \dfrac{\partial \phi}{\partial x_3} & 0 \\[2mm] Q & 0 & 0 & 0 \\[2mm] 0 & 0 & Q & 0 \\[2mm] 0 & 0 & 0 & \dfrac{Q}{Pr} \end{vmatrix} = 0,$$

$$(1.5.12)$$

where now

$$Q = -\frac{1}{Re}\left(\frac{\partial \phi}{\partial x_2}\right)^2,$$

$$(1.5.13)$$

and therefore the characteristic equation reduces to

$$\left[\frac{1}{Pr}\frac{\partial\phi}{\partial x_2}\right]\left[\frac{1}{Re}\left(\frac{\partial\phi}{\partial x_2}\right)^2\right]^3 = 0. \tag{1.5.14}$$

The variables t, x_1, and x_3 do not appear in the characteristic equation and therefore the system is parabolic. Wang notes that with the elimination of pressure, a step that physically corresponds to the incompressibility condition, our system looses its ability of upstream influence. The same is true with the second factor of Eq. (1.5.14). This factor derives from the diffusion terms of the original equations and indicates that within the boundary-layer approximation, diffusion is confined only in the x_2 direction and disturbances propagate instantly only perpendicular to the wall.

Consider now the boundary-layer approximation imposed hypothetically on the Euler equations or, alternatively, the assumption of inviscid flow imposed on the boundary-layer equation:

$$\frac{\partial u_1}{\partial x_1} + \frac{\partial u_2}{\partial x_2} + \frac{\partial u_3}{\partial x_3} = 0, \tag{1.5.15}$$

$$\frac{\partial u_1}{\partial t} + u_1\frac{\partial u_1}{\partial x_1} + u_2\frac{\partial u_1}{\partial x_2} + u_3\frac{\partial u_1}{\partial x_3} = -\frac{\partial p}{\partial x_1}, \tag{1.5.16}$$

$$\frac{\partial u_3}{\partial t} + u_1\frac{\partial u_3}{\partial x_1} + u_2\frac{\partial u_3}{\partial x_2} + u_3\frac{\partial u_3}{\partial x_3} = -\frac{\partial p}{\partial x_3}, \tag{1.5.17}$$

$$\frac{\partial T}{\partial t} + u_1\frac{\partial T}{\partial x_1} + u_2\frac{\partial T}{\partial x_2} + u_3\frac{\partial T}{\partial x_3} = 0. \tag{1.5.18}$$

The characteristic determinant and the characteristic equation then become

$$\begin{vmatrix} \dfrac{\partial\phi}{\partial x_1} & \dfrac{\partial\phi}{\partial x_2} & \dfrac{\partial\phi}{\partial x_3} & 0 \\ S & 0 & 0 & 0 \\ 0 & 0 & S & 0 \\ 0 & 0 & 0 & \dfrac{S}{Pr} \end{vmatrix} = 0, \tag{1.5.19}$$

or, if expanded,

$$\frac{1}{Pr}\frac{\partial\phi}{\partial x_2}S^3 = 0, \tag{1.5.20}$$

where

$$S = \frac{\partial\phi}{\partial t} + u_1\frac{\partial\phi}{\partial x_1} u_2\frac{\partial\phi}{\partial x_2} + u_3\frac{\partial\phi}{\partial x_3}. \tag{1.5.21}$$

This indicates that the particle paths are subcharacteristics of the boundary-layer equations. For steady flow the subcharacteristics coincide with the streamlines since there is no distinction between streamlines and particle paths. The contour of the body is in fact both a streamline and a particle path and therefore a subcharacteristic of our problem. The problem then falls in the special category described in the previous section: The boundary on which the boundary layer develops is a subcharacteristic.

1.6 Zones of Influence and Dependence

The concept of influence and dependence is most easily demonstrated by considering steady three-dimensional boundary-layer flow. In this case the system of equations retains some diffusion terms and therefore the elliptic character in the direction x_2, the direction perpendicular to the wall. A disturbance is therefore propagated instantly across the boundary layer. On the other hand, since the streamlines are subcharacteristics, information may travel along the streamlines with the speed of convection, that is, the component of the velocity parallel to the wall at the point of consideration (Raetz, 1957; Der and Raetz, 1962; Wang, 1971). The region of influence of a certain point P, (x_1, x_2, x_3), therefore, will include a wedge-shaped volume with generators the verticals that touch the outermost subcharacteristics emanating from the vertical at P, as shown schematically in Fig. 1.8. To be more explicit, consider all the velocity vectors of points on the vertical through P. Let the outermost directions correspond to the points P' and P''. A message may be convected parallel to the wall in any direction included in the two vertical planes that contain the velocity vectors of P' and P''. These

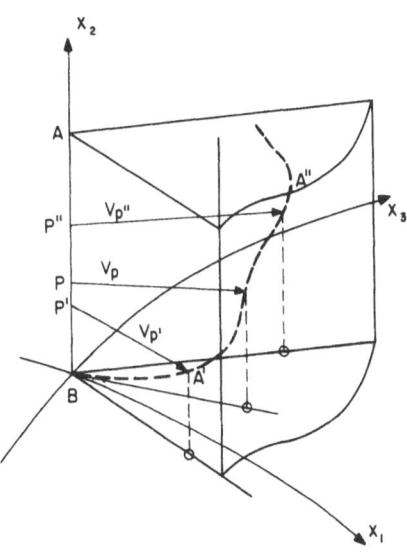

Fig. 1.8 Region of dependence in the immediate neighborhood of a point P in a three-dimensional boundary layer. The dashed line is the locus of the tips of velocity vectors.

planes are characteristic planes. Note that the locus of the tips of the velocity vectors touches the side walls of the wedge only at the points A' and A'', the tips of the velocities of points P' and P'', respectively. Once a disturbance arrives at a specific point within the wedge, it instantly propagates across the boundary layer, all the way down to the wall and up to the edge of the boundary layer. The speed of propagation is infinitely large but the magnitude of the disturbance is exponentially small.

The slope of the velocity vector in the x_1, x_3 plane is given by

$$\frac{dx_3}{dx_1} = \frac{u_3}{u_1},$$
(1.6.1)

and the largest and smallest values of this function for fixed x_1 and x_3 define the two outermost directions of the subcharacteristics. This is essentially due to the hyperbolic character of the operator $u_1\partial/\partial x_1 + u_3\partial/\partial x_3$.

Since the streamlines are the subcharacteristics of our problem, the outermost streamlines may be used to extend the region of influence as far downstream as necessary into a finite wedged-shaped configuration shown schematically in Fig. 1.9 (Wang, 1971). The lateral surfaces of the wedge-shaped volume are characteristic surfaces that contain the outermost subcharacteristics crossing AB. In the same figure we show schematically the zone of dependence of points on AB, which is defined in a similar way.

Let us repeat the arguments for another system of equations, defined again in a three-dimensional space, the space x_1, x_2, t instead of the space x_1, x_2, x_3. The role of x_3 will now be undertaken by time, a convenient interchange that has been identified and utilized by a great number of

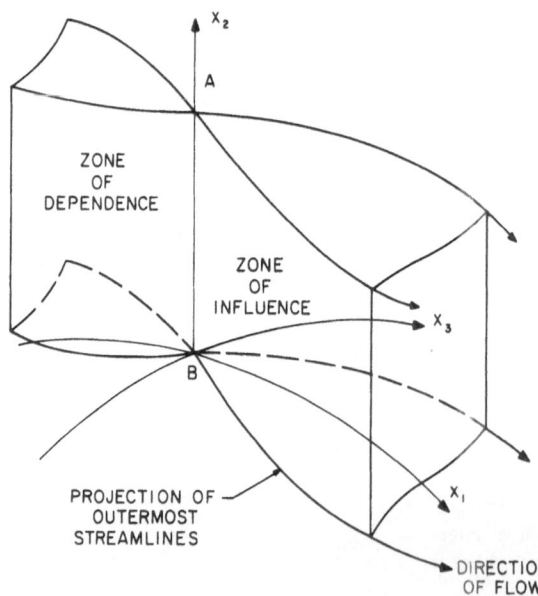

Fig. 1.9 Zones of influence and dependence of points on the line AB.

investigators. The boundary-layer equations then take the form

$$\frac{\partial u_1}{\partial x_1} + \frac{\partial u_2}{\partial x_2} = 0, \tag{1.6.2}$$

$$\frac{\partial u_1}{\partial t} + u_1 \frac{\partial u_1}{\partial x_1} + u_2 \frac{\partial u_1}{\partial x_2} = \frac{\partial U_{1e}}{\partial t} + U_{1e} \frac{\partial U_{1e}}{\partial x_1} + \frac{1}{Re} \frac{\partial^2 u_1}{\partial x_2^2}. \tag{1.6.3}$$

The characteristic and subcharacteristic equations of the above system are

$$\left(\frac{\partial \phi}{\partial x_2}\right)^3 = 0, \tag{1.6.4}$$

$$\frac{\partial \phi}{\partial t} + u_1 \frac{\partial \phi}{\partial x_1} + u_2 \frac{\partial \phi}{\partial x_2} = 0. \tag{1.6.5}$$

Once again we may define a wedge-shaped domain of influence in the three-dimensional space t, x_1, x_2 based on the largest and smallest value of the slope

$$\frac{dx_1}{dt} = u_1(x_1, x, t). \tag{1.6.6}$$

If the maximum and minimum slopes are given by U_{1e} and 0, respectively, which is true for steady flows (Nickel, 1958), the domains of influence and dependence can be represented by two wedges shown schematically in Fig. 1.10. Notice that as time increases, the influence of P extends further away

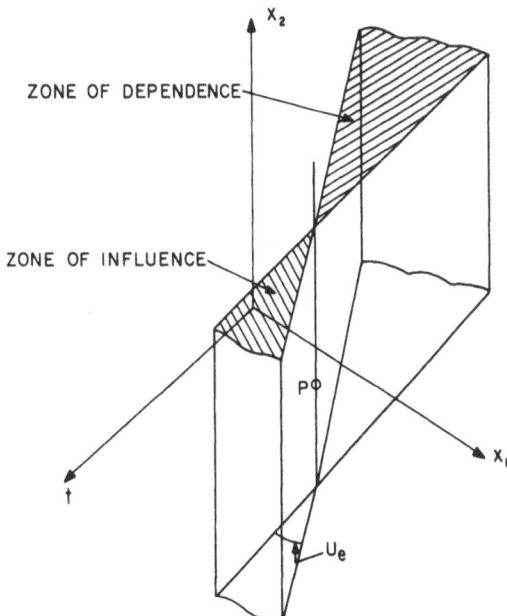

Fig. 1.10 Zones of influence and dependence for unsteady nonreversing flows.

into larger and larger distances x_1. However, this is not generally the case for unsteady boundary-layer flows. The basic theorems derived by Nickel (1958) are not valid for time-dependent equations and hence maxima and minima may be found away from the boundaries. The maximum velocity at an x_1, t station therefore is not necessarily U_{1e}. An account of Nickel's results and the extensions to unsteady flows reported by Velte (1960) will be found in the next section. A number of recent numerical investigations have indeed proved that in unsteady boundary-layer flows the velocity profiles may reverse or overshoot their outer flow limit. In this case the domains of influence and dependence extend both upstream and downstream as shown schematically in Fig. 1.11. In fact the slope of the characteristic planes is given by

$$\tan \theta_1 = \min(u), \qquad \tan \theta_2 = \max(u),$$

where positive angles are measured counterclockwise, that is, in the upstream direction. The significance of this finding becomes apparent if the outer flow velocity is in the direction of x_1, but the velocity u changes sign within the boundary layer. The properties of the flow at P are then influenced by points upstream as well as downstream of P. This will play a significant role in developing a proper differencing scheme and a stable method of integration of the boundary-layer equations.

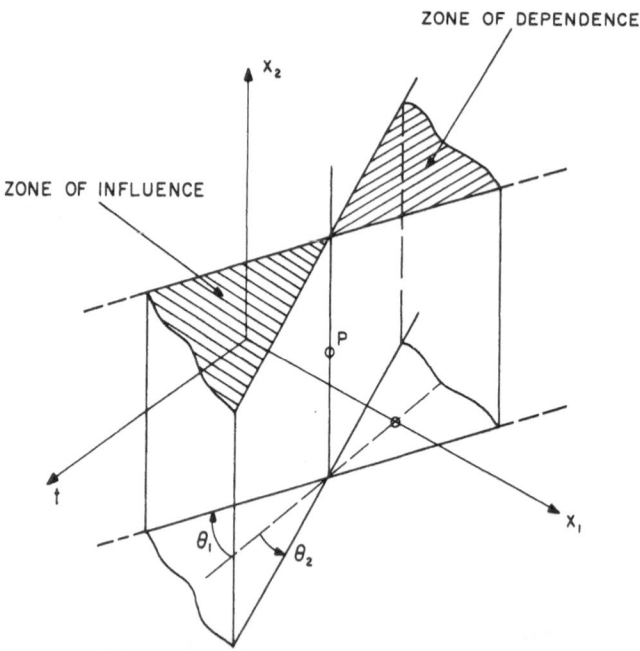

Fig. 1.11 Zones of influence and dependence for reversing flows.

1.7 Existence, Uniqueness, and Bounds

Existence and uniqueness of solutions of the full Navier–Stokes and the boundary-layer equations were not investigated until recently and Russian mathematicians have been by far the protagonists in this area (Ladyzhenskaya, 1969; Oleinic, 1969). This author, however, has found that few of the competent investigators who made significant contributions to the field of computations of unsteady viscous flows are familiar with such rigorous mathematical theorems. A formal development of the theory that leads to final proofs of existence and uniqueness is beyond the scope of this monograph. However, it was felt that a mere presentation of the theorems that pertain exclusively to unsteady viscous flows and a few representative references may be appropriate here, at least for the sake of completeness. The material of this section, therefore, is not in the mainstream of the contents of this monograph.

More interesting to numerical analysts but also rather unnoticed are the works of Nickel (1958) and Velte (1960) pertaining to bounds on various properties of boundary-layer flow. Such constraints could be incorporated in computer codes to prevent the calculations from a totally erroneous path. Unfortunately, little has been done in this direction for unsteady flows.

Consider the two-dimensional problem to which Eqs. (1.2.16)–(1.2.19) and the boundary conditions (1.2.20)–(1.2.26) reduce, if the flow is assumed two-dimensional and the temperature field is not taken into consideration.

The domain of integration $\mathcal{D}$ is defined by the inequalities

$$0 \leqslant x_1 \leqslant x_m, \qquad 0 \leqslant t \leqslant t_m, \qquad 0 \leqslant x_2 \leqslant \infty, \tag{1.7.1}$$

and the boundaries are

$$R' : x_1 = x_m, \qquad 0 < x_2 < \infty, \qquad 0 \leqslant t \leqslant t_m;$$
$$t = t_m, \qquad 0 \leqslant x_2 < \infty, \qquad 0 \leqslant x_1 \leqslant x_m, \tag{1.7.2}$$

R : the rest of the boundary of the domain $\mathcal{D}$.

The existence and uniqueness of this problem have been investigated by Oleinic (1966). It was proven that under certain smoothness conditions on the functions u_{11}, u_{12}, u_{20}, U_{1e}, and with $u_{1w} = 0$, as well as under some compatibility conditions on the initial profiles, there exists a unique solution to the problem, for any $x_m \leqslant \infty$, when t_m is sufficiently small, or for any $t_m \leqslant \infty$, when x_m is sufficiently small. Nonexistence is attributed, for larger x_m and t_m, to separation. Oleinic (1966) also proves that the solution to this problem depends continuously on the given functions $u_{1i}(t_1, x_2)$, $u_{20}(x_1, x_2, x_3)$, $u_{2w}(t)$, and $U_{1e}(t, x_1)$, and that this solution is stable for $t \to \infty$. Moreover, if these functions have limits for $t \to \infty$, then the solution tends asymptotically to a steady solution corresponding to steady flow with stationary boundary conditions, the limits of the above conditions.

For two-dimensional steady flows, Nickel (1958) has used a maximum–minimum principle to derive some basic but very interesting properties of the velocity profile of boundary layers. Very briefly, these are (Nickel, 1973):

(a) If $u_1(0, x_2) \leqslant U_{1e}(0)$, then $u_1(x_1, x_2) \leqslant U_{1e}(x_1)$ for $x_1 > 0$. In the terminology of fluid mechanics, the velocity profile at a certain station x_1 will never exceed the outer flow velocity at the same station, if the initial velocity at $x_1 = 0$ does not exceed the initial outer-flow velocity.

(b) If $dU_{1e}/dx_1 \geqslant 0$ and if $u_1(0, x_2) \leqslant U_{1e}(0) + e$, where e is a positive quantity, then $u_1(x_1, x_2) \leqslant U_{1e}(x_1) + e$ for $x_1 > 0$. Namely, for a favorable pressure gradient, the overshoot of the profile at a certain station is not larger than the overshoot of the initial profile. The term "overshoot" is used to denote the amount by which the velocity within the boundary layer exceeds the outer-flow velocity at the same station.

(c) If $dU_{1e}/dx_1 \leqslant 0$ and if $u_1(0, x_2) \leqslant U_{1e}(0) + e$, then $u_1(x_1, x_2) \leqslant U_{1e}(0) + e$ for $x_1 > 0$. Namely, for adverse pressure gradients, the maximum of the velocity profile at a certain station does not exceed the maximum of the initial velocity profile.

These three theorems were schematically presented by Nickel (1973) as shown in Fig. 1.12. An uphill sloping wall represents flow with favorable pressure gradients, and the opposite is true for a downhill sloping wall.

Moreover, Nickel discovered the following:

(d) The slope of a profile does not grow in the interior of the flow.

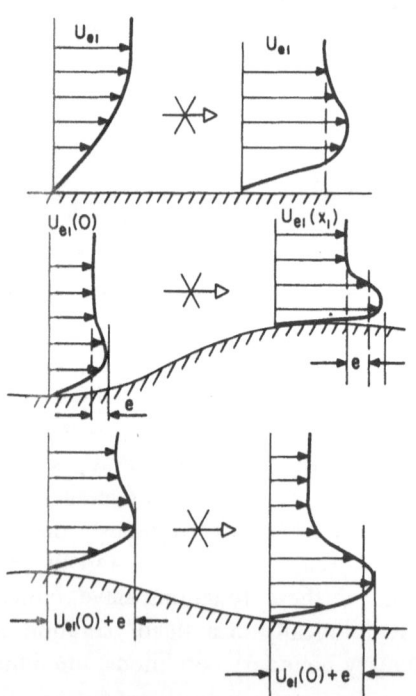

Fig. 1.12 Symbolic representation of Nickel's constraining properties.

(e) Monotonic profiles remain monotonic.

(f) The number of extreme points of a profile does not increase downstream.

With regard to reversed flow, for $dU_{1e}/dx_1 \geqslant 0$,

(g) In the interior of the flow, the velocity component cannot vanish; that is, no reversed flow may develop.

(h) If $du_1/dx_2 > 0$ at $x_1 = x_2 = 0$, then $du_1/dx_2 > 0$ at $x_2 = 0$ for all x_1; that is, no reversed flow and, hence, for steady flows, no separation may occur.

Finally,

(i) If $dU_{1e}/dx_1 \geqslant 0$, then convex profiles remain convex.

(j) If the function dU_{1e}/dx_1 changes its sign n times, then the final profile at $x_1 = x_m$ has $n + 1$ more turning points than the initial profile. Here $n + 1$ may be replaced by n if the initial profile is convex at the wall and in case U_{1e} has a relative minimum at $x_1 = 0$, or, alternatively, if the initial profile is concave at the wall and if U_{1e} at $x_1 = 0$ has a relative minimum.

All these statements give a very good insight into the character of steady boundary layers and were given here as a reference, because almost all are violated in unsteady flows. Unfortunately there is no formal proof or rigorous information on bounds of unsteady boundary layers. In this monograph we shall only refer to classes of problems whose numerical solutions were shown to violate the properties that Nickel discovered for steady flow. This is essentially due to the acceleration term $\partial U_{1e}/\partial t$. Oscillatory flow over a flat plate as we shall discuss in Chapter 4 violates condition (a). For adverse pressure gradients that lead to separation, we shall describe in Chapter 7 solutions that involve overshoots of the order of 3 to 4 times the outer-flow velocity downstream of initial velocity profiles that have no overshoot at all, in violation of Nickel's condition (c).

It is worthwhile to enumerate here a few more cases of unsteady flows that violate Nickel's properties in order to emphasize the variety of the characteristics of unsteady flows and contrast them with those of steady flow. In Chapter 5 we discuss the response of laminar boundary layers to pulsating external flows and describe how the nonlinearities may give rise to nonmonotonic profiles of both the velocity and temperature, which in fact may exhibit three or even five extrema even though the initial profiles are monotonic. This is opposite to Nickel's steady-flow properties (e) and (f). For the same case, it has been known for a long time that reverse flow may temporarily appear even for favorable pressure gradients in violation of property (h). Finally, in Chapter 7 we describe how steady flow over moving walls separates according to the MRS pattern, that is, a stagnation point develops away from the wall, in violation of property (g).

In an attempt to generalize the theory to unsteady flow, Velte (1960) considered the linearized version of the problem by assuming that the coefficients in the convective part of the equations do not vary with space or

Fig. 1.13 Domain of integration for
the two-dimensional unsteady equations.

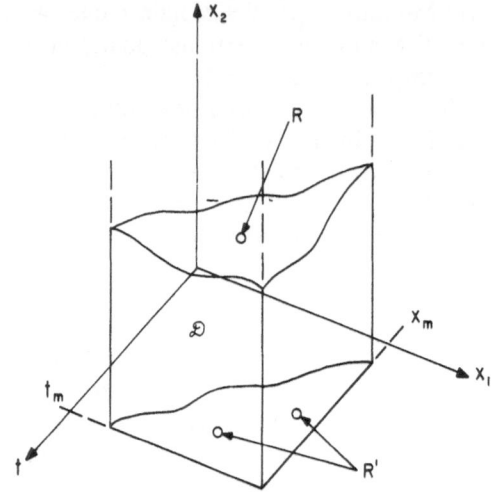

time. He first derived similar conclusions with Nickel, for three-dimensional
steady flow and for less-stringent conditions by employing Nirenberg's
maximum principle. He then converted his equations from three-dimensional
steady to two-dimensional unsteady. However the boundary conditions
assumed by Velte are time-independent. The problems he examined therefore belong to a very small class of rather unimportant character with regard
to physical applications.

Velte (1960) assumed a more general form of the domain of integration.
He considers the space $\mathfrak{D}$ defined by the inequalities

$$x_i < x_1 < x_m, \qquad 0 < x_2 < \infty, \qquad t_i < t < t_m$$

with the boundaries

$$R' : x_1 = x_m, \qquad 0 < x_2 < \infty, \qquad t_i < t \leqslant t_m;$$
$$t = t_m, \qquad 0 < x_2 < \infty, \qquad x_i < x_1 \leqslant x_m;$$

R : the points on the boundary of $\mathfrak{D}$, that do not belong to R'.

This space is shown schematically in Fig. 1.13. Velte then proceeded to
prove theorems, some of which are listed below.

Theorem 1. *Let $u_1(x_1, x_2, t) \geqslant 0$ in $\mathfrak{D} + R + R'$, $U_{1e}(x_1) > 0$ and dU_{1e}/dx_1
> 0 in $x_i \leqslant x_1 \leqslant x_m$. Then $u_1(x_1, x_2, t) > 0$ in $\mathfrak{D} + R'$. That is for all times,
the velocity component u_1 will vanish at most at the wall or at an initial
station.*

Theorem 2. *Let $u_1 \leqslant 0$, $U_{1e} < 0$, $dU_{1e}/dx \leqslant 0$ in $x_i \leqslant x_1 \leqslant x_m$. Then (a)
$u_1(x_1, x_2, t) \leqslant \sup u$ in $\mathfrak{D} + R'$, where $\sup u$ is the supremum of all u_1s
evaluated at points in R. If no reversed flow exists and U_{1e} does not increase
anywhere, that is, there is no adverse pressure gradient, then the function u_1*

cannot grow larger than the constant $\sup u$ *in the interval* $t_i \leqslant t \leqslant t_m$. (b) *If, moreover,* $u_1 < \sup u$ *at the wall, then,* $u_1(x_1, x_2, t) < \sup u$ *in* $\mathcal{D} + R'$.

Theorem 3. *Let* $u_1 \geqslant 0$ *on* R: *then* (a) $u_1^2 \geqslant U_{1e}^2 + M$ *in* $\mathcal{D} + R'$ *and, moreover, if this inequality is satisfied at the wall with only the "less than" sign, then* (b) $u_1^2 < U_e^2 + M$ *in* $\mathcal{D} + R'$, *where* M *is the supremum of* $u_1^2 - U_{1e}^2$.

REFERENCES

Cole, J. D., 1968. *Perturbation Methods in Applied Mathematics*, Blaisdell, Boston, Massachusetts.

Courant, R. and Hilbert, D., 1962. *Methods of Mathematical Physics*, Interscience, New York.

Der, J., Jr., and Raetz, G. S., 1962. I.A.S. Paper No. 62–70.

Ladyzhenskaya, O. A., 1969. *The Mathematical Theory of Viscous Incompressible Flow*, Glasgow, The University Press.

Nayfeh, A. H., 1973. *Perturbation Methods*, Wiley, New York.

Nickel, K., 1958. *Arch. Rat. Mech. Anal.*, **2**, 1–31.

Nickel, K., 1973. *Annual Review of Fluid Mechanics*, eds. Van Dyke, M., Vincenti, W. G., and Wehausen, J. V., Annual Reviews, Inc., **5**, 405–428.

Oleinic, O., 1966. Linsei-Rendiconti, *Sci. Fis. Mat. Nat.*, **XLI**, 32–40.

Oleinic, O., 1969. "Mathematical Problems of Boundary Layer Theory," Lecture Notes, Univ. of Minnesota Press.

Petrovskii, I. G., 1967. *Partial Differential Equations*, Saunders, Philadelphia, Pennsylvania.

Raetz, G. S., 1957. "A Method of Calculating Three-Dimensional Laminar Boundary Layers of Steady Compressible Flows," Northrop Corp., Report No. NA1-58-73.

Schlichting, H., 1968. *Boundary-Layer Theory*, McGraw-Hill, New York.

Van Dyke, M., 1964. *Perturbation Methods in Fluid Mechanics*, Academic Press, New York.

Velte, W., 1960. *Arch. Rat. Mech. Anal.*, **5**, 420–431.

Wang, K. C., 1971. *J. Fluid Mech.*, **48**, 397–404.

Wang, K. C., 1975. *Phys. Fluids*, **18**, 951–955.

White, F. M., 1974. *Viscous Fluid Flow*, McGraw-Hill, New York.

CHAPTER 2

Numerical Analysis

2.1 Introduction

This monograph is addressed to physicists and engineers with background in fluid mechanics and some familiarity with numerical analysis. No special sections are included to introduce the reader to basic concepts of fluid mechanics. However, since the monograph appears as a volume of a series in Computational Physics, it is perhaps pertinent here to include some fundamental concepts, formulas, and theorems on numerical methods.

We start with definitions and provide some elementary examples, with parabolic equations mostly in mind. The presentation is compact and the reader should not expect to educate himself on the topic by reading this chapter. To some, the material may appear trivial and elementary. The writer has nevertheless found it necessary in a graduate course on unsteady viscous flows to review such basic concepts. It is felt that others, too, may find the following outline helpful when teaching a similar course. The intent of this chapter is, in short, to introduce elementary ideas in a comprehensive way, introduce a notation convention, state elementary theorems, refer the reader to more appropriate references for more details, and, finally, concentrate on schemes necessary for the solution of unsteady viscous flows, mainly the unsteady boundary-layer equations.

The last few sections are devoted to modern developments in numerical analysis inspired by and used exclusively in problems of unsteady flows. The results of such methods and their physical significance will be discussed in later chapters. Most of the model examples are linear, and convergence and stability are discussed rigorously only for linear equations. Such methods are merely indicative when applied to nonlinear equations, and, to the knowledge of the author, no extensions of the theory to nonlinear systems are available.

Approximate numerical solutions of differential equations can be derived by solving the corresponding finite-differencing equations. Classical texts on the topic are concerned with the relations between the differential equations, the difference equations and their solutions (Richtmyer, 1957; Smith, 1965; Richtmyer and Morton, 1967; Carnahan *et al.*, 1969; Roache, 1972). The

Fig. 2.1 A uniform grid
and notation for grid
points.

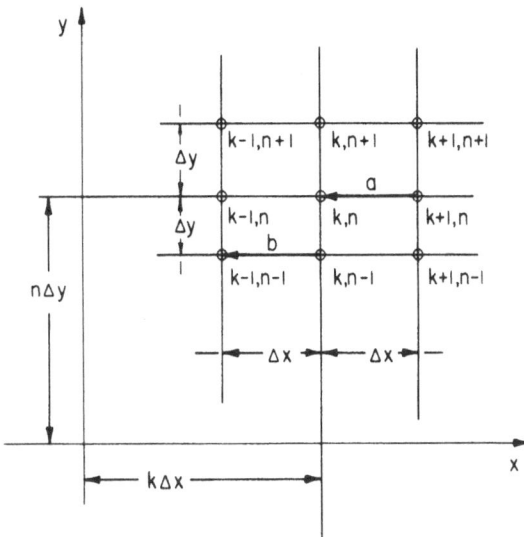

domain of integration is covered by a *net* (or *grid* or *lattice*) whose mesh size
is at the disposal of the numerical analyst. Consider for example the
two-dimensional space x, y (Fig. 2.1) covered with a net of uniform mesh
sizes Δx by Δy.

The point $(k\Delta x, n\Delta y)$ in the two-dimensional space is denoted by the pair
(k, n). A continuous function u of x and y is denoted by $u(x, y)$, whereas a
discrete function defined at the grid points (k, n) is denoted by $u_{k,n}$. Differences are denoted by the symbols Δ and δ and represent differences of the
continuous function $u(x, y)$, or the discrete function $u_{k,n}$, respectively. For
example, differences in the x-direction are

$$\Delta u(x, y) = u(x + \Delta x, y) - u(x, y), \tag{2.1.1}$$

$$\delta u_{k,n} = u_{k+1,n} - u_{k,n}. \tag{2.1.2}$$

It is preferable to use a more complete notation, whereby a subscript on the
symbol δ represents the direction in which the difference is taken:

$$(\delta_x u)_{k,n} = u_{k+1,n} - u_{k,n}. \tag{2.1.3}$$

Quite often such differences will be sketched in the graphical representation of the net by a vector and will be denoted by a plain symbol. An
average of two differences shown in Fig. 2.1, for example, can then quickly
be identified as

$$\frac{1}{2}\left[(\delta_x u)_{k+1,n} + (\delta_x u)_{k,n-1}\right] = \frac{a+b}{2}. \tag{2.1.4}$$

Consider now Taylor's expansions about the point (x, y):

$$u(x + \Delta x, y) = u(x, y) + \frac{\partial u(x, y)}{\partial x} \Delta x + \frac{1}{2!} \frac{\partial^2 u(x, y)}{\partial x^2} (\Delta x)^2 + \cdots,$$

$$(2.1.5)$$

$$u(x - \Delta x, y) = u(x, y) - \frac{\partial u(x, y)}{\partial x} \Delta x + \frac{1}{2!} \frac{\partial^2 u(x, y)}{\partial x^2} (\Delta x)^2 + \cdots,$$

$$(2.1.6)$$

$$u(x, y + \Delta y) = u(x, y) + \frac{\partial u(x, y)}{\partial y} \Delta y + \frac{1}{2!} \frac{\partial^2 u(x, y)}{\partial y^2} (\Delta y)^2 + \cdots.$$

$$(2.1.7)$$

Rearranging these equations, adding, or subtracting, etc., we can express derivatives in terms of differences. For example,

$$\frac{\partial u}{\partial x} = \frac{u(x + \Delta x, y) - u(x, y)}{\Delta x} + O(\Delta x), \qquad (2.1.8)$$

$$\frac{\partial u}{\partial y} = \frac{u(x, y + \Delta y) - u(x, y)}{\Delta y} + O(\Delta y), \qquad (2.1.9)$$

$$\frac{\partial^2 u}{\partial x^2} = \frac{u(x + \Delta x, y) - 2u(x, y) + u(x - \Delta x, y)}{(\Delta x)^2} + O\left[(\Delta x)^2\right], \quad (2.1.10)$$

where the symbol $O(\ )$ represents the order of magnitude of the terms omitted and, in the terminology of numerical analysis, the *truncation error*. The corresponding difference ratios of the discrete function are expressed accordingly:

$$\frac{\delta u}{\delta x} = \frac{u_{k+1,n} - u_{k,n}}{\Delta x}, \qquad (2.1.11)$$

$$\frac{\delta u}{\delta y} = \frac{u_{k,n+1} - u_{k,n}}{\Delta y}. \qquad (2.1.12)$$

It is common in literature to denote central differencing of the second order in a compact way:

$$\left(\delta_x^2 u\right)_{k,n} = u_{k+1,n} - 2u_{k,n} + u_{k-1,n} \qquad (2.1.13)$$

and thus,

$$\frac{\delta^2 u}{\delta x^2} = \frac{u_{k+1,n} - 2u_{k,n} + u_{k-1,n}}{(\Delta x)^2}. \qquad (2.1.14)$$

We can always define a differencing equation corresponding to a differential equation by replacing all differences of the form of (2.1.8)–(2.1.10) with differences of the form of (2.1.11)–(2.1.14). The boundary conditions can be expressed in terms of the discrete functions. The problems of how well the solution of the difference equations for the discrete function $u_{k,n}$ approximates the function $u(k\Delta x, n\Delta y)$ are the concern of the theory of computational methods and will be outlined in the sections that follow. There are essentially two types of errors that may appear:

(i) Errors due to the fact that derivatives are approximated by differences. These are essentially created when we drop the terms of order Δx, Δy, etc., in the expressions (2.1.8)–(2.1.10). They are called *discretization errors* and are associated with the concept of *convergence*.

(ii) Errors due to the fact that calculations can be carried out only to a finite number of decimal places or significant figures and other numerical errors that can propagate and possibly grow. These are called *round-off errors* and are connected with the concept of *stability*. The next two sections deal in more detail with convergence and stability.

2.2 Convergence

Let $u(x,t)$ represent the exact solution of a differential equation and $u_{k,n}$ the exact solution of the corresponding difference equation. The finite-difference equation is said to be *convergent*, if

$$u_{k,n} \rightarrow u(k\Delta x, n\Delta t) \quad \text{as } \Delta x \rightarrow 0,$$

$$\text{and } \Delta t \rightarrow 0. \tag{2.2.1}$$

The definition can be extended to hold for systems of partial differential equations with respect to three or more independent variables. The concept of convergence is more subtle than one may think, as Roache (1972) notes. Clearly in the limit of $\Delta x \rightarrow 0$, the expression (2.1.8), for example, reduces to Newton's definition of a derivative. However, the limits denoted in Eq. (2.2.1) imply the limits of the solutions of the differential and the difference equation, respectively, and not merely the individual terms.

A simple numerical example

Let us consider here the one-dimensional heat equation

$$\frac{\partial u}{\partial t} = \alpha \frac{\partial^2 u}{\partial x^2}, \tag{2.2.2}$$

where α is a dimensionless diffusivity and u is the temperature. This is a classical example, quite pertinent here, since our main interest is in parabolic equations. A finite-difference form of Eq. (2.2.2) is

$$\frac{u_{k,n+1} - u_{k,n}}{\Delta t} = \alpha \frac{u_{k+1,n} - 2u_{k,n} + u_{k-1,n}}{(\Delta x)^2}, \tag{2.2.3}$$

where the time derivative has been expressed by a forward difference. This equation can be solved by sweeping the domain of integration along the time axis. If all the data are known for all k at the station n, then each of Eqs. (2.2.3) can be solved independently to generate the temperature profile at the column $n + 1$. Calculations should start at $n = 0$, where the initial profile $u_{k,0}$ is given for all k. This is an *explicit* scheme of differencing. The parameter

$$\lambda = \alpha \frac{\Delta t}{(\Delta x)^2} \tag{2.2.4}$$

is at our disposal and will play a significant role in the development that follows.

In almost all classical texts a simple numerical example is presented to demonstrate how this scheme may converge or diverge. We chose here to describe an example proposed by Carnahan et al. (1969). Equation (2.2.3) may be rewritten in the form

$$u_{k,n+1} = \lambda u_{k+1,n} + (1 - 2\lambda)u_{k,n} + \lambda u_{k-1,n}. \tag{2.2.5}$$

Let the boundary and initial conditions be

$$u = 0, \quad \text{for } t \leqslant 0, 0 < x < 5, \tag{2.2.6}$$

$$u = 100, \quad \text{for } x = 0 \text{ and } 5, t > 0. \tag{2.2.7}$$

For simplicity let $\alpha = 1$. Then two sets of calculations can be easily carried out by hand, one for $\Delta t = 0.01$, $\Delta x = 0.2$ and another for $\Delta t = 0.04$, $\Delta x = 0.2$. They correspond to $\lambda = \frac{1}{4}$ and $\lambda = 1$, respectively. The results are plotted in Figs. 2.2a and 2.2b.

Physically the problem may represent an infinite slab of thickness 5, at a uniform temperature, whose walls are suddenly given a change in temperature. Our mathematical model describes the way in which the temperature of the slab will increase, asymptotically approaching the temperature of the surface. A closed-form solution exists and is given in terms of a Fourier series:

$$u = 100\left[1 - \frac{4}{\pi} \sum_{n=1,3}^{\infty} \frac{1}{n} e^{-(n\pi/5)^2 \alpha t} \sin \frac{n\pi x}{5}\right] \tag{2.2.8}$$

It is known that the solution for $t > 0$ has a finite and negative curvature and remains monotonic on each side of the middle of the slab. It can easily be shown by comparison with the exact solution that the curves of Fig. 2.2a calculated with $\lambda = \frac{1}{4}$ represent a crude but reasonable approximation of the solution for the first few n's. In this figure we show the exact solution only for $t = 0.05$. Clearly the erratic jumps of the solution for $\lambda = 1$, shown in Fig. 2.2b are unacceptable. The latter solution is nevertheless an exact solution of the difference equation and its values are given in Table 2.1 for the first few n's. The exact solution to the difference equation with $\lambda = 1$ is therefore

Fig. 2.2(a) Solution of the heat equation for $\Delta t = 0.01$, $\Delta x = 0.2$, i.e., $\lambda = \frac{1}{4}$. The exact solution for $t = 0.05$ is plotted with a dashed line. (b) Solution of the heat equation for $\Delta t = 0.04$, $\Delta x = 0.2$, i.e., $\lambda = 1$.

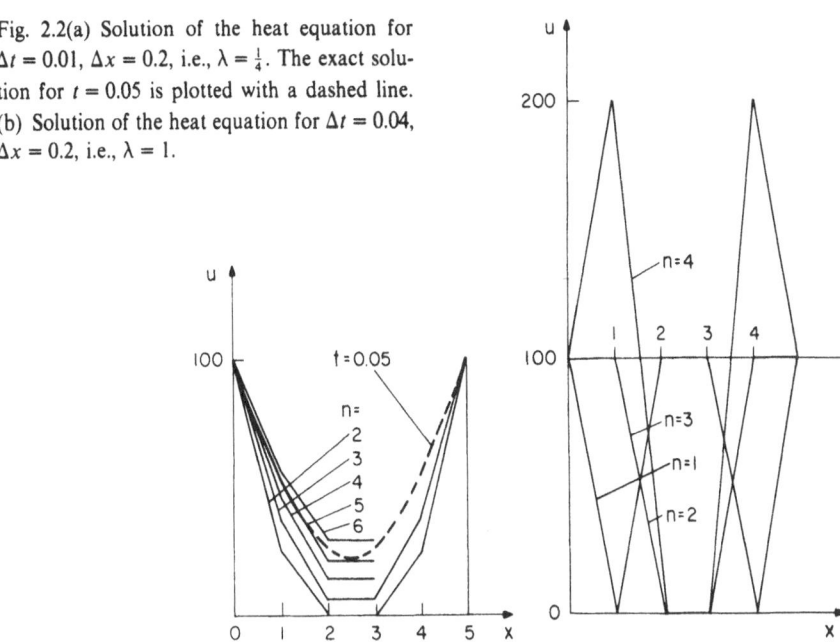

(a)   (b)

shown to deviate more and more from the exact solution of the differential equation as time increases.

No round-off errors can be blamed for the failure of this scheme to come even close to the solution of the differential equation. For $\lambda = 1$, the solution of the difference equation would exhibit this erratic behavior, even if the mesh sizes Δt and Δx chosen were much smaller. This is a typical example of a diverging numerical scheme.

Some authors describe this behavior as "unstable." This is a poor choice, since the term *stability* is an explicitly defined term in numerical analysis. In fact the use of the term here is most unfortunate, because this example is a typical example of *nonconvergence* and not an example of instability. The fact is that Eq. (2.2.3) is also unstable for $\lambda = 1$. This should not confuse the issue here because in the numerical example, the difference equation has been solved exactly.

Table 2.1

Time Subscript	Space Subscript k					
n	0	1	2	3	4	5
0	0	0	0	0	0	0
1	100	0	0	0	0	100
2	100	100	0	0	100	100
3	100	0	100	100	0	100
4	100	200	0	0	200	100
5	100	− 100	200	200	− 100	100

General approach

In the spirit of Eqs. (2.1.8)–(2.1.10), derivatives of the function $u(x,t)$ can be expressed in terms of higher-order derivatives. For example, the time derivative becomes

$$\frac{\partial u(x,t)}{\partial t} = \frac{u(x,t+\Delta t) - u(x,t)}{\Delta t} - \frac{\Delta t}{2}\frac{\partial^2 u(x,t)}{\partial t^2} + 0\left[(\Delta t)^2\right]. \qquad (2.2.9)$$

Adding Eqs. (2.1.5) and (2.1.6) expanded to order $(\Delta x)^6$ yields

$$\frac{\partial^2 u(x,t)}{\partial x^2} = \frac{u(x+\Delta x,t) - 2u(x,t) + u(x-\Delta x,t)}{(\Delta x)^2}$$

$$+ \frac{1}{12}\frac{\partial^4 u(x,t)}{\partial x^4}(\Delta x)^2 + 0\left[(\Delta x)^4\right]. \qquad (2.2.10)$$

Substitution of (2.2.9) and (2.2.10) into (2.2.2) with $\alpha = 1$ yields

$$u(x,t+\Delta t) = \lambda u(x+\Delta x,t) + (1-2\lambda)u(x,t)$$

$$+ \lambda u(x-\Delta x,t) + \frac{(\Delta t)^2}{2}\frac{\partial^2 u(x,t)}{\partial t^2}$$

$$- \frac{\Delta t(\Delta x)^2}{12}\frac{\partial^4 u(x,t)}{\partial x^4} + 0\left[(\Delta t)^3\right]$$

$$+ 0\left[\Delta t(\Delta x)^4\right]. \qquad (2.2.11)$$

The discretization error is defined as the difference

$$\epsilon_{k,n} = u(k\Delta x, n\Delta t) - u_{k,n}. \qquad (2.2.12)$$

We can readily derive an equation for $\epsilon_{k,n}$ if we evaluate Eq. (2.2.11) at $x = k\Delta x$ and $t = n\Delta t$. We shall then simply subtract Eq. (2.2.5) from (2.2.11) to arrive at almost the same form as the equation that governs $u(x,t)$, i.e., (2.2.11):

$$\epsilon_{k,n+1} = \lambda\epsilon_{k+1,n} + (1-2\lambda)\epsilon_{k,n} + \lambda\epsilon_{k-1,n}$$

$$+ \frac{(\Delta t)^2}{2}\frac{\partial^2 u}{\partial t^2} - \frac{\Delta t(\Delta x)^2}{12}\frac{\partial^4 u}{\partial x^4} + 0\left[(\Delta t)^3\right] \qquad (2.2.13)$$

$$+ 0\left[\Delta t(\Delta x)^4\right].$$

If R is a sufficiently large number, and $\lambda \leqslant 1/2$, then

$$|\epsilon_{k,n+1}| \leqslant \lambda|\epsilon_{k+1,n}| + (1-2\lambda)|\epsilon_{k,n}| + \lambda|\epsilon_{k-1,n}| + R\Delta t(\Delta x)^2. \qquad (2.2.14)$$

Let ϵ_n be the upper bound of the absolute values of $\epsilon_{k,n}$ at the time level n

$$\epsilon_n = \sup|\epsilon_{k,n}|, \qquad (2.2.15)$$

then Eq. (2.2.14) becomes

$$|\epsilon_{k,n+1}| \leqslant \lambda\epsilon_n + (1 - 2\lambda)\epsilon_n + \lambda\epsilon_n + R\Delta t(\Delta x)^2 = \epsilon_n + R\Delta t(\Delta x)^2. \quad (2.2.16)$$

Equation (2.2.16) holds for all $\epsilon_{k,n+1}$, hence for their maximum as well:

$$\epsilon_{n+1} \leqslant \epsilon_n + R\Delta t(\Delta x)^2. \quad (2.2.17)$$

By evaluating Eq. (2.2.17) for all n and taking into account the fact that $\epsilon_0 = 0$, we conclude that

$$\epsilon_n \leqslant Rn\Delta t(\Delta x)^2. \quad (2.2.18)$$

Thus the discretization error is bounded and can be made smaller if smaller values of Δt and Δx are chosen. The explicit scheme is therefore a converging scheme if $\lambda \leqslant \frac{1}{2}$.

Going back to Eq. (2.2.13), we note that, for the special case of $\lambda = \frac{1}{6}$, two more terms cancel, because $\partial^2 u/\partial t^2 = \partial^4 u/\partial x^4$, a relationship which can be easily derived from Eq. (2.2.2). Thus, the discretization error becomes

$$\epsilon_n \leqslant Rn\Delta t(\Delta x)^4. \quad (2.2.19)$$

2.3 Stability

We reserve the term "stability" in numerical analysis to describe the decay or amplification of numerical errors that somehow penetrated our calculations. The most common of such errors are rounding errors; that is, errors committed because our actual calculations are carried out to a finite number of decimal places. In an actual calculation we shall start accumulating both errors, round-off and discretization errors, and it will be impossible to find a dichotomy between the two.

The concept of *convergence* is easier to grasp and more convenient to express symbolically than the concept of *stability*. If a specific scheme of numerical analysis is chosen, then the truncation error of each term in the equation is uniquely defined. There are two very well-defined solutions: the solution to the differential equation $u(x,t)$, and the solution to the finite-difference equation $u_{k,n}$. True, the last may depend on the choice of Δt and Δx. Moreover, the final expression for the discretization error is generally given in terms of unknown derivatives for which no upper and lower bounds can be estimated. Nevertheless, convergence is not foreign to minds trained in Newtonian calculus. It is intimately connected with the accuracy involved in trying to approximate a derivative with a quotient of finite differences.

Let $\tilde{u}_{k,n}$ be an actual numerical solution that, unfortunately, may never be the same if two different investigators were to attempt to derive it. Then the *round-off* error is defined as

$$E_{k,n} = u_{k,n} - \tilde{u}_{k,n}. \tag{2.3.1}$$

In general, a set of finite-difference equations is *stable* when the cumulative effect of all the rounding errors is negligible. This is essentially the definition of O'Brien, *et al.* (1951) and Eddy (1949). Lax and Richtmyer (1956) define stability more generally by requiring a bounded extent to which any component of the initial data can be amplified in the numerical procedure.

Let us assume that a certain number of errors were committed at the station $\tilde{n}$ and we need to calculate their propagation for larger n. We shall then substitute Eq. (2.3.1) into the finite difference equation. We see immediately that since $u_{k,n}$ satisfies the finite-difference equation, the errors $E_{k,n}$ do so as well, provided the coefficients of the equation do not depend on the unknown functions. To be more specific, we assume that we are asked to carry out two calculations from the time plane $\tilde{n}$ on; one, starting with the exact values $u_{k,\tilde{n}}$ and another, starting with a set of values $\tilde{u}_{k,\tilde{n}}$ that deviate a little from $u_{k,\tilde{n}}$. We further assume that from then on, we shall be able to solve the difference equation with no further numerical errors, i.e., that both $u_{k,n}$ and $\tilde{u}_{k,n}$ for $n > \tilde{n}$ satisfy exactly the difference equation. We need to know now whether for increasing n, the solution $\tilde{u}_{k,n}$ will deviate further from $u_{k,n}$ or whether such deviations will be eventually smoothed out.

A simple numerical example

A very simple example of the decay of an error can be given if we consider the explicit scheme for the solution of the heat equation for $\lambda = \frac{1}{2}$ (see Eq. (2.2.5)).

$$u_{k,n+1} = \tfrac{1}{2}(u_{k-1,n} + u_{k+1,n}), \tag{2.3.2}$$

which can be written in terms of the error

$$E_{k,n+1} = \tfrac{1}{2}(E_{k-1,n} + E_{k+1,n}). \tag{2.3.3}$$

If an error equal to 2 is committed at $k = 5$ and $n = 0$, (Smith, 1965), then it propagates at later times according to Eq. (2.3.3) as shown schematically in Fig. 2.3. It appears that the scheme is stable, which is in fact the case for $\lambda \leqslant \frac{1}{2}$ as will be shown later.

General approach

A wide range of methods for the investigation of stability are available today as, for example, the *matrix* method (Smith, 1965), the *discrete perturbation analysis* method (Thom and Apelt, 1961), and Hirt's method (Hirt, 1965). The most common method was developed by von Neumann in the early

Fig. 2.3 The decay of an
error in the heat equation
with $\lambda = \frac{1}{2}$.

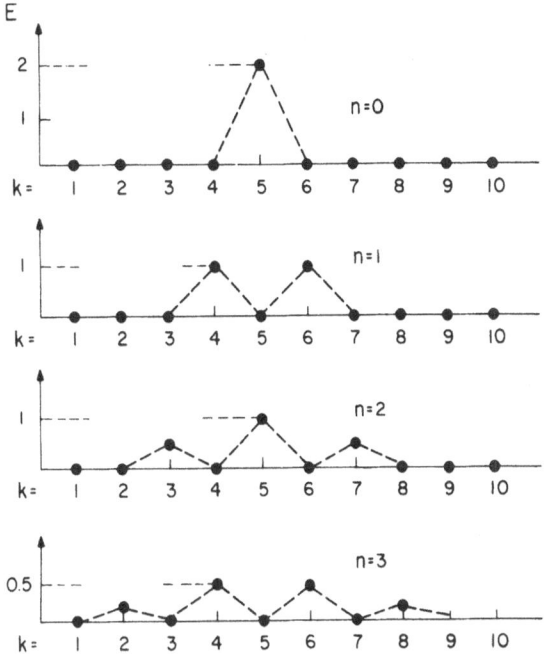

1940s but was finally presented for the first time and discussed in detail by
O'Brien *et al.* (1951). Roache (1972) gives a detailed description as well as a
thorough comparison and evaluation of most of these methods. In this
section we only present an outline of the von Neumann analysis. An initial
line of errors is expressed in terms of a finite Fourier series. The growth of a
function that for $t = 0$ reduces to the Fourier series is then studied by the
method of separation of variables. Von Neumann's analysis is therefore
applicable only for linear problems.

Consider a set of errors $E_{k,n}$ that have somehow creeped into our calculations at the time step n. This arbitrary distribution of values can be Fourier
analyzed and finally expressed in terms of a sine and a cosine series in the
form

$$E_{k,n} = \sum a_m \cos \frac{m\pi x}{L} + \sum b_m \sin \frac{m\pi x}{L}, \qquad (2.3.4)$$

where L is the domain of integration in the x direction. It is more convenient
to work here with complex notation, replacing Eq. (2.3.4) by

$$E_{k,n} = \sum_{m=0}^{N} (A_m)_n e^{im\pi x/L}, \qquad (2.3.5)$$

where i is the imaginary unit and the A_m's are arbitrary complex coefficients.
Equations (2.3.5), with $x = k\Delta x$ are sufficient for the determination of the
$N + 1$ unknowns $(A_0)_n, (A_1)_n, \ldots, (A_N)_n$. An arbitrary distribution of errors
in the x-direction can thus be studied. However, since the problem is linear,

we shall consider only one of the Fourier components. The notation will become less cumbersome if we drop the index m, which defines the Fourier component, and replace the factor $m\pi/L$ by the quantity k_x. Our error then simply becomes

$$E_{k,n} = A_n e^{ik_x x}, \qquad k_x = \frac{m\pi}{L}.$$

(2.3.6)

Moreover, in terms of the phase angle $\theta = k_x \Delta x$, we have

$$E_{k,n} = A_n e^{ik\theta}.$$

(2.3.7)

It is unfortunate that the symbol k_x may be confused with the counter k. However we decided to retain it, since it represents in Eq. (2.3.6) a wave number of a wave propagating in the x direction, and k is traditionally used to denote wave numbers.

We need to investigate now how the coefficient A_n will behave as we proceed to higher values of n. In other words, the Fourier analysis has been used to determine all the A_n's at the nth station. We need to determine whether the amplitude of the error will grow or decay as we proceed to the time level $n + 1$. Note that if we express A_n in terms of an exponential

$$A_n = e^{\alpha t} = e^{\alpha n \Delta t},$$

(2.3.8)

then Eq. (2.3.6) takes the more familiar form of a propagating wave.

The amplification factor is defined as the ratio

$$\xi = \frac{A_{n+1}}{A_n}$$

(2.3.9)

and our error will decrease with n if

$$|\xi| = \left| \frac{A_{n+1}}{A_n} \right| \leqslant 1.$$

(2.3.10)

This is the von Neumann criterion for stability.

Let us consider again our example, Eq. (2.2.5), and recall that the same equation governs the propagation of the errors $E_{k,n}$. Substituting in this equation, expressed in terms of E's, the quantity (2.3.7) and

$$E_{k,n+1} = A_{n+1} e^{ik\theta},$$

(2.3.11)

$$E_{k\pm1,n} = A_n e^{i(k\pm1)\theta},$$

(2.3.12)

we arrive at

$$A_{n+1} e^{ik\theta} = \lambda A_n e^{i(k+1)\theta} + (1 - 2\lambda) A_n e^{ik\theta} + \lambda A_n e^{i(k-1)\theta}$$

(2.3.13)

This can be easily brought into the form

$$A_{n+1} = A_n \left[1 - 2\lambda(1 - \cos\theta) \right].$$

(2.3.14)

The amplification factor is now the ratio

$$\xi = \frac{A_{n+1}}{A_n} = 1 - 2\lambda(1 - \cos\theta), \tag{2.3.15}$$

and the stability requirement becomes

$$-1 \leqslant 1 - 2\lambda(1 - \cos\theta) \leqslant 1 \tag{2.3.16}$$

and must be satisfied for all possible θ, that is, for all possible Fourier terms. The right-hand side is always satisfied while the left-hand side will be valid for all θ, if

$$\lambda = \frac{\alpha\Delta t}{(\Delta x)^2} \leqslant \frac{1}{2}, \tag{2.3.17}$$

that is, if

$$\Delta t \leqslant \frac{1}{2} \frac{(\Delta x)^2}{\alpha}. \tag{2.3.18}$$

We therefore arrive by mere chance at the same restriction imposed on us by the convergence requirements.

Application to implicit schemes

Expressing the time derivative in Eq. (2.2.3) in terms of a backward differencing scheme, we arrive at a system of equations that are coupled:

$$\frac{u_{k,n+1} - u_{k,n}}{\Delta t} = \alpha \frac{u_{k+1,n+1} - 2u_{k,n+1} + u_{k-1,n+1}}{(\Delta x)^2}. \tag{2.3.19}$$

The round-off error satisfies the same equation and therefore substitution of expression (2.3.7) yields

$$e^{ik\theta}A_{n+1} - e^{ik\theta}A_n = \lambda\Big[e^{i(k+1)\theta}A_{n+1} - 2e^{ik\theta}A_{n+1} \\ + e^{i(k-1)\theta}A_{n+1}\Big]. \tag{2.3.20}$$

Thus

$$A_{n+1} - A_n = \lambda A_{n+1}(e^{i\theta} + e^{-i\theta} - 2) \tag{2.3.21}$$

and the amplification factor becomes

$$\xi = \frac{A_{n+1}}{A_n} = \frac{1}{1 + 4\lambda\sin^2(\theta/2)}. \tag{2.3.22}$$

Since $|\xi| < 1$ for all values of the phase θ and the parameter λ, this differencing scheme is stable for any choice of the values of Δt and Δx. The importance of such schemes is now apparent.

There is a large number of implicit schemes available that are unconditionally stable. Richtmyer and Morton (1967) provide an inclusive table of such schemes with their truncation error and the appropriate references.

One of the most important of such schemes is a weighted average between the purely implicit and purely explicit schemes. A finite difference form of Eq. (2.2.2) expressed in this scheme is

$$\frac{u_{k,n+1} - u_{k,n}}{\Delta t} = \alpha \frac{\sigma(\delta_x^2 u)_{k,n+1} + (1 - \sigma)(\delta_x^2 u)_{k,n}}{(\Delta x)^2} , \qquad (2.3.23)$$

where σ may take any value between 0 and 1. For $\sigma = \frac{1}{2}$, we recover the well-known Crank–Nickolson scheme, which is again unconditionally stable.

It can be easily proved (Richtmyer and Morton, 1967) that this scheme of integration is unconditionally stable if

$$\tfrac{1}{2} < \sigma \leqslant 1, \qquad (2.3.24)$$

and that for $0 \leqslant \sigma \leqslant \frac{1}{2}$, the stability criterion is

$$2\lambda \leqslant \frac{1}{1 - 2\sigma} . \qquad (2.3.25)$$

2.4 Consistency and the Equivalence Theorem

The term *truncation error* refers to errors involved when the terms of a differential equation are replaced by finite differences. It should be borne in mind that convergence analysis goes one step further. It investigates the difference of the solutions of the difference and the differential equation. The *discretization* error was defined as the difference $u(k\Delta x, n\Delta t) - u_{k,n}$.

The truncation error for the scheme given by Eq. (2.2.3) is defined as the difference:

$$T = \left[\frac{u(x, t + \Delta t) - u(x, t)}{\Delta t} - \alpha \frac{u(x + \Delta x, t) - 2u(x, t) + u(x - \Delta x, t)}{(\Delta x)^2} \right]$$

$$- \left[\frac{\partial u(x, t)}{\partial t} - \alpha \frac{\partial^2 u(x, t)}{\partial x^2} \right]. \qquad (2.4.1)$$

Using again a Taylor series expansion, it can be easily proved that the truncation error for this explicit scheme is

$$T = O(\Delta t) + O\left[(\Delta x)^2 \right]. \qquad (2.4.2)$$

If the truncation error of a numerical scheme tends to zero as the mesh sizes tend to zero, then the scheme is called *consistent* or *compatible* with the differential equation. It is emphasized that consistency does not imply that the solution of the numerical scheme will converge to the solution of the differential equation. In other words, it is possible that the truncation error tends to zero as the mesh sizes approach zero but the discretization error does not. That is, if the stability condition is violated, then the exact solution of the differential equations comes closer to satisfying the difference equations, but the exact solution of the difference equations departs more and more from the true solution of the problem.

Of fundamental importance here is Lax' equivalence theorem (Richtmyer and Morton, 1967), which states essentially that for linear equations *consistency* and *stability* are necessary and sufficient conditions for *convergence*. This theorem is valid only under relatively stringent conditions and although it has been extended to quasilinear problems (Thompson, 1964), it is far from being a useful tool in engineering applications. However, proving that the numerical scheme of a linearized problem is stable may be considered at least an indication that the solution will converge. The only other method to check for the stability and convergence of a scheme would be to compare the numerical solution with a known exact solution of the problem.

The stability and, therefore most often, the convergence of a numerical scheme is also restricted by the Courant condition (Courant et al., 1928), often referred to in literature as the Courant–Friedrichs–Lewy criterion. This condition states that the finite-difference domain of influence should include the continuum domain of influence. Some clarification here is required. Finite-difference equations are written for each grid point in the mesh. For a particular point then, a specific difference scheme extends to some of the neighboring points, thus forming a local net. The Courant–Friedrichs–Lewy criterion states that the local numerical net of influence of each grid point should contain the domain of influence of the point as defined by the differential equation. Consider for example point A in Fig. 2.4 and assume in

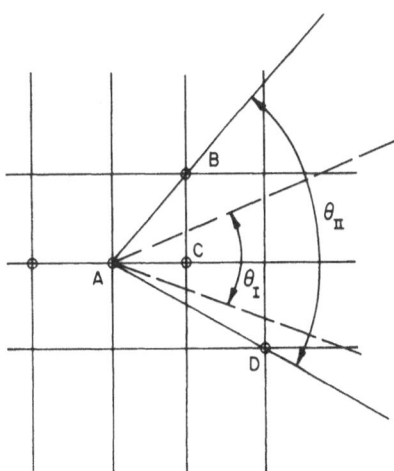

Fig. 2.4 Schematic example of domain of influence for the numerical scheme θ_{II}, and the differential scheme (θ_I) at point A.

a schematic representation that a specific differencing scheme involves the points B, C, and D. The domain of influence of the finite scheme is marked in the figure by the wedge θ_{II}. Let θ_{I} be the wedge of influence of the differential equation determined according to the theory of Chapter 1. The Courant–Friedrichs–Lewy criterion then simply requires that $\theta_{\text{II}} > \theta_{\text{I}}$.

2.5 Unsteady Flow Equations

A great variety of numerical schemes have been proposed for the solution of the unsteady Navier–Stokes equations and their approximate form, the boundary-layer equations. Critically describing all of these schemes would be a tremendous task. We shall be discussing here a more narrow class of problems, mostly based on the unsteady boundary-layer equation. In fact, emphasis will be given to numerical solutions that shed light on the important physical aspects of the problems.

Stability and convergence analysis of the complete system of equations is not possible due to the nonlinearities. In this section we shall investigate the stability of model linear equations and we shall describe various numerical schemes proposed for the solution of the unsteady boundary-layer equations.

The effect of convection

As a first step let us include a first-order derivative in Eq. (2.2.2), which simulates the convective part of the time derivative

$$\frac{\partial u}{\partial t} + U \frac{\partial u}{\partial x} = \alpha \frac{\partial^2 u}{\partial x^2} . \tag{2.5.1}$$

In this equation, U is a constant, which may be the known and constant velocity of a field. Equation (2.5.1) then describes the way heat is conducted and convected in the direction of x.

An explicit finite-difference scheme for Eq. (2.5.1) is

$$u_{k,n+1} - u_{k,n} + \frac{\kappa}{2} (u_{k+1,n} - u_{k-1,n}) = \lambda(u_{k+1,n} - 2u_{k,n} + u_{k-1,n}), \tag{2.5.2}$$

where

$$\kappa = U \frac{\Delta t}{\Delta x} , \qquad \lambda = \alpha \frac{\Delta t}{(\Delta x)^2} , \tag{2.5.3}$$

and the first x derivative has been expressed in terms of a central difference. The truncation error of this scheme is thus $0[\Delta t + (\Delta x)^2]$.

Assuming as before the round-off error in the form $A_n e^{ik\theta}$ and recalling that the propagation of such errors is governed by the same difference equation, we get

$$A_{n+1} = A_n \left[1 - \frac{\kappa}{2} (e^{i\theta} - e^{-i\theta}) + \lambda(e^{i\theta} + e^{-i\theta} - 2) \right], \tag{2.5.4}$$

and therefore the amplification factor becomes

$$\xi = \frac{A_{n+1}}{A_n} = 1 - 2\lambda(1 - \cos\theta) - i\kappa\sin\theta. \tag{2.5.5}$$

It is very interesting to note that now the amplification factor is complex. The imaginary part on the right-hand side derives from the convection term of the original equation. The stability condition is expressed in terms of the modulus of ξ. In the complex plane, therefore, the vector ξ should lie within the unit circle. This is expressed by the condition

$$|\xi|^2 = \xi\bar{\xi} < 1, \tag{2.5.6}$$

where an overbar denotes the complex conjugate.

Roache (1972) suggests that Eq. (2.5.5) be rewritten in the form

$$\xi = (1 - 2\lambda) + 2\lambda\cos\theta - i\kappa\sin\theta. \tag{2.5.7}$$

This then represents an ellipse with major and minor axes parallel to the real and the imaginary axis in the Argand plane, respectively. The center of the ellipse is at the point $1 - 2\lambda$ and half-axis lengths are equal to κ and 2λ, respectively. Notice that the point $\xi = 1$ is always a point of the ellipse.

The absolute values of the real and the imaginary parts of (2.5.7) should be less than 1, independently of each other and for all values of θ. This implies that necessary conditions for stability are

$$\kappa \leqslant 1, \qquad \lambda \leqslant \tfrac{1}{2}. \tag{2.5.8}$$

The more general condition, Eq. (2.5.6), requires that the modulus of ξ be less than 1 for all possible values of θ. Fromm (1964) has carried out involved calculations to find that one more restriction should be added to (2.5.8), namely,

$$\kappa^2 \leqslant 2\lambda. \tag{2.5.9}$$

From Eqs. (2.5.8) and (2.5.9), we further conclude that

$$\Delta t \leqslant \frac{2\alpha}{U^2}, \qquad \Delta x \leqslant \frac{2\alpha}{U}. \tag{2.5.10}$$

Figures 2.5a and 2.5b depict schematically the polar diagram of the amplification ξ for a case that satisfies conditions (2.5.8) and (2.5.9) and

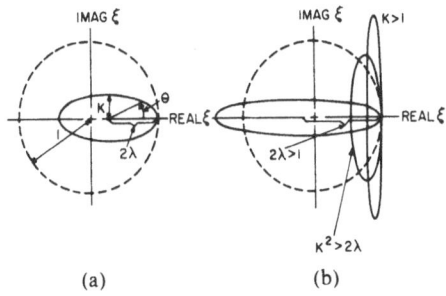

Fig. 2.5 Plot of $\xi = $ const. curves (solid line). The dotted line is the unit circle shown here for comparison (Eq. (2.5.7)). (a) Stable; (b) unstable.

three cases in which these conditions are violated. Note that stability requires that the entire ellipse is contained in the unit circle, i.e., Eq. (2.5.6) should be satisfied.

Two-dimensional problems

Consider a model that includes two-dimensional convection and diffusion:

$$\frac{\partial u}{\partial t} + U \frac{\partial u}{\partial x} + V \frac{\partial u}{\partial y} = \nu \left(\frac{\partial^2 u}{\partial x^2} + \frac{\partial^2 u}{\partial y^2} \right), \tag{2.5.11}$$

where U and V are known functions and the coefficient of the second derivatives is a dimensionless viscosity ν to signify the fact that the term represents diffusion of momentum.

The unknown function u depends on the space variables x, y and time t. The integers m, k, and n will be running in the directions x, y, and t, respectively,

$$u(x, y, t) = u(m\Delta x, k\Delta y, n\Delta t), \tag{2.5.12}$$

and the corresponding discrete function will be $u_{m,k}^n$.

A straightforward explicit scheme for the solution of this equation is (Roache, 1972)

$$\frac{u_{m,k}^{n+1} - u_{m,k}^n}{\Delta t} + U \frac{u_{m+1,k}^n - u_{m-1,k}^n}{2\Delta x} + V \frac{u_{m,k+1}^n - u_{m,k-1}^n}{2\Delta y}$$

$$= \nu \left[\frac{\left(\delta_x^2 u \right)_{m,k}^n}{(\Delta x)^2} + \frac{\left(\delta_y^2 u \right)_{m,k}^n}{(\Delta y)^2} \right]. \tag{2.5.13}$$

This scheme is shown schematically in Fig. 2.6. In terms of the differencing vectors shown in this figure, it can be put in the following compact form:

$$\frac{a}{\Delta t} + U \frac{(c+b)}{2\Delta x} + V \frac{(d+f)}{2\Delta y} = \nu \left[\frac{c-b}{(\Delta x)^2} + \frac{f-d}{(\Delta y)^2} \right]. \tag{2.5.14}$$

An alternative scheme, implicit in time and space, is shown in Fig. 2.7. Its finite-difference form is

$$\frac{u_{m,k}^{n+1} - u_{m,k}^n}{\Delta t} + U \frac{u_{m+1,k}^{n+1} - u_{m,k}^{n+1}}{\Delta x} + V \frac{u_{m,k+1}^{n+1} - u_{m,k-1}^{n+1}}{2\Delta y}$$

$$= \nu \left[\frac{\left(\delta_x^2 u \right)_{m,k}^{n+1}}{(\Delta x)^2} + \frac{\left(\delta_y^2 u \right)_{m,k}^{n+1}}{(\Delta y)^2} \right], \tag{2.5.15}$$

and in compact form (see Fig. 2.7),

$$\frac{a}{\Delta t} + U \frac{b}{\Delta x} + V \frac{d+f}{2\Delta y} = \nu \left[\frac{c-b}{(\Delta x)^2} + \frac{f-d}{(\Delta y)^2} \right]. \tag{2.5.16}$$

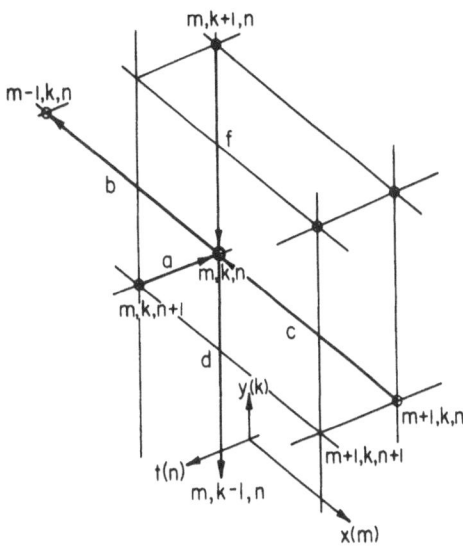

Fig. 2.6 Mesh configuration for an explicit and implicit scheme of integration of Eq. (2.5.11).

Scheme (2.5.14) has a truncation error $0[\Delta t + (\Delta x)^2 + (\Delta y)^2]$, whereas scheme (2.5.15) has a truncation error $0[\Delta t + \Delta x + (\Delta y)^2]$.

Collecting terms, we may rewrite Eqs. (2.5.13) and (2.5.15) in this general form:

$$T_k u_{m+1,k}^{n+1} = E_k u_{m,k}^{n+1} + F_k u_{m+1,k}^n + G_k u_{m,k}^n + H_k u_{m-1,k}^n, \qquad (2.5.17)$$

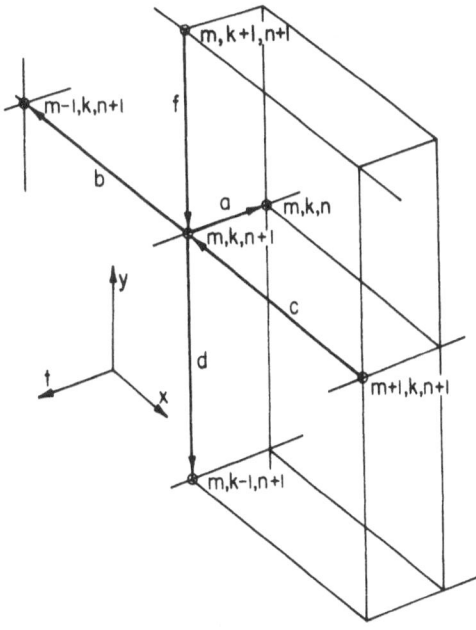

Fig. 2.7 Mesh configuration for a scheme implicit in time and space for integration of Eq. (2.5.11).

where for the scheme (2.5.13) we have

$$T_k = 0, \qquad E_k = 1, \qquad F_k = -U\frac{\Delta t}{2\Delta x},$$

$$G_k = 1 + \nu\Delta t\left[\frac{\delta_x^2}{(\Delta x)^2} + \frac{\delta_y^2}{(\Delta y)^2}\right] - V\Delta t\frac{\delta_{2y}}{2\Delta y}, \qquad (2.5.18)$$

$$H_k = U\frac{\Delta t}{2\Delta x},$$

and for the scheme (2.5.15), we have

$$T_k = 1, \qquad E_k = 1 - \frac{1}{U}\frac{\Delta x}{\Delta t} + \frac{\nu\Delta x}{U}\left[\frac{\delta_x^2}{(\Delta x)^2} + \frac{\delta_y^2}{(\Delta y)^2} - V\frac{\delta_{2y}}{2\Delta y}\right],$$
$$(2.5.19)$$

$$G_k = 0, \qquad G_k = \frac{1}{U}\frac{\Delta x}{\Delta t}, \qquad H_k = 0.$$

In these equations, δ_{2y} and δ_y^2 are centered first- and second-order differencing operators in the y direction, respectively.

Let us consider now the stability of scheme (2.5.13). The round-off error takes the form

$$E_{m,k}^n = A_n e^{i(m\theta_x + k\theta_y)}, \qquad (2.5.20)$$

where all or some of the integers m, k, and n may be replaced by $m \pm 1$, $k \pm 1$, and $n \pm 1$, respectively, and the phase angles θ_x and θ_y vary independently. After some straightforward steps, we may readily derive the expression for the amplification factor

$$\xi = 1 - 2(\lambda_x + \lambda_y) + 2\lambda_x \cos\theta_x + 2\lambda_y \cos\theta_y$$
$$- i(\kappa_x \sin\theta_x + \kappa_y \sin\theta_y). \qquad (2.5.21)$$

In the above equation the constants κ and λ are equivalent to the ones defined by (2.5.3)

$$\kappa_x = U\frac{\Delta t}{\Delta x}, \qquad \kappa_y = V\frac{\Delta t}{\Delta y}, \qquad (2.5.22)$$

$$\lambda_x = \nu\frac{\Delta t}{(\Delta x)^2}, \qquad \lambda_y = \nu\frac{\Delta t}{(\Delta y)^2}. \qquad (2.5.23)$$

Again the requirement that the absolute values for the real and the imaginary parts of ξ be less than 1 reduce to

$$\lambda_x + \lambda_y \leqslant \tfrac{1}{2}, \qquad \kappa_x + \kappa_y \leqslant 1. \qquad (2.5.24)$$

These two conditions were derived for $\theta_x = \theta_y = 90°$.

Fromm (1964) further shows that for $\Delta x = \Delta y = \Delta$ and $\theta_x = \theta_y$,

$$|U| + |V| \leqslant \frac{4\nu}{\Delta}. \qquad (2.5.25)$$

For this uniform space net we have $\lambda_x = \lambda_y$ and the first of (2.5.24) reduces to

$$\lambda_x = \lambda_y \leqslant \tfrac{1}{4}, \tag{2.5.26}$$

which is twice as restrictive as the corresponding condition for one-dimensional flow. Similarly, if $\kappa_x = \kappa_y$, we have

$$\kappa = \kappa_x = \kappa_y \leqslant \tfrac{1}{2}. \tag{2.5.27}$$

For a boundary-layer type of an equation, the term $\partial^2 u / \partial x^2$ is missing from the right-hand side of Eq. (2.5.11), and condition (2.5.24) reduces to its earlier form

$$\lambda_y = \nu \frac{\Delta t}{\Delta y^2} \leqslant \frac{1}{2}. \tag{2.5.28}$$

Extensions of this analysis to three-dimensional flow is straightforward and can be found in Roache (1972).

Farn and Arpaci (1966) have proposed and used an explicit scheme of calculation almost identical to the scheme given by Eq. (2.5.15) or (2.5.16). Their actual numerical results will be discussed in later chapters.

2.6 Implicit Schemes

A variety of implicit schemes have been employed for the solution of unsteady boundary-layer equations. The first two methods that appeared (Dwyer, 1968; Hall, 1969), although developed independently, employ a very similar implicit scheme as described in the next chapter. Let us return to a model equation in order to examine the stability conditions of such schemes. Consider the equation

$$\frac{\partial u}{\partial t} + U \frac{\partial u}{\partial x} = \nu \frac{\partial^2 u}{\partial y^2}. \tag{2.6.1}$$

A very simple scheme, implicit in both the coordinates x and t and therefore essentially a Laasonnen scheme, is shown schematically in Fig. 2.8. Backward-facing differencing is used for x and t derivatives and the y differencing is implemented at the station with the largest integer m and n, i.e., the station $m + 1$, k, $n + 1$. This scheme was used by Telionis et al. (1973). The finite-difference equation becomes

$$\frac{u_{m+1,k}^{n+1} - u_{m+1,k}^{n}}{\Delta t} + U \frac{u_{m+1,k}^{n+1} - u_{m,k}^{n+1}}{\Delta x} = \nu \frac{\left(\delta_y^2 u\right)_{m,k}^{n+1}}{(\Delta y)^2}, \tag{2.6.2}$$

or simply

$$\frac{a}{\Delta t} + U \frac{b}{\Delta x} = \nu \frac{f - d}{(\Delta y)^2}. \tag{2.6.3}$$

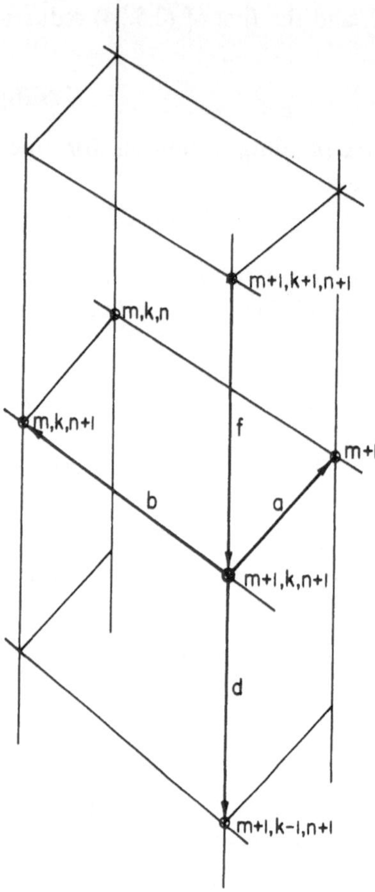

Fig. 2.8 A Laasonnen scheme for
the solution of Eq. (2.6.1).

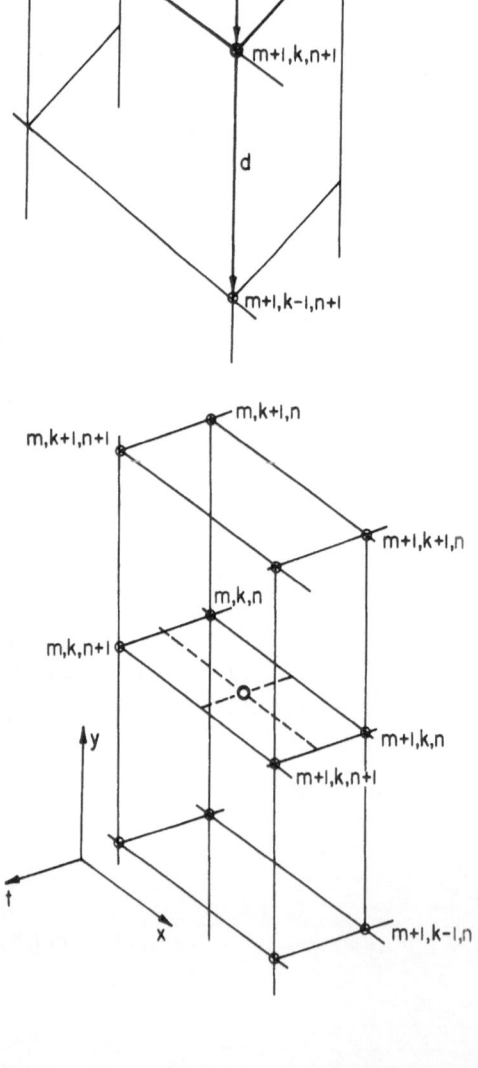

Fig. 2.9 A Crank–Nicholson
scheme for the solution of
Eq. (2.6.1).

This equation can be brought into the form of Eq. (2.5.17), that is,

$$T_k u^{n+1}_{m+1,k} = E_k u^{n+1}_{m,k} + F_k u^n_{m+1,k} + G_k u^n_{m,k} + H_k u^n_{m-1,k} \tag{2.6.4}$$

with coefficients

$$\begin{aligned} T_k &= 1 + \kappa_x - \lambda_y \delta_y^2, & E_k &= \kappa_x, \\ F_k &= 1, & G_k &= 0, & H_k &= 0, \end{aligned} \tag{2.6.5}$$

where the symbols κ_x and λ_y are equal to $U\Delta t/\Delta x$ and $\nu\Delta t/(\Delta y)^2$, respectively, and were defined by (2.5.22) and (2.5.23), respectively.

Wirz (1975) proposes two more implicit schemes: a Crank–Nicholson and a Mehrstellen. These schemes are shown schematically in Figs. 2.9 and 2.10. They can be brought again into the form of Eq. (2.6.4) with coefficients given by

Crank–Nicholson

$$\begin{aligned} T_k &= 1 + \kappa_x - \tfrac{1}{2}\lambda_y \delta_y^2, & E_k &= \kappa_x - 1 + \tfrac{1}{2}\lambda_y \delta_y^2, \\ F_k &= 1 - \kappa_x + \tfrac{1}{2}\lambda_y \delta_y^2, & G_k &= 1 + \kappa_x + \tfrac{1}{2}\lambda_y \delta_y^2 \end{aligned} \tag{2.6.6}$$

Mehrstellen

$$\begin{aligned} T_k &= S_y + S_y^* - \tfrac{1}{2}\lambda_y \delta_y^2, & E_k &= S_y^* - S_y + \tfrac{1}{2}\lambda \delta_y^2, \\ F_k &= S_y - S_y^* + \tfrac{1}{2}\lambda_y \delta_y^2, & G_k &= S_y + S_y^* + \tfrac{1}{2}\lambda_y \delta_y^2 \end{aligned} \tag{2.6.7}$$

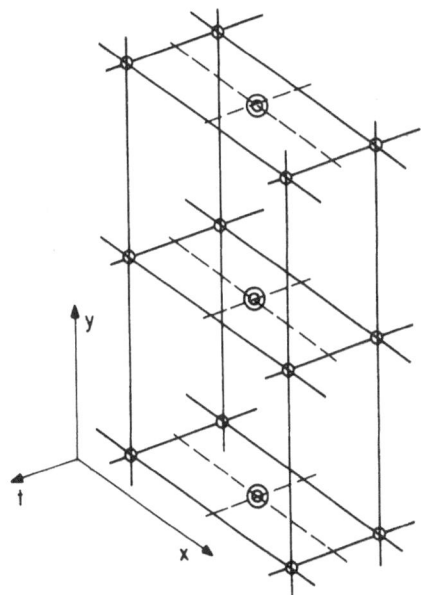

Fig. 2.10 A Mehrstellen scheme for the solution of Eq. (2.6.1).

and with the operators S_y and S_y^*, defined as follows:

$$S_y(\ )_k = \tfrac{1}{12}\left[(\ )_{k+1} + 10(\ )_k + (\ )_{k-1}\right] \tag{2.6.8}$$

$$S_y^*(\ )_k = \tfrac{1}{12}\left[\kappa_{x,k+1}(\ )_{k+1} + 10\kappa_{x,k}(\ )_k + \kappa_{x,k-1}(\ )_{k-1}\right]. \tag{2.6.9}$$

The truncation error of the Laasonnen scheme [(Eq. (2.6.5)] is of order $\Delta t + \Delta x + (\Delta y)^2$, whereas the truncation errors of the Crank–Nicholson and the Mehrstellen schemes are of order $(\Delta t)^2 + (\Delta x)^2 + (\Delta y)^2$ and $(\Delta t)^2 + (\Delta x)^2 + (\Delta y)^4$, respectively. Wirz (1975) investigated the stability of the last three schemes. He calculated the moduli of the amplification factors that are given below:

For the Laasonnen scheme of Eq. (2.6.5),

$$\xi\bar\xi = \left[\left(1 + 4\lambda_y \sin^2 \frac{\theta_y}{2}\right)^2 + 4\kappa_x\left(1 + \kappa_x + 4\lambda_y \sin^2 \frac{\theta_y}{2}\right)^2 \sin^2 \frac{\theta_x}{2}\right]^{-1}. \tag{2.6.10}$$

For the Crank–Nicholson scheme of Eq. (2.6.6),

$$\xi\bar\xi = \frac{\cos^2(\theta_x/2)\left[1 - 2\lambda_y \sin^2(\theta_y/2)\right] + \kappa_x^2 \sin^2(\theta_x/2)}{\cos^2(\theta_x/2)\left[1 + 2\lambda_y \sin^2(\theta_y/2)\right] + \kappa_x^2 \sin^2(\theta_x/2)}. \tag{2.6.11}$$

For the Mehrstellen scheme of Eq. (2.6.7),

$$\xi\bar\xi = \frac{\cos^2(\theta_x/2)\left[1 - 2\lambda_y \sin^2(\theta_y/2)\right]^2 + \kappa_x^2 \sin^2(\theta_x/2)\left(1 - \tfrac{1}{3}\sin^2(\theta_y/2)\right)^2}{\cos^2(\theta_x/2)\left[1 + 2\lambda_y \sin^2(\theta_y/2)\right]^2 + \kappa_x^2 \sin^2(\theta_x/2)\left(1 - \tfrac{1}{3}\sin^2(\theta_y/2)\right)^2}. \tag{2.6.12}$$

All quotients are less than 1 for any values of the phases θ_x and θ_y and therefore all schemes are unconditionally stable.

In all the implicit schemes described in this section, three unknown quantities spaced vertically at grid points $k - 1$, k, and $k + 1$ are contained in each equation. The system of algebraic equations thus formed has a tridiagonal structure and efficient algorithms are today available for its solution (Isaacson and Keller, 1966, pp. 55–61). To give an example, let us assume that all the unknown functions at grid points with indices less than $m + 1$ and $n + 1$ have been factored into coefficients and that the system of equations for u_{m+1}^{n+1} can be written as

$$A_k u_{m+1,k+1}^{n+1} + B_k u_{m+1,k}^{n+1} + C_k u_{m+1,k-1}^{n+1} = D_k. \tag{2.6.13}$$

If instead of the model equation (2.6.1) the proper momentum equation is considered, then nonlinearities are introduced that we incorporate in the coefficients A_k, B_k, and C_k as well. Moreover, one more first-order differential equation is introduced, the continuity equation

$$\frac{u_{m+1,k}^{n+1} - u_{m,k}^{n+1}}{\Delta x} + \frac{v_{m+1,k+1}^{n+1} - v_{m+1,k}^{n+1}}{\Delta y} = 0. \tag{2.6.14}$$

The dependent variable v is also contained in the coefficients A_k, B_k, C_k, and D_k, which are assumed known at each iteration. Small modifications of the classical method of solution have been proved quite efficient. One of the fastest methods is due originally to Davis and is described in Davis *et al.* (1970), Werle and Davis (1972) and Werle and Bertke (1972). A variation of the classical Thomas algorithm is introduced, which uses Newton–Raphson linearization and has quadratic convergence. The main idea behind such schemes is described briefly here.

Assuming that a set of values of u for $k = K - 1$ are given, where K is the total number of grid points in the y direction, the coefficients A_k, B_k, and C_k can be calculated. The assumed values of u may either be an arbitrary guess or the latest iterate in the solution. Werle and Bertke start their calculation by assuming a uniform profile for u except for $k = 1$, the wall, where $u = 0$. Equation (2.6.14) can then be solved by marching from $k = 1$, where the boundary condition $v_{m,1}^{n} = 0$ is also imposed. The set of Eqs. (2.6.13) is then solved by any of the available methods, subject to appropriate conditions at $k = 1$ and K. The new values of u are then used to update the coefficients of the equation, A_k, B_k, C_k, and D_k. Moreover, continuity equation is solved once more to yield updated values of the quantity v. The process is repeated at each station until convergence is achieved.

The method has been extended to transient and oscillatory flows by Telionis *et al.* (1973) and Telionis and Tsahalis (1974a). More references and more details will be found in later sections together with the description of the physical problem and the interpretation of the results.

2.7 The Keller–Box Method

An alternative method for solving boundary-layer equations is based on Keller's "box method" or "midpoint scheme" (Keller, 1969, 1971). It was first employed in solving boundary-layer equations by Keller and Cebeci (1971). Information on applications of the method to a variety of boundary-layer flow problems can be found in two recent articles by Keller (1975, 1978).

The Keller-box method is based on the idea of expressing all functions and derivatives in terms of quantities at the corners of one computational block. To this end new variables are introduced in order to eliminate all second-order derivatives. In particular, with τ a new variable, defined by $\tau = v \, \partial u / \partial y$, the two-dimensional version of Eqs. (1.2.27) and (1.2.28) becomes equivalent to a set of three first-order differential equations:

$$\frac{\partial u}{\partial x} + \frac{\partial v}{\partial y} = 0,$$ *(2.7.1)

$$\frac{\partial u}{\partial t} + u\frac{\partial u}{\partial x} + v\frac{\partial u}{\partial y} = -\frac{1}{\rho}\frac{\partial p}{\partial x} + \frac{\partial \tau}{\partial y},$$ *(2.7.2)

$$\tau = \nu\frac{\partial u}{\partial y}.$$ *(2.7.3)

All derivatives can now be approximated by simple centered differences and two-point averages, using only values at the corners of the three-dimensional box. Brackets are used here to denote averaged values of independent variables as well as finite differences. For example, the expressions

$$[u]_{m,k-1/2}^{n-1} = \tfrac{1}{2}\left(u_{m,k}^{n-1} + u_{m,k-1}^{n-1}\right),$$ (2.7.4)

$$[u]_{m,k-1/2}^{n-1/2} = \tfrac{1}{2}\left\{[u]_{m,k-1/2}^{n} + [u]_{m,k-1/2}^{n-1}\right\},$$ (2.7.5)

represent averages of the independent variable u. The first is an average between the values at points B' and C' in Fig. 2.11, whereas the second is an average between all four points, $BB'C'C$, in the plane $x = x_m$. Averaging among points A', B', C', D' would give the approximate value of the function at the center E of the rectangle on the time plane n, whereas similarly, averaging among all corners of the three-dimensional box, that is, the

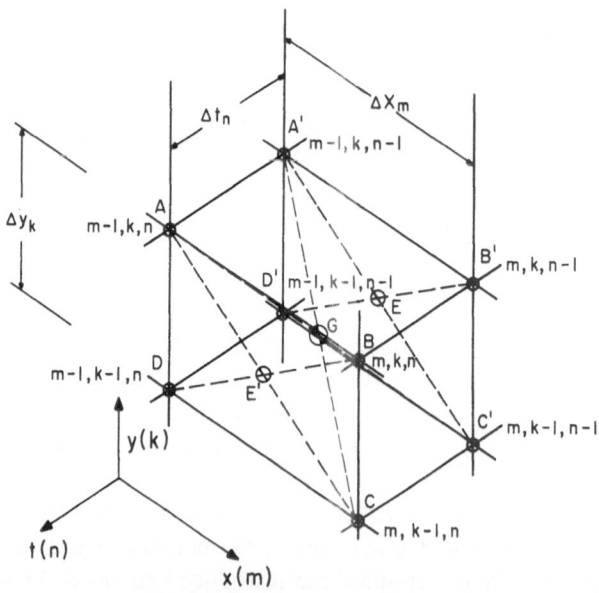

Fig. 2.11 The "Keller-box" scheme.

quantity $[u]^{n-1/2}_{m-1/2,k-1/2}$ is an approximation of the function at the center of the box, G.

In a similar way derivatives are expressed as follows:

$$\left[\frac{\delta u}{\delta x} \right]^{n-1}_{m-1/2,k} = \frac{1}{\Delta x_m} \left(u^{n-1}_{m,k} - u^{n-1}_{m-1,k} \right), \tag{2.7.6}$$

$$\left[\frac{\delta u}{\delta x} \right]^{n-1}_{m-1/2,k-1/2} = \frac{1}{\Delta x_m} \left(u^{n-1}_{m,k-1/2} - u^{n-1}_{m-1,k-1/2} \right), \tag{2.7.7}$$

where Δx_m is one of the dimensions of the box that could vary in space. The method can be used directly in computations with variable mesh sizes since all the averaging and finite differencing is done within the box and evaluations at midpoints in a plane or the midpoint of the box are always unambiguously defined.

Keller and Cebeci (1971) describe a method for solving the steady form of these equations. They impose the classical boundary conditions to arrive again at a system of coupled nonlinear equations, which they solve by Newton's method. Newton's iterates are defined and the equations are formulated in terms of the corrections δu, which have the form of Eq. (2.6.13). The system has a tridiagonal structure but due to the special zero pattern of the coefficients, it can be solved by a very efficient method, which Keller (1978) claims would be faster if compared with Crank–Nicholson schemes. Moreover since Newton's method converges quadratically, two Newton iterates are usually enough for convergence.

The method has been employed for oscillatory laminar boundary-layer flows by Phillips and Ackerberg (1973) and for oscillatory turbulent flows by Cebeci (1977). Keller (1978) proposes to solve the problem by centering the continuity equation at the time plane t_n, i.e., the point $E'(t_n, x_{m-1/2}, y_{k-1/2})$ in Fig. 2.11.

$$\left[\frac{\delta u}{\delta x} \right]^{n}_{m-1/2,k-1/2} + \left[\frac{\delta v}{\delta y} \right]^{n}_{m-1/2,k-1/2} = 0 \tag{2.7.8}$$

Similarly, the definition of the shear is centered at the midpoint between B and C, i.e., the point $(t_n, x_m, y_{k-1/2})$.

$$[\tau]^{n}_{m,k-1/2} = \left[\nu \frac{\delta u}{\delta y} \right]^{n}_{m,k-1/2}. \tag{2.7.9}$$

The momentum equation can be centered in the middle of the box, i.e., the point G. Keller (1978) proposes to center this equation at the midpoint of the subcharacteristic facing backward, in order to satisfy the Courant–Friedrichs–Lewy condition. The direction of the subcharacteristic depends on the value and the sign of the quantity u, a point which is the topic of the next section. Discussion of applications of the Keller-box method to reversing flows will also be found at the end of the following section.

2.8 Upwind Differencing

In the late fifties, Sears (1956), Rott (1956), and Moore (1957) argued that the vanishing of the wall shear may not necessarily imply separation in unsteady flows. Their arguments were based on the similarities between steady flows over moving walls and unsteady flows over fixed walls. Their work and subsequent developments, theoretical and experimental, will be described in Chapter 7. Even though the physical significance of the point of zero skin friction has been deemphasized, it plays a vital role in the development of computational methods, and this is the topic of the present section.

It is well known that for steady flow, the boundary-layer equations are singular at the point of zero skin friction. The behavior of the equations in the neighborhood of this point has been studied extensively by both analytical and numerical methods. A thorough review on the topic was prepared by Brown and Stewartson (1969). It is well established (Dean, 1950) that the full Navier–Stokes equations are nonsingular at the point of zero skin friction. However, it was only after 20 years that boundary-layer calculations were successful in passing through the point of zero skin friction and into an area of reversed flow (Catherall and Mangler, 1966; Klemp and Acrivos, 1971; Klineberg and Steger, 1974; Carter, 1974). These authors discovered that if instead of the outer-flow pressure distribution they were to prescribe any other flow property, as for example, the displacement thickness, then their calculations would proceed through the point of zero skin friction without the apperance of any singularity. However the continuation of the calculations would require boundary conditions on the downstream side of the outer flow.

The situation is depicted schematically in Fig. 2.12. The problem of steady two-dimensional boundary-layer flow is well posed if boundary conditions

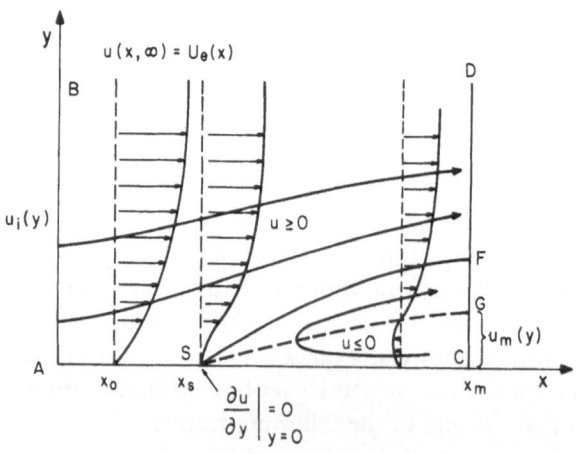

Fig. 2.12 The neighborhood of a vanishing wall shear for steady boundary-layer flow. Initial conditions $u_i(y)$ and $u_m(y)$ are required along AB and GC, respectively.

are imposed at the wall and at infinity, $y = 0$ and ∞, and if an initial profile is given at $x = 0$. For each station x_0 then, the domain of influence extends to all points $x \geqslant x_0$. Disturbances travel in the y direction instantly and in the x direction they are convected with speed equal to the local value of the velocity. In this way a disturbance travels always downstream, as long as there is no reversed flow, that is, for $x < x_s$, with the speed of the outer flow U_e. This is the largest speed at a station x_0, as Nickel (1958) pointed out, provided that the initial profile is monotonic.

These concepts have been discussed in the first chapter in the spirit of the theory of characteristics. Stewartson (1960) has also given a lucid description of how disturbances propagate. Disturbances travel up in the y direction at the station $x = x_0$ immediately after they have been made. Then they travel through the outer part of the boundary layer, where $u = U_e$, in the direction of increasing x with velocity U_e. As soon as they reach a new station, they immediately diffuse to all heights y. Although such disturbances travel with a finite velocity, there is no discontinuity at the wave front because the disturbance at a station downstream of x_0 is exponentially small to begin with so that continuity of all derivatives is ensured.

The situation is drastically changed beyond the point $x = x_s$, that is, downstream of the point where $\partial u/\partial y|_w = 0$. A region of reversed flow sets in beneath the dotted curve SG. In this region, since $u \leqslant 0$, disturbances propagate in the direction of decreasing x. To complete such a calculation, an initial profile $u_m(y)$ is needed at the station x_m. Only then would the problem be well posed, as Giraud (1965) points out. However, numerical integration is very difficult, since a straightforward marching in the x direction is not possible any more. This is due to the fact that the finite-differencing mesh should include the regions of dependence in order to meet the Courant–Friedrichs–Lewy criterion. The calculations should start with initial profiles at AB and GC and march to the right and left of AB and GC, respectively. However, the locus of points with zero velocity u, that is, the line SG, is not known a priori and some type of an iteration scheme is necessary to complete the calculation. Needless to say that this approach makes sense physically only if the separated region remains thin and laminar. All such methods developed up to now do not provide any information about the character of the separated region. On many occasions, as, for example, downstream of a relatively abrupt change in the boundary slope, the flow enters into a region of violent, although slow-moving, turbulent wake flow.

The situation is a little different in unsteady flows. To begin with, the appearance of a zero wall shear at the time t_s and the station x_s say, is not accompanied by a singularity as Telionis et al. (1973) discovered by numerical experiments. This was corroborated by other investigators as well (Phillips and Ackerberg, 1973; Williams and Johnson, 1974 a and b; Nash et al., 1975; Shen and Nenni, 1975). At later times, and if the pressure gradient

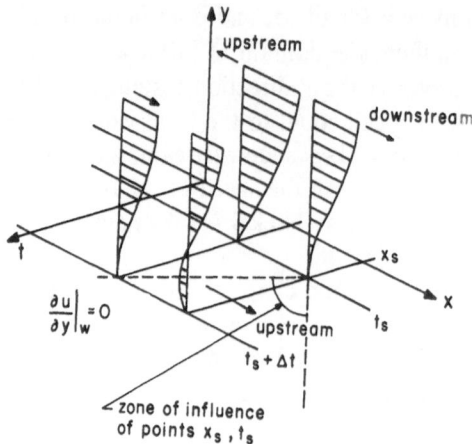

Fig. 2.13 The neighborhood of zero wall shear for unsteady boundary-layer flow.

becomes more adverse, the point of zero skin friction moves upstream, leaving behind a region of partially reversed flow. The flow is depicted schematically in Fig. 2.13. It turns out that now it is very possible that boundary-layer calculations of such fields can be completed with initial conditions of the form of (1.2.20)–(1.2.26), that is, with information on the initial planes xy and ty. This is indeed the case if the regions of dependence of the points of interest, as defined in the first chapter, contain only known initial information in both the x and t initial planes. This is true if condition (1.6.6) is satisfied.

Wang (1975) notes that numerical calculations of unsteady boundary-layer calculations reported earlier (Phillips and Ackerberg, 1973; Telionis et al., 1973; Telionis and Tsahalis, 1974a), did not recognize the concepts of zones of influence and dependence. Wang concedes that nevertheless, such calculations could satisfy the rules of influence and dependence and they were indeed successfully carried out. It appears that Phillips and Ackerberg (1973) were aware of the implication of zones of influence and dependence and their Eq. (3.22) is nothing but the statement of Eq. (1.6.6). Moreover, Telionis and Tsahalis (1974b) in a later publication describe explicitly the situation. The calculations of Phillips and Ackerberg could proceed to the right until an asymptotic solution valid for large x was met. The calculations of Telionis and Tsahalis were open on the right-hand side. The last authors argue that their domain of integration should collapse at each time plane by at least one grid point. Telionis and Tsahalis explicitly mention that disturbances in parabolic equations of the boundary layer form travel with a velocity equal to the maximum flow velocity. Thus in their case they had to guarantee that their calculation does not enter into regions that have been influenced by points downstream.

Keller (1978) approaches the problem in terms of the projection of the subcharacteristics on planes perpendicular to the y axis of Fig. 2.13

$$\frac{dx(s)}{ds} = u\frac{dt(s)}{ds} \, ,\tag{2.8.1}$$

where s is the variable along such projections. It is clear that if u is positive, then all such curves run to the right in t–x planes, viewed as shown in Fig. 2.14, and therefore all points simply influence their downstream domain. The situation changes drastically if a region is doubly covered, a fact indicating the possible formation of discontinuities. However, Keller points out that the diffusion term in the boundary-layer equation eliminates the possibility of shock formations. Consider, for example, the case whereby a sub-characteristic emanating from a point A, $0 < x_A \leqslant x_m$ at $t = 0$, eventually hits the time axis at a point B, $0 < t_B < t_m$ for $x = 0$. Obviously then, data on these two points could not be arbitrarily described. Moreover, the calculation could not be continued to the right of this subcharacteristic (see Fig. 2.14a). In a similar manner, if a subcharacteristic cuts across the domain of integration $0 < t < t_m$ and $0 < x < x_m$ as shown in Fig. 2.14b, then again the triangular zone CDE is cut off from the domain of possible calculations.

Proper numerical schemes for the solution of such problems should satisfy the Courant–Friedrichs–Lewy criterion. The computation mesh should enclose the dependence zone of the differential equation. This requires that if reversed flow exists, information be received from some station downstream of the point of interest and in such a way that the Courant–Friedrichs–Lewy criterion is met. Telionis et al. (1973) and Phillips and Ackerberg (1973) developed independently very similar upwind-differencing schemes to solve reversed flow problems. Similar schemes have been employed earlier for a variety of other situations as described extensively by Roache (1972). It appears that proper differencing for reversing boundary-layer equations was actually employed for the first time by Robins (1970) (see also Belcher et al.,

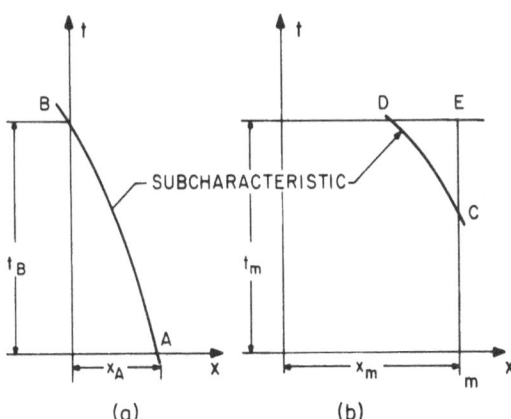

Fig. 2.14 Projection of subcharacteristics on planes y = constant.

(a)   (b)

1972), a fact that the above authors seem to be unaware of. Belcher *et al.* have used values of u at two previous time planes and both upstream and downstream neighboring stations of the axial location x.

Telionis *et al.* (1973) propose two alternative schemes that satisfy the Courant–Friedrichs–Lewy criterion. For regions of nonreversing flow, they use a regular implicit scheme, very similar to scheme (2.3.23). When at a station a zero or negative value of the velocity is encountered, then information from downstream is included in the x derivative according to the following differencing scheme:

$$\left(\frac{\delta u}{\delta x} \right)^n_{m,k} = \frac{1}{\Delta x} \left(u^{n-1}_{m+1,k} - u^n_{m,k} \right). \tag{2.8.2}$$

In this way the vector $b/\Delta x$ (see Fig. 2.15), is used as a first guess for the x derivative at the point m, k, n. An alternative method that was used by Telionis *et al.* (1973) is based on a zig-zag scheme that is essentially an average of the x derivative evaluated at the stations m and $m-1$ for the

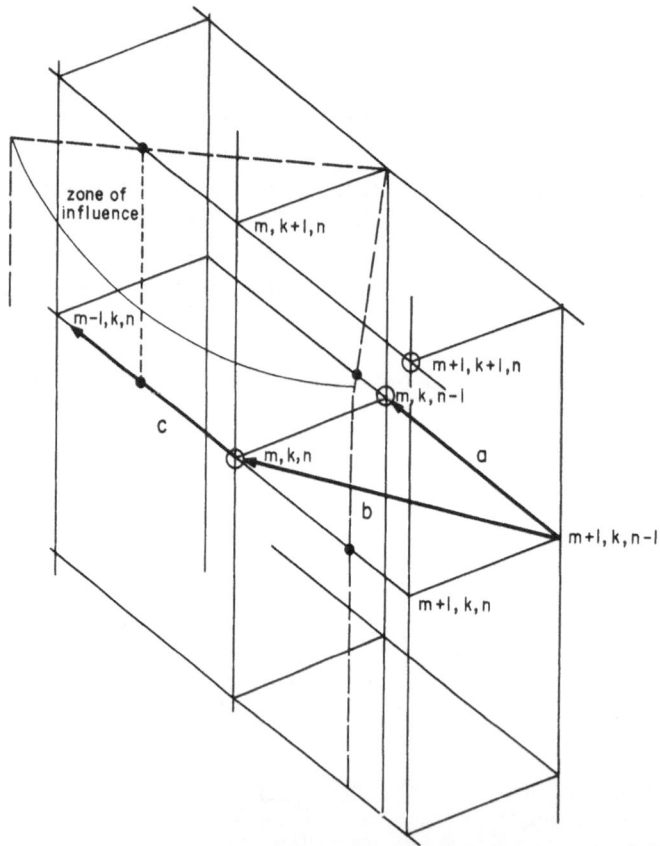

Fig. 2.15 Zig-zag and upwind-differencing schemes.

time plane n, and the stations $m + 1$ and m for the time plane $n - 1$.

$$\left(\frac{\delta u}{\delta x}\right)^n_{m,k} = \frac{1}{\Delta x}\left[\frac{u^n_{m,k} - u^n_{m-1,k}}{2} + \frac{u^{n-1}_{m+1,k} - u^{n-1}_{m,k}}{2}\right]. \tag{2.8.3}$$

This is written in terms of the differencing vectors of Fig. 2.15 as follows

$$\left(\frac{\delta u}{\delta x}\right)^n_{m,k} = \frac{1}{2\Delta x}(a + c). \tag{2.8.4}$$

Both techniques have been tested and later used for the calculation of certain flow fields. The advantage of (2.8.4) is that it may be used throughout the calculation for reversing or nonreversing flows. The results were compared with closed-form analytical solutions or numerical solutions of the full Navier–Stokes equation. Details will be found in later chapters.

Phillips and Ackerberg (1973) employ a more involved differencing scheme based on the Keller-box method (Keller and Cebeci, 1971). The basic idea of the Keller-box scheme, as described in the previous section, is to introduce a new dependent variable, a dimensionless shear stress $\tau = \partial u/\partial y$ and a new equation, the definition of τ. Phillips and Ackerberg (1973) express their functions and first derivatives in terms of averages of differences in the present time plane n, averages in the past time plane $n - 1$, and the same quantities calculated at a previous x station. The basic principles of the method are demonstrated below in terms of a few examples expressed for simplicity with uniform spacing. All quantities are calculated at the center of the box (see Fig. 2.16), i.e., the point $p = m - \frac{1}{2}$, $q = k - \frac{1}{2}$, $r = n - \frac{1}{2}$. The functions themselves at p, q, r are expressed in terms of an average of corner

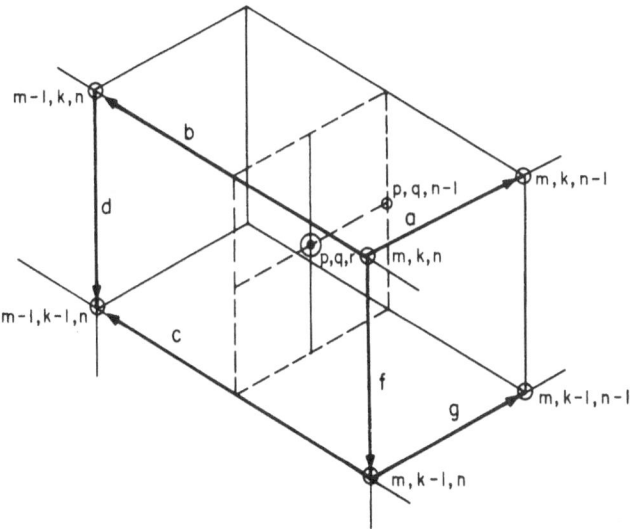

Fig. 2.16 The Keller-box scheme as modified by Phillips and Ackerberg.

values

$$u_{p,q}^r = \frac{1}{8}\left[u_{m,k}^n + u_{m,k-1}^n + u_{m-1,k-1}^n + u_{m-1,k}^n \right] + \frac{1}{2} u_{p,q}^{n-1}. \tag{2.8.5}$$

Derivatives with respect to the parabolic variables are also expressed in terms of averages:

$$\left(\frac{\delta u}{\delta x} \right)_{p,q}^r = \frac{1}{2}\left[\frac{u_{m,k}^n - u_{m-1,k}^n}{\Delta x} + \frac{u_{m,k-1}^n - u_{m-1,k-1}^n}{\Delta x} \right] + \frac{1}{2}\left(\frac{\delta u}{\delta x} \right)_{p,q}^{n-1},$$

$$\tag{2.8.6}$$

$$\left(\frac{\delta u}{\delta x} \right)_{p,q}^r = \frac{1}{4\Delta t}\left[u_{m,k}^n + u_{m,k-1}^n + u_{m-1,k-1}^n + u_{m-1,k}^n \right] - \frac{u_{p,q}^{n-1}}{\Delta t}. \tag{2.8.7}$$

Note that the x derivative is an average of differences at the time planes n and $n-1$, whereas the time derivative is expressed in terms of backward differences emanating from the single point p, q, $n-1$. In terms of our compact notation (see also Fig. 2.17), these derivatives can be written as

$$\left(\frac{\delta u}{\delta x} \right)_{p,q}^r = \frac{1}{2}\left(\frac{b+c}{\Delta x} \right) + \frac{1}{2}\left(\frac{\delta u}{\delta x} \right)_{p,q}^{n-1}, \tag{2.8.8}$$

$$\left(\frac{\delta u}{\delta t} \right)_{p,q}^r = \frac{1}{4\Delta t}(a_1 + a_2 + a_3 + a_4). \tag{2.8.9}$$

Derivatives in the normal direction are also written in terms of averages (Fig. 2.16)

$$\left(\frac{\delta u}{\delta y} \right)_{p,r}^q = \frac{1}{2\Delta t}\left(\frac{f+d}{\Delta y} \right) + \frac{1}{2}\left(\frac{\delta u}{\delta y} \right)_{p,q}^{n-1}. \tag{2.8.10}$$

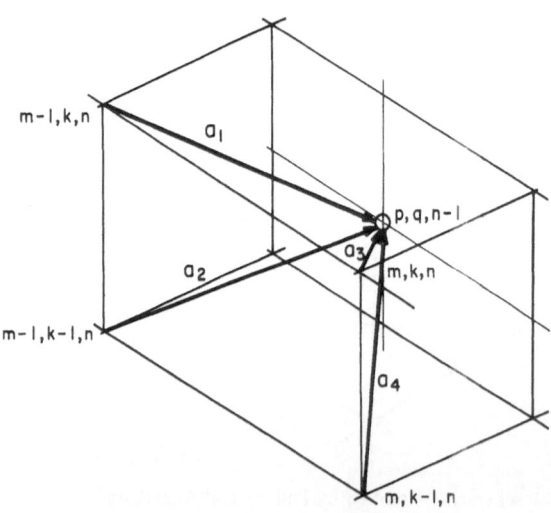

Fig. 2.17 Schematic representation of the Phillips and Ackerberg scheme.

In regions of backward flow, Phillips and Ackerberg (1973) employ a zig-zag differencing scheme. To provide for upwind influence, they calculate their dependent variables and their derivatives of the earlier time plane by averaging over stations $m - 1$, m, and $m + 1$. For example,

$$u_{p,k-1}^{n-1} = \frac{1}{2}\left(u_{p+1,q}^{n-1} + u_{p-1,q}^{n-1}\right), \tag{2.8.11}$$

$$\left(\frac{\delta u}{\delta x}\right)_{p,q-1}^{n} = \frac{1}{2}\left[\left(\frac{\delta u}{\delta x}\right)_{p+1,q}^{n-1} + \left(\frac{\delta u}{\delta x}\right)_{p-1,q}^{n-1}\right]. \tag{2.8.12}$$

They switch to this scheme only when negative velocities are encountered, although this could be used throughout the calculation as mentioned earlier.

A better understanding of the differencing schemes discussed in this section and their effectiveness in integrating through regions of reversed flow can be obtained by decomposing difference vectors as suggested by Shen (1978). Consider, for example, the decomposition of the upwind scheme represented by the vector a in Fig. 2.18 into the sum of the vectors b and c, namely,

$$u_{m+1,k}^{n-1} - u_{m,k}^{n} = \left(u_{m+1,k}^{n-1} - u_{m,k}^{n-1}\right) + \left(u_{m,k}^{n-1} - u_{m,k}^{n}\right). \tag{2.8.13}$$

The x derivative expressed by upwind differencing can then be approximated by

$$\left(\frac{\delta u}{\delta x}\right)_{m,k}^{n} = \frac{u_{m+1,k}^{n-1} - u_{m,k}^{n-1}}{\Delta x} - \frac{u_{m,k}^{n} - u_{m,k}^{n-1}}{\Delta t}\frac{\Delta t}{\Delta x}. \tag{2.8.14}$$

The second ratio of differences on the right-hand side is, to a first-order, nothing but $\delta u / \delta t$ and therefore the equation becomes

$$\left(\frac{\delta u}{\delta x}\right)_{m,k}^{n} = \left(\frac{\delta u}{\delta x}\right)_{m,k}^{n-1} - \left(\frac{\delta u}{\delta t}\right)_{m,k}^{n}\frac{\Delta t}{\Delta x}, \tag{2.8.15}$$

or, if solved in terms of $\delta u / \delta t$,

$$\left(\frac{\delta u}{\delta t}\right)_{m,k}^{n} = \lambda\left[\left(\frac{\delta u}{\delta x}\right)_{m,k}^{n-1} - \left(\frac{\delta u}{\delta x}\right)_{m,k}^{n}\right], \tag{2.8.16}$$

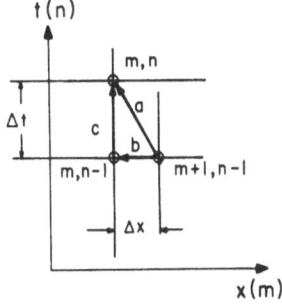

Fig. 2.18 Shen's decomposition of upwind differencing.

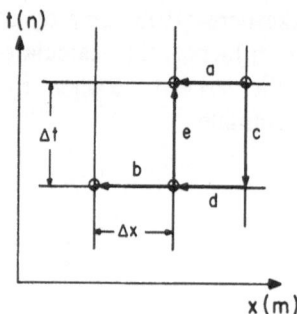

Fig. 2.19 Shen's decompos'tion of a zig-zag scheme.

where $\lambda = \Delta x / \Delta t$. Shen (1978) substitutes this expression in the momentum equation, the two-dimensional version of Eq. (1.2.28), to arrive at a quasi-steady form. To emphasize the fact that only the time is discretized at this point, we may write the differencing ratios again in the form of partial derivatives and retain only the superscripts, which define the time plane:

$$(u^n + \lambda)\left(\frac{\delta u}{\delta x}\right)^n + v^n \left(\frac{\delta u}{\delta y}\right)^n$$

$$= \left(\frac{\delta U_e}{\delta t} + U_e \frac{\delta U_e}{\delta x}\right)^n + \lambda \left(\frac{\delta u}{\delta x}\right)^{n-1} + \nu \left(\frac{\delta^2 u}{\delta y^2}\right)^n. \tag{2.8.17}$$

Following a similar decomposition for the zig-zag scheme depicted schematically in Fig. 2.19, namely, $a + b = c + d + e + b$, we arrive at an alternative form of the momentum equation

$$(u^n_{m,k} + 2\lambda)\left(\frac{\delta u}{\delta x}\right)^n_{m,k} + v^n_{m,k}\left(\frac{\delta u}{\delta y}\right)^n_{m,k}$$

$$= \left(\frac{\partial U_e}{\partial t} + U_e \frac{\partial U_e}{\partial x}\right)^n_{m,k} + \left(\frac{\delta u}{\delta t}\right)^n_{m-1,k} + \nu \left(\frac{\delta^2 u}{\delta y^2}\right)^n_{m,k}. \tag{2.8.18}$$

Note that now a time derivative evaluated at an earlier space station is unavoidable. The quasi-steady form of the equation expressed in terms of partial derivatives in the plane x and y is not any more permissible.

Both Eqs. (2.8.17) and (2.8.18) are parabolic in character with respect to the parabolic variable x. Their integration therefore can be continued until the coefficient of $\delta u / \delta x$ changes signs. It becomes immediately apparent that the zig-zag scheme, incorporated in Eq. (2.8.18), is more stable than the upwind-differencing scheme since the coefficient $u^n_{m,k} + 2\lambda$ will permit smaller negative values of u than the coefficient $u^n_{m,k} + \lambda$.

A more direct approach to the problem is proposed by Keller (1978), who uses differences along the subcharacteristics themselves. The method has not yet been tested and will require interpolation, which is often the source of bookkeeping errors. However, it is conceptually the only exact method. Working again in a plane perpendicular to the y axis, Keller (1978) identifies the subcharacteristic of the operator $\partial/\partial t + u\,\partial/\partial x$, i.e., Eq. (2.8.1), where s is the coordinate along the subcharacteristic. He then proposes to express the first two terms in the momentum equation in terms of an s derivative

$$\frac{\partial}{\partial t} + u\frac{\partial}{\partial x} = \frac{1}{dt/ds}\frac{d}{ds}\, , \qquad (2.8.19)$$

which is essentially nothing but a projection of the Lagrangian derivative in the plane of consideration

$$\frac{\partial}{\partial t} + u\frac{\partial}{\partial x} = \frac{d}{dt}\bigg|_{s=\text{const}}. \qquad (2.8.20)$$

We may arrive at a similar conclusion using Shen's method of analysis. If PQ is a backward facing subcharacteristic (see Fig. 2.20), then by definition $u \approx \Delta x_t/\Delta t$. The reader is cautioned to the fact that Δx_t is not the same as the width of the mesh, Δx. Euler's operator now becomes

$$\frac{\delta u}{\delta t} + u\frac{\delta u}{\delta x} \approx \frac{u_m^n - u_m^{n-1}}{\Delta t} + \frac{\Delta x_t}{\Delta t}\frac{u_m^{n-1} - u_Q^{n-1}}{\Delta x_t} = \frac{u_m^n - u_Q^{n-1}}{\Delta t}. \qquad (2.8.21)$$

Keller evaluates this derivative at the midplane, $k - \frac{1}{2}$ of the box, and writes the finite-differencing form of the momentum equation as follows:

$$\frac{u_{m,k-1/2}^n - u_{Q,k-1/2}^n}{\Delta t} = -\left[\frac{\partial p}{\partial x}\right]_m + \left[\frac{\partial \tau}{\partial y}\right]_{k-1/2}^{PQ/2} - [v]_{k-1/2}^{PQ/2}. \qquad (2.8.22)$$

In this equation the superscript $PQ/2$ denotes the evaluation at the midpoint of the segment PQ.

This method has not yet been tested; however, it appears to be mathematically the most appropriate model. There is only one point that is not mentioned in the original paper of Keller and may need some clarification.

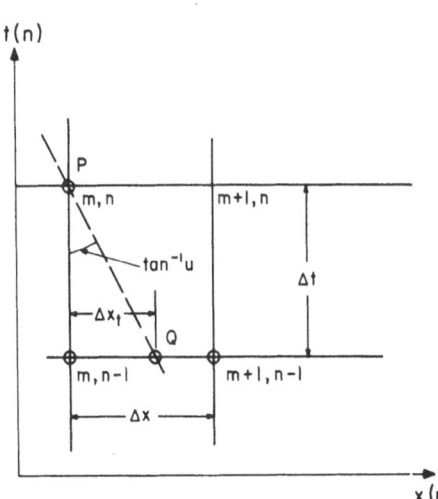

Fig. 2.20 Projection of a sub-characteristic on a $y = \text{constant}$ plane.

The suggestion is that the finite difference should be formulated along the local subcharacteristic. However, the regions of influence of a point are defined by the outermost subcharacteristics along the vertical at the station of interest, as described in the previous chapter. For regions of partially reversed flow, therefore, the point $Q(x_Q, y_Q, t_Q)$ should be defined for all vertical locations, along the characteristic $x = x_Q$, $t = t_Q$, inclined by a slope equal to max(u). Implementation of the method will also require careful interpolations and iterations. Interpolations are necessary for evaluating the functions at point Q between the grid points $(m, n - 1)$ and $(m + 1, n - 1)$ (Fig. 2.20). Moreover, the slope of PQ can be guessed first, but should be updated since it depends on the unknown function u.

2.9 Navier–Stokes—Lattices and Boundary Conditions

Numerical methods for the solution of elliptic equations have been extensively discussed in literature. They are most demanding with respect to computer time and space, because the algebraic system of all equations involved is coupled and the equations must be solved simultaneously. Methods have been proposed for the improvement of the computation speed as, for example, the Jacobi method, the Gauss–Seidel method, the extrapolated method, and others.

A method most closely related to the techniques described in this monograph is the alternating-direction implicit method (ADI). An initial guess is made for all mesh points. Equations are then solved, say for a two-dimensional mesh in the x direction first, assuming necessary values in the y direction are known. The computed values serve as known quantities in the next step, which involves solution in the y direction. The equations are solved alternately by sets of rows and sets of columns and the method may be considered as a line method with alternating directions. The references of Chapter 1 describe adequately the work in this area.

Some of the most characteristic difficulties in the solution of elliptic problems center around the size and shape of the domain of integration and the appropriate boundary conditions. Elliptic problems require Dirichlet or von Neumann-type of conditions all along the periphery of the domain of integration. In aerodynamics one should extend the domain to infinity in order to impose the condition of the undisturbed free stream. This of course is impossible since the space of our computing machines is finite. One may invoke an approximation by imposing the free stream conditions at the periphery of a finite but large domain of integration. More sophisticated methods are available involving variable mesh sizes or appropriate transformations that map infinity onto a finite position. Some of these methods will be discussed here.

The most problematic area for a finite size domain of integration is the downstream boundary, which is overrun by the wake of a fully submerged body moreover, the boundary conditions on the surface of the body may introduce bookkeeping difficulties depending on the slope of the body. In this section we concentrate the discussion on the choice of the mesh configuration and the appropriate boundary conditions. The problems of interest involve the full Navier–Stokes equations.

Numerical solutions of the full Navier–Stokes equations are not uncommon today. To do justice to a complex topic like this, one should devote to it a monograph at least twice as long as the present. The equations and finite differencing involved are very lengthy and complex, and space limitations would not permit here detailed descriptions of different methods. In this section we describe briefly some of the principles involved, discuss the different forms of the boundary conditions and outline methods of solution. More contributions are referenced in this monograph and in later chapters with respect to specific applications. The emphasis again is on the unsteady effects, although this aspect of the problem is not a critical element of the solution. In fact, many solutions of steady-state problems were framed as transients, asymptotically arriving at the steady-state form.

Lattices

Almost all of the computation schemes are formulated in terms of the equations of vorticity and stream function and deal strictly with two-dimensional flow. For incompressible, two-dimensional flow, the stretching term $\Omega_j \partial u_i / \partial x_j$ in Eq. (1.2.7) vanishes and, in a more traditional notation, this equation reads

$$\frac{\partial \Omega}{\partial t} + u \frac{\partial \Omega}{\partial x} + v \frac{\partial \Omega}{\partial y} = \nu \left[\frac{\partial^2 \Omega}{\partial x^2} + \frac{\partial^2 \Omega}{\partial y^2} \right], \tag{2.9.1}$$

with Ω the single component of vorticity in the z direction. By virtue of the continuity equation, Eq. (2.9.1) may also be written as

$$\frac{\partial \Omega}{\partial t} + \frac{\partial(u\Omega)}{\partial x} + \frac{\partial(v\Omega)}{\partial y} = \nu \left[\frac{\partial^2 \Omega}{\partial x^2} + \frac{\partial^2 \Omega}{\partial y^2} \right]. \tag{2.9.2}$$

This equation together with Eq. (1.2.9) repeated here for completeness

$$\Omega = - \left(\frac{\partial^2 \psi}{\partial x^2} + \frac{\partial^2 \psi}{\partial y^2} \right) \tag{2.9.3}$$

form a closed system, provided the velocity components u and v are expressed in terms of the stream function ψ. For a rectangular nonuniform mesh, Thoman and Szewczyk (1969) propose to express the u-component of velocity at the horizontal interface boundaries of the cell centered at the point (k, m)

$$u_{m,k-1/2} = 2 \frac{\psi_{m,k} - \psi_{m,k-1}}{\Delta y_k + \Delta y_{k-1}}, \tag{2.9.4}$$

$$u_{m,k+1/2} = 2 \frac{\psi_{m,k+1} - \psi_{m,k}}{\Delta y_k + \Delta y_{k+1}}, \tag{2.9.5}$$

and, similarly, the v component of the velocity is computed at the vertical interface boundaries. The subscript of the variable grid size Δy indicates the position of the grid point. The velocity components at the center of the cell are then taken as the average of boundary values:

$$u_{m,k} = \tfrac{1}{2}(u_{m,k-1/2} + u_{m,k+1/2}), \tag{2.9.6}$$

$$v_{m,k} = \tfrac{1}{2}(v_{m-1/2,k} + v_{m+1/2,k}). \tag{2.9.7}$$

The u component of velocity on vertical boundaries and the v component of velocity on horizontal boundaries are expressed as interpolations of the velocities calculated in (2.9.4)–(2.9.7). In terms of these expressions, the finite-difference form of the vorticity equation becomes

$$
\begin{aligned}
\frac{\Omega_{m,k}^{n+1} - \Omega_{m,k}^{n}}{\Delta t} &+ \frac{(u\Omega)_{m+1/2,k}^{n} - (u\Omega)_{m-1/2,k}^{n}}{\Delta x_m} \\
&+ \frac{(v\Omega)_{m,k+1/2}^{n} - (v\Omega)_{m,k-1/2}^{n}}{\Delta y_k} \\
&= \frac{2\nu}{\Delta x_m} \left[\frac{\Omega_{m+1,k}^{n} - \Omega_{m,k}^{n}}{\Delta x_{m+1} + \Delta x_m} - \frac{\Omega_{m,k}^{n} - \Omega_{m-1,k}^{n}}{\Delta x_m + \Delta x_{m-1}} \right] \\
&+ \frac{2\nu}{\Delta y_k} \left[\frac{\Omega_{m,k+1}^{n} - \Omega_{m,k}^{n}}{\Delta y_{k+1} + \Delta y_k} - \frac{\Omega_{m,k}^{n} - \Omega_{m,k-1}^{n}}{\Delta y_k + \Delta y_{k-1}} \right].
\end{aligned}
\tag{2.9.8}
$$

Equation (2.9.3) expressed in terms of finite differences centered at the point (m, k) and solved for $\psi_{m,k}$ becomes

$$
\begin{aligned}
\psi_{m,k} = \frac{2}{d_{m,k}} &\left[\frac{\Omega_{m,k}}{2} + \frac{\psi_{m-1,k}}{\Delta x_m(\Delta x_{m-1} + \Delta x_m)} + \frac{\psi_{m+1,k}}{\Delta x_m(\Delta x_m + \Delta x_{m+1})} \right. \\
&\left. + \frac{\psi_{m,k-1}}{\Delta y_k(\Delta y_{k-1} + \Delta y_k)} + \frac{\psi_{m,k+1}}{\Delta y_k(\Delta y_k + \Delta y_{k+1})} \right],
\end{aligned}
\tag{2.9.9}
$$

where

$$
\begin{aligned}
d_{m,k} = &\left[\frac{2}{\Delta x_m(\Delta x_{m-1} + \Delta x_m)} + \frac{2}{\Delta x_m(\Delta x_m + \Delta x_{m+1})} \right. \\
&\left. + \frac{2}{\Delta y_k(\Delta y_{k-1} + \Delta y_k)} + \frac{2}{\Delta y_k(\Delta y_k + \Delta y_{k+1})} \right].
\end{aligned}
\tag{2.9.10}
$$

Equations (2.9.8) and (2.9.9) form a system to be solved for the quantities $\Omega^n_{m,k}$ and $\psi^n_{m,k}$. Pressures are then computed by forward integration of the momentum equation.

The von Neumann stability criterion for the diffusion terms of the vorticity equation requires that the coefficient $\Omega^n_{m,k}$ be greater than or equal to zero:

$$1 - (\Delta t)\nu\, d_{m,k} \geqslant 0 \quad \text{or} \quad \Delta t \leqslant \frac{1}{\nu\, d_{m,k}}. \tag{2.9.11}$$

Similarly the convection terms of the equation dictate the inequality

$$\Delta t \leqslant \left\{ \frac{|u^n_{m,k}|}{\Delta x_m} + \frac{|v^n_{m,k}|}{\Delta y_k} \right\}^{-1}. \tag{2.9.12}$$

Let $(\Delta t)_d$ be the minimum of the time increment imposed by the diffusion terms as given by Eq. (2.9.11) and $(\Delta t)_c$ the minimum increment imposed by the convection terms and given by Eq. (2.9.12). Then the criterion for stability of the complete equation becomes

$$\Delta t \leqslant \frac{1}{\left[1/(\Delta t)_d\right] + \left[1/(\Delta t)_c\right]}. \tag{2.9.13}$$

Note that Δt is not less than or equal to the minimum of Δt_d and Δt_c but rather to the equivalent value of the resistance of two electrical resistances connected in parallel, with magnitudes Δt_c and Δt_d, respectively.

A crucial step in the numerical solution of these equations is the appropriate choice of the mesh configuration. Variable mesh sizes are a necessity, because in the regions next to the solid boundaries large values of the dependent variables are expected. This is especially true for impulsive changes of the flow. On the other hand, very far from the solid surfaces, the variations are very weak and considerable efficiency and storage savings can be accomplished by increasing the size of the mesh. Working with a purely rectangular scheme to solve the problem of impulsively started flow around a cylinder, Thoman and Szewczyk (1969) propose a rectangular mesh configuration shown in Fig. 2.21. This generates considerable difficulties at the intersection of the lattice with the solid boundary and the expression of the wall-boundary conditions. It appears that a mixed version, with a cylindrical lattice formation in the vicinity of the cylinder and a rectangular lattice away from the cylinder is more appropriate for efficient computations of the developing boundary layer. This mesh configuration is shown in Fig. 2.22. The difficulties now are concentrated at the interface of the two schemes. This is a purely bookkeeping problem and the details can be found in Thoman and Szewczyk (1969). The problem is not trivial because the finite-difference equations must be expressed in terms of cylindrical coordinates for the inner region, and at the interface of the two meshes care should be taken to define two sequences of cells of the two systems, respectively,

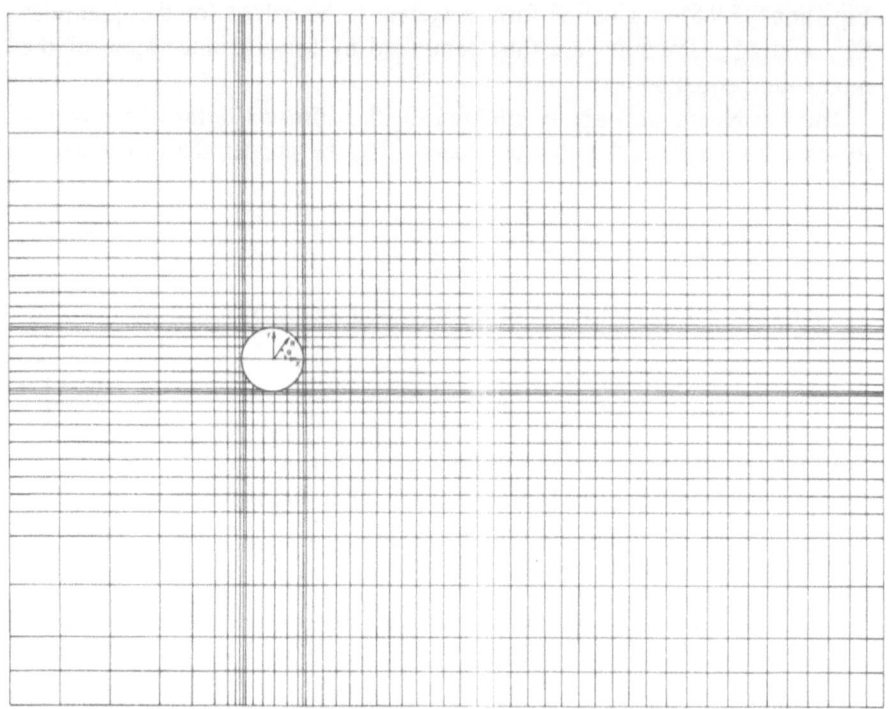

Fig. 2.21 Rectangular mesh configuration with variable mesh sizes.

with common centers. This is shown schematically in Fig. 2.23. In this figure the cells belonging to the polar system of coordinates are denoted by solid lines and the cells of the rectangular mesh are denoted by broken lines.

Moreover, ir and $i\theta$ are the indices that vary in the r and θ direction, whereas i and j run along the two directions of the rectangular mesh. Extrapolations are necessary to convert from the cylindrical to the rectangular system. The mesh is thick in the immediate vicinity of the wall, where large gradients are expected. The mesh size is progressively reduced away from the wall. The scale for the variable mesh is controlled by the Reynolds number since thinner boundary layers and, therefore, larger gradients perpendicular to the wall are expected for larger Reynolds numbers.

The idea of searching for an appropriate mesh configuration to conform with the shape of the particular solid body under consideration, and to cope with the regions where computations will encounter particular difficulties, has been pursued vigorously in the 1970s. Mehta and Lavan (1975) propose to use the classical Joukowski transformation in order to map the flow around a Joukowski airfoil into the interior of a unit circle. Consider the transformation

$$z = \frac{1}{\zeta} + \gamma + \frac{\zeta c^2}{1 + \gamma \zeta} ,$$

(2.9.14)

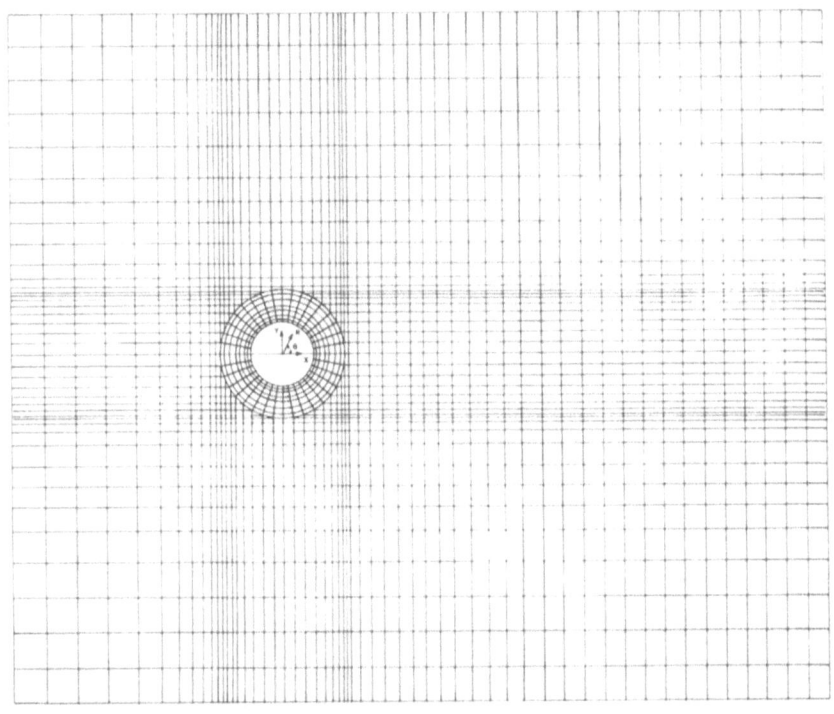

Fig. 2.22 Mixed system of coordinates with variable mesh sizes.

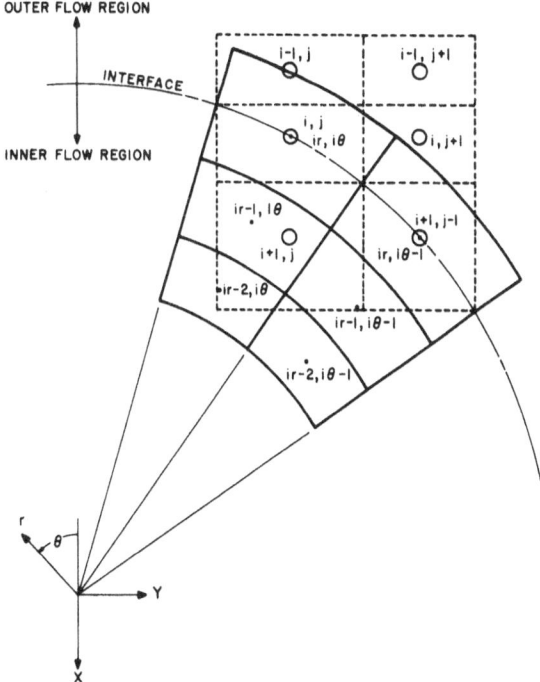

Fig. 2.23 Interface of the rectangular and the polar mesh of Fig. 2.22.

where z and ζ are complex numbers

$$z = x + iy, \qquad \zeta = re^{i\theta} \tag{2.9.15}$$

and γ and c are constants. This transformation maps the interior of the circle of radius c in the ζ plane onto a cambered airfoil in the z plane. The more familiar form of the transformation, namely, $z = \zeta + c^2/\zeta$, requires the shifting of the center of the circle from the origin of the coordinates in order to produce airfoil-like shapes in the transformed plane. In Eq. (2.9.14) the shift represented by the complex constant $\gamma = \xi_0 + i\eta_0$ is incorporated in the transformation and, therefore, the singular points are not symmetric with respect to the origin. To avoid the cusp at the trailing edge of the airfoil, the circle is not passed through the singular point but very near to it. This is accomplished by allowing

$$c = \left(\xi_0 + \sqrt{1 - \eta_0^2} \right)(1 - \delta), \tag{2.9.16}$$

where δ is a very small number.

If the governing equations are expressed in terms of the new coordinates r and θ, then the entire flow field around the airfoil maps in the domain $r_0 \leqslant r < 1$ and $0 \leqslant \theta \leqslant 2\pi$. A uniform mesh in the ζ-plane then maps on a lattice in the z-plane, which is thicker in the neighborhood of the airfoil (Fig. 2.24). In fact in areas of sharp variation of the geometry as, for example, at the trailing edge, the grid spacing is even further condensed next to the surface.

Mehta and Lavan introduce one further stretching in order to thicken further the mesh in the neighborhood of the solid boundary. They stretch the variable r according to an arc tanh function

$$\rho = \frac{\tanh^{-1}(rk_3 - k_4) + k_2}{k_1 + k_2}. \tag{2.9.17}$$

The domain of the new variable is $\rho_0 < \rho < 1$ and the value $\rho = 1$ corresponds to the skin of the body. In terms of these coordinates the equations of vorticity and stream function, Eqs. (2.9.1) and (2.9.3), become

$$H^2 r^2 \frac{R}{L} \frac{\partial \Omega}{\partial t} = \left(\frac{d\rho}{dr} \right)^2 r^2 \frac{\partial^2 \Omega}{\partial \rho^2} + \left(\frac{d\rho}{dr} r + \frac{d^2\rho}{dr^2} r^2 \right) \frac{\partial \Omega}{\partial \rho}$$

$$+ \frac{\partial^2 \Omega}{\partial \theta^2} - r \frac{d\rho}{dr} \frac{R}{L} J\left(\frac{\Omega, \psi + y}{\rho, \theta} \right), \tag{2.9.18}$$

$$- H^2 r^2 \Omega = r^2 \left(\frac{d\rho}{dr} \right)^2 \frac{\partial^2 \psi}{\partial \rho^2} + \left(\frac{d\rho}{dr} r + \frac{d^2\rho}{dr^2} r^2 \right) \frac{\partial \psi}{\partial \rho} + \frac{\partial^2 \psi}{\partial \theta^2}, \tag{2.9.19}$$

where H is essentially the Jacobian of the transformation

$$rH^2 = J\left(\frac{x, y}{r, \theta} \right) = \left(\frac{\partial x}{\partial r} \frac{\partial y}{\partial \theta} - \frac{\partial x}{\partial \theta} \frac{\partial y}{\partial r} \right). \tag{2.9.20}$$

Fig. 2.24 Curviliar mesh about a
Joukowski airfoil defined by Joukowski's transformation.

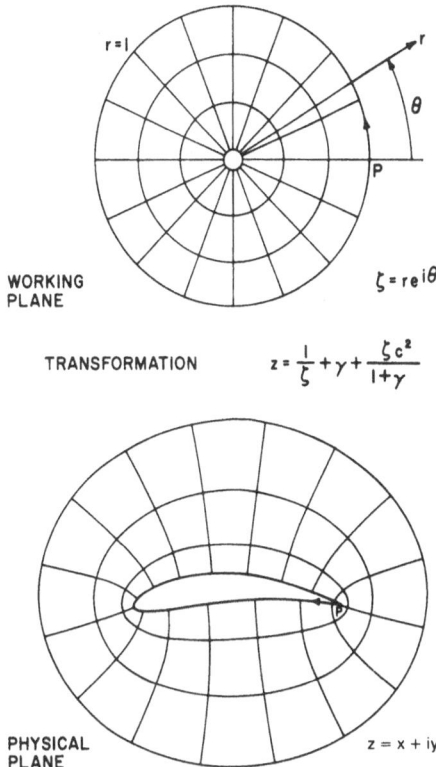

WORKING
PLANE

$\zeta = r e^{i\theta}$

TRANSFORMATION $z = \dfrac{1}{\zeta} + \gamma + \dfrac{\zeta c^2}{1+\gamma}$

PHYSICAL
PLANE

$z = x + iy$

ψ is a disturbance stream function, and R, L are the Reynolds number and
the dimensionless chord of the airfoil, respectively.

The proper choice of the coordinates may considerably improve the
efficiency of the numerical calculations. Ghia and Davis (1974) clearly
demonstrate this fact in their effort to solve the problem of steady flow about
a semiinfinite rectangular slab shown in Fig. 2.25. They introduce another
conformal transformation, the Schwarz–Christoffel transformation

$$\frac{dz}{d\zeta} = \left[Re - \zeta^2 \right]^{1/2}, \tag{2.9.21}$$

where again $z = x + iy$ and $\zeta = \xi + i\eta$, and Re is the Reynolds number
based on the thickness l, $Re = 2\pi U_\infty l / \nu$. Integration of Eq. (2.9.21) yields

$$z = \frac{1}{2} \left[\zeta (Re - \zeta^2)^{1/2} + Re \sin^{-1}(\zeta / \sqrt{Re}) \right]. \tag{2.9.22}$$

The transformation maps the body surface on a surface $\eta = \eta_w = $ constant.
The problem is therefore transformed to the problem of the flow about a flat
plate. Numerical integration is implemented on a (ξ, η) mesh, which in the
physical plane takes the form shown in Fig. 2.25. Note that for most of the
domain of integration, the new coordinates are more or less aligned with the
direction of the flow. Moreover, they are thickly spaced around the corner

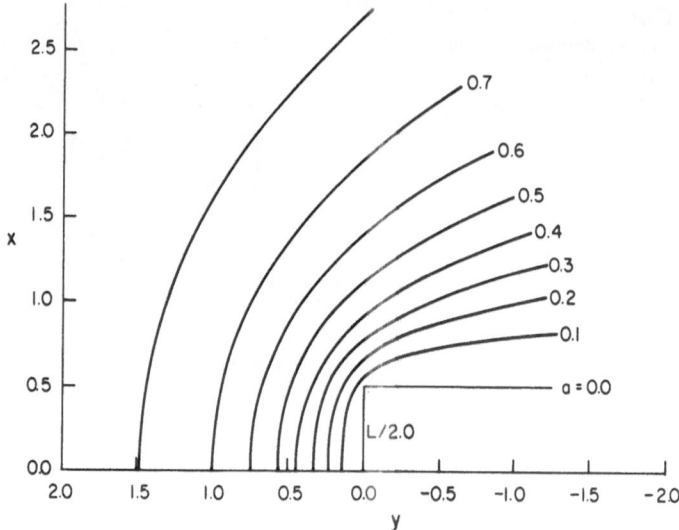

Fig. 2.25 Curvilinear mesh about a rectangular body defined by Schwarz–Cristoffel transformation.

where one would expect larger gradients. One more interesting observation is pertinent here. As Re approaches zero, the transformation yields the parabolic coordinates that are the optimal coordinates for boundary-layer flow past a thin semiinfinite flat plate (Kaplun, 1954, Davis, 1974). If the boundary-layer equations are expressed in terms of the optimal coordinates, they yield a uniformly valid solution for all η. The correct outer flow including displacement effects is then approached.

To facilitate integration in a finite domain, a further transformation is introduced

$$S = 1 - \frac{A}{\xi} \ln\left[1 + (\xi/A)\right], \tag{2.9.23}$$

$$N = \frac{\eta - \eta_w}{C + (\eta - \eta_w)}, \tag{2.9.24}$$

with A and C constants that can be chosen so as to obtain a reasonably smooth variation of the vorticity function in terms of the coordinate S and N. The domain of integration maps now into the finite domain $0 \leqslant S \leqslant 1$ and $0 \leqslant N \leqslant 1$. Ghia and Davis (1974) solve the full Navier–Stokes equations in terms of the independent variables S and N.

Boundary conditions

The proper choice of the boundary conditions to be imposed is a crucial step in the solution of the full Navier–Stokes equations. The problem under consideration is the flow around an obstacle immersed in a stream of infinite extent. It is desirable therefore to demand that the flow merges into a uniform stream, but this condition should be imposed at infinity. For integration in a physical plane, the approximation involved in imposing free-stream conditions on the edges of the finite domain is not always acceptable. A particularly problematic region is the downstream boundary of the domain of integration because a viscous wake usually extends quite far downstream of the body.

For a rectangular mesh (see Fig. 2.21) with dimensions $M \times K$, that is, $m = 1, \ldots, M$ and $k = 1, \ldots, K$, the stream function is required to satisfy the conditions (Thoman and Szewczyk, 1969)

$$\left. \frac{\partial \psi}{\partial y} \right|_{k=1+1/2} = U_\infty, \qquad \left. \frac{\partial \psi}{\partial x} \right|_{k=2} = 0, \tag{2.9.25}$$

$$\left. \frac{\partial \psi}{\partial y} \right|_{k=K-1/2} = U_\infty, \qquad \left. \frac{\partial \psi}{\partial x} \right|_{k=K-1} = 0, \tag{2.9.26}$$

so that the flow is forced to merge with the undisturbed stream on the upper and lower boundaries of the domain of integration. On the upstream boundary, the vorticity and the v component of the velocity are constrained:

$$\left. \frac{\partial \psi}{\partial x} \right|_{m=1+1/2} = 0, \qquad \Omega_{2,k} = 0. \tag{2.9.27}$$

This permits the inflow streamlines to shift to accommodate the circulation produced by a rotating cylinder or a lifting body.

The crucial conditions are the boundary conditions on the downstream boundary. The boundary conditions there should not permit accumulation of vorticity that may feed through the system in the upstream direction. Thoman and Szewczyk found by numerical experiments that the most effective conditions were

$$\left. \frac{\partial \Omega}{\partial x} \right|_{m=M-1/2} = 0, \qquad \left. \frac{\partial v}{\partial x} \right|_{m=M-1/2} = 0, \tag{2.9.28}$$

which can be written in terms of ψ and Ω in a finite-difference form as follows:

$$\psi_{M,k} = \mu_{M-1,k} + \frac{x_M - x_{M-1}}{x_{M-1} - x_{M-2}} (\psi_{M-1,k} - \psi_{M-2,k}), \tag{2.9.29}$$

$$\Omega_{M,k} = \Omega_{M-1,k}. \tag{2.9.30}$$

On the walls of a cylinder, the no-penetration and no-slip conditions are imposed, namely, that the velocity components parallel and perpendicular to the wall vanish on a fixed wall or, in general, that they are equal to the velocity of the wall.

Mehta (1972) observes that far from the body the dominant term in the governing equations is convection. In the phase plane ρ, θ, he proposes to express the fact that diffusion is negligible in Eq. (2.9.18) and imposes the following condition:

$$H^2 r \frac{\partial \Omega}{\partial t} = - \frac{d\rho}{dt} J\left(\frac{\Omega, \psi + y}{\rho, \theta} \right) \quad \text{at } \rho = \rho_0. \tag{2.9.31}$$

Similarly, dropping diffusion and the pressure gradient from the θ component of the momentum equation yields

$$\frac{\partial u_\theta}{\partial t} = - \left[\frac{1}{2rH} \frac{\partial (u_r^2 + u_\theta^2)}{\partial \theta} + u_r \Omega \right], \tag{2.9.32}$$

with

$$u_r = \frac{1}{rH} \left(\frac{\partial \psi}{\partial \theta} + \frac{\partial y}{\partial \theta} \right), \qquad u_\theta = - \frac{1}{H} \left(\frac{d\rho}{dr} \frac{\partial \psi}{\partial \rho} + \frac{\partial y}{\partial r} \right). \tag{2.9.33}$$

Equation (2.9.31) then gives

$$\frac{\partial \psi}{\partial \rho} = - \frac{dr}{d\rho} \left(u_\theta H + \frac{\partial y}{\partial r} \right), \quad \text{at } \rho = \rho_0. \tag{2.9.34}$$

These boundary conditions permit eddies or vortices to pass through the downstream boundary. Moreover, since the velocity itself is not specified, circulation can change with time. The method thus circumvents the difficulty of specifying downstream boundary condition across a wake that is actually part of the solution and therefore not known a priori.

On the surface of the body the no-penetration and no-slip conditions read

$$\psi = -y, \qquad \partial \psi / \partial \rho = 0. \tag{2.9.35}$$

Implementation of the methods described is of course a lot more involved than implied by the simplified version of the formulas provided. In the chapters that follow we describe in some detail features of the computational methods that are important only in the calculation of unsteady flows. The reader will find more details in the original works referenced here.

REFERENCES

Belcher, B. J., Burggraf, O. R., Cooke, J. C., Robins, A. J., and Stewartson, K., 1972. In *Recent Research of Unsteady Boundary Layers*, ed. Eichelbrenner, E. A., 1444–1465.

Brown, S. N., and Stewartson, K., 1969. In *Annual Review of Fluid Mechanics*, ed. Sears, W. R., Annual Reviews, Inc., **1**, 45–72.

Carnahan, B., Luther, H. A., and Wilkes, J. O., 1969. *Applied Numerical Methods*, Wiley, New York.

Carter, J. E., 1974. AIAA Paper No. 74-583.

Catherall, D., and Mangler, K. W., 1966. *J. Fluid Mech.*, **26**, 163–182.

Cebeci, T., 1977. *Proc. R. Soc. Lond.*, **A355**, 225–238

Courant, R., Friedrichs, K. O., and Lewy, H., 1928. *Math. Annalen.*, **100**, 32–74.

Davis, R. T., Whitehead, R. E., and Wornom, S. F., 1970. "The Development of an Incompressible Boundary-Layer Theory Valid to Second Order," Va. Poly. Inst. & State Univ. Eng. Report, VPI-E-70-1.

Davis, R. T., 1974. "A Study of the Use of Optimal Coordinates in the Solution of the Navier–Stokes Equations," Univ. of Cincinnati, Report No. AFL 74-12-14.

Dean, W. R., 1950. *Proc. Cambridge Philos. Soc.*, **46**, 293–306.

Dwyer, H. A., 1968. *AIAA J.*, **6**, 2447–2448.

Eddy, E. P., 1949, "Stability in the Numerical Solution of Initial Value Problems in Partial Differential Equations," NOLM 10232, Naval Ordinace Laboratory, White Oak, Silver Spring, Maryland.

Farn, C. L. S., and Arpaci, V. S., 1966. *AIAA J.*, **4** 730–732.

Fromm, J. E., 1964. *Meth. Comp. Physics*, **3**, 345–382.

Ghia, U., and Davis, R. T., 1974. *AIAA J.*, **12**, 1659–1665.

Guirand, J. P., 1969. *C. R. Acad. Sci. Paris*, **268**, 239–241.

Hall, M. G., 1969. *Proc Roy. Soc. Lond.* **A310**, 401–414.

Hirt, C. W., 1965. AIAA Paper No. 65-3.

Isaacson, E., and Keller, H. B., 1966. *Analysis of Numerical Methods*, Wiley, New York.

Kaplun, S., 1954. *ZAMP* **5**, 111–135.

Keller, H. B., 1969. *SIAM J. Num. Anal.*, **6**, 8–30.

Keller, H. B., 1971. In *Numerical Solution of Partial Differential Equations*, ed. Hubbard, B., **2**, 327–350.

Keller, H. B., 1975. In *Lecture Notes in Physics, Proceedings of Fourth International Conference on Numerical Methods in Fluid Dynamics*, 1–21, Springer, Colorado.

Keller, H. B. 1978. *Ann. Rev. Fluid. Mech.*, **10**, 417–433.

Keller, H. B., and Cebeci, T., 1971. In *Lecture Notes in Physics, Proceedings of Second International Conference on Numerical Methods in Fluid Dynamics*, Springer, Berlin.

Klemp, J. B., and Acrivos, A., 1971. *J. Fluid Mech.*, **53**, 177–191.

Klineberg, J. M., and Steger, J. L., 1974. AIAA Paper No. 74-94.

Lax, P. D., and Richtmyer, R. D., 1956. *Comm. Pure Appl. Math.*, **9**, 267–293.

Mehta, U. B., 1972. "Starting Vortex, Separation Bubbles and Stall-0 Numerical Study of Laminar Unsteady Flow Around an Airfoil," Ph.D. Thesis, Illinois Institute of Technology.

Mehta, U. B., and Lavan, Z., 1975. *J. Fluid Mech.*, **67**, 227–256.

Moore, F. K., 1957. In *Boundary Layer Research*, ed. Görtler, H., 296–311, Springer, Berlin.

Nash, J. F., Carr, L. W., and Singleton, R. E., 1975. *AIAA J.*, **13**, 167–172.

Nickel, K., 1958. *Arch. Rat. Mech. Anal.*, **2**, 1–31.

O'Brien, G. G., Hyman, M. A., and Kaplan, S., 1951. *J. Math. Physics*, **29**, 223–251.

Phillips, J. H., and Ackerberg, R. C., 1973. *J. Fluid Mech.*, **58**, 561–579.

Robins, A. J., 1970. Ph.D. Thesis, Bristol University.

Richtmyer, R. D., 1957. *Difference Methods for Initial-Value Problems*, Interscience, New York.

Richtmyer, R. D., and Morton, K. W., 1967. *Difference Methods for Initial Value Problems*, second edition, Wiley, New York.

Roache, P. J., 1972. *Computational Fluid Dynamics*, Hermosa Publishers, Albuquerque.

Rott, N., 1956. *Q. Appl. Math.*, **13**, 444–451.

Sears, W. R., 1956. *J. Aerosol Sci.*, **23**, 490–499.

Shen, S. F., 1978. *Adv. Appl. Mech.*, **18**, 177–220.

Shen, S. F., and Nenni, J. P., 1975. In *Unsteady Aerodynamics*, ed. Kinney, R. B., **1**, 245–259.

Smith, G. D., 1965. *Numerical Solution of Partial Differential Equations*, Oxford Univ. Press, New York and London.

Stewartson, K., 1960. In *Advances in Applied Mechanics*, eds. Dryden, H. L., and Von Kármán, T., **6**, 1–37, Academic Press, New York.

Telionis, D. P., and Werle, M. J., 1973. *J. Appl. Mech.*, **95**, 369–374.

Telionis, D. P., and Tsahalis, D. Th., 1974a. *AIAA J.*, **12**, 1469–1476.

Telionis, D. P., and Tsahalis, D., Th. 1974b. *Acta Astron.*, **1**, 1487–1505.

Telionis, D. P., Tsahalis, D. Th., and Werle, M. J., 1973. *Phys. Fluids*, **16**, 968–973.

Thom, A., and Apelt, C. J., 1961. *Field Computations in Engineering and Physics*, C. Van Nostrand, Princeton, New Jersey.

Thoman, D. C., and Szewczyk, A. A., 1969. *Phys. Fluids*, Suppl. II, 76–86.

Thompson, R. J., 1964. *J. Soc. Indust. Appl. Math.*, **12**, 189.

Wang, K. C., 1975. *Phys. Fluids*, **18**, 951–955.

Werle, M. J., and Bertke, S. D., 1972. *AIAA J.*, **10**, 1250–1252.

Werle, M. J., and Davis, R. T., 1972. *J. Appl. Mech.*, **39**, 7–12.

Williams, J. C., and Johnson, W. D., 1974a. *AIAA J.*, **12**, 1388–1393.

Williams, J. C., and Johnson, W. D., 1974b. *AIAA J.*, **12**, 1427–1429.

Wirz, H. J., 1975. In *Progress in Numerical Fluid Dynamics*, *Lecture Notes in Physics*, ed. Wirz, H. J., **41**, 442–476, Springer, New York.

CHAPTER 3

Impulsive Motion

3.1 Introduction

Impulse is the application of concentrated or distributed forces of large magnitude F for very short periods of time Δt. In mechanics we study the effects of impulses by assuming that the duration of force application tends to zero but the product $F\Delta t$ remains finite. The motion that results from the application of an impulse is an impulsive motion.

Impulsive fluid motions do not exist in reality. However, many practical engineering problems can be solved approximately, by assuming that the external disturbance is impulsive. Evidence collected most recently supports the idea that viscous phenomena display a characteristic inertia. Indeed, most abrupt changes of the boundary conditions as, for example, the sudden change of the angle of attack of an airfoil, are followed by relatively slower adjustments of the viscous field. Changes of the boundary conditions may therefore be considered, with a reasonable accuracy, as impulsive. This is especially true in the case of unsteady separation, a topic that will be discussed in detail in a later chapter.

The assumption of impulsive motion results in significant simplifications of the mathematical formulation, because time dependence is eliminated from the boundary conditions. Due to their simplicity, impulsive motion problems have attracted the interest of many investigators. It is very interesting to note here that some basic concepts of boundary-layer theory were actually presented and adequately discussed, in conjunction with a typical impulsive viscous flow problem (Rayleigh 1911). The Rayleigh problem is in fact a classical example, quite popular among instructors of fluid mechanics, because its properties are characteristic of impulsive viscous flows.

Impulsive motions are of particular interest to students of viscous flows, because for very small times following the impulse, the field changes are practically inviscid. In other words, the impulsive generation of the flow field of a viscous fluid is at first "inviscid" and, similarly, the impulsive change of an established viscous flow field can be derived, for small times, by superimposing an "inviscid correction." The early stages of impulsive viscous flows contain valuable information of flow properties that control the subsequent development of the phenomenon, and thorough understanding of its properties is essential in the study of unsteady flows. Moreover, the character of the solution and its asymptotic behavior for small times is necessary input for the numerical calculations, as described in detail in the following sections.

In this chapter we briefly review early and classical contributions of Blasius, Lamb, Goldstein and Rosenhead, and Stewartson, typical examples of impulsive viscous flows. In particular we examine the growth of the boundary layer over a cylinder or a flat plate started impulsively from rest. In the first case the thickening of the boundary layer is accompanied by the appearance of zero-shear profiles, reversing flow, and eventually separation. A semiinfinite flat plate started impulsively from rest requires careful consideration of the region where the elementary solutions of Rayleigh and Blasius overlap. Such analytical solutions have been chosen as test cases for the numerical calculations that followed.

In this chapter we also describe some of the numerical methods employed to solve impulsive-flow problems. A variety of methods have been used and we describe here a representative of each category, namely: exact solutions of an approximate equation, numerical solutions in the reduced variable space of semisimilar equations, and straightforward numerical computations in the physical space of variables. Comparison of the relative performance of such methods is provided together with some arguments and conclusions on the physical aspects of the problem.

3.2. The Flow Immediately After an Impulsive Start

It has been known since the 1930's that impulsively generated flows are initially inviscid. The main idea is contained in the classical text of Lamb (1932), but we found later presentations (Sears, 1949) more tractable. The subsequent derivation is following closely the steps outlined in an unpublished work of Professor W. R. Sears (private communication).

Imcompressible flows will be assumed throughout to avoid the complication of acoustic disturbances. Let $\mathbf{V}_1$, p_1 be the velocity and pressure fields respectively of a steady flow that satisfies the Navier–Stokes equation for $t < 0$:

$$\mathbf{V}_1 \cdot \nabla \mathbf{V}_1 = -\frac{1}{\rho} \nabla p_1 + \nu \nabla^2 \mathbf{V}_1. \qquad *(3.2.1)$$

It is emphasized again that an asterisk is being used in this monograph to denote dimensional velocities, pressures, space coordinates, and time, u^*, v^*, p^*, x^*, y^*, and t^*, respectively. However, if all the quantities that appear in the equation are dimensional and there is no danger of confusion, only one asterisk will be used in front of the equation number.

Let us assume that at $t = 0$ a new pressure distribution p_2 is superposed on our flow field and generates a new velocity field $\mathbf{V} = \mathbf{V}_1 + \mathbf{V}_2$, where $\mathbf{V}_2$ represents the vector change of the velocity due to the pressure change. The new flow $(\mathbf{V}, p)$ satisfies again the Navier–Stokes equation

$$\frac{\partial \mathbf{V}_2}{\partial t} + (\mathbf{V}_1 + \mathbf{V}_2) \cdot \nabla(\mathbf{V}_1 + \mathbf{V}_2)$$
$$= -\frac{1}{\rho} \nabla(p_1 + p_2) + \nu \nabla^2(\mathbf{V}_1 + \mathbf{V}_2). \qquad \text{*(3.2.2)}$$

Subtracting Eq. (3.2.1) from (3.2.2), we readily arrive at

$$\frac{\partial \mathbf{V}_2}{\partial t} + \mathbf{V}_1 \cdot \nabla \mathbf{V}_2 + \mathbf{V}_2 \cdot \nabla(\mathbf{V}_1 + \mathbf{V}_2) = -\frac{1}{\rho} \nabla p_2 + \nu \nabla^2 \mathbf{V}_2. \qquad \text{*(3.2.3)}$$

We now assume, that the pressure p_2 is applied impulsively and therefore generates a change of the flow field in a very small time t_ϵ. One then may integrate Eq. (3.2.3) with respect to time and note that for $t_\epsilon \to 0$, the convective and viscous terms tend to zero since the generated velocity is finite. Our assumption here is that the dominant terms in our momentum equation are the impulsive-pressure changes and the resulting acceleration $\partial \mathbf{V}_2/\partial t$. The generated changes of the velocity field are finite and the nonlinear terms of the equation, as well as the viscous diffusion terms, may be neglected if integrated over a vanishingly small period of time. Hence,

$$\mathbf{V}_2(t_\epsilon) = -\frac{1}{\rho} \nabla\left(\int_0^{t_\epsilon} p_2 \, dt \right). \qquad \text{*(3.2.4)}$$

We define the integral on the right-hand side for $t_\epsilon \to 0$, as the impulsive-pressure distribution $P = -\int_0^{t_\epsilon} p_2 \, dt$ and thus

$$\mathbf{V}_2(t_\epsilon) = -\frac{1}{\rho} \nabla P. \qquad \text{*(3.2.5)}$$

We immediately conclude that the field $\mathbf{V}_2(t_\epsilon)$ is irrotational, since $\nabla \times \nabla P$ vanishes by the well-known identity that holds for any scalar P. We have therefore proved that if the motion is started impulsively from rest, that is, $\mathbf{V}_1 = 0$, the generated field $\mathbf{V}_2$ is irrotational and, similarly, if an impulsive pressure produces a change $\mathbf{V}_2$ on an established viscous flow $\mathbf{V}_1$, then this change is also irrotational.

The process of starting or stopping an irrotational flow impulsively should not be thought of as requiring the application of impulsive pressures throughout the flow field (Sears, 1949). It will only be necessary to supply the boundary values of P, for the pressures at interior points are internal forces exerted between adjacent fluid particles. Mathematically this becomes obvious if we take the divergence of Eq. (3.2.5). The left-hand side then vanishes by the equation of continuity, $\nabla \cdot \mathbf{V} = 0$, and hence,

$$\nabla^2 P = 0. \qquad \text{*(3.2.6)}$$

The solution of this equation, the glorified Laplace equation, depends only on values of P or its derivatives on the boundaries of the domain.

Let us now consider what will happen after the time $t = t_\epsilon$. The reader should recall that the impulse is assumed to consist of an infinitely large pressure exerted for an infinitely small time, so that the integral $\int_0^{t_\epsilon} p_2\, dt$ is finite. At the end of this phase of the motion, or for $t_\epsilon \to 0$, at $t = 0^+$ as traditionally denoted in literature, the flow consists of the field $\mathbf{V}_1$ and the distrubance field $\mathbf{V}_2$. The latter, being irrotational, does not satisfy the no-slip condition on the solid boundaries. Alternatively we can say that at $t = 0^+$, all the solid boundaries are covered with vortex sheets or, equivalently, that the new field $\mathbf{V}_2$ introduces new amounts of vorticity which are confined to regions infinitely close to the solid boundaries, thus leaving the field $\mathbf{V}_2$ essentially irrotational.

After $t = 0^+$ the effect of the impulsive pressure disappears and the most dominant characteristic of the flow is the concentrated vorticity that needs to be diffused into the flow. The diffusion terms will then outweigh all other terms in the momentum equation and will balance the unsteady term $\partial \mathbf{V}/\partial t$, at least for times $0^+ < t < t_1$ where $t_1 \ll L/U_\infty$

$$\frac{\partial \mathbf{V}}{\partial t} = \nu \nabla^2 \mathbf{V} \qquad\qquad\qquad\qquad *(3.2.7)$$

This phenomenon will be studied more carefully in the following sections.

Rayleigh's problem is a very simple example of impulsive motion displaying most of the characteristic features of impulsive flows. Consider the flow started impulsively over an infinite flat plate. The symmetry of the problem requires that no quantities vary with x and we may deduce immediately from the continuity equation that $\partial v/\partial y = 0$. Since v vanishes at the wall, it must vanish everywhere and therefore the flow has streamlines straight and parallel to the wall. The second momentum equation then simply gives $\partial p/\partial y = 0$, and the pressure is constant throughout the flow field, say p_∞. The x component of the momentum equation then takes the form

$$\frac{\partial u}{\partial t} = \nu \frac{\partial^2 u}{\partial y^2}, \qquad\qquad\qquad\qquad *(3.2.8)$$

and a solution that meets the initial and boundary conditions, $u(x, y, 0) = 0$, $u(x, 0, t) = 0$, and $u(x, \infty, t) \to U_\infty$ (see Fig. 3.1) is

$$u = U_\infty \operatorname{erf}\left(y/2\sqrt{\nu t}\right), \qquad\qquad\qquad *(3.2.9)$$

where erf stands for the error function.

Notice that for $t \to 0$, the velocity profile approaches plug flow, that is, a uniform flow $u = U_\infty$ for all $y > 0$ and $u = 0$ for $y = 0$ (see Fig. 3.2). This is essentially the inviscid solution, i.e., a uniform velocity throughout, which violates the no-slip condition at the wall. The discontinuity at $y = 0$, which immediately develops to very large velocity gradients $\partial u/\partial y$ at $t > 0$, is

Fig. 3.1 The Rayleigh problem; boundary conditions.

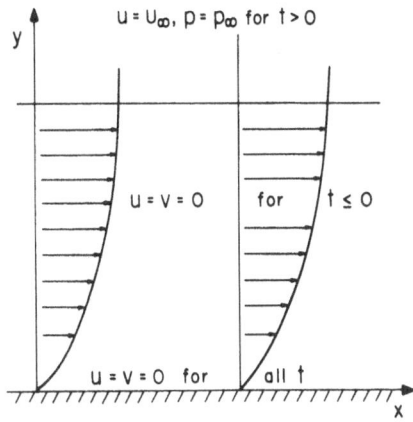

essentially a vortex sheet that wraps around the solid body at the beginning of the motion. In this simple and well-known example we therefore see that for $t = 0^+$, we recover an inviscid velocity distribution in accordance with the theorem we proved. In fact, since here the full Navier–Stokes equations reduce to the form given in Eq. (3.2.8), no approximation is necessary and our solution is exact. If the thickness of the viscous region is defined as the domain where $u < (0.99) U_\infty$, then $\delta \approx 3.64\sqrt{\nu t}$. The viscous region therefore is proportional to the square root of viscosity.

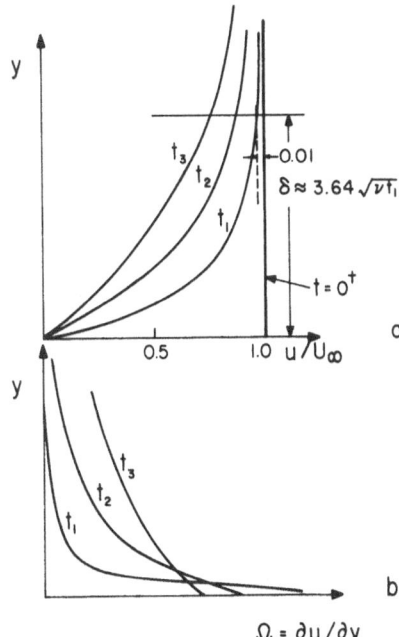

Fig. 3.2 Velocity and vorticity profiles of the Rayleigh flow; schematic representation for $t_1 < t_2 < t_3$.

One final comment is pertinent here. The reader may easily verify that the "flux of vorticity," that is, the integral $\int \mathbf{\Omega} \cdot d\mathbf{s}$, with $\mathbf{\Omega}$ and $\mathbf{s}$ the vorticity and the area, respectively, is independent of time, if the integation extends over the entire two-dimensional space. The area under the curves of Fig. 3.2b is constant. In other words the "amount" of vorticity generated instantly by the impulsive motion is conserved, even though vorticity slowly diffuses away from the wall.

The reader will find detailed descriptions of the Rayleigh problem in classical texts (Schlichting, 1968; White, 1974). Let us now proceed further to describe qualitatively the situation in the case of an impulsive change superimposed on an established viscous flow.

Let us assume that the field $\mathbf{V}_1$ represents the flow about a circular cylinder in a stream of velocity $\mathbf{U}_{\infty 1}$. The well-known solution for potential flow gives at $\theta = 90°$ (see Fig. 3.3a):

$$v_{\theta 1} = - U_{\infty 1}\left(1 + \frac{R^2}{r^2}\right),$$

*(3.2.10)

which yields $v_{\theta 1} = -2 U_{\infty 1}$ at $r = R$.

The boundary layer modifies this profile so that the velocity vanishes at $r = R$ (dashed line in Fig. 3.3b) and is approximately equal to $2U_{\infty 1}$ at the edge of the boundary layer. This of course is true in the ideal situation of an infinite Reynolds number. For finite Reynolds numbers and boundary-layer thicknesses, the velocity at $\theta = 90°$ never reaches the value $2U_{\infty 1}$. The maximum approaches the value $- U_{\infty 1}[1 + (R + \delta^*)^2(r + \delta)^{-2}]$ where δ and δ^* are the thickness and the displacement thickness respectively.

Let us now assume that the cylinder is given impulsively a speed $- U_{\infty 2}$ or, equivalently, that the flow accelerates impulsively from $U_{\infty 1}$ to $U_{\infty 1} + U_{\infty 2}$. According to our theorem, the viscous flow field at $t = 0^+$ can be derived by superimposing the potential solution due to $U_{\infty 2}$. At $t = 0^+$ the correction to the radial velocity component at $\theta = 90°$ will be

$$v_{\theta 2} = - U_{\infty 2}\left(1 + \frac{R^2}{r^2}\right).$$

*(3.2.11)

The boundary-layer profile, within the boundary-layer approximation, is uniformly displaced to the right by the quantity $v_{\theta 2}(R, 90°) = 2 U_{\infty 2}$. As a result, its wall value is $2U_{\infty 2}$ and the no-slip condition is violated (see Fig. 3.3b).

At $t = 0^+$, there is a discontinuity in the velocity profiles around the cylinder. The no-slip condition requires that the velocity vanishes at $r = R$, but our solution at $t = 0^+$ and $r \to R$ gives a finite velocity $v_\theta = 2U_{\infty 2}\sin\theta$. Very large values of vorticity are confined to very small distances from the wall of the cylinder. For $t > 0^+$, the vorticity generated by the impulsive change will start diffusing into the flow. The boundary-layer velocity profiles will then indicate ever-decreasing velocity gradients as shown schematically

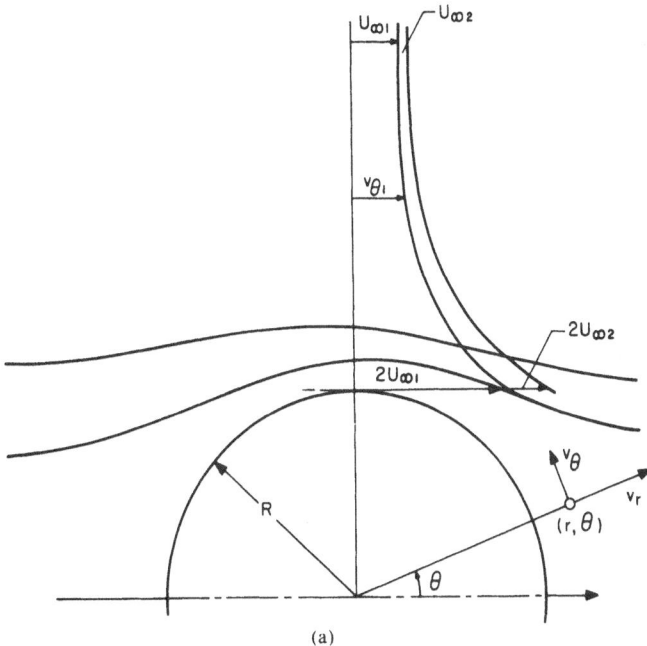

Fig. 3.3a Schematic of velocity profiles immediately after an impulsive change of the free stream from $U_{\infty 1}$ to $U_{\infty 1} + U_{\infty 2}$.

in Fig. 3.3c. This description is based on the classical features of the Euler and boundary-layer equations. It is within the accuracy of these solutions that, for example, the edge velocities are $2U_{\infty 1}$ and $2(U_{\infty 1} + U_{\infty 2})$ before and after the impulse, respectively. However, the qualitative description is certainly valid for the true behavior of a fluid as well as the exact solution of the full Navier–Stokes equations.

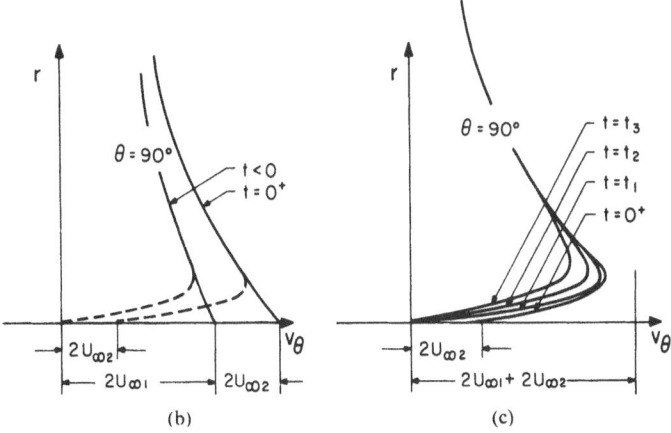

Fig. 3.3b Detail of the profile at $\theta = 90°$ of Fig. 3.3a for $t < 0$ and $t = 0^+$. (c) Detail of the profile at $\theta = 90°$ of Fig. 3.3a for $t > 0$.

3.3 An Order of Magnitude Analysis

We are now interested in analyzing more carefully the motion immediately after an impulsive start. In fact we would be content to solve the problem for the early stages of the flow, that is, for small times. Let t_0 be a small time representing the order of magnitude of our interest. The arguments that follow hold for any type of unsteady flows that possess a characteristic small-time scale, as, for example, periodic flows with large frequencies. In the last case the parameter t_0 can be defined as the inverse of the frequency. Let lengths, velocities, and time be scaled with a typical length L, velocity U_0, and the quantity t_0, respectively.

$$\left.\begin{array}{ll} x = x^*/L, & u = u^*/U_0 \\ y = y^*/L, & v = v^*/U_0 \end{array}\right\} t = t^*/t_0. \tag{3.3.1}$$

Here again, except for the quantities U_0, L, and t_0, symbols with and without asterisks represent dimensional and dimensionless quantities, respectively.

The continuity and momentum equations (1.2.4) and (1.2.5) for two-dimensional incompressible flow then become

$$\frac{\partial u}{\partial x} + \frac{\partial v}{\partial y} = 0, \tag{3.3.2}$$

$$\frac{L}{U_0 t_0} \frac{\partial u}{\partial t} + u \frac{\partial u}{\partial x} + v \frac{\partial u}{\partial y} = \frac{1}{\rho U_0^2} \frac{\partial p^*}{\partial x} + \frac{\nu}{U_0 L}\left(\frac{\partial^2 u}{\partial x^2} + \frac{\partial^2 u}{\partial y^2}\right), \tag{3.3.3}$$

$$\frac{L}{U_0 t_0} \frac{\partial v}{\partial t} + u \frac{\partial v}{\partial x} + v \frac{\partial v}{\partial y} = \frac{1}{\rho U_0^2} \frac{\partial p^*}{\partial y} + \frac{\nu}{U_0 L}\left(\frac{\partial^2 v}{\partial x^2} + \frac{\partial^2 v}{\partial y^2}\right). \tag{3.3.4}$$

The coordinate system is aligned with the skin of the body.

Based on physical intuition, engineering experience, and the arguments of the previous section, we expect that for small times, the viscous effects will be confined in regions very close to the solid boundaries. After all, we know that even as $t \to \infty$, when the steady-state field is approached, the viscous effects will remain confined in thin layers, at least as long as the flow is attached. Therefore, following the line of arguments employed in the development of the classical boundary-layer theory, we would like to look again into a thin region next to the wall, in which the v component of the velocity is expected to be very small. The stretching factor in the present case should contain the parameter t_0. All these arguments can be also motivated by inspecting Rayleigh's solution. According to this solution, the boundary layer starts off with zero thickness at $t = 0$ and grows with the square root of time. It seems reasonable to search for a similar solution in the case of impulsive flow about a body of arbitrary configuration. Rayleigh's solution further indicates that a proper dimensionless small number, characteristic of the phenomenon could be $\sqrt{\nu t_0}/L$. Let us therefore stretch the dimensionless quantities y and v as follows:

$$Y = y\left(L/\sqrt{\nu t_0}\right), \qquad V = v\left(L/\sqrt{\nu t_0}\right), \tag{3.3.5}$$

and replace the pressure p^* with the dimensionless quantity $p = p^*/\rho U_0^2$. Equations (3.3.2)–(3.3.4) then become

$$\frac{\partial u}{\partial x} + \frac{\partial V}{\partial Y} = 0, \tag{3.3.6}$$

$$\frac{\partial u}{\partial t} + \frac{U_0 t_0}{L}\left(u\frac{\partial u}{\partial x} + V\frac{\partial u}{\partial Y}\right) = -\frac{U_0 t_0}{L}\frac{\partial p}{\partial x} + \frac{\nu t_0}{L^2}\left(\frac{\partial^2 u}{\partial x^2} + \frac{L^2}{\nu t_0}\frac{\partial^2 u}{\partial Y^2}\right),$$

$$\tag{3.3.7}$$

$$\frac{\partial V}{\partial t} + \frac{U_0 t_0}{L}\left(u\frac{\partial V}{\partial x} + V\frac{\partial V}{\partial Y}\right)$$

$$\tag{3.3.8}$$

$$= -\frac{U_0 t_0}{L}\frac{L}{\sqrt{\nu t_0}}\frac{\partial p}{\partial Y} + \frac{\nu t_0}{L^2}\left(\frac{\partial^2 V}{\partial x^2} + \frac{L^2}{\nu t_0}\frac{\partial^2 V}{\partial Y^2}\right).$$

There are two dimensionless parameters appearing (Stuart, 1964), the quantities $U_0 t_0/L$ and $\nu t_0/L^2$. With t_0 a very small number, we expect that both these dimensionless quantities are also small. However, in the limit of $\nu \to 0$, we expect that $\nu t_0/L^2 \ll U_0 t_0/L$, or, equivalently, that $\nu/L \ll U_0$. The last inequality is expressing the fact that the Reynolds number is large, which is certainly true for most practical applications. It appears therefore reasonable as a first approximation to drop the terms of order $\nu t_0/L^2$ but retain terms of order $U_0 t_0/L$. We immediately conclude from Eq. (3.3.8) that to this level of approximation, $\partial p/\partial Y \approx 0$, that is, the pressure does not vary across the boundary layer. In dimensionless form and after the terms of order $\epsilon = \nu t_0/L^2$ have been eliminated, Eqs. (3.3.6)–(3.3.7) and the appropriate boundary conditions read

$$\frac{\partial u}{\partial x} + \frac{\partial V}{\partial Y} = 0, \tag{3.3.9}$$

$$\frac{\partial u}{\partial t} + \frac{U_0 t_0}{L}\left(u\frac{\partial u}{\partial x} + V\frac{\partial u}{\partial Y}\right) = -\frac{U_0 t_0}{L}\frac{\partial p}{\partial x} + \frac{\partial^2 u}{\partial Y^2} + O(\epsilon), \tag{3.3.10}$$

$$u = v = 0, \qquad \text{at } Y = 0, \tag{3.3.11}$$

$$u \to U_e, \qquad \text{as } Y \to \infty, \tag{3.3.12}$$

$$u = u_0, v = v_0, \quad \text{at } t = 0. \tag{3.3.13}$$

For the record, we provide the equivalent dimensional form of the preceding equations:

$$\frac{\partial u^*}{\partial x^*} + \frac{\partial v^*}{\partial y^*} = 0, \tag{3.3.14}$$

$$\frac{\partial u^*}{\partial t^*} + u^*\frac{\partial u^*}{\partial x^*} + v^*\frac{\partial v^*}{\partial y^*} = -\frac{1}{\rho}\frac{\partial p^*}{\partial x^*} + \nu\frac{\partial^2 u^*}{\partial y^{*2}}. \tag{3.3.15}$$

The reader must have noticed that we have rederived Prandtl's boundary-layer equations by using different scaling parameters. The preceding analysis has also indicated that for very small times, a further approximation could be derived from Eq. (3.3.10) by dropping the terms that multiply the small parameter $U_0 t_0 / L$.

$$\frac{\partial u}{\partial t} = \frac{\partial^2 u}{\partial Y^2} .$$
(3.3.16)

Finally the dimensionless quantitites appearing in Eqs. (3.3.7) and (3.3.8) are estimates of the relative "weight" carried by the terms that appear in these equations. That is,

$$\frac{\nu t_0}{L^2} = \frac{\text{rate of diffusion parallel to the wall}}{\text{rate of diffusion across the boundary layer}} ,$$

$$\frac{U_0 t_0}{L} = \frac{\text{rate of convection parallel to the wall}}{\text{rate of diffusion across the boundary layer}} .$$

Our estimates indicate that for small times, the diffusion is a very strong effect in the direction perpendicular to the wall but practically nonexistent in the direction parallel to the wall; further that diffusion far outweighs convection, a fact that Blasius (1908) and later Goldstein and Rosenhead (1936) have used in their analyses.

The dimensionless parameter $L / U_0 t_0$ is often referred to in literature as the Strouhal number. The reader should be cautioned to the fact that a large group of investigations of blunt-body aerodynamics refer to the Strouhal number defined by $L \omega^* / U_0$, where ω^* is the frequency of vortex shedding. In this case the Strouhal number is a consequence of the particular geometrical configuration and the hydrodynamic effects and, therefore, an unknown parameter. The coefficient of the diffusion terms in Eqs. (3.3.7) and (3.3.8) is the product of the Strouhal number and the Reynolds number. In terms of the dimensionless parameters

$$St = \frac{L}{U_0 t_0} , \qquad Re = \frac{U_0 L}{\nu} ,$$
(3.3.17)

the full Navier–Stokes equations can be expressed in dimensionless variables, as follows:

$$\frac{\partial u}{\partial t} + \frac{1}{St}\left(u \frac{\partial u}{\partial x} + v \frac{\partial u}{\partial y} \right) = \frac{1}{St} \frac{\partial p}{\partial x} + \frac{1}{ReSt}\left(\frac{\partial^2 u}{\partial x^2} + \frac{\partial^2 u}{\partial y^2} \right),$$
(3.3.18)

$$\frac{\partial v}{\partial t} + \frac{1}{St}\left(u \frac{\partial u}{\partial x} + v \frac{\partial u}{\partial y} \right) = \frac{1}{St} \frac{\partial p}{\partial y} + \frac{1}{ReSt}\left(\frac{\partial^2 v}{\partial x^2} + \frac{\partial^2 v}{\partial y^2} \right).$$
(3.3.19)

The two-dimensional form of all equations is given here for brevity but their extension to three-dimensional flow is straightforward.

3.4 Infinite Bodies—Rayleigh Problems

There is a large class of problems that are often categorized as Rayleigh problems. They involve impulsive changes of the flow parallel to the generators of infinite cylinders. Schlichting (1968) points out that Stokes (1851) fully discussed and solved the problem of the impulsive start of an infinite plate. He therefore correctly refers to them as Stokes problems of the first kind. However, such problems are familiar to most investigators, perhaps not very unjustly, as Rayleigh problems and we shall follow here this tradition.

Consider an infinite cylinder with generators that touch a closed or open curve. Let the x axis be aligned with the generators of the body (Fig. 3.4). Since the geometry of the solid surface is independent of the coordinate x, it is reasonable to assume that the flow properties do not depend on x as well. For a two-dimensional configuration, one may invoke the continuity equation and argue that since $\partial u/\partial x = 0$, v may at most be a constant. However, a constant other than zero would violate the nonpenetration condition. The extension of the argument to three-dimensional flows is not straightforward. To be sure and with no loss of generality, we may start with the assumption that $v = w = 0$. The continuity equation then is automatically satisfied. The second and third components of the momentum equation reduce to

$$\frac{\partial p}{\partial y} = \frac{\partial p}{\partial z} = 0. \qquad\qquad *(3.4.1)$$

Thus p may vary only with x and it can be at most a linear function of x. With $\partial u/\partial x = v = w = 0$ now, the x component of momentum equation becomes

$$\frac{\partial u}{\partial t} = -\frac{1}{\rho}\frac{\partial p}{\partial x} + \nu\left(\frac{\partial^2 u}{\partial y^2} + \frac{\partial^2 u}{\partial z^2}\right). \qquad\qquad *(3.4.2)$$

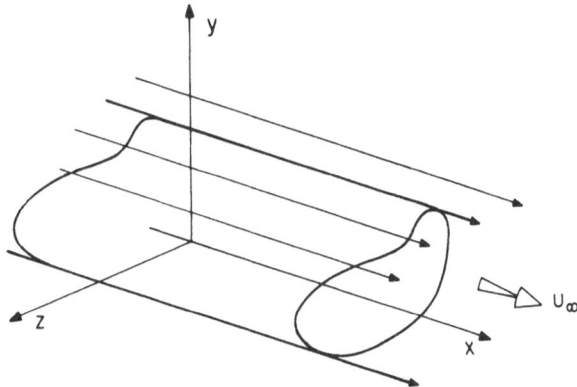

Fig. 3.4 Flow parallel to the generators of a cylindrical surface.

This is the heat equation with a forcing term and its solutions have been widely discussed for a variety of boundary conditions (Carlsaw and Jaeger, 1947).

Some interesting special cases of Rayleigh's problem are the impulsive starts of a semiinfinite flat plate or a right-angle wedge parallel to their edges (see Figs. 3.5a and b). The first problem was considered by Howarth (1950) who assumed that the flow and the plate are at rest for $t < 0$ and that the plate is given a velocity U_∞ in the direction of x at $t = 0$. Howarth develops the solution in terms of new independent variables:

$$Y = y/(\nu t)^{1/2}, \qquad Z = z/(\nu t)^{1/2}, \qquad R = r/(\nu t)^{1/2}, \qquad \text{*(3.4.3)}$$

with

$$r^2 = y^2 + z^2 \qquad\qquad \text{*(3.4.4)}$$

and in two forms:

   (i) a solution in series
   (ii) a solution by operational methods.

His results are shown in Fig. 3.6 expressed in terms of the ratio u/U_∞. As $Z \to \infty$, that is, as we move away from the edge of the plate, we recover the classical Rayleigh solution for an infinite plate, whereas for $Z \to -\infty$ the flow remains undisturbed. This solution therefore describes in a closed form the edge effects of a plate with a boundary parallel to the direction of the flow. This is a pure convection-versus-diffusion problem and no separation effects may arise. It can be seen from Fig. 3.6 that significant changes of the velocity are confined approximately to $|Z| < 2$, and it can be shown that this is also true approximately for

$$y^2 + z^2 < 2\sqrt{\nu t} \ . \qquad\qquad \text{*(3.4.5)}$$

Howarth further extends Rayleigh's arguments by replacing x with Ut and

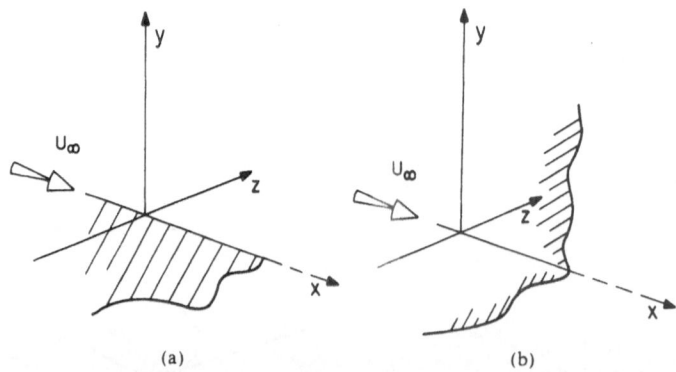

Fig. 3.5(a) Flow parallel to the edge of a semiinfinite plate. (b) Flow parallel to the edge of a 90° infinite wedge.

Fig. 3.6 Velocity distribution for various dimensionless heights in the neighborhood of the edge $Z = 0$ of a plate started impulsively in a direction parallel to the edge (Howarth, 1950).

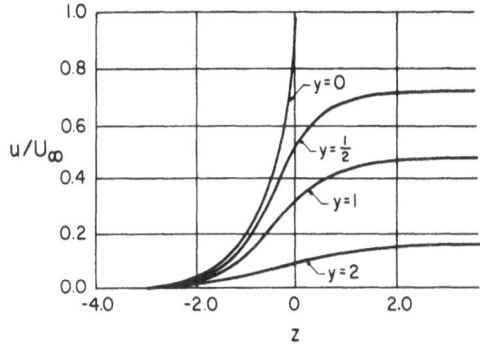

thus reducing the problem to the steady flow over a rectangular flat plate placed in a flow parallel to two of its sides.

Problems of infinite wedges were considered by Sowerby (1951), Hasimoto (1951), Sowerby and Cooke (1953), and others. For the right-angle wedge shown in Fig. 3.5b, the solution is given in closed form

$$\frac{u}{U_\infty} = 1 - \text{erf } \frac{y}{2(\nu t)^{1/2}} \text{ erf } \frac{z}{2(\nu t)^{1/2}} \; . \qquad \text{*(3.4.6)}$$

Turning now to cylinders of a finite cross section, let us study the impulsive start of the flow in a circular cylinder. Let the fluid and the cylinder be at rest. At $t = 0$, we apply a uniform and constant pressure gradient dp/dx. In polar coordinates the problem can be stated as follows

$$\frac{\partial u}{\partial t} = -\frac{1}{\rho}\frac{dp}{dx} + \nu\left(\frac{\partial^2 u}{\partial r^2} + \frac{1}{r}\frac{\partial u}{\partial r}\right), \qquad \text{*(3.4.7)}$$

$$u = 0, \quad \text{at } r = a \text{ and } t = 0, \qquad \text{*(3.4.8)}$$

and $u < \infty$ everywhere,

where a is the radius of the pipe. The solution can be written in terms of an infinite series of Bessel functions (Szymanski, 1932)

$$\frac{u}{u_{\max}} = 1 - \left(\frac{r}{a}\right)^2 - \sum_{n=1}^{\infty} \frac{8 J_0(\lambda_n r/a)}{\lambda_n^3 J_1(\lambda_n)}\exp\left(-\lambda_n^2\frac{\nu t}{a^2}\right), \qquad \text{*(3.4.9)}$$

$$u_{\max} = -\frac{dp}{dx}\frac{a^2}{4\mu}, \qquad \text{*(3.4.10)}$$

where J_0 and J_1 are the Bessel functions of zeroth- and first-order, respectively and λ_n are the zeros of J_0. Figure 3.7 shows some profiles of this velocity distribution. The familiar features of Rayleigh flows are evident. A thin boundary layer is generated at first, while the core of the flow is almost uniform. There is a characteristic difference however, because the flow is here started via an impulsive pressure. As a result all velocities and velocity gradients are always finite. The core velocity starts from zero and smoothly

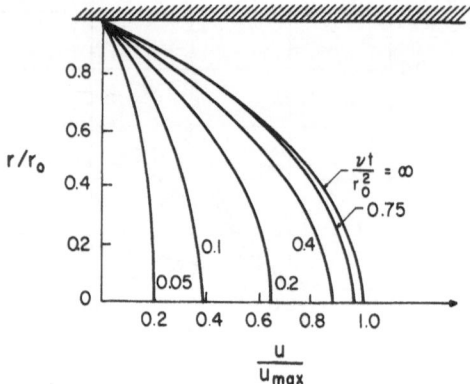

grows. As time increases, the velocity profiles approach quickly the Poiseuille profile

$$u = u_{max}\left[1 - \left(\frac{r}{a}\right)^2\right]. \qquad *(3.4.11)$$

A typical time required for the asymptotic solution to be reached within 1% of its final value is $t \cong a^2/\nu$. White (1974) notes that for air in a 30-cm-diameter pipe, this would require more than 25 min. For large pipes therefore it would be reasonable to assume boundary-layer flow for all practical applications of impulsive-flow starts. In contrast, the flow in a pipe 6-mm in diameter will achieve the Poiseuille profile in 0.5 sec. These points are further discussed by Ducoffe and White (1964).

3.5 Pointed Bodies Started Impulsively

The impulsive start of an infinite flat plate, that is, Rayleigh's problem, is a typical example of diffusion of vorticity and boundary-layer growth in time. On the other hand, steady flow over a semiinfinite plate, commonly known as the Blasius problem, is a typical case of self-similar flow and boundary-layer growth with distance from the leading edge of the plate. Both problems are parabolic in nature, with time t and axial distance x as parabolic variables. The similarities of the two problems are most clearly exemplified by the following observation. At the origin of both coordinate systems, that is, at the beginning of the motion all along the skin of the plate for the Rayleigh problem, or at the leading edge of the flat plate for the Blasius problem, an infinite amount of vorticity is introduced into the flow. Subsequently, at later times, or downstream points for the two problems, vorticity starts diffusing into the flow. However, the amount of vorticity introduced at $t = 0$ or $x = 0$ remains constant. It appears that the combined problem of the flow over a semiinfinite plate started impulsively in a direction perpendicular to its edge introduces unexpected mathematical difficulties that were not encountered in cases of blunt bodies.

The condition that a signal from the origin $x = 0$, traveling with the velocity of the outer flow $U_e(x, t)$, cannot reach any $x > 0$ in a finite time is (Stewartson, 1960)

$$\lim_{x \to 0} \left| \frac{U_e t}{x} \right| < \infty, \quad \text{for } t > 0. \qquad *(3.5.1)$$

This is true for blunt bodies where the flow at $x = 0$ is nothing but Hiemenz' stagnation flow and U_e varies linearly with x. It is not true, though, for sharp leading edges and, in particular, for flat plates. In the latter case, the message that a leading edge exists travels downstream with the speed U_∞, the speed of the free stream.

Let us consider a semiinfinite plate started impulsively from rest. At the instant t, points on the plate downstream of the point $x = U_\infty t$, are not aware of the fact that the plate has a leading edge and, for $x > U_\infty t$, a solution independent of x, that is, the classical Rayleigh solution is valid. In his first publication on the topic, Stewartson (1951) used a Stokes approximation in order to bridge the two regions. The problem is thus reduced to a linear parabolic equation:

$$\frac{\partial u^*}{\partial t^*} + U_\infty \frac{\partial u^*}{\partial x^*} = \nu \frac{\partial^2 u^*}{\partial y^{*2}}. \qquad (3.5.2)$$

For the flow about a semiinfinite flat plate that is suddenly given a velocity $- U_\infty$, the proper initial and boundary conditions are

$$u = v = 0, \quad \text{for } y = 0, \quad x > 0, t > 0, \qquad *(3.5.3a)$$

$$u = U_\infty, \quad \text{for } y \to \infty, \quad x \geqslant 0, t \geqslant 0, \qquad *(3.5.3b)$$

$$u = U_\infty, \quad \text{for } y > 0, \quad x \geqslant 0, t = 0, \qquad *(3.5.3c)$$

$$\text{and} \qquad x = 0, \qquad t \geqslant 0. \qquad *(3.5.3d)$$

The solution of Eq. (3.5.2) subject to the conditions (3.5.3) is

$$u^* = U_\infty \operatorname{erf} \eta, \qquad (3.5.4)$$

where

$$\eta = \frac{y^*}{2\sqrt{\nu x^*/U_\infty}}, \quad \text{for } x^* \leqslant U_\infty t^*, \qquad (3.5.4a)$$

and

$$\eta = \frac{y^*}{2\sqrt{\nu t^*}}, \quad \text{for } x^* > U_\infty t^*. \qquad (3.5.4b)$$

Stewartson found an essential singularity at $x^* = U_\infty t^*$, such that even though all derivatives with respect to the variable $\xi = U_\infty t^*/x^*$ are continuous there, the solution for $x^* > U_\infty t^*$ is not an analytic continuation of the solution for $x^* < U_\infty t^*$.

It is intuitively obvious that the metamorphosis of the boundary layer from its Rayleigh to its Blasius form cannot introduce any unexpected appreciable changes in features like the boundary-layer thickness, the skin friction, or even the boundary-layer profile. The latter for example are monotonic functions that vary between 0 and U_∞ for all x's and all t's. We may therefore argue that the problem bears rather little significance with respect to engineering applications, although very similar problems arise in the study of compressible flows, as, for example, the flow over a semiinfinite plate following the passage of a shock wave or the boundary-layer flow on the wall of a shock tube. There is no doubt that a solution of the full Navier–Stokes equations should not indicate any pecularities in the transition region where the error function profile changes over to the Blasius profile. It is only necessary to examine whether our simplified model, the boundary-layer equations can be used to calculate the flow through this region of transition. It would be in fact imperative for the numerical analyst tackling a more complex problem with a sharp leading edge to be aware of and familiar with the features of a traveling singularity in his domain of integration.

It is instructive at this point to recast the equations in terms of similarity variables

$$\xi = U_\infty t^*/x^*, \qquad \eta = y^*(U_\infty/\nu x^*)^{1/2}. \tag{3.5.5}$$

A straightforward application of the chain rule and substitution in Eqs. (3.3.14) and (3.3.15) with $\partial p/\partial x = 0$ yields

$$\frac{\xi}{x^*}\frac{\partial u^*}{\partial \xi} + \frac{\eta}{2x^*}\frac{\partial u^*}{\partial \eta} - \left(\frac{U_\infty}{\nu x^*}\right)^{1/2}\frac{\partial v^*}{\partial \eta} = 0, \tag{3.5.6}$$

$$\frac{U_\infty}{x^*}\frac{\partial u^*}{\partial \xi} - \frac{\xi}{x^*}u^*\frac{\partial u^*}{\partial \xi} - \frac{\eta}{2x^*}\frac{\partial u^*}{\partial \eta}$$

$$+ v^*\left(\frac{U_\infty}{\nu x^*}\right)^{1/2}\frac{\partial u^*}{\partial \eta} = \nu\left(\frac{U_\infty}{\nu x^*}\right)\frac{\partial^2 u^*}{\partial \eta^2}. \tag{3.5.7}$$

A quick observation indicates that proper stretching of the v component of velocity is enough to eliminate altogether the variable x^*,

$$u^* = uU_\infty, \qquad v^* = \left(\frac{\nu}{U_\infty x^*}\right)^{1/2}V. \tag{3.5.8}$$

In this way the number of independent variables is reduced from 3 to 2 and the semisimilar character of the equations is brought about:

$$\xi \frac{\partial u}{\partial \xi} + \frac{\eta}{2} \frac{\partial u}{\partial \eta} - \frac{\partial V}{\partial \eta} = 0, \tag{3.5.9}$$

$$(1 - \xi u) \frac{\partial u}{\partial \xi} - \left(\frac{\eta u}{2} - V \right) \frac{\partial u}{\partial \eta} = \frac{\partial^2 u}{\partial \eta^2}. \tag{3.5.10}$$

The singular parabolic nature of the problem is now clear. The change of the sign of the factor $1 - \xi u$ at $\xi = 1/u$, requires that the proper direction of integration of the parabolic equations is reversed. Equation (3.5.10) is nothing but the heat equation with a coefficient of conductivity that changes sign across the boundary layer and in the process it goes through infinity. Disturbances will travel in both directions and the domains of influence and dependence are interwoven. It is therefore no longer possible to start the integration at a certain station and march uniformly in one direction.

Many authors have subsequently attacked the problem but with rather limited success. Shuh (1953) and Oudart (1953) obtained approximate solutions. Cheng (1957) considered an outer mean flow accelerated with a velocity $U_e = At^n$, where A is a constant and $n > 0$, but ran into the same difficulties at $x = Ut$. Lam and Crocco (1959) considered a formulation in two independent variables and sought a solution of the boundary-value problem by iteration. However their computations ultimately diverged.

The problem of an impulsively started wedge was considered only a few years ago by Williams and Rhyne (1980). In terms of a new scaling of the coordinates, the resulting equation includes both the Rayleigh solution, for short time and the Falkner Skan solution for large time. In addition, the scaling reduces an infinite interval $(0 < x < \infty, 0 < t < \infty)$ to a finite interval. It is concluded that unsteady attached wedge type flows can be obtained for adverse pressure gradients that would induce separation if the flow were steady.

Direct numerical methods

Hall (1968) and Dwyer (1968a) attempted for the first time to solve the problem, integrating the partial differential equations in the three-coordinate space t, x, y. Dwyer converted a method developed for steady three-dimensional boundary layers (Dwyer, 1968b) to solve the unsteady two-dimensional flow problem. He imposed the following boundary conditions:

$$u = v = 0, \quad \text{at } y = 0, \text{ for } t > 0, x > 0, \tag{3.5.11}$$

$$u = U_\infty, \quad \text{at } t = 0,$$

$$u \to U_\infty, \quad \text{as } y \to \infty, \tag{3.5.12}$$

and integrated directly the differential equations.

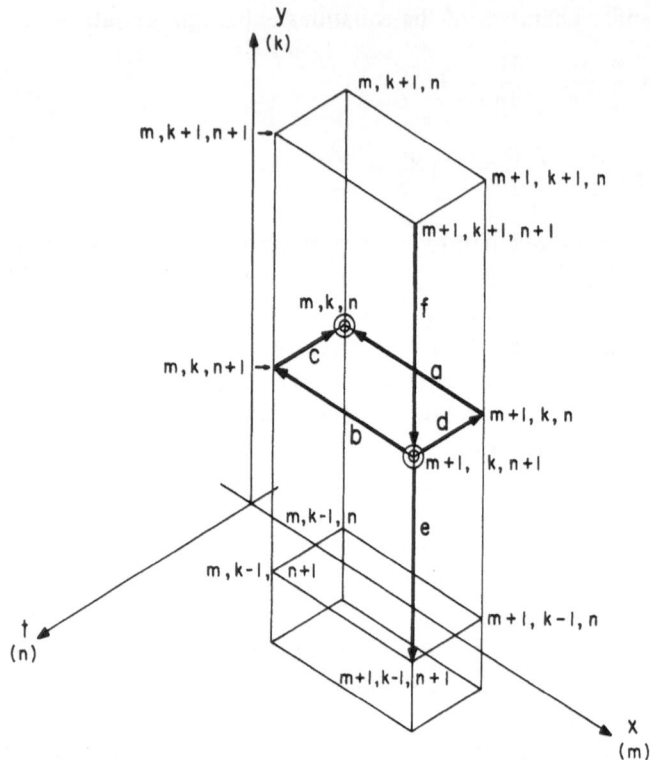

Fig. 3.8 Mesh and finite-difference vectors for Dwyer's (1968b) scheme.

Dwyer's differencing is based on a mixed Crank–Nicholson scheme outlined briefly below. Averaged differences in the direction of x and t are used (see Fig. 3.8).

$$\frac{\delta u}{\delta x} = \frac{u^n_{m+1,k} - u^n_{m,k}}{2\Delta x} + \frac{u^{n+1}_{m+1,k} - u^{n+1}_{m,k}}{2\Delta x}, \tag{3.5.13}$$

$$\frac{\delta u}{\delta t} = \frac{u^{n+1}_{m,k} - u^n_{m,k}}{2\Delta t} + \frac{u^{n+1}_{m+1,k} - u^n_{m+1,k}}{2\Delta t}. \tag{3.5.14}$$

In terms of the differencing vectors shown in Fig. 3.8, these derivatives are written as follows:

$$\frac{\partial u}{\partial x} \cong \frac{a+b}{2\Delta x}, \tag{3.5.15}$$

$$\frac{\partial u}{\partial t} \cong \frac{c+d}{2\Delta t}. \tag{3.5.16}$$

The y derivatives are expressed in terms of central differences, and the expressions are combined with weighted averages of the same derivatives calculated at earlier stations:

$$\frac{\delta u}{\delta y} = \frac{1}{4} \frac{u_{m+1,k+1}^{n+1} - u_{m+1,k-1}^{n+1}}{2\Delta y} + \frac{1}{4}\left(\frac{\delta u}{\delta y}\right)_{m+1,k}^{n}$$

$$+ \frac{1}{4}\left(\frac{\delta u}{\delta y}\right)_{m,k}^{n} + \frac{1}{4}\left(\frac{\delta u}{\delta y}\right)_{m,k}^{n+1},$$

(3.5.17)

$$\frac{\delta^2 u}{\delta y^2} = \frac{1}{4} \frac{u_{m+1,k+1}^{n+1} - 2u_{m+1,k}^{n+1} + u_{m+1,k-1}^{n+1}}{(\Delta y)^2}$$

$$+ \frac{1}{4}\left[\left(\frac{\delta^2 u}{\delta y^2}\right)_{m+1,k}^{n} + \left(\frac{\delta^2 u}{\delta y^2}\right)_{m,k}^{n} + \left(\frac{\delta^2 u}{\delta y^2}\right)_{m,k}^{n+1}\right].$$

(3.5.18)

In terms of the differencing vectors of Fig. 3.8 these become

$$\frac{\partial u}{\partial y} \cong \frac{1}{4} \frac{f+e}{\Delta y} + \frac{3}{4}\left(\overline{\frac{\partial u}{\partial y}}\right),$$

(3.5.19)

$$\frac{\partial^2 u}{\partial y^2} \cong \frac{1}{4} \frac{f-e}{(\Delta y)^2} + \frac{3}{4}\left(\overline{\frac{\partial^2 u}{\partial y^2}}\right),$$

(3.5.20)

where the bar denotes averaging at previous stations where the quantities have been already calculated. Dwyer found no difficulties or any evidence of singularity in the neighborhood of the point $\xi = U_\infty t/x = 1$. His calculated value of a reduced wall shear stress is shown in Fig. 3.9. The shear stress τ is divided by the value estimated by the Rayleigh theory τ_R. The ratio therefore should be equal to 1 for $\xi < 1$ and should approach the function τ_B/τ_R

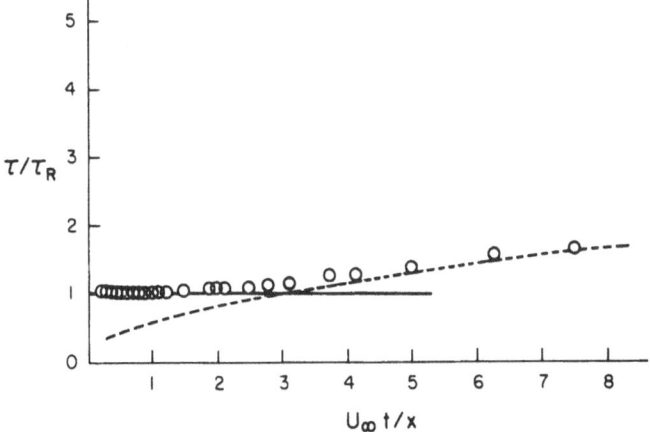

Fig. 3.9 The shear stress reduced with Rayleigh's shear stress τ/τ_R as a function of the parameter $\xi = U_\infty t/x$. ---, τ_B/τ_R, the Blasius solution; O, points computed by Dwyer (1968b).

where τ_B is the Blasius shear stress, as ξ grows. The two curves that represent the Rayleigh and Blasius solutions are shown with solid and dashed lines in the figure, respectively. The numerical data indicate a smooth transition from the one solution to the other. Minor inaccuracies for $\xi < 1$ are attributed to the method employed for starting the calculations at the leading edge of the plate.

It is interesting to note that Hall (1969a, b) in his simultaneous but independent effort employs mixed differencing schemes identical to those given by Eqs. (3.5.13) and (3.5.14). However, Hall's boundary conditions are entirely different and essentially designed to capture the transition from the Rayleigh to the Blasius solution. The domain of integration is given by the inequalities

$$1 < x < 2, \quad t < 1, \quad y > 0, \tag{3.5.21}$$

where x, t, and y are dimensionless quantities:

$$x = \frac{x^*}{L}, \quad t = \frac{U_\infty t^*}{L}, \quad y = Re^{1/2}\frac{y^*}{L}. \tag{3.5.22}$$

At the wall and the outer edge of the boundary, the classical conditions are specified:

$$u = v = 0, \quad \text{at } y = 0, \tag{3.5.23}$$

$$u \to U_\infty, \quad \text{as } y \to \infty. \tag{3.5.24}$$

The initial field at $t = 1$ is known in closed form. It is nothing but the Rayleigh solution, since for $t = 1$ the domain $x > 1$ is unaware of the existence of the leading edge.

$$u = \operatorname{erf}\frac{y}{2t^{1/2}}, \quad \text{at } t = 1. \tag{3.5.25}$$

However, the initial profile at $x = 1$ is not known a priori, since for $t > 1$ it will have to participate in the transition from the Rayleigh to the Blasius flow. Hall makes use of a well-known property of the solution, the existence of the similarity variables defined by Eq. (3.5.5), i.e., $\xi = U_\infty t^*/x^*$ and $\eta = y^*(U_\infty/\nu x^*)^{1/2}$. A profile of the velocity calculated by numerical integration at the station $x = 2$ and for $t = t_2$ can then be used to estimate the initial profile at $x = 1$, but for $t = t_2/2$. That is, if the profile $u(t/x, y/\sqrt{x})$ is known for $x = 2$ and $t = t_2$, then

$$u\left(\frac{t_2}{2}, \frac{y}{\sqrt{2}}\right) = u\left(\frac{t_2/2}{1}, \frac{y/\sqrt{2}}{1}\right), \tag{3.5.26}$$

and the profile at $x = 1$ at $t = t_2/2$ can be derived via stretching in the direction perpendicular to the wall by the factor $2^{-1/2}$. Hall then employs an iteration scheme, whereby the profile at $x = 1$ is first assumed arbitrarily:

$$u = u_B + (u_R - u_B)\exp\left[-\tfrac{1}{4}(t - 1)^2\right], \quad \text{at } x = 1, \tag{3.5.27}$$

with u_B and u_R the Blasius and the Rayleigh solutions. This expression simply represents an arbitrary smooth variation from the Rayleigh to the Balsius solution. With the initial profiles (3.5.25) and (3.5.27) for $t = 1$ and $x = 1$, respectively, calculations can be carried out by sweeping the domain of integration in the positive x direction and then in the positive t direction. Equation (3.5.26) can then be used to convert information derived by direct integration at $x = 2$ to initial profiles at $x = 1$ corresponding to earlier times. These updated profiles progressively replace the expression (3.5.27) and the calculations are repeated until convergence is achieved.

Hall finds that the transition to the steady Blasius solution is very rapid. His results appear to be very close to those of Dwyer shown in Fig. 3.9. He notes that even for $\xi = 4$, the solution is already identical—to within four significant figures—to the steady-state solution. In fact, he finds it impossible to check his numerical results with the asymptotic solution given by Stewartson (1951). When ξ is large enough for Stewartson's solution to be valid, the perturbation ends up being smaller than the errors in the computation.

Von Neumann's theory for the numerical stability is not applicable to nonlinear problems as described in the previous chapter. However, if we were to disregard the fact that the coefficients in the differential equations are varying with u, we can prove that the schemes of Dwyer (1968a) and Hall (1969a, b) are both unconditionally stable. This of course provides nothing more but an indication of stability. Such methods can be tested for stability only by comparison with exact solutions. Dwyer actually performed some numerical experiments with different step sizes and found that such changes do not seem to be restricted by stability considerations.

Integration in terms of similarity variables

The methods of Dwyer and Hall require numerical integration in a domain of three independent variables and, as a result, storage of information in all points of a two-dimensional grid. Integration in the space of the two similarity variables of Eqs. (3.5.5) was attempted later by Dennis (1972). In this way the number of independent variables is reduced from 3 to 2. The required computer space and time is thus considerably reduced.

Dennis introduces a stream function

$$u^* = \frac{\partial \psi}{\partial y^*}, \qquad v^* = -\frac{\partial \psi}{\partial x^*} \tag{3.5.28}$$

and, in terms of the Blasius stream function Ψ defined by

$$\psi = (U_\infty \nu x^*)^{1/2} \Psi, \tag{3.5.29}$$

he writes the momentum equation as

$$\frac{\partial^2 u}{\partial \eta^2} + \left[\frac{1}{2} \Psi - \xi \frac{\partial \Psi}{\partial \xi} \right] \frac{\partial u}{\partial \eta} = (1 - u\xi) \frac{\partial u}{\partial \xi}, \tag{3.5.30}$$

where u is the dimensionless velocity

$$u = \frac{u^*}{U_\infty} = \frac{\partial \Psi}{\partial \eta}. \tag{3.5.31}$$

The boundary conditions (3.5.11)–(3.5.12) are imposed, which are expressed in terms of the variables u and Ψ, as follows:

$$\Psi = 0, \quad \text{at } \eta = 0, \tag{3.5.32}$$

$$u = 0, \quad \text{at } \eta = 0, \tag{3.5.33}$$

$$u \to 1, \quad \text{as } \eta \to \infty. \tag{3.5.34}$$

Initial conditions are specified for $\xi_0 \leqslant 1$ so that the Rayleigh solution holds, that is, (3.5.4) with η given by (3.5.4b). The initial condition on the stream function then is

$$\Psi = 2\xi^{1/2} \left[\zeta \operatorname{erf} \zeta + \frac{1}{\sqrt{\pi}} (e^{-\zeta^2}) - 1 \right], \tag{3.5.35}$$

where $\zeta = y^*/2(\nu t_0)^{1/2}$. The system of Eqs. (3.5.30) and (3.5.31) is parabolic and can be integrated by any of the marching techniques in the direction of increasing ξ, provided that all the coefficients remain positive. Let these coefficients be denoted by

$$p(\xi, \eta) = \frac{1}{2} \Psi - \xi \frac{\partial \Psi}{\partial \xi}, \quad q(\xi, \eta) = 1 - u\xi. \tag{3.5.36}$$

For $u > 1/\xi$, q changes sign and the proper direction of integration changes as described earlier in this chapter. To account for this fact, Dennis is using backward- and forward-differencing schemes to express the ξ derivatives. With

$$u(k\Delta\xi, m\Delta\eta) = u_{k,m}, \tag{3.5.37}$$

he writes

$$\frac{\partial u}{\partial \xi} \cong \frac{u_{k,m} - u_{k-1,m}}{\Delta\xi}, \quad \text{for } q > 0, \tag{3.5.38}$$

$$\frac{\partial u}{\partial \xi} \cong \frac{u_{k+1,m} - u_{k,m}}{\Delta\xi}, \quad \text{for } q < 0. \tag{3.5.39}$$

With first and second η derivatives expressed by central differences, the finite-difference form of Eq. (3.5.30) becomes

$$\left(1 + \frac{\Delta\eta}{2} p\right) u_{k,m+1} + \left(1 - \frac{\Delta\eta}{2} p\right) u_{k,m-1} - \left(2 - \frac{(\Delta\eta)^2}{\Delta\xi} q\right) u_{k,m}$$

$$= -q \frac{(\Delta\eta)^2}{\Delta\xi} u_{k-1,m}, \quad \text{for } q > 0, \tag{3.5.40}$$

$$\left(1 + \frac{\Delta\eta}{2} p\right) u_{k,m+1} + \left(1 - \frac{\Delta\eta}{2} p\right) u_{k,m-1} - \left(2 + \frac{(\Delta\eta)^2}{\Delta\xi} q\right) u_{k,m}$$

$$= q \frac{(\Delta\eta)^2}{\Delta\xi} u_{k+1,m}, \quad \text{for } q < 0. \tag{3.5.41}$$

The problem can be viewed now as a boundary-value problem and a condition at some $\xi > \xi_0$ should be used. It is known of course that for large times, the solution should approach the Balsius function (Schlichting, 1968)

$$\Psi(\xi, \eta) \sim f(\eta), \quad \text{as } \xi \to \infty. \tag{3.5.42}$$

A large enough ξ_m is chosen as a right-hand side limit and the problem is integrated by the successive over-relaxation method [see, for example, Roache (1972), p. 117].

To start the calculation, an arbitrary distribution $u(\xi, \eta)$ is assumed throughout the domain of integration. The coefficients p and q are then calculated from (3.5.36) for all the grid points and the system of Eqs. (3.5.36) and (3.5.40) or (3.5.41) is solved by marching from $\xi = 1 + \Delta\xi$ toward increasing ξ. During this process the coefficients p and q calculated initially are used and therefore the system is rendered linear. If q is positive at a station k, then the quantity $u_{k-1,m}$ from the previous station is used in Eq. (3.5.40). However, if $q < 0$, the quantity $u_{k+1,m}$ from the earlier iteration is used. This is strongly reminiscent of the zig-zag scheme described in the last section of the previous chapter (Fig. 2.15). The only difference is that now each two-dimensional plane corresponds to an updated iterate, whereas in the straightforward calculation described before, these planes represent subsequent time stations. In this way, the entire plane ξ, η is swept and an updated function $u(\xi, \eta)$ is calculated. The coefficients p and q are again calculated in terms of the function u, and the process is repeated until convergence is achieved as depicted schematically in Fig. 3.10.

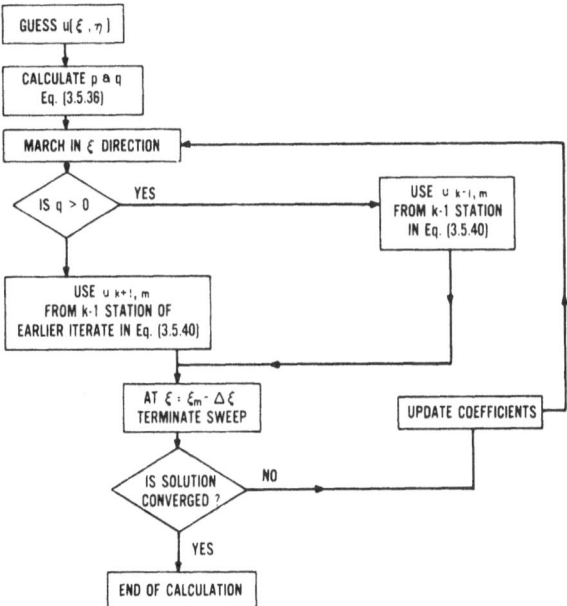

Fig. 3.10 Flow chart for integration of Eq. (3.5.30).

It is very interesting to note that since

$$\frac{u(\xi + \Delta\xi) - u(\xi)}{\Delta\xi} = \frac{\partial u}{\partial \xi} + \frac{1}{2}\Delta\xi \frac{\partial^2 u}{\partial \xi^2} + O(\Delta\xi)^2, \tag{3.5.43}$$

Eqs. (3.5.40) and (3.5.41) can be viewed as the finite difference form of the differential equation

$$\frac{\partial^2 u}{\partial \eta^2} + p\frac{\partial u}{\partial \eta} + \frac{1}{2}\Delta\xi|q|\frac{\partial^2 u}{\partial \xi^2} = q\frac{\partial u}{\partial \xi} \tag{3.5.44}$$

accurate to order $(\Delta\xi)^2$ and $(\Delta\eta)^2$. The elliptic behavior of this parabolic problem can now be explained.

Dennis (1972) shows very good agreement between his numerical results and those of Hall (1968). He also finds that the velocity component in the direction normal to the plate is continuous at $\xi = 1$, and the general features of the solution in the neighborhood of $\xi = 1$ are consistent with Stewartson's essential singularity.

The problem is certainly a challenge to numerical analysts and, in each of the investigations described, the difficulties are shifted to different parts of the method of solution. Dwyer (1968a) integrates in the physical plane from $x = 0$ and is faced with the leading edge singularity for all times. Moreover in the initial steps of the integration, very large gradients are encountered. Hall (1969a, b) integrates again in the physical plane but only from $t = 1$ and $x = 1$. His initial profiles at $t = 1$ for all $x > 1$ are given by the Rayleigh solution, since the message of the existence of the leading edge has not yet entered the domain of integration. However, an initial profile is needed at $x = 1$ for $t > 1$ and Hall introduces an iteration scheme making use of the self-similar nature of the solution. In this way he integrates a parabolic equation with nonnegative coefficients and captures the transition region within his domain of integration. Dennis (1972) instead integrates in the similarity variable domain (ξ, η), from $\xi = 1$ to some large value of ξ's, by converting the problem to a boundary-value problem and making use of the Blasius solution as a right-hand side boundary condition.

3.6 Blunt Bodies Started Impulsively

We consider now flows around bodies that meet condition (3.5.1). These are blunt bodies where the flow stagnates. If the origin is located at the point of stagnation, then the outer-flow velocity varies linearly with x in the neighborhood of this point.

The fact that messages from the origin can never reach a station x at a finite time have some very interesting implications for the numerical integration of such problems. The numerical solution is rather insensitive to the initial profiles of the velocity. This is essentially due to the Courant–

Friedrichs–Lewy condition, which requires that the domain of dependence
of the numerical scheme includes the domain of dependence of the partial
differential equation.

In this section we describe first the classical analysis of Blasius (1908) and
Goldstein and Rosenhead (1936). This seems to be a straightforward extension of the Rayleigh problem to bodies with finite thickness. The description
of approximate results due to later investigations follows. Finally we provide
a short account of numerical methods and the most recent results that
extend into the regions of reversed flow. However, to adhere to the historical
development of the theory, we confine the methods described here to
boundary-layer models and reserve discussions of more general solutions for
the next section.

We shall gain a good understanding of the physical concepts involved by
studying the development of the flow immediately after the impulsive start.

The Blasius expansion

Let us assume that at $t = 0$ the motion is generated impulsively, so that at
$t = 0^+$, a steady potential flow exists but the solid boundaries of our blunt
body are covered with vortex sheets. We intend to study how the vorticity
will diffuse into the flow or, equivalently, how the boundary layer will grow,
immediately after $t = 0^+$. We seek a solution to the boundary-layer equations. The outer flow is potential and therefore Bernoulli's equation is valid
at the edge of the boundary layer and the pressure term can be replaced by
$U_e \partial U_e / \partial x$. Equations (3.3.14) and (3.3.15) can thus be written:

$$\frac{\partial u^*}{\partial x^*} + \frac{\partial v^*}{\partial y^*} = 0, \tag{3.6.1}$$

$$\frac{\partial u^*}{\partial t^*} + u^* \frac{\partial u^*}{\partial x^*} + v^* \frac{\partial v^*}{\partial y^*} = U_e^* \frac{dU_e^*}{dx} + v \frac{\partial^2 u^*}{\partial y^{*2}} . \tag{3.6.2}$$

The appropriate initial and boundary conditions are

$$\text{for } t^* = 0^+ \text{ and } y^* > 0: \quad \begin{cases} u^* = U_e^*(x^*), \\ v^* = 0, \end{cases} \tag{3.6.3}$$

$$\text{for } t^* > 0: \quad \begin{cases} u^* = 0, \quad v^* = 0 \quad \text{at } y^* = 0, \\ u^* = U_e^*(x^*) \quad \text{as } y^* \to \infty. \end{cases} \tag{3.6.4}$$

In a preceding section we have seen by an order of magnitude analysis
that diffusion in the y direction is the most violent effect and that a first
approximation to Eq. (3.6.2) is the equation

$$\frac{\partial u^*}{\partial t^*} = v \frac{\partial^2 u^*}{\partial y^{*2}} . \tag{3.6.5}$$

The above equation, together with the boundary conditions (3.6.3) and

(3.6.4), is essentially the Rayleigh problem. Its solution is

$$u^*(x^*, y^*, t^*) = U_e^*(x^*) \operatorname{erf} \frac{y^*}{2\sqrt{\nu t^*}} . \tag{3.6.6}$$

The variable x^* enters as a parameter via the boundary condition. The shape of the particular body under consideration enters parametrically via the function $U_e^*(x^*)$. Motivated by the Rayleigh solution and in an attempt to generate a better approximation, we introduce the following dimensionless independent variables:

$$\xi = \frac{x^*}{L}, \qquad u = \frac{u^*}{U_0}, \qquad U_e = \frac{U_e^*}{U_0},$$

$$\eta = \frac{y^*}{2\sqrt{\nu t^*}}, \qquad v = \frac{v^*}{U_0}, \qquad t = \frac{t^* U_0}{L}. \tag{3.6.7}$$

The coordinate η has a form reminiscent of the stretched coordinate Y that appeared in Eq. (3.3.5). The development in this section indicates that we need some stretching of the v component of velocity as well. This will become more obvious very soon. It is important only to emphasize that the coordinate Y is only scaled by a small time t_0, whereas the coordinate η is a combination of y^* and t^* in the spirit of the familiar similarity solution of the heat equation.

Substitution of Eq. (3.6.7) in (3.6.1) and (3.6.2) yields

$$\frac{\partial u}{\partial \xi} + \frac{L}{2\sqrt{\nu t^*}} \frac{\partial v}{\partial \eta} = 0, \tag{3.6.8}$$

$$-\frac{\eta}{2t} \frac{\partial u}{\partial \eta} + \frac{\partial u}{\partial t} + u \frac{\partial u}{\partial \xi} + \frac{L}{2\sqrt{\nu t^*}} v \frac{\partial u}{\partial \eta} = U_e \frac{\partial U_e}{\partial \xi} + \frac{1}{4t} \frac{\partial^2 u}{\partial \eta^2} . \tag{3.6.9}$$

The necessity of the stretching of the v component of the velocity is now clear. The stretching factor is $L/2\sqrt{\nu t^*}$, which becomes

$$\frac{L}{2\sqrt{\nu t}} = \left(\frac{U_0 L}{\nu t} \right)^{1/2} = \frac{1}{2} \left[\frac{Re}{t} \right]^{1/2}, \tag{3.6.10}$$

where $Re = U_0 L/\nu$ is the Reynolds number.

There is a fundamental difference in the development of the present section as compared to the analysis of the previous sections. In Section 3.3 we assumed that a small time constant exists and we formulated dimensionless parameters that contain it. In the present section we define a dimensionless time t and propose to investigate the behavior of the solution for small values of this independent variable. Notice that now the appearance of the Reynolds number is inevitable. Both processes are appropriate if the behavior of the mathematical model for small times is to be studied.

Let us seek a solution in the form of an expansion in powers of the coordinate t:

$$u = u_1 + tu_2 + t^2u_3 + \cdots, \qquad (3.6.11)$$

$$v = t^{1/2}Re^{-1/2}(v_1 + tv_2 + t^2v_3 + \cdots), \qquad (3.6.12)$$

where the unknown functions u_i and v_i $(i = 1, 2, \ldots)$ are functions of the variables ξ and η. The reader should be cautioned to the fact the coordinates ξ, η, and t are independent variables. Thus, notice that in the time derivative

$$\frac{\partial}{\partial t^*} = \frac{\eta}{2t}\frac{\partial}{\partial \eta} + \frac{\partial}{\partial t} \qquad (3.6.13)$$

the dominant term is the first, since it multiplies the factor $\eta/2t$, while the term $\partial/\partial t$ will be finite if the expansions of Eqs. (3.6.11) and (3.6.12) are assumed. Substitution of expressions (3.6.11)–(3.6.12) into Eqs. (3.6.8) and (3.6.9) and collection of equal powers of t yields:

From the continuity equation:

$$O(t^0), \qquad \frac{\partial u_1}{\partial \xi} + \frac{\partial v_1}{\partial \eta} = 0, \qquad (3.6.14)$$

$$O(t^1), \qquad \frac{\partial u_2}{\partial \xi} + \frac{\partial v_2}{\partial \eta} = 0. \qquad (3.6.15)$$

From the momentum equation:

$$O(t^{-1}), \qquad \frac{\partial^2 u_1}{\partial \eta^2} + 2\eta\frac{\partial u_1}{\partial \eta} = 0, \qquad (3.6.16)$$

$$O(t^0), \qquad \frac{\partial^2 u_2}{\partial \eta^2} + 2\eta\frac{\partial u_2}{\partial \eta} - u_2 = 4\left(u_1\frac{\partial u_1}{\partial \xi} + v_1\frac{\partial u_1}{\partial \eta} - U_e\frac{dU_e}{d\xi}\right).$$

$$(3.6.17)$$

Blasius (1908) arrived at these equations in a slightly different form and solved them in closed form. Later Goldstein and Rosenhead (1936) carried the expansion to one more order of accuracy. At this point let us only note that the above system of differential equations can be solved successively for u_1, v_1, then u_2, v_2, etc., where the first term is indeed given by Eq. (3.6.6). The solution to the first equation that satisfies the boundary conditions (3.6.3) and (3.6.4) is the error function

$$u_1 = U_e\,\mathrm{erf}(\eta). \qquad (3.6.18)$$

The next term in the expansion, calculated by Blasius (1908), is

$$u_2(\xi, \eta) = U_e\frac{dU_e}{dx}f_2(\eta), \qquad (3.6.19)$$

where

$$f_2(\eta) = \frac{1}{2}(2\eta^2 - 1)\operatorname{erf}^2\eta + \frac{3}{\sqrt{\pi}}\eta e^{-\eta^2}\operatorname{erf}\eta + 1$$

$$- \frac{4}{3\pi}e^{-\eta^2} + \frac{2}{\pi}e^{-2\eta^2} - \left(1 + \frac{2}{3\pi}\right)(2\eta^2 + 1) \tag{3.6.20}$$

$$+ \frac{1}{\sqrt{\pi}}\left(1 + \frac{4}{3\pi}\right)\left[\frac{1}{2}\sqrt{\pi}(2\eta^2 + 1)\operatorname{erf}\eta + \eta e^{-\eta^2}\right].$$

Goldstein and Rosenhead (1936) tabulated the function $f_2(\eta)$ and proceeded to calculate the next term in the expansion. Wundt (1955) later repeated the calculations to provide more accurate values of all the functions up to the third power in the expansion.

One of the most interesting features of the flow, for small times, is the appearance of a point of zero skin friction, which marks the onset of flow reversal. Equating the derivative $\partial u/\partial y$ to zero for $y = 0$, Blasius arrived, using two terms in his expansion, at the condition

$$1 + \left(1 + \frac{4}{3\pi}\right)\frac{dU_e}{dx}t = 0. \tag{3.6.21}$$

From the above equation it appears that a vanishing wall shear is possible only if $dU_e/dx < 0$, that is, if there exists a region with adverse pressure gradient. In fact, the first instant t_s that the skin friction will vanish is given by the smallest value of dU_e/dx and

$$t_s = -\left(1 + \frac{4}{3\pi}\frac{dU_e}{dx}\right)^{-1}. \tag{3.6.22}$$

The most common example that investigators consider as a test case is the circular cylinder. For such a cylinder, the smallest value of dU_e/dx, that is, the largest adverse-pressure gradient occurs at the rear stagnation point. Equation (3.6.20) gives $t_s = 0.35$, while Goldstein and Rosenhead estimated, with one more term in their expansion, $t_s = 0.32$. From then on the point of zero skin friction starts moving upstream, thus generating a thin region of recirculating flow.

Modern methods

The problem described in this section has interested a large number of investigators in the last few decades and is considered today as a classical test case. Modern methods, analytical and numerical, have made it possible to extend the calculations and investigate the upstream motion of the point of flow reversal and the rate of thickening of the recirculating bubble. Görtler (1946) extended the theory to take into account different variations of the free-stream velocity. Shuh (1953) developed a method based on the

Kármán–Polhausen technique to integrate the unsteady boundary-layer equations about a circular cylinder, whereas Watson (1955) and Wundt (1955) expanded again in powers of time to provide more accurate tabulations of the results for a wider choice of accelerations and cylinder shapes. In particular, Görtler (1946) investigated a free-stream velocity increasing with a power of time

$$U_\infty = At^n,$$ (3.6.23)

with n a positive integer and A a constant, whereas Watson (1955) considered a velocity given by

$$U_\infty = At^\alpha,$$ (3.6.24)

where α is a real and positive number. For the last case, and with a stream function defined by the equations

$$u = U_\infty \frac{\partial \psi}{\partial \eta},$$ *(3.6.25)

$$v = -2U_\infty \sqrt{\nu t / L^2} \, \frac{\partial \psi}{\partial \eta},$$ *(3.6.26)

Watson (1955) expanded in powers of time as follows:

$$\psi = \psi_0(\xi, \eta) + t^{\alpha+1}\psi_1(\xi, \eta) + t^{2\alpha+2}\psi_2(\xi, \eta) + \cdots .$$ (3.6.27)

Expansions in powers of a normalized time were proposed also by Collins and Dennis (1973a), who match their solution with the uniform flow far from the cylinder. The first approximation was derived analytically, but Collins and Dennis proceed to compute numerically up to the seventh-order terms including the effect of finite but high values of the Reynolds number. In the following subsection we shall describe a subsequent work of the same authors which solves the same problem by numerical integration of the Navier–Stokes equations.

Exact numerical solutions of the boundary-layer equations treating the impulsive start on a flat plate appeared in the late 1960s as described in the previous section. Numerical solutions of the boundary-layer equations for flows over blunt bodies became available only in the early 1970s. It is interesting to note that such solutions followed the numerical integration of the full Navier–Stokes equations. Yet, even with today's computational capabilities, it is prohibitive to seek solutions to the Navier–Stokes equations with realistic Reynolds numbers and flows that involve turbulent boundary layers. The exact behavior of the boundary-layer model and the extent of its validity in time, space, and variety of problems is still of great interest. It has been proved, for example, that boundary-layer models can capture fields that contain thin regions of recirculating flows, and work is being continued to model separated regions in terms of Kirchhoff's free-streamline theory and free-shear layer calculations.

Stewartson (Belcher *et al.*, 1971) reports a purely numerical calculation of the unsteady boundary-layer equations in the three-coordinate space. With spacings $\Delta t = 0.025$, $\Delta x = \pi/18$, and $\Delta y = 0.1$, integration proceeds via a Crank–Nicholson scheme by marching in the positive x and t direction until regions of negative velocities are encountered. At this point values of $\partial u/\partial x$ are calculated using values of u at the two previous stations of t and neighboring upstream and downstream points in the x grid. An explicit formula is not given in this reference but it appears that the scheme must be similar to the "zig-zag scheme" described in the earlier chapter. Stewartson suggests an alternative scheme that may be similar to the "upwind differencing" of the previous chapter. The numerical results of this work will be commented upon together with the results of methods that follow. In this section we provide a more detailed description of a numerical method, developed by the author's group, for solving the unsteady boundary-layer equations (Telionis, *et. al.*, 1973).

A variety of transformations based on Prandtl's (1904) similarity variables has proved useful in the analysis of boundary layers. Self-similar solutions were studied first by Falkner and Skan (1931) for flat plate and wedge flows, but later the ideas were extended in terms of coordinate expansions to include more general situations (Howarth, 1938; Görtler, 1946). It turns out that such transformations are useful even if exact numerical solutions are sought. This is due mainly to the fact that in terms of the new coordinates, the boundary-layer thickness and the domain of integration in the normal direction grow very mildy with the downstream distance. Moreover, for situations that are not far from the flat plate or the wedge flow, that is, for flows that are nearly self-similar, the dependence on the coordinate x is very mild and, therefore, a coarser mesh configuration can be used.

As an example we consider the generalization of Görtler's (1957) transformation to unsteady flow (Telionis, *et al.*, 1973):

$$\xi(x, y, t) = \int_0^x U_e(x, t)\, dx, \tag{3.6.28}$$

$$\eta(x, y, t) = \frac{U_e(x, t)}{(2\xi)^{1/2}}\, Y, \tag{3.6.29}$$

$$\tau = t. \tag{3.6.30}$$

All variables in these equations are dimensionless and Y is the stretched boundary-layer coordinate normal to the wall. New dependent variables are also introduced (Werle and Davis, 1972):

$$F = \frac{u}{U_e}, \tag{3.6.31}$$

$$V = \frac{(2\xi)^{1/2}}{U_e} \left[v + \frac{\partial \eta}{\partial s} (2\xi)^{1/2} F \right]. \tag{3.6.32}$$

In terms of the above variables the differential equations (3.6.1) and (3.6.2) and the boundary conditions (3.6.3) and (3.6.4) read

$$2\xi \frac{\partial F}{\partial \xi} + F + \frac{\partial V}{\partial \eta} = 0, \tag{3.6.33}$$

$$\frac{2\xi}{U_e^2} \frac{\partial F}{\partial t} + 2\xi F \frac{\partial F}{\partial \xi} + F \frac{\partial F}{\partial \eta} + \beta(F^2 - 1) = \frac{\partial^2 F}{\partial \eta^2}, \tag{3.6.34}$$

$$F(\xi, 0, t) = V(\xi, 0, t) = 0, \tag{3.6.35}$$

$$F(\xi, \infty, t) = 1, \tag{3.6.36}$$

where β is the pressure gradient function,

$$\beta = \frac{2\xi}{U_e} \frac{dU_e}{d\xi}. \tag{3.6.37}$$

The initial conditions for flows started impulsively from rest are

$$F(\xi, \eta, 0^+) = 1, \quad \text{for } \eta \to 0, \tag{3.6.38}$$

$$V(\xi, \eta, 0^+) = 0. \tag{3.6.39}$$

To facilitate numerical integration, Eq. (3.6.34) is written in the form

$$F_{\eta\eta} + A_1 F_\eta + A_2 F + A_3 + A_4 F_\xi + A_5 F_t = 0, \tag{3.6.40}$$

where the coefficients A_i depend on V, F, and β. A grid is introduced with variable mesh sizes $\Delta\xi$, $\Delta\eta$, and $\Delta\tau$. Let m, k, and n represent the running indices in the ξ, η, and τ direction, respectively. The zero value of the independent variables, that is, the station of stagnation, the wall, and the initial plane are denoted by $m = 1$, $k = 1$, and $n = 1$, respectively. The time derivative is written in a difference form

$$\left(\frac{\delta F}{\delta \tau}\right)_{m,k}^n = \frac{F_{m,k}^n - F_{m,k}^{n-1}}{\Delta\tau} + O(\Delta\tau) \tag{3.6.41}$$

and (3.6.40) reduces to

$$F_{\eta\eta} + \alpha_1 F_\eta + \alpha_2 F + \alpha_3 + \alpha_4 F_\xi = 0, \tag{3.6.42}$$

where the coefficients now depend on the value of the function F at the previous time step as well:

$$\alpha_1 = A_1, \tag{3.6.43}$$

$$\alpha_2 = A_2 + \frac{A_5}{\Delta\tau}, \tag{3.6.44}$$

$$\alpha_3 = A_3 - \frac{A_5 F_{m,k}^{n-1}}{\Delta\tau}, \tag{3.6.45}$$

$$\alpha_4 = A_4. \tag{3.6.46}$$

Equation (3.6.42) is essentially in the form of the steady-state boundary-layer equation and a code developed by Werle and Davis (1972) is used as part of the solution.

In the early stages of the computations, considerable numerical difficulties are encountered because of extreme values of velocity gradients next to the walls. In the terminology of fluid mechanics, the vortex sheet that initially covers the skin of the body would require an extremely fine mesh in the vicinity of the wall. To avoid this difficulty, it is more practical (Telionis and Tsahalis, 1974a, b) to make use of an asymptotic solution for small times and start the numerical calculation at an instant t_e, where t_e is a very small number. The Blasius solution is of course readily available for this purpose, since it provides a good approximation of the flow for small times. The solution represented by Eq. (3.6.6) is evaluated at $t = t_e$ and this serves as an initial condition for the numerical calculations that follow.

3.7 Solutions to the Full Navier–Stokes Equations

With the development of the method of inner and outer expansions, it became apparent that higher-order contributions to the boundary-layer solutions from the outer flow may have to be considered, before one would proceed to the third term of the expansion (3.6.11) and (3.6.12). In other words, expansions in powers of the small time parameter will certainly provide a more accurate solution to the boundary-layer equation. However, it will become clear from the subsequent development due to Wang (1967a) that the boundary-layer approximation itself will have to be corrected before one would proceed to higher-order terms in the expansion. It will therefore be necessary to reconsider the problem in terms of the full Navier–Stokes equations. The development that follows is also a nice example of reworking the problem in terms of a parameter expansion, rather than a coordinate expansion (Nayfeh, 1973). Blasius' analysis is based on expansions in powers of the variable t, whereas Wang's solution proceeds in powers of the small parameter ε, the inverse of the Strouhal number.

The first purely numerical solution of the Navier–Stokes equations for the impulsively started cylinder appeared in the late 1950s (Payne, 1958), but the bulk of the contributions came almost a decade later (Hirota and Miyakoda, 1965; Kawaguti and Jain, 1966; Ingham, 1968; Son and Hanratty, 1969; Jain and Rao, 1969; Thoman and Szewczyk, 1969; Dennis and Staniforth, 1971) and simultaneously with the work of Wang (1967a). In this section we outline two procedures for the numerical integration of the Navier–Stokes equations. In the first, integration proceeds in the physical plane; the second involves a conformal transformation and integration in a phase plane.

Inner and outer expansions

Consider again the two-dimensional form of continuity and the full Navier–Stokes equations in the dimensionless form given by Eqs. (3.3.18) and (3.3.19) but expressed in terms of polar coordinates:

$$\frac{\partial u}{\partial \theta} + \frac{\partial}{\partial r}(rv) = 0, \tag{3.7.1}$$

$$\frac{\partial u}{\partial t} + \frac{1}{St}\left(\frac{u}{r}\frac{\partial u}{\partial \theta} + v\frac{\partial u}{\partial r} + \frac{uv}{r}\right)$$

$$= -\frac{1}{r}\frac{\partial p}{\partial \theta} + \frac{1}{ReSt}\left(\frac{\partial^2 u}{\partial r^2} + \frac{1}{r}\frac{\partial u}{\partial r} + \frac{1}{r^2}\frac{\partial^2 u}{\partial \theta^2}\right. \tag{3.7.2}$$

$$\left. + \frac{2}{r^2}\frac{\partial v}{\partial \theta} - \frac{u}{r^2}\right),$$

$$\frac{\partial v}{\partial t} + \frac{1}{St}\left(\frac{u}{r}\frac{\partial v}{\partial \theta} + v\frac{\partial v}{\partial r} - \frac{u^2}{r}\right)$$

$$\tag{3.7.3}$$

$$= -\frac{\partial p}{\partial r} + \frac{1}{ReSt}\left(\frac{\partial^2 v}{\partial r^2} + \frac{1}{r}\frac{\partial v}{\partial r} + \frac{1}{r^2}\frac{\partial^2 v}{\partial \theta^2} - \frac{v}{r^2} - \frac{2}{r^2}\frac{\partial u}{\partial \theta}\right).$$

In the above equations r, θ and v, u are the coordinates and the velocity components in the radial and transverse direction, respectively. For an impulsive start, the analysis of Section 3.2 indicates that at $t = 0^+$, potential flow has been established and, therefore, the proper initial condition for the flow around a circular cylinder of radius 1 is

$$u = \left(1 + \frac{1}{r^2}\right)\sin\theta, \quad \text{at } t = 0^+, \tag{3.7.4}$$

$$v = -\left(1 - \frac{1}{r^2}\right)\cos\theta, \quad \text{at } t = 0^+. \tag{3.7.5}$$

The familiar no-slip and no-penetration condition will now be imposed on the skin of the cylinder,

$$u = v = 0, \quad \text{at } r = 1, \tag{3.7.6}$$

and the flow should merge smoothly into the undisturbed free stream at infinity. Wang (1967a) assumes that the Reynolds number and the Strouhal number are of the same order, say ε^{-1}.

$$Re = O(\varepsilon^{-1}), \; St = O(\varepsilon^{-1}) \tag{3.7.7}$$

An asymptotic expansion in powers of ε is then possible. Expecting a boundary-layer type of flow, in the spirit of the development of Section 3.3, let us introduce a stretched variable in the direction normal to the wall

$$r = 1 + \varepsilon\tilde{r} \tag{3.7.8}$$

and asymptotic expansions for the inner region that incorporate the assumption that v is one order of magnitude smaller than u:

$$u = u_0 + \varepsilon u_1 + \varepsilon^2 u_2 + \ldots, \quad v = \varepsilon v_0 + \varepsilon^2 v_1 + \ldots, \tag{3.7.9}$$

$$p = p_0 + \varepsilon p_1 + \varepsilon^2 p_2 + \ldots . \tag{3.7.10}$$

Substituting these expressions in Eqs. (3.7.1)–(3.7.3) and collecting the coefficients of like powers of ε, we arrive at

Order ε^0

$$\frac{\partial u_0}{\partial \theta} + \frac{\partial v_0}{\partial \tilde{r}} = 0, \tag{3.7.11}$$

$$\frac{\partial u_0}{\partial t} - \frac{\partial^2 u_0}{\partial \tilde{r}^2} = -\frac{\partial p_0}{\partial \theta}, \tag{3.7.12}$$

$$\frac{\partial p_0}{\partial \tilde{r}} = 0. \tag{3.7.13}$$

Order ε

$$\frac{\partial u_1}{\partial \theta} + \frac{\partial v_1}{\partial \tilde{r}} = -\frac{\partial}{\partial \tilde{r}}(\tilde{r} v_0), \tag{3.7.14}$$

$$\frac{\partial u_1}{\partial t} - \frac{\partial^2 u_1}{\partial \tilde{r}^2} = -v_0 \frac{\partial u_0}{\partial \tilde{r}} - u_0 \frac{\partial u_0}{\partial \theta} - \frac{\partial p_1}{\partial \theta} + \tilde{r}\frac{\partial p_0}{\partial \theta} + \frac{\partial u_0}{\partial \tilde{r}}, \tag{3.7.15}$$

$$\frac{\partial p_1}{\partial \tilde{r}} = 0. \tag{3.7.16}$$

Note that the equations of order ε^0 are identical to the Blasius equations, but the equations of order ε already contain terms due to curvature effects. Such terms would have been eliminated only if the Reynolds number were substantially larger than the Strouhal number. The solution to the problem of order ε^0 is well known. It is Eq. (3.6.18) recast in polar variables:

$$u_0 = 2\,\mathrm{erf}\left(\frac{\tilde{r}}{2\sqrt{t}}\right)\sin\theta. \tag{3.7.17}$$

Introducing the familiar similarity variable

$$\eta = \frac{\tilde{r}}{2\sqrt{t}}, \tag{3.7.18}$$

we write the inner expansion in terms of a stream function

$$\psi = \varepsilon\psi_0(\eta, \theta, t) + \varepsilon^2\psi_1(\eta, \theta, t) + \ldots, \tag{3.7.19}$$

where from (3.7.17) and the continuity equation,

$$\psi_0 = 4\sqrt{t}\sin\theta\left[\eta - \eta\,\mathrm{erfc}\,\eta + \frac{1}{\sqrt{\pi}}(e^{-\eta^2} - 1)\right], \tag{3.7.20}$$

and erfc is the complementary error function.

Reserving capital symbols, U, V, P, and Ψ for the outer-flow velocity components, pressure, and stream function, respectively, we expand

$$\Psi = \Psi_0(r,\theta,t) + \varepsilon\Psi_1(r,\theta,t) + \varepsilon^2\Psi_2(r,\theta,t) + \cdots, \tag{3.7.21}$$

where

$$\Psi_0 = \left(r - \frac{1}{r}\right)\sin\theta. \tag{3.7.22}$$

To match the inner and outer solutions at this level, Wang introduces an intermediate variable

$$\bar{\eta} = \sqrt{\varepsilon}\,\eta. \tag{3.7.23}$$

Matching requires that the limit of the inner solution for $\bar{r}\to\infty$ or $\eta\to\infty$ equals the limit of the outer solution for $r\to 1$

$$\lim_{\eta\to\infty}\psi(\eta,\theta,t)\to\lim_{r\to 1}\Psi(r,\theta,t). \tag{3.7.24}$$

A quick inspection of (3.7.23) indicates that the first limiting procedure $\eta\to\infty$ is equivalent to $\varepsilon\to 0$, keeping the intermediate variable $\bar{\eta}$ fixed. If Eq. (3.7.8) is rewritten in terms of η or $\bar{\eta}$

$$r = 1 + \varepsilon(2\sqrt{t}\,\eta) = 1 + \sqrt{\varepsilon}\,(2\sqrt{t}\,\bar{\eta}), \tag{3.7.25}$$

it becomes clear that the same is true for the second limiting procedure. In other words, $r\to 0$ is equivalent to $\varepsilon\to 0$, keeping $\bar{\eta}$ fixed. Matching therefore requires that

$$\lim_{\varepsilon\to 0}\psi(\bar{\eta},\theta,t)\Big|_{\eta=\mathrm{const}} = \lim_{\varepsilon\to 0}\Psi(\bar{\eta},\theta,t)\Big|_{\bar{\eta}=\mathrm{const}}. \tag{3.7.26}$$

Expressing the inner solution (3.7.20) in terms of the intermediate variable and expanding for small ε gives

$$\lim_{\varepsilon\to 0}\psi = \varepsilon\lim_{\varepsilon\to 0}\psi_0 + \varepsilon^2\lim_{\varepsilon\to 0}\psi_1 + \cdots$$
$$= 4\sqrt{t}\sin\theta\left(\sqrt{\varepsilon}\,\bar{\eta} - \frac{\varepsilon}{\sqrt{\pi}}\right) + O(\varepsilon^{3/2}\bar{\eta},\varepsilon\bar{\eta}^2,\dots). \tag{3.7.27}$$

To express the outer expansions in terms of the intermediate variable, note that by virtue of (3.7.8) and (3.7.25),

$$\frac{1}{r} = \frac{1}{1+\sqrt{\varepsilon}\,(2\sqrt{t}\,\bar{\eta})} = 1 - \sqrt{\varepsilon}\,(2\sqrt{t}\,\bar{\eta}) + O(\varepsilon) \tag{3.7.28}$$

and thus,

$$r + \frac{1}{r} = 1 + \sqrt{\varepsilon}\,(2\sqrt{t}\,\bar{\eta}) - \left[1 - \sqrt{\varepsilon}\,(2\sqrt{t}\,\bar{\eta}) + \varepsilon 4t\bar{\eta}^2 + O(\varepsilon^{3/2})\right]. \tag{3.7.29}$$

The outer solution expanded for small ε thus becomes

$$\lim_{\varepsilon \to 0} \Psi = \lim_{\varepsilon \to 0} \Psi_0 + \varepsilon \lim_{\varepsilon \to 0} \Psi_1 + \cdots$$

$$= \sin\theta \left[\sqrt{\varepsilon} \,(4\sqrt{t}\,\bar{\eta}) + \varepsilon(4t\bar{\eta}^2) + O(\varepsilon^{3/2}) \right] \tag{3.7.30}$$

$$+ \varepsilon \Psi_1 \big|_{r=1} + (\varepsilon^{3/2}\bar{\eta})2\sqrt{t}\,\frac{\partial \psi_1}{\partial r}\bigg|_{r=1} + O(\varepsilon^2\bar{\eta}^2) + \cdots .$$

Matching according to (3.7.26) now yields

$$\Psi_1 \big|_{r=1} = -4\sqrt{\frac{t}{\pi}}\,\sin\theta. \tag{3.7.31}$$

This term represents the first phase of the interaction of the boundary layer with the outer flow. It is a correction proportional to the square root of time, to be added to the potential solution (3.7.4 and 5). This term will eventually feed back into second-order terms of the boundary-layer equation. The solution of the outer flow with (3.7.31) as an inner-boundary condition is

$$\Psi_1 = -4\sqrt{\frac{t}{\pi}}\,\frac{\sin\theta}{r}. \tag{3.7.32}$$

The first two orders of pressure distribution are related to the velocity field by

$$\frac{\partial p_0}{\partial \theta} = \frac{\partial P_0}{\partial \theta}\bigg|_{r=1} = -\frac{\partial U_0}{\partial t}\bigg|_{r=1} = 0, \tag{3.7.33}$$

$$\frac{\partial p_1}{\partial \theta} = \frac{\partial P_1}{\partial \theta}\bigg|_{r=1} = -\frac{\partial U_1}{\partial t}\bigg|_{r=1} + U_0\frac{\partial U_0}{\partial \theta}\bigg|_{r=1}$$

$$+ V_0\frac{\partial U_0}{\partial r}\bigg|_{r=1} + U_0 V_0\bigg|_{r=1}. \tag{3.7.34}$$

Wang (1967a) proceeds to calculate the second and third terms in the expansion by matching at each step the inner and outer solution. He finds

$$u_1 = 4\sqrt{t}\,(\sin\theta)\eta + 2\sqrt{\frac{t}{\pi}}\,\sin\theta(2 - 2e^{-\eta^2} + 3\sqrt{\pi}\,\mathrm{erfc}\,\eta)$$

$$+ 4(\sin\theta\cos\theta)tf(\eta), \tag{3.7.35}$$

$$\psi_1 = -4t(\sin\theta)\eta^2 + \frac{4t}{\sqrt{\pi}}\sin\theta\left[2\eta - \frac{\sqrt{\pi}}{4}\,\mathrm{erf}\,\eta + \frac{3\sqrt{\pi}}{2}\,\eta^2\,\mathrm{erfc}\,\eta - \frac{3\eta}{2}e^{-\eta^2}\right]$$

$$+ 8t^{3/2}\int_0^\eta f(\eta)\,d\eta\,\sin\theta\cos\theta, \tag{3.7.36}$$

where

$$
\int_0^\eta f(\eta)\,d\eta = \frac{11}{6\sqrt{\pi}}\,e^{-\eta^2}\operatorname{erfc}\eta - \frac{8}{3\sqrt{2\pi}}\operatorname{erfc}\sqrt{2}\,\eta + \frac{\eta^3}{3}\operatorname{erfc}^2\eta
$$

$$
- \frac{2}{3\sqrt{\pi}}\eta^2 e^{-\eta^2}\operatorname{erfc}\eta + \frac{1}{3\pi}\eta e^{-2\eta^2} - \frac{\eta}{2}\operatorname{erfc}^2\eta
$$

$$
+ \frac{1}{\sqrt{\pi}}\left(\frac{4}{9\pi} - \frac{3}{2}\right)e^{-\eta^2} - \left(1 + \frac{4}{9\pi}\right)\eta^3\operatorname{erfc}\eta
$$

$$
+ \frac{1}{\sqrt{\pi}}\left(1 + \frac{4}{9\pi}\right)\eta^2 e^{-\eta^2} + \frac{2}{3\sqrt{\pi}}\operatorname{erfc}\eta + \left(\frac{1}{2} - \frac{2}{3\pi}\right)\eta\operatorname{erfc}\eta
$$

$$
+ \frac{1}{\sqrt{\pi}}\left(\frac{8}{3\sqrt{2}} - \frac{4}{9\pi} - 1\right). \tag{3.7.37}
$$

It is interesting to note that the last term in Eq. (3.7.36) is the solution derived to this order by Blasius (1908).

A composite solution is also derived (Van Dyke, 1964; Nayfeh, 1973) by adding the inner and outer solutions and subtracting the common part

$$
U = \sin\theta\left(1 + \frac{1}{r^2} - 2\operatorname{erfc}\eta\right) + \varepsilon 2\sqrt{\frac{t}{\pi}}\,\sin\theta\left(\frac{2}{r^2} - 2e^{-\eta^2} + 3\sqrt{\pi}\,\eta\operatorname{erfc}\eta\right)
$$

$$
+ \varepsilon 4t\sin\theta\cos\theta\,f(\eta) + O(\varepsilon^2), \tag{3.7.38}
$$

$$
\Psi = \sin\theta\left(r - \frac{1}{r}\right) + 4\varepsilon\sqrt{\frac{t}{\pi}}\,\sin\theta\left(-\frac{1}{r} + e^{-\eta^2} - \sqrt{\pi}\,\eta\operatorname{erfc}\eta\right)
$$

$$
+ 4\varepsilon^2\frac{t}{\sqrt{\pi}}\sin\theta\left[-\frac{\sqrt{\pi}}{4}(1 - \operatorname{erfc}\eta) + \frac{3\sqrt{\pi}}{2}\eta^2\operatorname{erfc}\eta - \frac{3}{2}\eta e^{-\eta^2}\right]
$$

$$
+ 8\varepsilon^2 t^{3/2}\sin\theta\cos\theta\int_0^\eta f(\eta)\,d\eta + O(\varepsilon^3). \tag{3.7.39}
$$

Wang employed the same method (Wang, 1967b) to calculate flows around ellipses of various slenderness ratios at an angle of attack. For all problems he calculated and plotted instantaneous streamline patterns, very useful for comparisons with the results of other methods.

In an alternative approach, Collins and Dennis (1973a) reduced the number of independent variables by assuming the dependence on the azimuthal angle in the form of an asymptotic expansion. In this work the authors proceeded in a manner similar to Blasius by expanding again in powers of time. However, they generated solutions only for the inner region that were not matched with any outer solution but adjusted to match a uniform flow far from the cylinder. In the following subsection we begin by formulating the problem in a Fourier series in terms of the azimuthal angle.

Collins and Dennis employed this method in their boundary-layer calculations (Collins and Dennis, 1973a) as well as in their Navier–Stokes solutions
(Collins and Dennis, 1973b).

Bar-Lev and Yang (1975) formulated the problem in terms of the vorticity
in order to eliminate the pressure and expanded again in powers of a small
Strouhal number. Carrying their inner and outer expansions to a third order
and expanding in powers of the Reynolds number as well, they were able to
show that the outer flow is irrotational at least to the third order. They
obtained streamline configurations, pressure distributions, vorticity distributions, etc., for a wide range of Reynolds numbers. Some of their results will
be commented upon in the last section of this chapter as well as in the
chapter on unsteady separation.

Seminumerical solutions

Consider a modified radial coordinate in a polar system r, θ:

$$Y = \log\left(\frac{r}{a}\right), \tag{3.7.40}$$

where a is the radius of the cylinder. Consider further a dimensionless stream
function defined by the equation

$$u = e^{-Y}\frac{\partial \psi}{\partial \theta}, \qquad v = -e^{-Y}\frac{\partial \psi}{\partial Y}. \tag{3.7.41}$$

The problem can be formulated (Collins and Dennis, 1973b) in terms of the
stream function ψ and the dimensionless vorticity ζ. In its dimensionless
form the momentum equation becomes

$$e^{2Y}\frac{\partial \zeta}{\partial t} + \frac{\partial \psi}{\partial \theta}\frac{\partial \zeta}{\partial Y} - \frac{\partial \psi}{\partial Y}\frac{\partial \zeta}{\partial \theta} = \frac{2}{Re}\left(\frac{\partial^2 \zeta}{\partial Y^2} + \frac{\partial^2 \zeta}{\partial \theta^2}\right), \tag{3.7.42}$$

and the two functions are related by

$$\frac{\partial^2 \psi}{\partial Y^2} + \frac{\partial^2 \psi}{\partial \theta^2} = e^{2Y}\zeta. \tag{3.7.43}$$

Time is always made dimensionless according to the formula $t = t^* U/L$ as
in Eq. (3.6.7). The system of Eqs. (3.7.42) and (3.7.43) is a closed set of
equations for the unknown quantities ψ and ζ, with boundary conditions

$$\psi = \frac{\partial \psi}{\partial Y} = 0, \quad \text{at } Y = 0, \tag{3.7.44}$$

$$e^{-Y}\frac{\partial \psi}{\partial \theta} \to \cos \theta, \quad e^{-Y}\frac{\partial \psi}{\partial Y} \to \sin \theta, \quad \text{as } Y \to \infty. \tag{3.7.45}$$

Considering only flows symmetric about the x axis, i.e., the axis parallel to the mean flow, assume the solution in the form

$$\psi = \sum_{n=1}^{\infty} f_n(Y,t)\sin n\theta, \tag{3.7.46}$$

$$\zeta = \sum_{n=1}^{\infty} g_n(Y,t)\sin n\theta. \tag{3.7.47}$$

Substituting these expansions in the governing equations, Eqs. (3.7.42) and (3.7.43), multiplying each equation by $\sin n\theta$, and integrating with respect to θ from $\theta = 0$ to $\theta = \pi$, we obtain

$$e^{2Y}\frac{\partial g_n}{\partial t} = \frac{2}{R}\frac{\partial^2 g_n}{\partial Y^2} + nf_{2n}\frac{\partial g_n}{\partial Y} + \left(\frac{n}{2}\frac{\partial f_{2n}}{\partial Y} - \frac{2}{R}n^2\right)g_n + S_n, \tag{3.7.48}$$

$$\frac{\partial^2 f_n}{\partial Y^2} - n^2 f_n = e^{2Y}g_n, \tag{3.7.49}$$

with

$$S_n = \frac{1}{2}\sum_{\substack{m=1\\ m\neq n}}^{\infty}\left\{\left[(m+n)f_{m+n} - jf_j\right]\frac{\partial g_m}{\partial Y}\right.$$
$$\left. + m\left[\frac{\partial f_{m+n}}{\partial Y} - \mathrm{sgn}(m-n)\frac{\partial f_j}{\partial Y}\right]g_m\right\}, \tag{3.7.50}$$

where $j = |m - n|$ and $\mathrm{sgn}(\ )$ denotes the sign of the expression in the parenthesis. The boundary conditions at the wall require that

$$f_n = \partial f_n/\partial Y = 0, \quad \text{at } Y = 0, \tag{3.7.51}$$

while far from the body

$$g_n \to 0, \quad \text{as } Y \to \infty, \tag{3.7.52}$$

and conditions (3.7.45) imply that

$$e^{-Y}f_1 \to 1, \quad e^{-Y}\frac{\partial f_1}{\partial Y} \to 1, \quad \text{as } Y \to \infty, \tag{3.7.53}$$

but

$$e^{-Y}f_n \to 0, \quad e^{-Y}\frac{\partial f_n}{\partial Y} \to 0, \quad \text{as } Y \to \infty \text{ for } n > 1. \tag{3.7.54}$$

Collins and Dennis (1973a) point out that by multiplying Eq. (3.7.49) by e^{-nY} and integrating from $Y = 0$ to $Y = \infty$, we obtain the conditions

$$\int_0^{\infty} e^{Y}g_1(Y,t)\,dY = 2, \tag{3.7.55}$$

$$\int_0^{\infty} \exp\{(2-n)Y\}\,g_n(Y,t)\,dY = 0, \quad \text{for } n > 1, \tag{3.7.56}$$

which together with (3.7.51) and (3.7.52) are sufficient to solve the problem. In fact, if these conditions are satisfied and the quantity $e^{2Y}g_n$ is bounded for all n as $Y \to \infty$, then the flow is automatically adjusted to satisfy the external stream condition (3.7.54).

Seeking a solution in terms of expansions in powers of a small Strouhal number or an exact numerical solution of the above equations, the classical Blasius solution described in Section 3.6 is very useful. In the first case it becomes the first term of the asymptotic expansion. In the second it serves as an initial condition that allows the investigator to bypass the domain of very large derivatives encountered in the initial stages of the motion. To this end, boundary-layer independent and dependent variables are introduced:

$$Y = 2\left(\frac{2t}{R}\right)^{1/2}\eta, \tag{3.7.57}$$

$$f_n = 2\left(\frac{2t}{R}\right)^{1/2}F_n, \qquad g_n = \frac{1}{2}\left(\frac{R}{2t}\right)^{1/2}G_n. \tag{3.7.58}$$

A translation of (3.6.18) in terms of the present variables then gives

$$G_1 = \frac{4}{\sqrt{\pi}}\,e^{-\eta^2}, \tag{3.7.59}$$

$$F_1 = 2\left[\eta\,\mathrm{erf}\,\eta - \frac{1}{\sqrt{\pi}}\left(1 - e^{-\eta^2}\right)\right]. \tag{3.7.60}$$

With these expressions as initial profiles the parabolic set (3.7.48) and (3.7.49) can be integrated by any established method. Collins and Dennis describe a Crank–Nicolson scheme for integrating successively these equations from $n = 1$ to a final n ranging from 20 to 80. The time steps Δt start from small values of order 10^{-4} for the first 10 steps followed by 24 steps with $\Delta t = 10^{-3}$ and further increase for subsequent times. Solutions were derived for a wide range of Reynolds numbers, $Re = 5, 10, 40, 100, 200, 500, 1000, 5000$, and so on.

Exact numerical solutions

A good number of successful numerical integrations of the unsteady Navier–Stokes equations have been reported in literature. We have already outlined at the end of the previous chapter the need for appropriate lattices and boundary conditions. One of the first contributions that demonstrated historically the power of the modern computer in solving unsteady viscous flow problems is due to Fromm and Harlow (1963). We shall describe here very briefly the approach of Thoman and Szewczyk (1969) for integration of the finite-difference equations over the rectangular mesh of Fig. 2.21. This is one of the first works dealing with an impulsively started circular cylinder via the Navier–Stokes equations.

An initial guess is made for the stream function; for the case of a circular cylinder started from rest, let

$$\psi_{m,k} = U_\infty y \left[1 - \frac{a^2}{x_m^2 + y_k^2} \right], \tag{3.7.61}$$

with a the radius of the cylinder. The stream function on the cylinder is assigned the value zero and is not allowed to vary in subsequent iterations. Values are then calculated on the periphery of the domain of integration and an updated set of $\psi_{m,k}$'s are calculated by solving Eq. (2.9.9) with $\Omega_{m,k} = 0$, since initially vorticity is zero throughout the flow field.

$$\psi_{m,k}^0 = \frac{2}{d_{m,k}} \left[\frac{\psi_{m-1,k}}{\Delta x_m(\Delta x_{m-1} + \Delta x_m)} + \frac{\psi_{m+1,k}}{\Delta x_m(\Delta x_m + \Delta x_{m+1})} \right.$$
$$\left. \times \frac{\psi_{m,k-1}}{\Delta y_k(\Delta y_{k-1} + \Delta y_k)} + \frac{\psi_{m,k+1}}{\Delta y_k(\Delta y_k + \Delta y_{k+1})} \right]. \tag{3.7.62}$$

The process is repeated as shown in the flow chart in Fig. 3.11 until convergence is achieved. Then, velocity components are calculated via Eqs. (2.9.4)–(2.9.7) and the vorticity distribution on the skin of the body is computed in terms of the vorticity definition. For example, for $x < 0$ and $y < 0$,

$$\Omega_{m,k}^0 = 2 \frac{v_{m,k}^0 - v_{m-1,k}^0}{\Delta x_m + \Delta x_{m-1}} - 2 \frac{u_{m,k}^0 - u_{m,k-1}^0}{\Delta y_k + \Delta y_{k-1}}. \tag{3.7.63}$$

A flow chart of the steps required to march the calculation in time is shown in Fig. 3.12. Time is advanced by one step and the vorticity distribution is calculated from Eq. (2.9.8) with the velocity expressed in terms of the stream function of the previous time station

$$\frac{\Omega_{m,k}^1 - \Omega_{m,k}^0}{\Delta t} + \frac{(u^0 \Omega^1)_{m+1/2,k} - (u^0 \Omega^1)_{m-1/2,k}}{\Delta x_m}$$
$$+ \frac{(v^0 \Omega^1)_{m,k+1/2} - (v^0 \Omega^1)_{m,k-1/2}}{\Delta y_k} = \nu \nabla^2 \Omega_{m,k}^1. \tag{3.7.64}$$

The vorticity on the skin of the cylinder is maintained at zero in accordance with the wall-boundary condition. The stream-function is then calculated at the new time by an iteration identical to the one already described for the initial station. After convergence is achieved, velocity components at the new time plane are also calculated as before, and the vorticity on the skin of the body is calculated via equations like (3.7.63). Time is incremented and the process is repeated. Thoman and Szewczyk obtain streamlines by using a two-dimensional linear interpolation of the four nearest velocity components u and v. They also describe a scheme of integration over their hybrid lattice shown in Fig. 2.22.

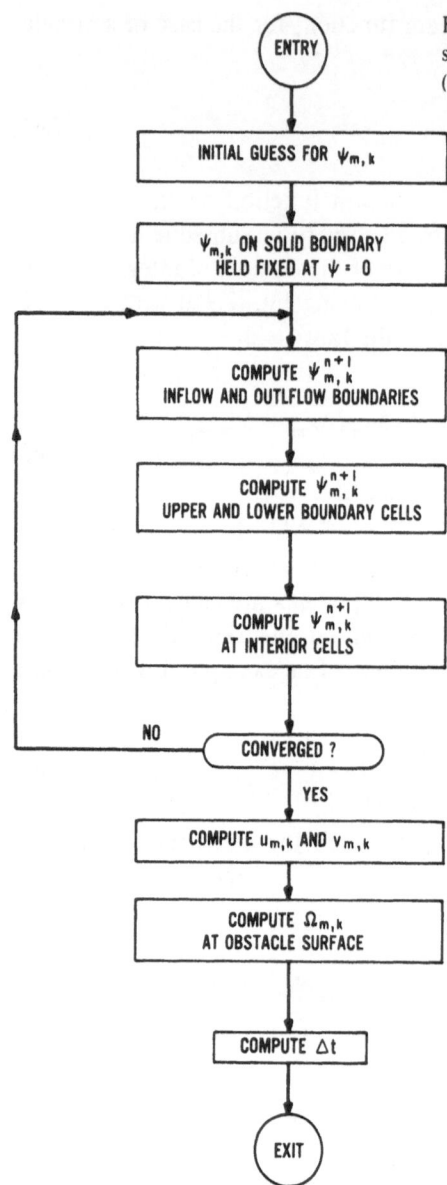

Fig. 3.11 Flow chart for the initial step of the solution of the full Navier–Stokes equations (Thoman and Scewczyk, 1969).

Numerical solutions of the full Navier–Stokes equations for impulsively started cylinders, plates, and airfoils have been reported in the 1970s (Wu and Thompson, 1973; Mehta, 1972, 1977; Schmall and Kinney, 1974; Kinney and Paolino, 1974; Thompson et al., 1974; Mehta and Lavan, 1975; Mehta, 1977; Wu et al., 1977). We chose to outline here the integration of the finite-difference form of the governing equations in the phase space (ρ, θ), as described at the end of the previous chapter.

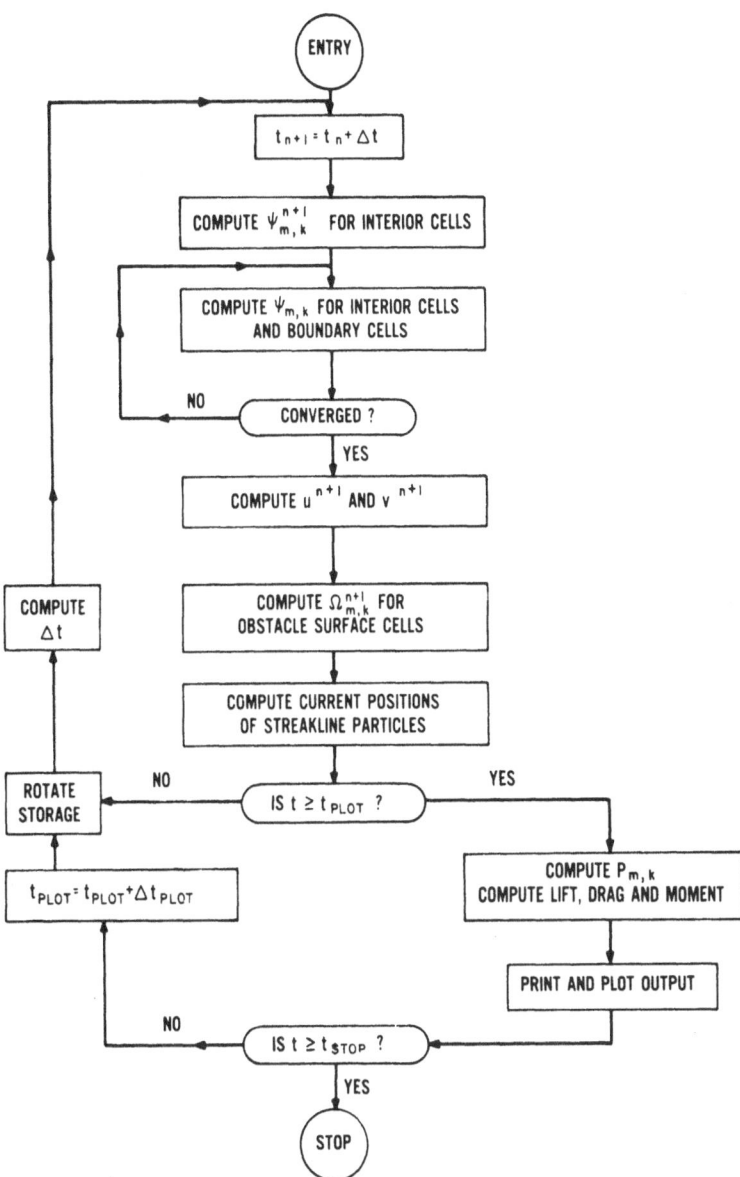

Fig. 3.12 Flow chart for the transient part of the solution of the full Navier–Stokes equations (Thoman and Scewczyk. 1969).

Consider a three-point backward difference for the time derivatives

$$\left(\frac{\partial \Omega}{\partial t}\right)^n_{m,k} \cong \frac{3\Omega^n_{m,k} - 4\Omega^{n-1}_{m,k} + \Omega^{n-2}_{m,k}}{2\Delta t} + \frac{(\Delta t)^2}{3} \overline{\frac{\partial^3 \Omega}{\partial t^3}}. \qquad (3.7.65)$$

An overbar denotes evaluation of the quantity at some mean value of time, $\bar{t}$, namely, $t - 2\Delta t < \bar{t} < t$. At $t = 0$ and $t = \Delta t$, a two-point backward difference is necessary (Richtmyer and Morton, 1967):

$$\left(\frac{\partial \Omega}{\partial t}\right)_{m,k}^{n} \cong \frac{\Omega_{m,k}^{n} - \Omega_{m,k}^{n-1}}{\Delta t} + \frac{\Delta t}{2} \overline{\frac{\partial^2 \Omega}{\partial t^2}}. \tag{3.7.66}$$

Mehta writes the Jacobian in Eq. (2.9.18) in terms of a second-order expression patterned according to a formula developed by Arakawa (1966) for equal step sizes. For $\Delta\theta \neq \Delta\rho$ the Jacobian is decomposed and expressed as follows (Mehta, 1972):

$$-J\left(\frac{\Omega, \psi + y}{\rho, \theta}\right)_{m,k} = \frac{1}{12\Delta\rho\Delta\theta}\left[\Omega_{m,k+1}(VE) + \Omega_{m-1,k+1}(VSE)\right.$$

$$+ \Omega_{m-1,k}(VS) + \Omega_{m-1,k-1}(VSW)$$

$$+ \Omega_{m,k-1}(VW) + \Omega_{m+1,k-1}(VNW) \tag{3.7.67}$$

$$\left. + \Omega_{m+1,k}(VN) + \Omega_{m+1,k+1}(VNE)\right] - E,$$

where in terms of

$$\Psi = \psi + y \tag{3.7.68}$$

the following quantities are defined:

$$VE = \Psi_{m-1,k} + \Psi_{m-1,k+1} - \Psi_{m+1,k} - \Psi_{m+1,k+1}, \tag{3.7.69}$$

$$VSE = \Psi_{m-1,k} - \Psi_{m,k+1}, \tag{3.7.70}$$

$$VS = \Psi_{m-1,k-1} + \Psi_{m,k-1} - \Psi_{m-1,k+1} - \Psi_{m,k+1}, \tag{3.7.71}$$

$$VSW = \Psi_{m,k-1} - \Psi_{m-1,k}, \tag{3.7.72}$$

$$VW = \Psi_{m+1,k-1} + \Psi_{m+1,k} - \Psi_{m-1,k-1} - \Psi_{m-1,k}, \tag{3.7.73}$$

$$VNW = \Psi_{m+1,k} - \Psi_{m,k-1}, \tag{3.7.74}$$

$$VN = \Psi_{m,k+1} + \Psi_{m+1,k+1} - \Psi_{m,k-1} - \Psi_{m+1,k-1}, \tag{3.7.75}$$

$$VNE = \Psi_{n,k+1} - \Psi_{m+1,k}, \tag{3.7.76}$$

and E is the truncation error

$$E = 0\left[(\Delta\rho)^2 + (\Delta\theta)^2\right]. \tag{3.7.77}$$

This expression of the Jacobian conserves the mean vorticity, the mean square vorticity, and the mean-square kinetic energy.

First and second derivatives of the diffusion terms in Eq. (2.9.19) are expressed in terms of central differences

$$\left(\frac{\partial\Omega}{\partial\rho}\right)_{m,k} \cong \frac{\Omega_{m,k+1} - \Omega_{m,k-1}}{2\Delta\rho} - \frac{(\Delta\rho)^2}{6}\,\overline{\frac{\partial^3\Omega}{\partial\rho^3}} , \tag{3.7.78}$$

$$\left(\frac{\partial^2\Omega}{\partial\rho^2}\right)_{m,k} \cong \frac{\Omega_{m,k+1} + \Omega_{m,k-1} - 2\Omega_{m,k}}{(\Delta\rho)^2} - \frac{(\Delta\rho)^2}{12}\,\overline{\frac{\partial^4\Omega}{\partial\rho^4}} , \tag{3.7.79}$$

$$\left(\frac{\partial^2\Omega}{\partial\theta^2}\right)_{m,k} \cong \frac{\Omega_{m+1,k} + \Omega_{m-1,k} - 2\Omega_{m,k}}{(\Delta\rho)^2} - \frac{(\Delta\theta)^2}{12}\,\overline{\frac{\partial^4\Omega}{\partial\theta^4}} . \tag{3.7.80}$$

In terms of the above finite-difference expressions, the finite-difference form of Eq. (2.9.19) becomes

$$\frac{H^2 r^2}{2\Delta t}\frac{R}{L}\left[T_1\Omega_{m,k}^n + T_2\Omega_{m,k}^{n-1} + T_3\Omega_{m,k}^{n-2}\right]$$

$$= \left(\frac{d\rho}{dr}\right)^2 r^2\left[\frac{\Omega_{m,k+1}^n + \Omega_{m,k-1}^n - 2\Omega_{m,k}^n}{(\Delta\rho)^2}\right]$$

$$+ \left(\frac{d\rho}{dr}r + \frac{d^2\rho}{dr^2}r^2\right)\left[\frac{\Omega_{m,k+1}^n - \Omega_{m,k-1}^n}{2\Delta\rho}\right] \tag{3.7.81}$$

$$+ \frac{\Omega_{m+1,k}^n + \Omega_{m-1,k}^n - 2\Omega_{m,k}^n}{(\Delta\theta)^2} - r\frac{d\rho}{dr}\frac{R}{L}J\left(\frac{\Omega,\psi}{\rho,\theta}\right) + E',$$

where the Jacobian is given by Eq. (3.7.67), the truncation error is

$$E' = -\left[H^2 r^2\frac{R}{L}\frac{(\Delta t)^m}{m+1}\overline{\frac{\partial^{m+1}\Omega}{\partial t^{m+1}}} + \left(\frac{d\rho}{dr}\right)^2 r^2\frac{(\Delta\rho)^2}{12}\overline{\frac{\partial^4\Omega}{\partial\rho^4}}\right.$$

$$\left. + \left(\frac{d\rho}{dr}r + \frac{d^2\rho}{dr^2}r^2\right)\frac{(\Delta\rho)^2}{6}\overline{\frac{\partial^3\Omega}{\partial\rho^3}} + \frac{(\Delta\theta)^2}{12}\overline{\frac{\partial^4\Omega}{\partial\theta^4}} + E\right], \tag{3.7.82}$$

and the constants T and m correspond to the two backward time-differencing schemes:

3-point scheme: $T_1 = 3, T_2 = -4, T_3 = 1, m = 2;$ \hfill (3.7.83)

2-point scheme: $T_1 = 2, T_2 = -2, T_3 = 0, m = 1.$ \hfill (3.7.84)

The finite-difference equation that governs the stream function can be derived by substituting the expressions of this section in Eq. (2.9.19)

$$
\begin{aligned}
- H^2 r^2 \Omega_{m,k}^n = r^2 \left(\frac{d\rho}{dr} \right)^2 & \left[\frac{\psi_{m,k+1}^n + \psi_{m,k-1}^n - 2\psi_{m,k}^n}{(\Delta\rho)^2} \right] \\
& + \left(\frac{d\rho}{dr} r + \frac{d^2}{dr^2} r^2 \right) \left[\frac{\psi_{m,k+1}^n - \psi_{m,k-1}^n}{2\Delta\rho} \right] \\
& + \left[\frac{\psi_{m+1,k}^n + \psi_{m-1,k}^n - 2\psi_{m,k}^n}{(\Delta\theta)^2} \right] + E'',
\end{aligned}
\tag{3.7.85}
$$

with a truncation error given by

$$
\begin{aligned}
E'' = r^2 & \left(\frac{d\rho}{dr} \right)^2 \frac{(\Delta\rho)^2}{12} \frac{\partial^4 \psi}{\partial \rho^4} + \left(\frac{d\rho}{dr} r + \frac{d^2\rho}{dr^2} r^2 \right) \\
& \times \frac{(\Delta\rho)^2}{6} \frac{\partial^3 \psi}{\partial \rho^3} + \frac{(\Delta\theta)^2}{12} \frac{\partial^4 \psi}{\partial \theta^4}.
\end{aligned}
\tag{3.7.86}
$$

The solution to the coupled system of equations is sought by iteration according to a relaxation procedure. An initial guess is made for the vorticity and stream function $\Omega_{m,k}^{n,0}$, $\psi_{m,k}^{n,0}$. The first iteration, $\Omega_{m,k}^{n,1}$, $\psi_{m,k}^{n,1}$, is then obtained by sweeping the domain of integration first in the free stream direction for $\rho = $ const, namely from $\theta = \pi$ to 0 and from $\theta = \pi$ to 2π, so that numerical errors are propagated downstream. In the following sweep, the domain of integration is covered from $\rho = 1$ to $\rho = \rho_0$, namely, from the solid surface to the outer boundary. The differences between the new and the old iterations are then calculated. For example, assuming that iterations $\Omega_{m,k}^{n,j-1}$ and $\psi_{m,k}^{n,j-1}$ are known, the jth iterations can be calculated by solving Eqs. (3.7.81) and (3.7.85) marching in the direction of decreasing k, that is from $k = JL$ to $k = 0$. The difference of the iterates then is

$$
(\Delta\Omega)_{m,k}^{n,j} = \Omega_{m,k}^{n,j} - \Omega_{m,k}^{n,j-1}
\tag{3.7.87}
$$

$$
(\Delta\psi)_{m,k}^{n,j} = \psi_{m,k}^{n,j} - \psi_{m,k}^{n,j-1}
\tag{3.7.88}
$$

where $\Omega_{m,k}^{n,j}$ and $\psi_{m,k}^{n,j}$ are found from Eqs. (3.7.81) and (3.7.85), namely

$$
\begin{aligned}
\Omega_{m,k}^{n,j} = & \left[2\Delta t \left\{ \Omega_{m,k+1}^{n,j} (a_4 + (VE)a_9) + \Omega_{m,k-1}^{n,j-1} (a_5 + (VW)a_9) \right. \right. \\
& + \Omega_{m+1,k}^n (a_0 + (VN)a_9) + \Omega_{m-1,k}^n (a_0 + (VS)a_9) \\
& + a_9 \Omega_{m+1,k-1}^{n,j-1} (VNW) + a_9 \Omega_{m-1,k-1}^{n,j-1} (VSW) \\
& + a_9 \Omega_{m-1,k+1}^{n,k} (VSE) + a_9 \Omega_{m+1,k+1}^{n,j} (VNE) \right\} \\
& \left. - a_8 H_{m,k}^2 (T_2 \Omega_{m,k}^{n-1} + T_3 \Omega_{m,k}^{n-2}) \right] (T_1 a_8 H_{m,k}^2 + a_6)^{-1}
\end{aligned}
\tag{3.7.89}
$$

$$\psi_{m,k}^{n,j} = a_2 \Big[a_4 \psi_{m,k+1}^{n,j} + a_5 \psi_{m,k-1}^{n,j-1} + a_0 (\psi_{m+1,k}^{n} + \psi_{m-1,k}^{n})$$
$$+ (\Delta\rho)^2 r_k^2 H_{m,k}^2 \Omega_{m,k}^{n,j} \Big]$$

(3.7.90)

with

$$a_0 = \left(\frac{\Delta\rho}{\Delta\theta}\right)^2, \quad a_1 = \left(\frac{d\rho}{d\theta}\right)_k^2 r_k^2, \quad a_2 = \frac{1}{2(a_0 + a_1)}$$

$$a_3 = \frac{1}{2}\left(r^2 \frac{d^2\rho}{dr^2} + r \frac{d\rho}{dr}\right)_k \Delta\rho$$

(3.7.91)

$$a_4 = a_1 + a_3, \quad a_5 = a_1 - a_3, \quad a_6 = 4\Delta t (a_0 + a_1)$$

$$a_7 = \Delta\rho \frac{R}{L} r_j, \quad a_8 = \Delta\rho r_j a_7, \quad a_9 = a_7 \frac{(d\rho/dr)_j}{12\Delta\theta}$$

Notice that on the right hand sides, the jth iteration appears only at $k + 1$ terms, since marching proceeds towards decreasing k's. Moreover, for $(m \pm 1, k)$ terms, no iteration counter is shown, because the θ direction of sweep is altered during the iteration sequence. More specifically, for j odd, sweeping proceeds first from $\theta = \pi$ to $\theta = 2\pi$ and then $\theta = \pi$ to $\theta = 0$; this order is interchanged for j even. In this way, accumulation of errors in some region is avoided.

In the actual calculations reported by Mehta and Lavan (1975) the jth iteration is calculated as a weighted average

$$\Omega^j = \Omega^{j-1} + \beta_1 (\Delta\Omega)^j$$

(3.7.92)

$$\psi^j = \psi^{j-1} + \beta_2 (\Delta\psi)^j$$

(3.7.93)

with β_1 and β_2 relaxation parameters. The process is repeated as shown in the flow chart of Fig. 3.13 until the residue at each grid point becomes smaller than a small prescribed quantity. The pressure coefficient on the surface can then be calculated.

3.8 Results and Physical Interpretation

The transient flow over solid bodies goes through various stages if started from rest. At first it is irrotational, a very rare case in real life, whereby the flow around a body is potential. Very soon the vortex sheets that are initially tightly wrapped around the body expand into boundary layers with growing thickness. Up to this point, the development of the flow can be successfully computed with any of the models described in this section.

There are two events that influence significantly the subsequent development of the flow. The first is the appearance of hydrodynamic instabilities

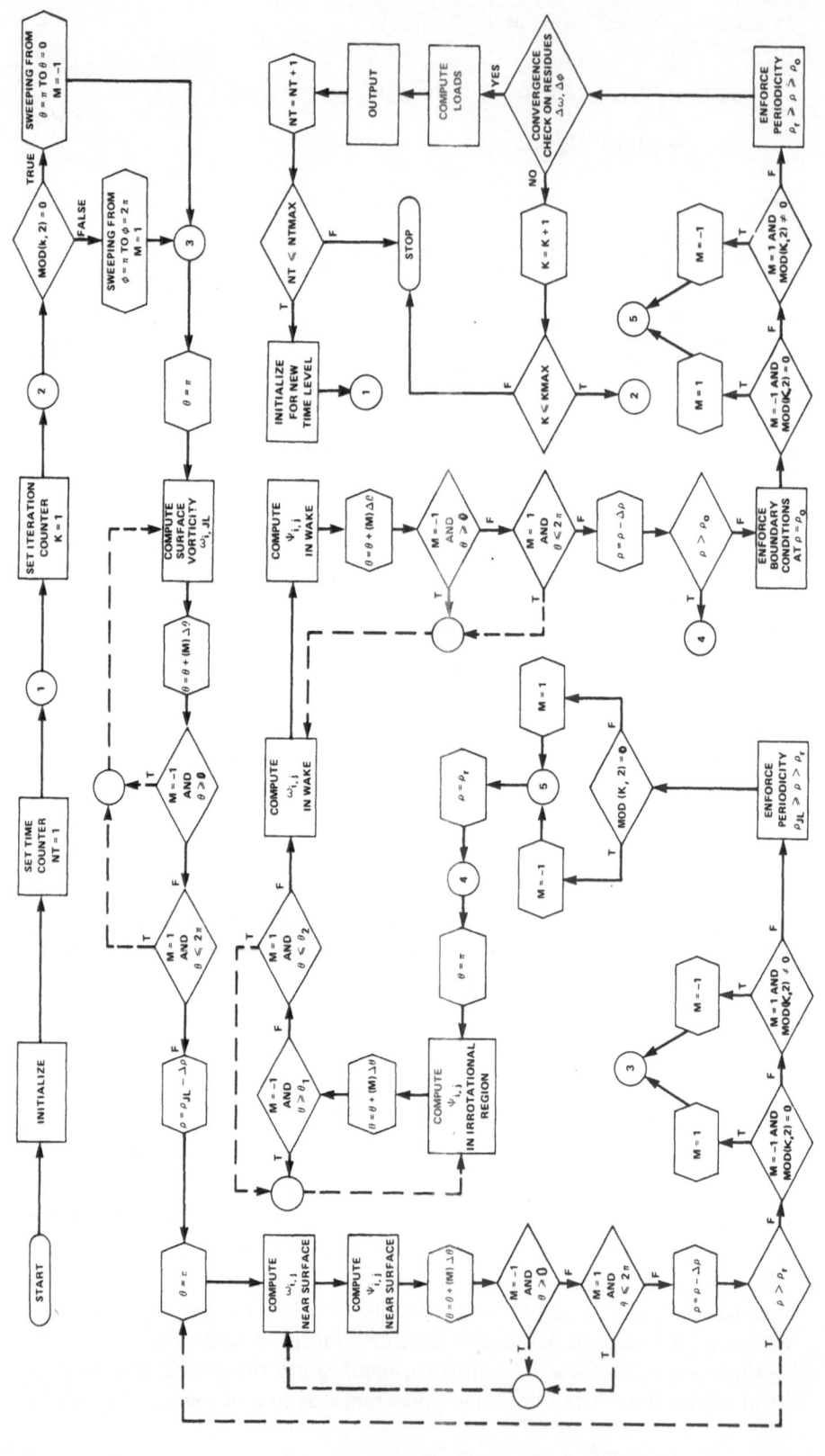

that may lead to transition, and subsequently the generation of boundary-layer turbulence. The second is the appearance of large-scale vortices in the aft region that eventually develop into a wake. Usually the second event precedes the first, but at this point very little information is available about the unsteady evolution of instabilities. In fact, to the knowledge of this author, there is no theoretical effort to model the transient development of large-scale vortices.

An equally important transient flow problem is the metamorphosis of an established viscous flow, involving perhaps turbulent boundary layers, separation and a wake to a new flow that corresponds to a new geometrical configuration, another angle of attack, or a new value of the oncoming mean flow velocity.

The methods described in this chapter have demonstrated, with variable success, ability to capture correctly the early phases of the boundary-layer development, the appearance of separated bubbles and secondary vortices, and the early stages of a fully developed wake. The problem of unsteady separation and the formation of an unsteady wake introduces considerable computational but mostly conceptual difficulties and a special chapter is devoted to it. In this section, therefore, we shall describe results pertaining to the early stages of the development of the flow. Results of the methods described in this chapter pertaining to unsteady separation are discussed in the chapter on separation.

Blasius (1908) and Goldstein and Rosenhead (1936) calculated the time required for the appearance of a point of zero skin friction. For a few decades following the latter work, it was believed that the calculations downstream of the point of zero skin friction would be meaningless. It was Proudman and Johnson (1962) who indicated that a thin layer of reversed flow is compatible with rear stagnation attached flow. They actually proved that for any fixed time t_0, no matter how large, the thickness of the rotational domain of the flow becomes arbitrarily small as $Re \to \infty$.

Earlier numerical calculations, using mostly integral methods, have been successfully employed to estimate the upstream propagation of the point of zero skin friction. There is a very large number of results today and in Fig. 3.14 we collected only a few, representative of different methods, namely, boundary-layer calculations (Presz and Heiser, 1968; Telionis and Tsahalis, 1974b; Cebeci, 1979), full Navier–Stokes solutions (Thoman and Schzew-czyk, 1969), and some experimental points (Weinberg, 1967). All the calculations represented in this figure correspond to very large or infinite Reynolds numbers. Comparison with experimental data with $Re < 10^3$ therefore is not very appropriate, but no other data are available.

What seems to be most interesting physically is that it takes some finite time for the first zero skin friction to appear on the skin of the cylinder. This

---

Fig. 3.13 (*opposite*) Flowchart for the solution of the full Navier–Stokes equations by the method of Mehta and Lavan (1975).

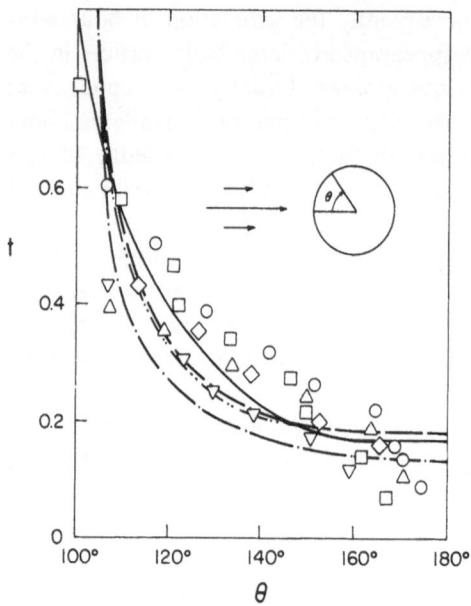

Fig. 3.14 The temporal path of the point of zero skin friction for a circular cylinder started impulsively from rest. $-\cdot-\cdot-$, Presz and Heiser (1968); ———, Thoman and Szewczyk (1969); – – –, Telionis and Tsahalis (1974a); ----, Cebeci (1979). Experimental points of Weinberg (1967) for Reynolds numbers: 405 ($\triangle, \nabla, \bigcirc,$); 819 ($\lozenge$).

is the time t_s that Blasius and Goldstein and Rosenhead originally calculated. There are some discrepancies on the calculated values of t_s. This is most clearly demonstrated in Table 3.1. It appears that most investigators have obtained a value of $t_s = 0.32$. All the theoretical data indicate that the point of zero skin friction travels upstream with a very large rate after $t = t_s$. In other words, once a thin recirculating bubble appears, it shoots upstream quickly, covering essentially all the aft part of the cylinder. The rate of upstream propagation slows down considerably but the bubble may extend to $\theta < 120°$ before any other signs of secondary vortices or separation become evident. The boundary-layer calculations approach asymptotically a value of θ larger than 100°, which corresponds to steady boundary-layer results with potential outer flow.

The experimental data in Fig. 3.14 do not indicate any specific Reynolds number effects. However all modern methods of analysis clearly show that for smaller Reynolds numbers, the transient flow phenomena require longer

Table 3.1 Times for initial appearance and upstream advancement of the point of zero skin friction in the aft region of a circular cylinder impulsively started from rest.

$\theta =$	180	166	146	138	124	110
Collins and Dennis (1973a)	0.322	0.331	0.39	0.43	0.589	0.90
Collins and Dennis (1973b)	0.322	0.33	0.39	0.42	0.59	1.10
Telionis and Tsahalis (1974a)	0.35	0.36	0.40	0.45	0.60	1.11
Bar-Lev and Yang (1975)	0.322	0.330	0.389	0.438	0.602	1.089
Cebeci (1979)	0.320	0.330	0.390	0.436	0.596	1.10
Wang (1979)	0.320	0.358	0.421	0.472	0.646	1.186

Fig. 3.15 The temporal path of the
point of zero skin friction over a
circular cylinder for different Reynolds numbers (Bar-Lev and Yang,
1975).

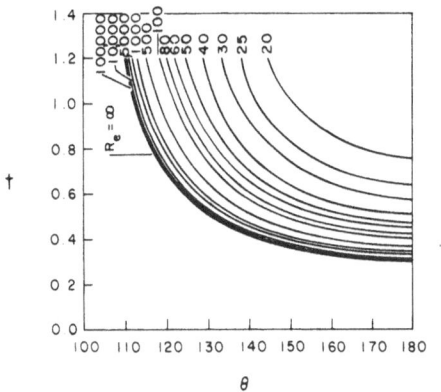

times to develop as shown in Fig. 3.15. This is an unexpected behavior. It is
well known that for unsteady impulsive changes, time is scaled with kinematic viscosity. Flows with larger viscosity, therefore, should develop in time
faster than their less viscous counterparts. For example, the velocity of the
impulsively generated flow over an infinite plate or wedge (see Eq. 3.4.6) will
reach a certain value twice as fast for a fluid with twice the viscosity of
another fluid. In the case of the appearance of a zero skin friction and the
development of a recirculation bubble, the opposite is true. The point of zero
skin friction will reach a specific location on the skin of the cylinder faster
for a less viscous fluid.

Some of the computational methods described in this section have been
used for the calculation of the impulsively generated flow about ellipses and
airfoils. One of the most interesting aspects of flows about asymmetric bodies
is the fact that the first point of zero-skin friction may not necessarily appear
at the rear stagnation point. True, for an impulsively started circular cylinder
and in general for relatively slender symmetric bodies, it is intuitively
obvious and a well documented fact that zero skin friction emerges at the
point of rear stagnation, as shown schematically in the first sketch of Fig.
3.16. The extreme counterexample involves a flat plate positioned perpendicular to the flow as shown schematically in the last sketch of Fig. 3.16. This in
fact is very similar to a problem discussed originally by Prandtl in his
pioneering work (Prandtl, 1904). Separated vortices now originate at the
edges of the plate. Howarth (1938) pointed out that for an elliptical cylinder,
the point of zero skin friction is away from the rear stagnation point if the
ratio of the major to the minor axes is less than 6.

In Fig. 3.17 we show schematically the flow pattern for the flow about an
elliptical cylinder started in a fluid otherwise at rest but at an angle of attack.
This problem has been solved by the method of inner and outer expansions
with the full Navier–Stokes equations as a model (Wang, 1967a) and by
direct numerical integration of the boundary-layer equations (Telionis and
Tsahalis, 1974a). The results are presented in terms of an azimuthal angle θ
measured from the point of stagnation. Consider the elliptic coordinates

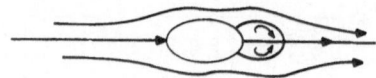

Fig. 3.16 Schematic representation of early stages of separation over impulsively started bodies.

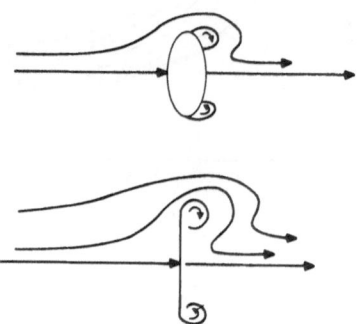

(σ, τ) defined by the equations

$$x = C(\cosh \tau)(\cos \sigma), \tag{3.8.1}$$

$$y = C(\sinh \tau)(\sin \sigma), \tag{3.8.2}$$

where C is the focal length (see Fig. 3.18). The equation of the ellipse is given by

$$\tau = \beta \equiv \coth^{-1}(a_1/b_1), \tag{3.8.3}$$

with a_1 and b_1 the major and minor axes of the ellipse. In terms of these coordinates, the velocity distribution on the skin of the ellipse is

$$u = U_\infty \sin(\theta + \alpha)e^{\beta}(\sinh^2 \beta + \sin^2 \theta)^{-1/2}, \tag{3.8.4}$$

$t = 0$

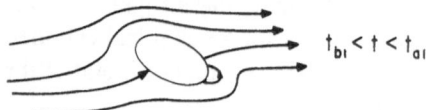

$t_{b1} < t < t_{a1}$

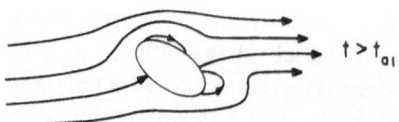

$t > t_{a1}$

Fig. 3.17 Schematic representation of streamline patterns for $t = 0$, $t_{b1} < t < t_{a1}$ and $t > t_{a1}$. Values t_{b1} and t_{a1} are defined in Fig. 3.19.

Fig. 3.18 Coordinate system for the
flow about an ellipse at an angle of
attack.

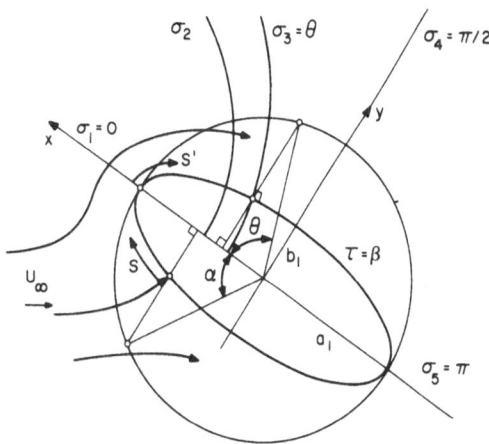

where α is the angle of attack of the oncoming flow. The numerical integration of the boundary-layer equation is marched from $\theta = -\alpha$ in increments of θ, along the boundary of the ellipse and is terminated at the rear stagnation point. For $\alpha = 0.3\pi$, the results shown in Fig. 3.19 indicate that the first zero skin friction appears first on the lower side of the ellipse and at a time approximately equal to $t_{b1} = 0.17$. At later times a recirculating bubble is generated on the upper side of the ellipse, but only after the lower bubble engulfs the rear stagnation point. It is interesting to note that the numerical calculation is marched by a zig-zag scheme through the upper side bubble, that is, through points of detachment and reattachment and into a second point of detachment that belongs to the lower bubble. The agreement of two entirely different methods is quite satisfactory.

The flow about a symmetric Joukowski airfoil started impulsively from

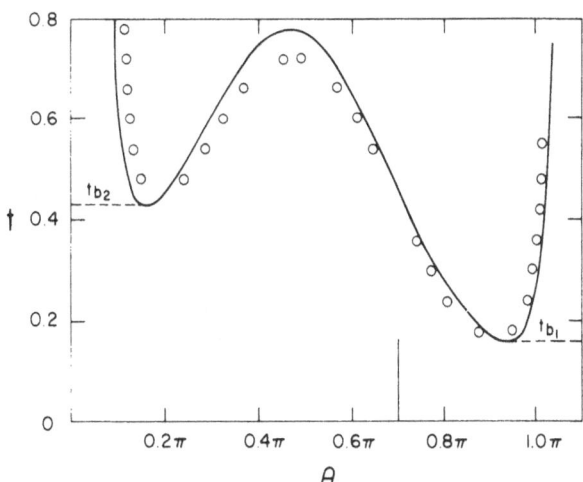

Fig. 3.19 The temporal
variation of skin friction
for flow over an ellipse at
an angle of attack; ——.
Telionis and Tsahalis
(1974a); O, Wang (1967b).

rest was studied by solving the full Navier–Stokes equations by Mehta and Lavan (1975) as described earlier in this section. In Eq. (2.9.16) the constants were chosen as follows:

$$\xi_0 = -0.05214, \eta_0 = 0 \quad \text{and} \quad \delta = 0.025,$$

which correspond to an 8.9998% thick symmetric profile with maximum thickness at 28.89% of the chord length. Calculations were performed with a mesh of 3840 total node points, 3024 of which were located in the rotational region, while Δt was equal to 0.001 at first but was increased for later times. Only the early stages of the flow are described here. A more complete description and a few computed streamlines and vorticity lines are included in the chapter on unsteady separation.

For an angle of attack of $\alpha = 15°$, Mehta and Lavan (1975) found that almost immediately after the impulsive start of the motion, that is for $t < 0.001$, the stagnation points displace quickly from their potential-flow position. The rear stagnation point moves from the upper side of the airfoil towards the trailing edge. This is accompanied by a small counterclockwise separation bubble between the rear stagnation point and the trailing edge. The rear dividing streamline leaves the airfoil practically at the trailing edge for $t > 0.006$. Simultaneously, the front stagnation point moves on the lower surface of the airfoil up and towards the leading edge, but soon after $t = 0.006$, it reverses its direction and displaces downstream, while the rear stagnation point remains practically at the trailing edge. This behavior is shown clearly in the plots of Fig. 3.20. It is characteristic that the displacement of the two stagnation points for $t < 0.02$, is almost symmetric. As a result, circulation is zero for $t \leqslant 0.02$.

Streamlines and equi-vorticity lines are shown in Fig. 3.21 for three characteristic instants. The plots of Fig. 3.20 indicate that for $t = 0.148$, the rear stagnation point is practially at the trailing edge. Figure 3.21 shows, however, that even though the trailing streamline emanates at the cusp, in a direction that appears to bisect the tangents to the body contours, it turns up a very short distance downstream. This is due to the starting vortex which is being convected in the direction of the flow. It is also interesting to note that the equivorticity lines, except for the immediate vicinity of the trailing edge, expand uniformly away from the airfoil. This is because, as described earlier, diffusion overpowers convection, and in the boundary layer the direction of diffusion is practically perpendicular to the solid boundary.

The stages shown in Fig. 3.21 incorporate all the important characteristics of impulsive motion and for low angles of attack, the flow arrives at a steady state without any significant qualitative change. The vorticity generated on the skin of the airfoil is shed at the trailing edge in the form of a free shear layer and the viscous effects are confined in the boundary layers and the free shear layer. The outer flow is very nearly potential. However, for the angle of attack discussed by Mehta and Lavan (1975), separation and recirculation bubbles are generated at later times, as described in Chapter VII.

Fig. 3.20 The temporal
excursions of the front
(O) and rear (□) stagna-
tion points for an impul-
sive start of an airfoil at
an angle of attack $\alpha =$
15° and Reynolds num-
ber $Re = 1000$. Open symbols, lower surface; solid
symbols, upper surface.

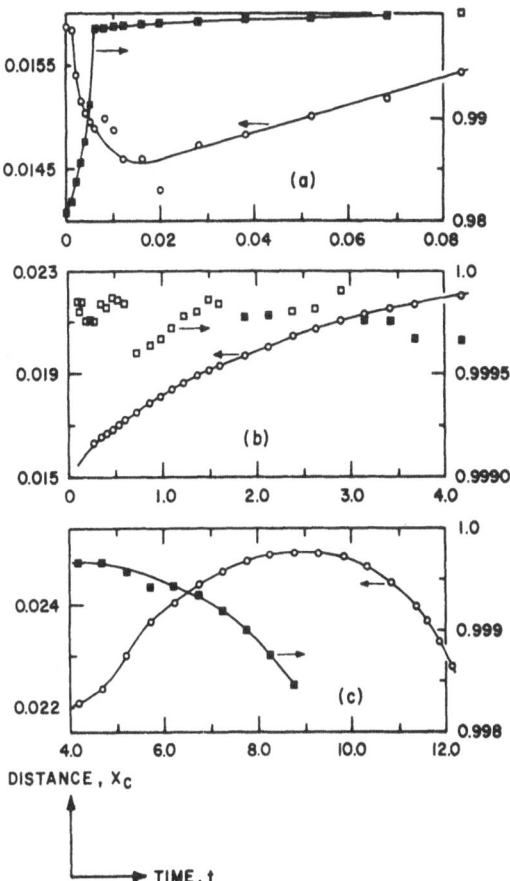

3.9 A Computer Program

As an example of a computer code for the calculation of an impulsively
started flow we present the program used by Telionis and Tsahalis (1974a).
The main elements were developed by Blottner and Flügge-Lotz (1963) but
the program was prepared by Werle and Davis (1972) and was later
modified and expanded for unsteady flow by Telionis, Tsahalis, and Werle
(1973). Descriptions of the finite differencing schemes are found and appro-
priately referenced in the previous and the present chapter. A flow chart
appears in Fig. 3.22.

   The computation is marched first in the downstream space direction. It
should be emphasized that the transformation relating x and ξ, Eq. (3.6.28),
depends upon time and therefore a certain x-station does not correspond to
the same ξ-station at different times. This necessitates marching in increments of ξ, so that grid points at the t_0 plane correspond to the same ξ values
with grid points at the $t_0 + \Delta t$ plane.

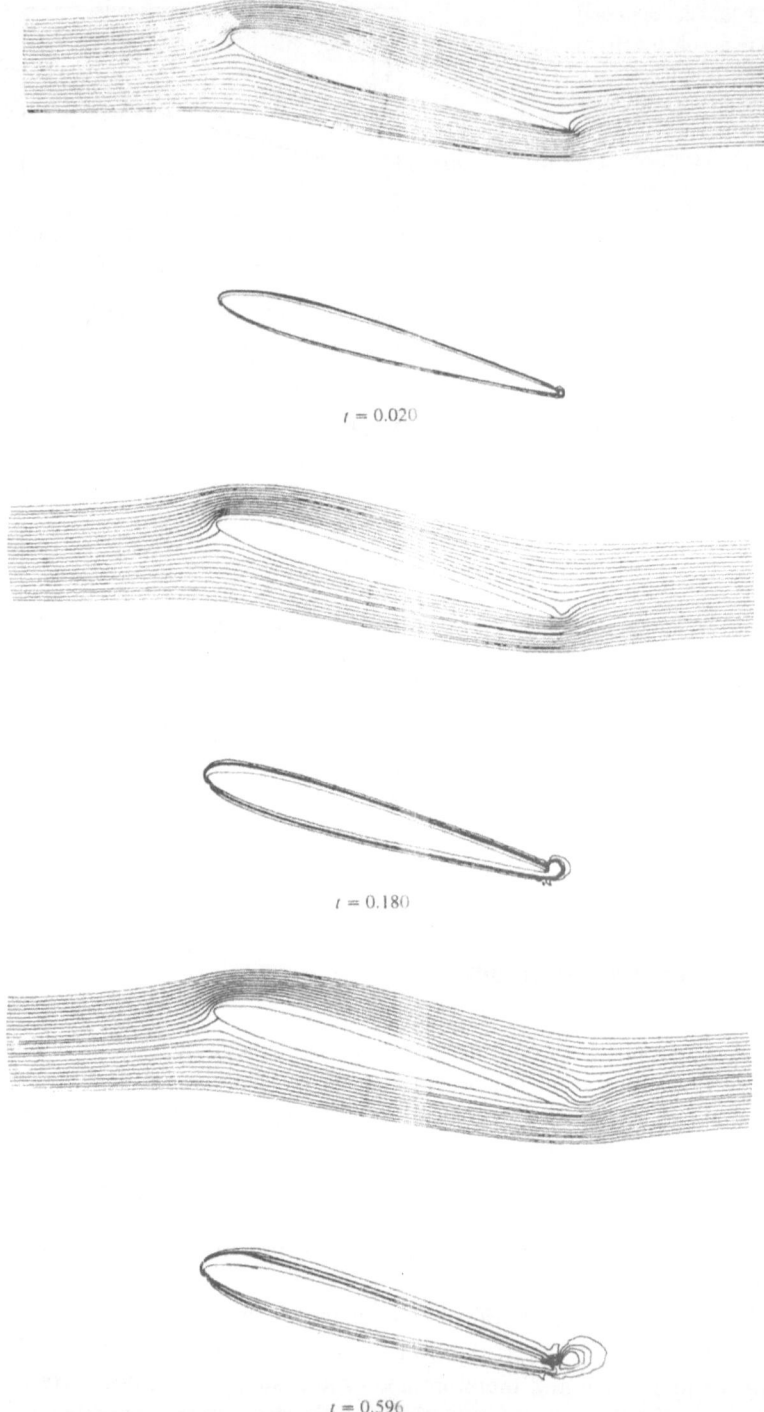

$t = 0.020$

$t = 0.180$

$t = 0.596$

Fig. 3.21 Streamlines and equivorticity lines for the airfoil of Fig. 3.20 at $t = 0.002$, 0.180 and 0.596.

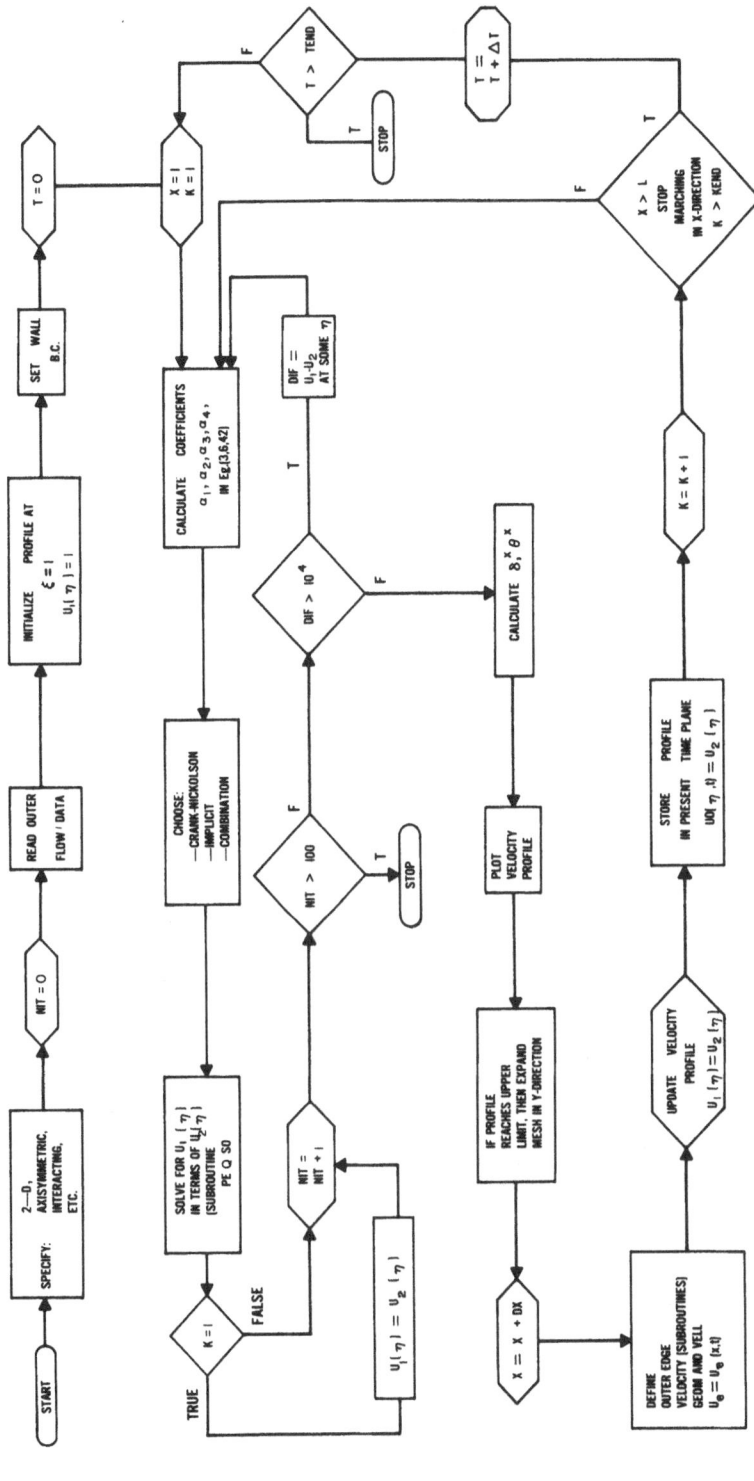

Fig. 3.22 Flowchart for the solution of the unsteady boundary layer-equations by the method of Telionis, Tsahalis and Werle (1973).

Very little information is necessary to start the calculation. Except for basic input data like mesh sizes and controlling flags one needs to specify the outer flow velocity distribution. This is done in the subroutine VELL. However, in the present case, the problem under consideration has a closed form outer flow solution given by Eq. (3.8.4). This is the potential flow about an ellipse at an angle of attack. If the outer flow is given in terms of experimental data, then an appropriate READ statement in the subroutine VELL and more data cards are necessary.

Input

There are five READ statements, all at the beginning of the main program. Each READ statement corresponds to one data card and the format is specified in the description below:

1st READ Statement
Read the following 4 real numbers and 2 integers in 4F10.6 and 2I5 format respectively.

DY   Initial $\Delta\eta$ spacing. In case of constant mesh size in the η-direction, DY is the grid spacing.

DXO   Axial distance spacing, $\Delta\xi$, or Δx. Unsteady calculations march in constant ξ-increments, $\Delta\xi = $ DXO. Notice that in the program $\Delta\xi$ is denoted by DS.

EPS   The inverse square root of the Reynolds number, necessary only for higher-order boundary layer calculations.

CRNB   Weight of implicit differencing in the x-direction. For example:
$= 1$, purely implicit differencing
$= 1/2$, Crank Nickolson scheme. This quantity may take any value between 0 and 1.

IE   Number of grid points in the η-direction.

IEND   Number of grid points in the ξ-direction.

IELD   Same as IEND but for asymptotic initial solution

2nd READ Statement
Read 6 real numbers in 6F10.6 format.

UO dU_e/dx at $x = 0$, the gradient of the outer flow velocity at the origin.

XJFAC   Controls two-dimensional or axisymmetric calculations
$= 0$, two-dimensional
$= 1$, axisymmetric

XNFAC   Power of outer flow velocity variation with axial distance at stagnation. For example,
$= 1$, stagnation flow
$= 0$, flat plate

XTCFO   Transverse curvature factor for higher order calculations
VW   Normal velocity at the wall, i.e., rate of blowing.
XPRI   Controls printing in the ξ-direction; velocity profiles are
           printed every XPRI space steps.

3rd READ Statement

Read in F10.6 format, YK a distance related to the angle of attack of a parabola. This is not related to the present problem.

4th READ Statement

Read in I5 format.

NITO   The minimum number of iterations that the program will go
           through regardless of convergence criteria.

Glossary and output

Some of the basic variables used in the program are

X x, physical downstream distance
DX Δx, x increment
XI ξ, transformed downstream distance
DS $\Delta\xi$, ξ increment
N Y, vertical coordinate
XN η, stretched variable
K, N k, m running indices in the x and y direction
                        respectively
F1(N) $F''_{k-1,m}$ the x-component of velocity at the station
                        $k - 1$.
F2(N) $F''_{k,m}$ the x-component of velocity at the station k
VC(N) $V''_{k,m}$ the y-component of velocity, V defined by
                        (3.6.33)
F1N(N) $\partial F/\partial\eta$, velocity derivative
F1NN(N) $\partial^2 F/\partial\eta^2$, second velocity derivative
A1(N), A2(N), . . .   Coefficients $\alpha_1, \alpha_2, \ldots$ of Eq. (3.6.42)

A typical output follows the program listing. The results printed correspond to the upper region and angular positions $\theta = 0°$ to $13°$. This is an impulsively started motion and therefore the outer flow is independent of time immediately after the beginning of the motion. Computations can thus be marched in increments of x, ξ, or θ. At each station the following quantities are given:

S x, physical distance along the ellipse measured from the point of
           stagnation
XI ξ, transformed coordinate
DEG θ, angular distance from the point of stagnation
BE β, pressure gradient parameter
A1 $\partial U/\partial\eta$ at $y = 0$, the velocity gradient at the wall

UE	U_e, outer flow velocity
KK	number of grid points within reversed flow region
DT	time increment
IK	number of grid points for zero net flow
NIT	total number of iterations required for convergence

At selected stations, various profiles are printed corresponding to the station that precedes this information. These profiles are

XN(N)	η, stretched distance from the wall
Y(N)	y, physical distance from the wall
F2N(N)	$\partial U/\partial \eta$, velocity gradient
VC(N)	V, modified velocity component in y-direction
UVEL	u, velocity component in x-direction
DU	$\partial u/\partial x$, the velocity gradient in the axial direction
VE	v, velocity component in the y-direction

```
      IMPLICIT REAL*8 (A-H, O-Z)
      COMMON /PECS/   DS,DN(151),IM,IE,A1(151),A2(151),A3(151),A4(151),
     1                XN(152),FO(181,151)
      COMMCN/GECME/X,DX,RO,UE,BETA,XI,XI2,PNC,PNC2,CPL,CPL2,
     1              CPT,CPT2,EPS,Z,XJFAC
      COMMON /VELL/ YK,T,T1,A,C,DEG,DT1
      DIMENSION DCIP(151),CIPP(151)
      DIMENSION H(151),HN(151),H2(151),HN2(151),B(151),ROR(151),
     1          ROH(151),ROHN(151),VC(151),FC(151),FCP(151),H1(151),
     2          CI(151),CI1(151),CI2(151),CIO(151),CIP(151),F2(151),
     3          F2N(151),F1(151),F1N(151),F1NN(151),F2NN(151),
     4          Y(151),Y2(151),R(151),YA(151),FERF(151),XA(151),DF(151)
      DIMENSION DU(151) , VE(151)
      READ(5,650)DY,EPS,CRNB,IE,IEND,IFLD
  650 FORMAT(3F10.6,3I5)
      READ(5,950)UO,XJFAC,XNFAC,XTCFO,VW,XPRI
  950 FORMAT(6F10.6)
      READ(5,5999) YK
 5999 FORMAT(F10.6)
      READ(5,7999) NITO
 7999 FORMAT(I5)
C     RATIO OF ELLIPSE AXES=2.0
C     ANGLE OF ATTACK 54 DEGREES
      A=0.3D0*3.14159DC
      C=0.549D0
      WRITE(6,902)EPS,CRNB,XJFAC,XNFAC
  902 FORMAT(2X,4HEPS=,F10.6,2X,5HCRNI=,F10.6,2X,2HHJ=,F10.6,2X,2HHN=,
     1F10.6//)
      UO=1.0D0/(1.0D0+YK**2 )
      WRITE(6,6999) UO,YK
 6999 FORMAT(2X,3HUO=,F10.6,2X,2HK=,F10.6//)
C     WRITE(6,6999) UC,YK,DX
C6999 FORMAT(2X,3HUO=,F10.6,2X,2HK=,F10.6,2X,3HDX=,F10.6//)
      NIT=0
      T=0.0D0
      DT=0.06D0
      IM=IE-1
 8000 PNC=0.0D0
      DT1=-3.14159D0/180.0D0
      XPR=0.0D0
      XKAP=1.0D0/(1.0D0+YK**2)**1.5D0
      XKAP2=1.0D0/(1.0D0+YK**2)**1.5D0
```

```
      XD=DABS(1.0D0-XNFAC)
      IF (XD.GT.0.001D0) GO TO 5
      PNC=DSQRT(1.0D0/(U0*(1.0D0+XJFAC)))
  5   BETA=2.0D0*XNFAC/(XNFAC+2.0D0*XJFAC+1.0D0)
      PNC2=PNC
      CPT=EPS*XJFAC*XTCF0
      CPT2=EPS*XJFAC*XTCF0
      CPL=EPS*XKAP
      CPL2=EPS*XKAP2
      DS=1.0D0
      CCRNB=1.0D0-CRNB
      E1=0.0D0
      FF=0.00D0
      T1=0.0D0
      DEG=0.0D0
      X=0.0D0
      XI=0.0D0
      XI2=2.0D0*XI
      UE=0.0D0
      R0=0.0D0
      Z=0.0D0
      UERO2=0.0DC
      XN(1)=0.0D0
      DN(1)=DY
      XN(2)=XN(1)+DN(1)
      Y(1)=PNC*XN(1)
      Y2(1)=Y(1)
      H(1)=1.0D0
      H1(1)=1.0D0
      H2(1)=1.0D0
      HN(1)=0.0D0
      HN2(1)=0.0CC
      P(1)=1.0D0
      ROR(1)=1.0CC
      ROH(1)=1.0DC
      ROHN(1)=0.0D0
      VC(1)=VW-XN(1)
      F1(1)=1.0C0
      F1N(1)=0.0DC
      F1NN(1)=0.0D0
      F2(1)=1.0D0
      F2N(1)=0.0D0
      F2NN(1)=0.0C0
      FC(1)=1.0D0
      FCP(1)=0.0D0
      CIC(1)=0.CDC
      CI(1)=0.0D0
      CI1(1)=0.0D0
      CI2(1)=0.0CC
      CIP(1)=0.0D0
      CIPP(1)=0.0D0
      R(1)=BETA
      DO 166 N=2,IE
      DN(N)=1.04D0*DN(N-1)
      XN(N+1)=XN(N)+DN(N)
      Y(N)=PNC*XN(N)
      Y2(N)=Y(N)
      H(N)=1.0D0
      H1(N)=1.0D0
      H2(N)=1.0D0
      HN(N)=0.0D0
      HN2(N)=0.CDC
      R(N)=1.0D0
      ROR(N)=1.0D0
      ROH(N)=1.0D0
      ROHN(N)=0.0D0
      VC(N)=VW-XN(N)
      F1(N)=1.0D0
```

```
      F1N(N)=0.0D0
      F1NN(N)=0.CC0
      F2(N)=1.0D0
      F2N(N)=0.0D0
      F2NN(N)=0.0D0
      FC(N)=1.0D0
      FCP(N)=0.0D0
      CIO(N)=0.0D0
      CI(N)=0.0D0
      CI1(N)=0.0D0
      CI2(N)=0.0D0
      CIP(N)=0.0D0
      CIPP(N)=0.0D0
      R(N)=BETA
 166  CONTINUE
1001  IF(T1.EQ.0.0D0) GO TO 174
      T=T+0.001D0
      DO 268 K=2,IELD
      DO 266 N=1,IE
      YA(N)=0.5D0*DSQRT(2.0D0*XI/T)*XN(N)/UE
      FO(K,N)=DERF(YA(N))
 266  CONTINUE
      T1=T1+DT1
      CALL GEOM(DS,XTCFO,CCRNI)
      IF(DEG.LT.-179.0D0) IELD=K
      IF(DEG.LT.-179.0D0) GO TO 268
 268  CONTINUE
      DO 269 N=1,IE
      FO(K+1,N)=2.0D0*FO(K,N)-FO(K-1,N)
 269  CONTINUE
      T=T+DT
      GO TO 8000
 174  CONTINUE
 541  CONTINUE
      F2(1)=0.0D0
      F1(1)=0.0D0
      FC(1)=0.0D0
      DO 1 K=1,IEND
 400  CONTINUE
      Y2(1)=0.0D0
      Y(1)=0.0D0
      YIE=Y(IE)
      DO 500 N=2,IE
      Y(N)=Y(N-1)+PNC*((1.D0+CPL*Y(N))/(1.D0+CPT*Y(N))+(1.D0+CPL*Y(N-1))
     1/(1.D0+CPT*Y(N-1)))*DN(N-1)*0.5D0
 500  Y2(N)=Y2(N-1)+PNC2*((1.D0+CPL2*Y2(N))/(1.DC+CPT2*Y2(N))+(1.D0+CPL2
     1*Y2(N-1))/(1.D0+CPT2*Y2(N-1)))*DN(N-1)*0.5D0
      CFAC=DABS(YIE-Y(IE))
      IF(CFAC.GT..00001D0)GO TO 400
      DO 601 N=1,IE
      H(N)=1.0D0+CPL*Y(N)
      H2(N)=1.0DC+CPL2*Y2(N)
      IF (X.GE.0.0001) GO TO 550
      H1(N)=H2(N)
 550  CONTINUE
      R(N)=1.0D0+CPT*Y(N)
      ROH(N)=(R(N)/H(N))**2
      ROR(N)=1.0D0/R(N)**2
 601  CONTINUE
      DO 600 N=2,IM
      HN(N)=(DN(N-1)*  H(N+1)/DN(N)-DN(N)*  H(N-1)/DN(N-1))/(DN(N)+
     1 DN(N-1))+(DN(N)-DN(N-1))*  H(N)/(DN(N)*DN(N-1))
      HN2(N)=(DN(N-1)* H2(N+1)/DN(N)-DN(N)* H2(N-1)/DN(N-1))/(DN(N)+
     1 DN(N-1))+(DN(N)-DN(N-1))* H2(N)/(DN(N)*DN(N-1))
      B(N)=BETA-XI2*(H2(N)-H1(N))/(H(N)*DS)
      ROHN(N)=(DN(N-1)*ROH(N+1)/DN(N)-DN(N)*ROH(N-1)/DN(N-1))/(DN(N)+
     1 DN(N-1))+(DN(N)-DN(N-1))*ROH(N)/(DN(N)*DN(N-1))
```

```
  600 ROHN(N)=ROHN(N)/ROH(N)
       HN(1)=- H(1)*(DN(2)+2.0D0*DN(1))/(DN(1)*(DN(2)+DN(1)))
      1       + H(2)*(DN(2)+DN(1))/(DN(2)*DN(1))
      2       - H(3)*DN(1)/(DN(2)*(DN(1)+DN(2)))
      HN2(1)=-H2(1)*(DN(2)+2.0D0*DN(1))/(DN(1)*(DN(2)+DN(1)))
      1       +H2(2)*(DN(2)+DN(1))/(DN(2)*DN(1))
      2       -H2(3)*DN(1)/(DN(2)*(DN(1)+DN(2)))
      HN2(IE)=H2(IE)*(DN(IM-1)+2.0D0*DN(IM))/(DN(IM)*(DN(IM)+DN(IM-1)))
      1        -H2(IE-1)*(DN(IM-1)+DN(IM))/(DN(IM)*DN(IM-1))
      2        +H2(IE-2)*DN(IM)/(DN(IM-1)*(DN(IM)+DN(IM-1)))
       F2N(1)=1.0D0
 1000 F2N1=F2N(1)
       F2N20=F2N(20)
       F2N40=F2N(40)
       F2N60=F2N(60)
       TS=UE
       DO 100 N=2,IM
       A1(N)=-VC(N)
       A2(N)=-B(N)*FC(N)
       A3(N)=B(N)
       IF(T1.EQ.0.0D0) GO TO 1112
       IF(T.GT.0.0D0) A2(N)=A2(N)-XI2/((TS**2)*DT)
       IF(T.GT.0.0D0) A3(N)=A3(N)+XI2*FC(K,N)/((TS**2)*DT)
 1112 A4(N)=-XI2*FC(N)
  100 IF(F2(N).LT.0.0D0) A4(N)=A4(N)+2.0D0*XI2*FC(N)
       IF(T1.LT.-0.0001D0) GO TO 130
       CRNI=1.0D0
       CCRNI=0.0D0
       GO TO 140
  130 CONTINUE
       CRNI=CRNB
       CCRNI=CCRNB
  140 CONTINUE
       CALL PEQSC(F1NN,F1N,F1,F2NN,F2N,F2,F1,FF,CRNI,T,K)
C      DIF=DABS(F2N(1)-F2N1)
       DIF1=DABS(F2N(1)-F2N1)
       DIF20=DABS(F2N(20)-F2N20)
       DIF40=DABS(F2N(40)-F2N40)
       DIF60=DABS(F2N(60)-F2N60)
       DO 150 N=1,IE
       FCP(N)=(F2(N)-F1(N))/DS
       IF(T.GT.0.0D0) FCP(N)=(FU(K+1,N)-FC(K,N))/(2.0D0*DS)+FCP(N)/2.0D0
  150 FC(N)=F2(N)*CRNI+F1(N)*CCRNI
       IF (T1.LE.-0.0001D0) GO TO 151
       DO 160 N=1,IE
       FCP(N)=0.0D0
       F1(N)=F2(N)
       FC(N)=F2(N)
       F1N(N)=F2N(N)
  160 F1NN(N)=F2NN(N)
  151 CONTINUE
       VC(1)=VW
       CI0(1)=0.D0
       DETA=0.0D0
       DO 200 N=2,IE
       DETA=DETA+(1.0D0-(FC(N)+FC(N-1))*0.5D0)*DN(N-1)
       VC(N)=VC(N-1)-(XI2*(FCP(N)+FCP(N-1))+FC(N)+FC(N-1))*DN(N-1)*.5D0
  200 CI0(N)=CI0(N-1)+(HN2(N)*(1.D0-F2(N)*F2(N))/(H2(N)**3)+HN2(N-1)*(1.
      1D0-F2(N-1)*F2(N-1))/(H2(N-1)**3))*DN(N-1)*0.5D0
       CIP1=CIP(1)
       DO 300 N=1,IE
       CI2(N)=(CI0(IE)-CI0(N))*H2(N)**2
       CI(N)=CI2(N)*CRNI+CI1(N)*CCRNI
  300 CIP(N)=(CI2(N)-CI1(N))/DS
       IF (T1.LE.-0.0001D0) GO TO 210
       DO 211 N=1,IF
       CI1(N)=CI2(N)
```

```
          CI(N)=CI2(N)
    211   CIP(N)=0.0D0
    210 CONTINUE
          CIP(N)=(CIP(N)+CIPP(N))/2.0D0
          NIT=NIT+1
          IF(NIT.LT.100) GO TO 220
C         IEND=K-1
C         K=IEND
C         GO TO 550C
          IF(NIT.GT.100) IEND=K
          IF(NIT.GT.100) GO TO 5500
    220 IF(NITO.GT.0) GO TO 20
          IF(T.LT.5.40D0) GO TO 555
          WRITE(6,2200) T1,XI,X,F2N(1),T
   2200 FORMAT(10X,5F15.6/)
C         WRITE(6,2200) F2N(1),A4(1),A4(2),A4(3)
C2200 FORMAT(10X,4F15.6/)
C         WRITE(6,2200) F2N(1),CI(1),CIP(1)
C2200 FORMAT(10X,3F15.6/)
C         IF (DIF.GT.0.0001D0) GO TO 1000
    555 IF(DIF1.GT.0.001D0) GO TO 1000
          IF(DIF20.GT.0.001D0) GO TO 1000
          IF(DIF40.GT.0.001D0) GO TO 1000
          IF(DIF60.GT.0.001D0) GO TO 1000
C         IF(X.LT.0.0001) GO TO 30
C         CDIF=DABS(CIP(1)/CIP1-1.0D0)
C         IF(CDIF.GT.0.001D0) GO TO 1000
C         GO TO 30
     20 IF(NIT.LT.NITO) GO TO 1000
     30 A1B=CRNI*F2N(1)+CCRNI*F1N(1)
          KDT=1
    943 FDIS=DETA-XN(KDT)
          IF (FDIS) 940,940,941
    941 KDT=KDT+1
          IF(KDT.GE.100) GO TO 5500
          GO TO 943
    940 DISP=Y(KDT-1)+(DETA-XN(KDT-1))*(Y(KDT)-Y(KDT-1))/DN(KDT-1)
          PFACW=1.0D0-2.0D0*CI(1)
          IKM=IE
          IKL=IE
          DOTA=0.0D0
          DITA=0.0D0
          DO 261 N=2,IKL
          IF(F2(N).GT.0.0D0) IKL=N
          IF(F2(N).GT.0.0D0) GO TO 261
          XA(N)=(XN(N+1)-XN(N-1))/2.0D0
          DF(N)=-XA(N)*F2(N)
          DITA=DITA+F2(N)
    261 CONTINUE
          DO 262 N=IKL,IKM
          IF(DOTA.GT.DITA) IKM=N
          IF(DOTA.GT.DITA) GO TO 262
          XA(N)=(XN(N+1)-XN(N-1))/2.0D0
          DF(N)=XA(N)*F2(N)
          DOTA=DOTA+DF(N)
    262 CONTINUE
          KK=IKM
C         CF=2.0D0*UE*A1B/PNC
C         VWU=VC(1)/PNC
C         IF(T.LT.0.8D0) GO TO 999
          WRITE(6,901)X,XI,DEG,T,BETA,A1B,UE,KK,DETA,IKL,NIT
    901 FORMAT(2X,2HS=,F8.5,1X,3HXI=,F8.5,1X,4HDEG=,F8.3,1X,2HT=,F8.5,1X,
         13HBE=,F9.5,1X,3HA1=,F8.5,1X,3HUE=,F8.5,1X,3HKK=,I3,1X,3HDT=,
          2F8.5,1X,3HIK=,I3,2X,4HNIT=,I2//)
C         WRITE(6,903)CF,VWU
C 903 FORMAT(2X,3HCF=,F8.5,1X,4HVWU=,F8.5//)
          PRIN=XPR-XI
```

```
C       IF(DEG.GT.0.376D0.AND.T.GT.0.0D0) PRIN=C.0D0
C       IF(T.GT.0.0D0) PRIN=0.0D0
        IF (PRIN.GE.0.0001D0) GC TO 999
        XPR=XPR+XPRI
        WRITE(6,695)
  695 FORMAT(9X,5HXN(N)10X,4HY(N),11X,6HF2N(N),10X,5HVC(N),10X,4HUVEL,
     111X,2HDU,10X,2HVE)
        DO 700 N=1,IE
        FH=FC(N)/H(N)
        PFAC=(1.0D0-2.0D0*CI(N))/(H(N)*H(N))
        IF(XI.GT..1D0) GC TO 111
        DU(N)=0.0D0
        VE(N)=0.0D0
        GO TO 788
  111 CONTINUE
        DU(N)=(1.D0/PNC**2)*BETA*F2(N)+(UE**2)*FCP(N)
     1+(1.D0/PNC**2)*(BETA-2.D0)*XN(N)*F2N(N)
        VE(N)=VC(N)/PNC-XN(N)/PNC*(BETA-1.D0)*F2(N)
  788 CONTINUE
C 700 CONTINUE
  700 WRITE(6,900) XN(N),Y(N),F2N(N),VC(N),FH,DU(N),VE(N)
  900 FORMAT(7F15.6)
        WRITE(6,980)
  980 FORMAT(1X,/)
C       WRITE(6,903)F2N(1),F2N(21),F2N(51),F2N(81),F2N(IE),VE(11),VE(21),
C     1VE(51),F2(81),F2(IE)
C 903 FORMAT(2X,4HFN1=,F8.5,1X,4HFN2=,F8.5,1X,4HFN5=,F8.5,1X,
C     14HFN8=,F8.5,1X,4HFNE=,F8.5,1X,4HVE1=,F8.5,1X,4HVE2=,F8.5,1X,
C     24HVE5=,F8.5,1X,4HF28=,F8.5,1X,4HF2E=,F8.5//)
  999 CONTINUE
        ABST=DABS(F2(IE)-F2(IE-4))
        IF (ABST.LE.C.0001D0) GO TO 997
        IE=IE+1
        IM=IE-1
        IF (IE.GE.200) GO TO 5000
        DN(IE)=DY
        XN(IE)=XN(IE-1)+DN(IE-1)
        Y(IE)=2.0D0*Y(IE-1)-Y(IE-2)
        Y2(IE)=2.0D0*Y2(IE-1)-Y2(IE-2)
        H2(IE)=2.0D0*H2(IE-1)-H2(IE-2)
        HN2(IE)=2.0D0*HN2(IE-1)-HN2(IE-2)
        F2(IE)=1.0D0
        F2N(IE)=0.0D0
        F2NN(IE)=C.0D0
        FC(IE)=1.0D0
        FCP(IE)=0.0D0
        VC(IE)=2.0D0*VC(IE-1)-VC(IE-2)
        CI(IE)=0.0D0
        CI2(IE)=0.CD0
        CIP(IE)=0.0D0
  997 CONTINUE
C       IF(XI.GE.0.35D0) DS=C.001D0
C       IF(XI.GE.0.36D0) DS=0.0002D0
  998 T1=T1+DT1
        CALL GEOM(DS,XTCFO,CCRNI)
        DO 175 N=1,IE
        F1(N)=F2(N)
        F1N(N)=F2N(N)
        F1NN(N)=F2NN(N)
        CI1(N)=CI2(N)
        DCIP(N)=CIP(N)-CIPP(N)
        CIPP(N)=CIP(N)
        CIP(N)=CIP(N)+DCIP(N)
        H1(N)=H2(N)
        FO(K,N)=F2(N)
  175 CONTINUE
        IF(T.LT.0.04D0) GO TO 1001
```

```
      IF(DEG.LT.-179.00D0) IEND=K
5500  NIT=0
   1  CONTINUE
      K=IEND
      IAND=IEND+1
      IF(DEG.LT.-179.00D0) GO TO 131
      DO 666 N=1,IE
      FO(K,N)=2.0D0*FO(K-1,N)-FO(K-2,N)
 666  CONTINUE
      GO TO 921
 131  CONTINUE
      DO 663 N=1,IE
      FO(K+1,N)=2.0D0*FO(K,N)-FO(K-1,N)
 663  CONTINUE
 921  T=T+DT
      IF(T.LT.0.8D0) GO TO 9743
      WRITE(7,667) T,IAND,IEND
 667  FORMAT(1F10.6,2I5)
      DO 664 K=1,IAND
      WRITE(7,662) (FC(K,N),N=1,151)
 662  FORMAT(7F10.6)
 664  CONTINUE
      GO TO 5000
9743  CONTINUE
      GO TO 8000
5000  CONTINUE
      STOP
      END
      SUBROUTINE PEQSO(W1NN,W1N,W1,W2NN,W2N,W2,E1,F1,CRNI,T,K)
      IMPLICIT REAL*8 (A-H, O-Z)
      COMMON /PEQS/  DS,DN(151),IM,IE,A1(151),A2(151),A3(151),A4(151),
     1               XN(152),FO(181,151)
      DIMENSION W1NN(151),W1N(151),W1(151)
      DIMENSION E(151),F(151),W2NN(151),W2N(151),W2(151)
      E(1)=E1
      F(1)=F1
      DO 10 N=2,IM
      A=(2.0D0-A1(N)*DN(N))/(DN(N-1)*(DN(N)+DN(N-1)))*CRNI
      B=((-2.0D0+A1(N)*(DN(N)-DN(N-1)))/(DN(N)*DN(N-1))+A2(N))*CRNI
     1   +A4(N)/DS
      IF(W2(N).LT.0.0D0.AND.T.GT.0.0D0) B=B-A4(N)/(2.0D0*DS)
      C=(2.0D0+A1(N)*DN(N-1))/(DN(N)*(DN(N)+DN(N-1)))*CRNI
      D=-(W1NN(N)+A1(N)*W1N(N)+A2(N)*W1(N))*(1.0D0-CRNI)
     1   -A3(N)+A4(N)*W1(N)/DS
      IF(W2(N).LT.0.0D0.AND.T.GT.0.0D0) D=D-A4(N)*W1(N)/(2.0D0*DS)
     1-A4(N)*(FO(K+1,N)-FO(K,N))/(2.0D0*DS)
      E(N)=-C/(B+A*E(N-1))
  10  F(N)=(D-A*F(N-1))/(B+A*E(N-1))
      W2(IE)=1.0D0
      KON=IM
      DO 20 N=2,IE
      W2(KON)=E(KON)*W2(KON+1)+F(KON)
  20  KON=KON-1
      DO 30 N=2,IM
      W2NN(N)=2.0D0*(W2(N+1)/DN(N)+W2(N-1)/DN(N-1))/(DN(N)+DN(N-1))
     1        -2.0D0*W2(N)/(DN(N)*DN(N-1))
  30  W2N(N)=(DN(N-1)/DN(N)*W2(N+1)/DN(N)-DN(N)*W2(N-1)/DN(N-1))/(DN(N)+
     1        DN(N-1))+(DN(N)-DN(N-1))*W2(N)/(DN(N)*DN(N-1))
      W2N(1)=-W2(1)*(DN(2)+2.0D0*DN(1))/(DN(1)*(DN(2)+DN(1)))
     1       +W2(2)*(DN(2)+DN(1))/(DN(2)*DN(1))
     2       -W2(3)*DN(1)/(DN(2)*(DN(1)+DN(2)))
      W2N(IE)=W2(IE)*(DN(IM-1)+2.0D0*DN(IM))/(DN(IM)*(DN(IM)+DN(IM-1)))
     1       -W2(IE-1)*(DN(IM-1)+DN(IM))/(DN(IM)*DN(IM-1))
     2       +W2(IE-2)*DN(IM)/(DN(IM-1)*(DN(IM)+DN(IM-1)))
      W2NN(1)=-W2N(1)*(DN(2)+2.0D0*DN(1))/(DN(1)*(DN(2)+DN(1)))
     1       +W2N(2)*(DN(2)+DN(1))/(DN(2)*DN(1))
     2       -W2N(3)*DN(1)/(DN(2)*(DN(1)+DN(2)))
      W2NN(IE)=W2N(IE)*(DN(IM-1)+2.0D0*DN(IM))/(DN(IM)*(DN(IM)+DN(IM-1)))
```

```
   1              -W2N(IE-1)*(DN(IM-1)+DN(IM))/(DN(IM)*DN(IM-1))
   2              +W2N(IE-2)*DN(IM)/(DN(IM-1)*(DN(IM)+DN(IM-1)))
   RETURN
   END
   SUBROUTINE GEOM(DS,XTCFO,CCRNI)
   IMPLICIT REAL*8 (A-H, O-Z)
   COMMON/GEOME/X,DX,RO,UE,BETA,XI,XI2,PNC,PNC2,CPL,CPL2,
   1              CPT,CPT2,EPS,Z,XJFAC
   COMMON /VELL/ YK,T,T1,A,C,DEG,DT1
   IF (RO.GE.0.0001D0) GO TO 10
   UERO2=0.0D0
10 CONTINUE
   CALL VEL(X,DX,UE,DUEDS,RO,XCOS,XKAP,Z)
   RP=RO
   HA=X-DX*0.5D0
   CALL VEL(HA,DXHA,UEHA,DUEHA,ROHA,XCCHA,XKHA,ZHA)
   X2=X+DX*CCRNI
   CALL VEL(X2,DX2,UE2,DUE2,RO2,XCCS2,XKAP2,Z2)
   X2M=X+DX*CCRNI*0.5D0
   CALL VEL(X2M,DXX2M,UEX2M,DUEM,ROX2M,XCOM,XKAM,Z2M)
   CPL=EPS*XKAP
   CPL2=EPS*XKAP2
   IF (XJFAC.LT.0.001D0) GO TO 20
   CPT=EPS*XCOS*XTCFO/RO
   CPT2=EPS*XCCS2*XTCFO/RO2
   GO TO 30
20 CONTINUE
   CPT=0.0D0
   CPT2=0.0D0
   RP=1.0D0
   RO2=1.0D0
   ROHA=1.0D0
   ROX2M=1.0D0
30 CONTINUE
   DS=(UERO2+4.0D0*UEHA*ROHA**2+UE*RP**2)*DX/6.0D0
   XI=XI+DS
   XI2=2.0D0*XI
   UERO2=UE*RP**2
   DXI2=(UERO2+4.0D0*UEX2M*ROX2M**2+UE2*RO2**2)*DX*CCRNI/6.0D0
   XI22=XI+DXI2
   XI222=2.0D0*XI22
   PNC=DSQRT(XI2)/(UE*RP)
   PNC2=DSQRT(XI222)/(UE2*RO2)
   BETA=XI2*DUEDS/(UE*RP)**2
   RETURN
   END
   SUBROUTINE VEL(X,DX,UE,DUEDS,RO,XCOS,XKAP,Z)
   IMPLICIT REAL*8 (A-H, O-Z)
   COMMON /VELL/ YK,T,T1,A,C,DEG,DT1
   DEG=T1*180.0D0/3.14159D0
   UE=-DSIN(T1)*DEXP(C)/DSQRT(DSINH(C)**2+DSIN(T1-A)**2)
   DUEDS=DEXP(C)*(DCOS(T1)/DSQRT(DSINH(C)**2+DSIN(T1-A)**2)
  1-DSIN(T1)*DSIN(T1-A)*DCOS(T1-A)/DSQRT(DSINH(C)**2+DSIN(T1-A)**2)
  2**3)*0.5DC*(1.0D0+2.0D0)/DSQRT((2.0D0*DSIN(T1-A))**2+DCOS(T1-A)
  3**2)
   DX=-2.0D0*DT1*DSQRT((2.0D0*DSIN(T1-A))**2+DCOS(T1-A)**2)/(1.0D0+
  12.0D0)
   X=X+DX
   RO=1.0D0
   Z=1.0D0
   XCOS=1.0D0/(1.0D0+1.0D0/2.0D0/Z)**0.5D0
   XKAP=1.0D0/(1.0D0+2.0D0*Z)**1.5D0
   RETURN
   END
```

EPS= 0.0   CRNI= 1.000000   J= 0.0   N=0.0

UO= 1.000000   K= 0.0

S= 0.0   X1= 0.0   DEG= 0.0   T= 0.0   BE= 0.0   A1=
0.4700

KK= 3   DT= 1.21510   IK= 2   NIT= 7

XN(N)	Y(N)	F2N(N)	VC(N)	UVEL
0.0	0.0	0.470080	0.0	0.0
0.002000	0.0	0.470080	-0.000001	0.000940
0.004080	0.0	0.470080	-0.000004	0.001918
0.006243	0.0	0.470080	-0.000009	0.002935
0.008493	0.0	0.470080	-0.000017	0.003992
0.010833	0.0	0.470080	-0.000028	0.005092
0.013266	0.0	0.470080	-0.000041	0.006236
0.015797	0.0	0.470080	-0.000059	0.007426
0.018428	0.0	0.470080	-0.000080	0.008663
0.021166	0.0	0.470080	-0.000105	0.009950
0.024012	0.0	0.470080	-0.000136	0.011288
0.026973	0.0	0.470080	-0.000171	0.012679
0.030052	0.0	0.470079	-0.000212	0.014127
0.033254	0.0	0.470079	-0.000260	0.015632
0.036584	0.0	0.470079	-0.000315	0.017197
0.040047	0.0	0.470078	-0.000377	0.018825
0.043649	0.0	0.470077	-0.000448	0.020519
0.047395	0.0	0.470076	-0.000528	0.022279
0.051291	0.0	0.470075	-0.000618	0.024111
0.055342	0.0	0.470074	-0.000720	0.026015
0.059556	0.0	0.470073	-0.000834	0.027996
0.063938	0.0	0.470071	-0.000961	0.030056
0.068496	0.0	0.470069	-0.001103	0.032198
0.073236	0.0	0.470066	-0.001261	0.034426
0.078165	0.0	0.470063	-0.001436	0.036744
0.083292	0.0	0.470059	-0.001631	0.039153
0.088623	0.0	0.470055	-0.001846	0.041660
0.094168	0.0	0.470050	-0.002084	0.044266
0.099935	0.0	0.470044	-0.002347	0.046977
0.105933	0.0	0.470037	-0.002638	0.049796
0.112170	0.0	0.470028	-0.002957	0.052727
0.118657	0.0	0.470019	-0.003309	0.055776
0.125403	0.0	0.470008	-0.003696	0.058947
0.132419	0.0	0.469995	-0.004121	0.062245
0.139716	0.0	0.469980	-0.004588	0.065674
0.147304	0.0	0.469963	-0.005100	0.069241
0.155197	0.0	0.469943	-0.005661	0.072950
0.163404	0.0	0.469920	-0.006276	0.076807
0.171941	0.0	0.469893	-0.006948	0.080818
0.180818	0.0	0.469862	-0.007684	0.084989
0.190051	0.0	0.469827	-0.008089	0.089327
0.199653	0.0	0.469787	-0.009368	0.093838
0.209639	0.0	0.469741	-0.010329	0.098530
0.220025	0.0	0.469688	-0.011378	0.103408
0.230826	0.0	0.469688	-0.012522	0.108481
0.242059	0.0	0.469627	-0.013770	0.113756
0.253741	0.0	0.469478	-0.015131	0.119241
0.265891	0.0	0.469388	-0.016614	0.124944
0.278526	0.0	0.469284	-0.018231	0.130874
0.291667	0.0	0.469167	-0.019991	0.137041
0.305334	0.0	0.469032	-0.021908	0.143452
0.319548	0.0	0.468879	-0.023994	0.150117
0.334329	0.0	0.468705	-0.026264	0.157047
0.349703	0.0	0.468705	-0.028734	0.164251
0.365691	0.0	0.468281	-0.031420	0.171740
0.382318	0.0	0.468025	-0.034340	0.179524
0.399611	0.0	0.467734	-0.037515	0.187615

0.417596	0.0	0.467404	-0.040964	0.196024
0.436299	0.0	0.467030	-0.044712	0.204763
0.455751	0.0	0.466605	-0.048784	0.213844
0.475981	0.0	0.466123	-0.053205	0.223279
0.497021	0.0	0.465578	-0.058006	0.223080
0.518901	0.0	0.464961	-0.063217	0.243261
0.541658	0.0	0.464262	-0.068873	0.253834
0.565324	0.0	0.463472	-0.075011	0.264812
0.589937	0.0	0.462578	-0.081669	0.276209
0.615534	0.0	0.461569	-0.088890	0.288038
0.642156	0.0	0.460429	-0.096722	0.300311
0.669842	0.0	0.459143	-0.105212	0.313041
0.698635	0.0	0.457693	-0.114416	0.326241
0.728581	0.0	0.456058	-0.124390	0.339924
0.759724	0.0	0.454218	-0.135197	0.354099
0.792113	0.0	0.452146	-0.146904	0.368779
0.825798	0.0	0.449817	-0.159582	0.383971
0.860830	0.0	0.447200	-0.173308	0.399685
0.897263	0.0	0.444264	-0.188166	0.415927
0.935153	0.0	0.440972	-0.204244	0.432700
0.974559	0.0	0.437285	-0.221636	0.450007
1.015542	0.0	0.433163	-0.240444	0.467847
1.058163	0.0	0.428558	-0.260775	0.486214
1.102490	0.0	0.423425	-0.282746	0.505101
1.148590	0.0	0.417711	-0.306478	0.524493
1.196533	0.0	0.411363	-0.332101	0.544372
1.246934	0.0	0.404327	-0.359751	0.564714
1.298250	0.0	0.396545	-0.389573	0.585485
1.352180	0.0	0.387962	-0.421719	0.606646
1.408267	0.0	0.378524	-0.456347	0.628148
1.466598	0.0	0.368179	-0.493623	0.649933
1.527262	0.0	0.356883	-0.533718	0.671934
1.590353	0.0	0.344597	-0.576808	0.694071
1.655967	0.0	0.331296	-0.623077	0.716253
1.724205	0.0	0.316970	-0.672708	0.738379
1.795174	0.0	0.301626	-0.725889	0.760337
1.868980	0.0	0.285295	-0.782806	0.782002
1.945470	0.0	0.268035	-0.843647	0.803243
2.025569	0.0	0.249936	-0.908595	0.823919
2.108592	0.0	0.231118	-0.977828	0.843886
2.194936	0.0	0.211739	-1.051518	0.862999
2.284733	0.0	0.191993	-1.129826	0.881114
2.378123	0.0	0.172107	-1.212906	0.898098
2.475247	0.0	0.152337	-1.300897	0.913829
2.576257	0.0	0.132959	-1.393929	0.928205
2.681308	0.0	0.114261	-1.492117	0.941149
2.790560	0.0	0.096527	-1.595566	0.952615
2.904182	0.0	0.080022	-1.704371	0.962588
3.022350	0.0	0.064976	-1.818620	0.971091
3.145244	0.0	0.051568	-1.938397	0.978183
3.273053	0.0	0.039912	-2.063787	0.983958
3.405975	0.0	0.030051	-2.194881	0.988535
3.544214	0.0	0.021951	-2.331779	0.992059
3.687983	0.0	0.015511	-2.474595	0.994685
3.837502	0.0	0.010567	-2.623461	0.996575
3.993002	0.0	0.006917	-2.778530	0.997883
4.154723	0.0	0.004332	-2.939978	0.998751
4.322911	0.0	0.002585	-3.108003	0.999299
4.497828	0.0	0.001462	-3.282825	0.999628
4.679741	0.0	0.000780	-3.464688	0.999815
4.868931	0.0	0.000389	-3.653852	0.999914
5.065688	0.0	0.000181	-3.850597	0.999963
5.270315	0.0	0.000077	-4.055219	0.999985
5.483128	0.0	0.000030	-4.268030	0.999995
5.704453	0.0	0.000011	-4.489354	0.999998
5.934631	0.0	0.000003	-4.719532	1.000000
6.174017	0.0	0.000001	-4.958917	1.000000
6.422977	0.0	0.000000	-5.207878	1.000000

6.681896	0.0	0.000000	-5.466797	1.000000
6.951172	0.0	0.000000	-5.736073	1.000000
7.231219	0.0	0.000000	-6.016120	1.000000
7.522468	0.0	0.000000	-6.307368	1.000000
7.825366	0.0	0.000000	-6.610267	1.000000
8.140381	0.0	0.000000	-6.925282	1.000000
8.467996	0.0	0.000000	-7.252897	1.000000
8.808716	0.0	0.000000	-7.593617	1.000000
9.163065	0.0	0.000000	-7.947966	1.000000
9.531588	0.0	0.000000	-8.316488	1.000000
9.914851	0.0	0.000000	-8.699752	1.000000
10.313445	0.0	0.000000	-9.098346	1.000000
10.727983	0.0	0.000000	-9.512883	1.000000
11.159102	0.0	0.000000	-9.944003	1.000000
11.607466	0.0	0.000000	-10.392367	1.000000
12.073765	0.0	0.000000	-10.858666	1.000000
12.558715	0.0	0.000000	-11.343616	1.000000
13.063064	0.0	0.000000	-11.848666	1.000000
13.587587	0.0	0.000000	-12.372487	1.000000
14.133090	0.0	0.000000	-12.917991	1.000000
14.700414	0.0	0.000000	-13.485314	1.000000
15.290430	0.0	0.000000	-14.075331	1.000000
15.904048	0.0	0.000000	-14.688948	1.000000
16.542209	0.0	0.000000	-15.327110	1.000000
17.205898	0.0	0.000000	-15.990798	1.000000
17.896134	0.0	0.000000	-16.681034	1.000000

S= 0.0   X1= 0.0   DEG= 0.0   T= 0.06100   BE= 0.0   A1= 0.47

UE= 0.0   KK= 3   DT= 1.21510   IK= 2   NIT= 14

0.0	0.0	0.470080	0.0	0.0
0.002000	0.0	0.470080	-0.000001	0.000940
0.004080	0.0	0.470080	-0.000004	0.001918
0.006243	0.0	0.470080	-0.000009	0.002935
0.008493	0.0	0.470080	-0.000017	0.003992
0.010830	0.0	0.470080	-0.000028	0.005092
0.013266	0.0	0.470080	-0.000041	0.006236
0.015797	0.0	0.470080	-0.000059	0.007426
0.018428	0.0	0.470080	-0.000080	0.008663
0.021166	0.0	0.470080	-0.000105	0.009950
0.024012	0.0	0.470080	-0.000136	0.011288
0.026973	0.0	0.470080	-0.000171	0.012679
0.030052	0.0	0.470079	-0.000212	0.014127
0.033254	0.0	0.470079	-0.000260	0.015632
0.036584	0.0	0.470079	-0.000315	0.017197
0.040047	0.0	0.470078	-0.000377	0.018825
0.043649	0.0	0.470077	-0.000448	0.020519
0.047395	0.0	0.470076	-0.000528	0.022279
0.051291	0.0	0.470075	-0.000618	0.024111
0.055342	0.0	0.470074	-0.000720	0.026015
0.059556	0.0	0.470073	-0.000834	0.027996
0.063938	0.0	0.470071	-0.000961	0.030056
0.068496	0.0	0.470069	-0.001103	0.032198
0.073236	0.0	0.470066	-0.001261	0.034426
0.078165	0.0	0.470063	-0.001436	0.036744
0.083292	0.0	0.470059	-0.001631	0.039153
0.088623	0.0	0.470055	-0.001846	0.041660
0.094168	0.0	0.470050	-0.002084	0.044266
0.099935	0.0	0.470044	-0.002347	0.046977
0.105933	0.0	0.470037	-0.002638	0.049796
0.112170	0.0	0.470028	-0.002957	0.052727
0.118657	0.0	0.470019	-0.003309	0.055776
0.125403	0.0	0.470008	-0.003696	0.058947
0.132419	0.0	0.469995	-0.004121	0.062245
0.139716	0.0	0.469980	-0.004588	0.065674
0.147304	0.0	0.469963	-0.005100	0.069241

0.155197	0.0	0.469943,	-0.005661	0.072950
0.163404	0.0	0.469920	-0.006276	0.076807
0.171941	0.0	0.469893	-0.006948	0.080818
0.180818	0.0	0.469862	-0.007684	0.084989
0.190051	0.0	0.469827	-0.008489	0.089327
0.199653	0.0	0.469787	-0.009368	0.093838
0.209639	0.0	0.469741	-0.010329	0.098530
0.220025	0.0	0.469688	-0.011378	0.103408
0.230826	0.0	0.469627	-0.012522	0.108481
0.242059	0.0	0.469558	-0.013770	0.113756
0.253741	0.0	0.469478	-0.015131	0.119241
0.265891	0.0	0.469388	-0.016614	0.124944
0.278526	0.0	0.469284	-0.018231	0.130874
0.291667	0.0	0.469167	-0.019991	0.137041
0.305334	0.0	0.469032	-0.021908	0.143452
0.319548	0.0	0.468879	-0.023994	0.150117
0.334329	0.0	0.468705	-0.026264	0.157047
0.349703	0.0	0.468507	-0.028734	0.164251
0.365691	0.0	0.468281	-0.031420	0.171740
0.382318	0.0	0.468025	-0.034340	0.179524
0.399611	0.0	0.467734	-0.037515	0.187615
0.417596	0.0	0.467404	-0.040964	0.196024
0.436299	0.0	0.467030	-0.044712	0.204763
0.455751	0.0	0.466605	-0.048784	0.213844
0.475981	0.0	0.466123	-0.053205	0.223279
0.497021	0.0	0.465578	-0.058006	0.233080
0.518901	0.0	0.464961	-0.063217	0.243261
0.541658	0.0	0.464262	-0.068873	0.253834
0.565324	0.0	0.463472	-0.075011	0.264812
0.589937	0.0	0.462578	-0.081669	0.276209
0.615534	0.0	0.461569	-0.088890	0.288038
0.642156	0.0	0.460429	-0.096722	0.300311
0.669842	0.0	0.459143	-0.105212	0.313041
0.698635	0.0	0.457693	-0.114416	0.326241
0.728581	0.0	0.456058	-0.124390	0.339924
0.759724	0.0	0.454218	-0.135197	0.354099
0.792113	0.0	0.452146	-0.146904	0.368779
0.825798	0.0	0.449817	-0.159582	0.383971
0.860830	0.0	0.447200	-0.173308	0.399685
0.897263	0.0	0.444264	-0.188166	0.415927
0.935153	0.0	0.440972	-0.204244	0.432700
0.974559	0.0	0.437285	-0.221636	0.450007
1.015542	0.0	0.433163	-0.240444	0.467847
1.058163	0.0	0.428558	-0.260775	0.486214
1.102490	0.0	0.423425	-0.282746	0.505101
1.148590	0.0	0.417711	-0.306478	0.524493
1.196533	0.0	0.411363	-0.332101	0.544372
1.246394	0.0	0.404327	-0.359751	0.564714
1.298250	0.0	0.396545	-0.389573	0.585485
1.352180	0.0	0.387962	-0.421719	0.606646
1.408267	0.0	0.378524	-0.456347	0.628148
1.466598	0.0	0.368179	-0.493623	0.649933
1.527262	0.0	0.356883	-0.533718	0.671934
1.590353	0.0	0.344597	-0.576808	0.694071
1.655967	0.0	0.331926	-0.623077	0.716253
1.724205	0.0	0.316970	-0.672708	0.738379
1.795174	0.0	0.301626	-0.725889	0.760337
1.868980	0.0	0.285295	-0.782806	0.782002
1.945740	0.0	0.268035	-0.843647	0.803243
2.025569	0.0	0.249936	-0.908595	0.823919
2.108592	0.0	0.231118	-0.977828	0.843886
2.194936	0.0	0.211739	-1.051518	0.862999
2.284733	0.0	0.191993	-1.129826	0.881114
2.378123	0.0	0.172107	-1.212906	0.898098
2.475247	0.0	0.152337	-1.300897	0.913829
2.576257	0.0	0.132959	-1.393929	0.928205
2.681308	0.0	0.114261	-1.492117	0.941149

2.790560	0.0	0.096527	-1.595566	0.952615
2.904182	0.0	0.080022	-1.704371	0.962588
3.022350	0.0	0.064976	-1.818620	0.971091
3.145244	0.0	0.051568	-1.938397	0.978183
3.273053	0.0	0.039912	-2.063787	0.983958
3.405975	0.0	0.030051	-2.194881	0.988535
3.544214	0.0	0.021951	-2.331779	0.992059
3.687938	0.0	0.015511	-2.474595	0.994685
3.837502	0.0	0.010567	-2.623461	0.996575
3.993002	0.0	0.006917	-2.778530	0.997883
4.154723	0.0	0.004332	-2.939978	0.998751
4.322911	0.0	0.002585	-3.108003	0.999299
4.497828	0.0	0.001462	-3.282825	0.999628
4.679741	0.0	0.000780	-3.464688	0.999815
4.868931	0.0	0.000389	-3.653852	0.999914
5.065688	0.0	0.000181	-3.850597	0.999963
5.270315	0.0	0.000077	-4.055219	0.999985
5.483128	0.0	0.000030	-4.268030	0.999995
5.704453	0.0	0.000011	-4.489354	0.999998
5.934631	0.0	0.000003	-4.719532	1.000000
6.174017	0.0	0.000001	-4.958917	1.000000
6.422977	0.0	0.000000	-5.207878	1.000000
6.681896	0.0	0.000000	-5.466797	1.000000
6.951172	0.0	0.000000	-5.736073	1.000000
7.231219	0.0	0.000000	-6.016120	1.000000
7.522468	0.0	0.000000	-6.307268	1.000000
7.825366	0.0	0.000000	-6.610267	1.000000
8.140381	0.0	0.000000	-6.925282	1.000000
8.467996	0.0	0.000000	-7.252897	1.000000
8.808716	0.0	0.000000	-7.593617	1.000000
9.163065	0.0	0.000000	-7.947966	1.000000
9.531588	0.0	0.000000	-8.316488	1.000000
9.914851	0.0	0.000000	-8.699752	1.000000
10.313445	0.0	0.000000	-9.098346	1.000000
10.727983	0.0	0.000000	-9.512883	1.000000
11.159102	0.0	0.000000	-9.944003	1.000000
11.607466	0.0	0.000000	-10.392367	1.000000
12.073765	0.0	0.000000	-10.858666	1.000000
12.558715	0.0	0.000000	-11.343616	1.000000
13.063064	0.0	0.000000	-11.847965	1.000000
13.587587	0.0	0.000000	-12.372487	1.000000
14.133090	0.0	0.000000	-12.917991	1.000000
14.700414	0.0	0.000000	-13.485314	1.000000
15.290430	0.0	0.000000	-14.075331	1.000000
15.904048	0.0	0.000000	-14.688948	1.000000
16.532209	0.0	0.000000	-15.327110	1.000000
17.205898	0.0	0.000000	-15.990798	1.000000
17.896134	0.0	0.000000	-16.681034	1.000000

S= 0.01986 XI= 0.00051 DEG= 1.000 T= 0.06100 BE= 1.68090 Al= 3.87291

UE= 0.03067 KK= 3 DT= 0.28429 IK= 2 NIT= 5

S= 0.03955 XI= 0.00162 DEG= 2.000 T= 0.06100 BE= 1.35580 Al= 3.52556

UE= 0.06187 KK= 3 DT= 0.25110 IK= 2 NIT= 3

S= 0.05906 XI= 0.00335 DEG= 3.000 T= 0.06100 BE= 1.25371 Al= 3.35733

UE= 0.09362 KK= 3 DT= 0.25690 IK= 2 NIT= 3

```
S= 0.07839  XI= 0.00568  DEG= 3.000  T= 0.06100  BE= 1.20760  A1= 3.25766
UE= 0.12594  KK=   3  DT= 0.26356  IK=  2  NIT= 3


S= 0.09754  XI= 0.00861  DEG= 5.000  T= 0.06100  BE= 1.18406  A1= 3.18819
UE= 0.15885  KK=   3  DT= 0.26892  IK=  2  NIT= 2


S= 0.11651  XI= 0.01216  DEG= 6.000  T= 0.06100  BE= 1.17197  A1= 3.13435
UE= 0.19237  KK=   3  DT= 0.27331  IK=  2  NIT= 2


S= 0.13529  XI= 0.01630  DEG= 7.000  T= 0.06100  BE= 1.16654  A1= 3.08961
UE= 0.22652  KK=   3  DT= 0.27707  IK=  2  NIT= 2


S= 0.15387  XI= 00.2105  DEG= 8.000  T= 0.06100  BE= 1.16536  A1= 3.05062
UE= 0.26132  KK=   3  DT= 0.26132  IK=  2  NIT= 2


S= 0.17227  XI= 0.02641  DEG= 9.000  T= 0.06100  BE= 1.16712  A1= 3.01552
UE= 0.29680  KK=   3  DT= 0.28349  IK=  2  NIT= 2


S= 0.19048  XI= 0.03236  DEG= 10.000  T= 0.06100  BE= 1.17100  A1= 2.98317
UE= 0.33296  KK=   3  DT= 0.28636  IK=  2  NIT= 2


S= 0.20848  XI= 0.03891  DEG= 11.000  T= 0.06100  BE= 1.17648  A1= 2.95285
UE= 0.36985  KK=   3  DT= 0.28910  IK=  2  NIT= 2


S= 0.22629  XI= 0.04605  DEG= 12.000  T= 0.06100  BE= 1.18320  A1= 2.92409
UE= 0.40745  KK=   3  DT= 0.29173  IK=  2  NIT= 2


S= 0.24391  XI= 0.05379  DEG= 13.000  T= 0.06100  BE= 1.19091  A1= 2.89656
UE= 0.44582  KK=   3  DT= 0.29429  IK=  2  NIT= 2
```

REFERENCES

Arakawa, A., 1966. *J. Comput. Phys.*, **1**, 119.

Bar-Lev, M., and Yang, H. T., 1975. *J. Fluid Mech.*, **48**, 33–55.

Belcher, B. J., Burggraf, O. R., Cooke, J. C., Robins, A. J., and Stewartson, K., 1971. In *Recent Research of Unsteady Boundary Layers*, ed. Eichelbrenner, E. A., 1444–1465.

Blasius, H., 1908. *Z. Math. Phys.*, **56**, 1–37.

Blottner, F. G. and Flügge-Lotz, I., 1963. *J. de Mecanique*, **2**, 397–423.

Carlsaw, H. S., and Jaeger, J. C., 1947. *Conduction of Heat in Solids*, Clarendon Press, London and New York.

Cebeci, T., 1979. *J. Comput. Phys.*, **31**, 153–172.

Cheng, S. I., 1957. *Q. Appl. Math.*, **14**, 337–352.

Collins, W. M., and Dennis, S. C. R., 1973a. *Q. J. Mech. Appl. Math.*, **26**, 53–75.

Collins, W. M., and Dennis, S. C. R., 1973b. *J. Fluid Mech.*, **60**, 105–127.

Dennis, S. C. R., 1972. *J. Inst. Math. Its Appl.*, **10**, 105–117.

Dennis, S. C. R., and Staniforth, A. N., 1971. *Lecture Notes in Physics*, **8**, 343.

Ducoffe, A. L., and White, F. M., 1964. *J. Basic Eng.*, **86**, 234–246.

Dwyer, H. A., 1968a. *AIAA J.*, **6**, 2447–2448.

Dwyer, H. A., 1968b. *AIAA J.*, **6**, 1336–1342.

Falkner, V. M., and Skan, S. W., 1931. *Philos. Mag.* **12**, 865–896.

Fromm, J. E., and Harlow, F. H., 1963. *Phys. Fluids*, **6**, 975–982.

Goldstein, S., and Rosenhead, L., 1936. *Proc. Cambridge Philos. Soc.*, **32**, 392–401.

Görtler, H., 1946. *Ing. Arch.*, **14**, 286–305.

Görtler, H., 1957. *Math. Mech.*, **6**, 1–66.

Hall, M. G., 1968. Unpublished Mintech Report.

Hall, M. G., 1969a. *Ing. Arch*, **38**, 97–106.

Hall, M. G., 1969b. *Proc. R. Soc. London*. A **310**, 401–414.

Hasimoto, H., 1951. *J. Phys. Soc. Jpn.*, **6**, 400–401.

Hirota, I., and Miyakoda, K., 1965. *Meteorol. Soc. Jpn. Ser. II*, **43**, 30.

Howarth, L., 1938. *Proc. R. Soc. London*, A **164**, 547–579.

Howarth, L., 1950. *Proc. Cambridge Philos. Soc.*, **46**, 127–140.

Ingham, D. B., 1968. *J. Fluid Mech.*, **31**, 815.

Jain, P. C., and Rao, K. S., 1969. *Phys. Fluids, Suppl. II*, **12**, 57.

Kawaguti, M., and Jain, P. C., 1966. *J. Phys. Soc. Jpn.*, **21**, 2055.

Kinney, R. B., and Paolino, M. A., 1974. *J. Appl. Mech.*, **41**, 919–924.

Lam, S. H., and Crocco, L., 1959. *J. Aerosol. Sci.*, **26**, 54–55.

Lamb, H., 1932. *Hydrodynamics*, Cambridge Univ. Press, London and New York.

Mehta, U. B., 1972. "Starting Vortex Separation Bubbles and Stall—a Numerical Study of Laminar Unsteady Flow Around an Airfoil," Ph.D. Thesis, Illinois Institute of Technology.

Mehta, U. B., 1977. In *Unsteady Aerodynamics*, AGARDOgraph, CP 227.

Mehta, U. B., and Lavan, Z., 1975. *J. Fluid Mech.*, **67**, 227–256.

Nayfeh, A. H., 1973. *Perturbation Methods*, Wiley, New York.

Oudart, A., 1953. *Rech. Aerosp.*, **31**, 7–12.

Payne, R. B., 1958. *J. Fluid Mech.*, **4**, 81–86.

Prandtl, L., 1904. "Über Flüssigkeitsbewegung bei sehr kleiner Reiburg," Proc. III Int. Math. Congr., Heidelberg, 484–491.

Presz, W. M., and Heiser, W. M., 1968. *Z. Flugwiss.*, **16**, 33–39.

Proudman, I., and Johnson, K., 1962. *J. Fluid Mech.*, **12**, 161–168.

Rayleigh, Lord, 1911. *Philos. Mag.*, **6**, 697–711.

Richtmyer, R. D. and Morton, K. W., 1967. *Difference Methods for Initial Value Problems*, Wiley, New York.

Roache, P. J., 1972. *Computational Fluid Dynamics*, Hermosa Publishers, Albuquerque.

Schlichting, H., 1968. *Boundary Layer Theory*, McGraw-Hill, New York.

Schmall, R. A., and Kinney, R. B., 1974. *AIAA J.*, **12**, 1566–1572.

Sears, W. R., 1949. *Introduction to Theoretical Hydrodynamics*, Cornell Univ. Press, Ithaca.

Schuh, H., 1953. *Z. Flugwiss*, **1**, 122–131.

Son, J. S. and Hanratty, T. J., 1969. *J. Fluid Mech.*, **35**, 369.

Sowerby, L., 1951. *Philos. Mag.*, **42**, 16.

Sowerby, L., and Cooke, J. C., 1953. *J. Mech. Appl. Math.*, **6**, 50–70.

Stewartson, K., 1951. *Q. Appl. Math. Mech.*, **4**, 182–198.

Stewartson, K., 1960. *Adv. Appl. Mech.*, **6**, 1–37.

Stokes, G. G., 1851. *Trans. Cambridge Philos. Soc.*, **9-11**, 8–106.

Stuart, J. T., 1964. In *Laminar Boundary Layers*, ed. L. Rosenhead, 349–406.

Szymanski, F., 1932. *J. Math. Pure Appl.*, **11**, 67.

Telionis, D. P., and Tsahalis, D. T., 1974a. *Acta Astron.*, 1, 1487–1505.

Telionis, D. P., and Tsahalis, D. T., 1974b. *AIAA J.*, 12, 614–619.

Telionis, D. P., Tsahalis, D. T., and Werle, M. J., 1973. *Phys. Fluids*, 16, 968–973.

Thoman, D. C., and Szewczyk, A. A., 1969. *Phys. Fluids Suppl.*, II, 12, 76.

Thompson, J. F., Shanks, S. P., and Wu, J. C., 1974. *AIAA J.*, 12, 787–794.

Van Dyke, M., 1964. *Perturbation Methods in Fluid Mechanics*, Academic Press, New York.

Wang, C. Y., 1967a. *J. Math. Phys.*, 46, 195–202.

Wang, C. Y., 1967b. *J. Appl. Mech.*, 34, 823–828.

Watson, E. J., 1955. *Proc. R. Soc. London*, A 231, 104–116.

Weinberg, S., 1967. "The Use of the Hydrogen Bubble Technique to Observe Unsteady Boundary Layers," MIT Master's Thesis.

Werle, M. J., and Davis, R. T., 1972. *J. Appl. Mech.*, 39, 7–12.

White, F. M., 1974. *Viscous Fluid Flow*, McGraw-Hill, New York.

Williams, J. C., and Rhyne, T. B., 1980. *SIAM J. Appl. Math.*, to appear.

Wu, J. C., and Thompson, J. R., 1973. *Comput. Fluids*, 1, 197–215.

Wundt, H., 1955. *Ing. Arch.*, 23, 212–230.

Chapter 4

Oscillations with Zero Mean

4.1 Introduction

To many fluid dynamicists the term "oscillation" and perhaps the term "fluctuation" usually imply the random unsteady motion found in turbulence. In this monograph we shall reserve these terms for unsteady periodic motions, generated externally by virtue of some imposed disturbing mechanism. Such disturbances can be generated by fluctuation of the solid boundaries themselves or unsteadiness contained in the free stream. In this and the next chapter, we examine the mathematical models that describe the response of viscous laminar flows to such external disturbances. Unsteadiness appears mathematically in the boundary and initial conditions. The development therefore is totally foreign to the study of hydrodynamic instabilities and their eventual conversion to turbulence, phenomena that are generated and sustained with almost no external interference.

These statements may appear to some as too uncompromising and, in a sense, inaccurate. External turbulence levels have been known for some time to have a direct effect on the stability characteristics of a boundary layer. Moreover, the response of viscous flows to deterministic external oscillations may not be so foreign to turbulence phenomena and may eventually help us understand the complicated interaction between free-stream turbulence and transition or boundary-layer turbulence. These topics will not, however, be addressed here; rather, we shall be exclusively concerned with oscillatory laminar viscous flows that respond to a discrete and usually harmonic external excitation.

There is another class of very important problems in fluid dynamics that involve periodic phenomena. These are problems dealing with wind loads on blunt bodies and structures and with flutter. In these cases periodicity is inherent, and the investigation focuses on the coupling between the natural oscillations of the solid and the fluid. This difficult problem is again not our present concern. The fluctuating flows investigated here are generated through imposed disturbances of known frequency and amplitude. Typical examples of such flows are the flow generated by bodies oscillating with a discrete frequency in a fluid at rest, such as the flow around a helicopter blade that experiences fluctuations in the magnitude and direction of the oncoming stream, or the flow about a cascade of blades that finds itself in the periodic wake of the previous stage. Problems involving disturbances of known frequency are encountered very often in natural or biological flows as well; for example, the motion of water waves over a shallow beach, the flow of blood in the arteries, and the mechanics of the cochlea in the human ear.

In this chapter we shall be concerned only with periodic disturbances imposed on a fluid otherwise at rest. Such disturbances generate oscillations in the fluid, and the most interesting effect of viscosity is the fact that the flow oscillations do not average to zero, but rather, a net steady flow is generated, conventionally labeled "steady streaming." Although such problems touch upon another very large class of phenomena, the generation and propagation of sound, these effects will also be disregarded here.

Lighthill (1964) states clearly the conditions that justify the neglect of compressibility effects. Assuming isentropic flow, with pressure and density related by $p\rho^{\gamma} = $ constant, where γ is a constant or, equivalently, that

$$\frac{dp}{dt} + \frac{\gamma p}{\rho} \frac{d\rho}{dt} = 0, \qquad \qquad *(4.1.1)$$

we may rewrite Eq. (1.2.1) as follows:

$$\frac{\partial u_i}{\partial x_i} = -\frac{1}{\rho} \frac{D\rho}{Dt} = \frac{1}{\gamma p} \frac{Dp}{Dt}. \qquad \qquad *(4.1.2)$$

The left-hand side of Eq. (4.1.2) represents the rate of increase of a volume of a fluid element per unit volume.

Consider an oscillatory disturbance with circular frequency ω. If p_ω is a typical value of the pressure disturbance and p_∞ is the level of the pressure, then the right-hand side of Eq. (4.1.2) can be approximated by $\omega p_\omega / p_\infty$. If U_ω is a typical velocity associated with these oscillations, then from Bernoulli's equation for unsteady irrotational flow, we conclude that p_ω is of order $\omega \rho U_\omega L$. Now, individual terms in $\partial u_i / \partial x_i$ have fluctuation with frequency ω of amplitude U_ω / L. Hence, the condition that $\omega p_\omega / p_\infty$ be negligible compared with U_ω / L is

$$\rho \omega^2 L^2 / p_\infty \ll 1. \qquad \qquad *(4.1.3)$$

Lighthill (1964) notes that if (4.1.3) is violated, then sound waves of wavelength comparable with the body size are formed, radiating with the speed of sound $a \sim \sqrt{p_\infty / \rho}$. If (4.1.3) is expressed in terms of the speed of sound, it takes the form

$$\frac{\omega L}{a} \ll 1. \qquad \qquad *(4.1.4)$$

Rayleigh (1894) derived (4.1.4) in his theory of sound as a condition that flow oscillations generated by the obstacle radiate negligible energy. Assuming that the flow velocities generated are of the same order of magnitude as the amplitude of the velocity fluctuations of the body U_∞, condition (4.1.4) can also be expressed in terms of the Mach number

$$M = \frac{U_\infty}{a} \ll 1. \qquad\qquad *(4.1.5)$$

These conditions are met in many real-life situations involving water waves or rigid bodies oscillating in liquids otherwise at rest.

Historically, the topic has interested investigators long before the development of boundary-layer concepts. Experimental evidence of steady streaming was reported by Faraday (1831) and Dvorak (1874). Rayleigh's (1894) work provided a theoretical explanation of the phenomenon described by earlier experimental investigations. The interest revived later with the experimental work of Carrière (1929), Andrade (1931), and Schlichting (1932). These authors gave contradictory results. For a circular cylinder immersed in an oscillating fluid, Carrière found that a steady-streaming layer is formed directing the flow toward the cylinder along the axis of the oscillation. Andrade and Schlichting found that jetlike layers are shot away from the cylinder along the same axis.

The controversy has long been resolved with the theoretical work of Schlichting and others; physical arguments are presented in the review articles of Rott (1964) and Stuart (1963) and are briefly outlined here. The reader will find the most comprehensive and inclusive account of earlier experimental works as well as theoretical analyses in Riley (1967) and his updated review (Riley, 1975).

In this chapter we shall be mostly concerned with the analytical development and the properties of the equations that will subsequently play a fundamental role in the computation of oscillatory flows. We start with a classical and well-known problem, oscillations over an infinite flat plate, and proceed with the study of oscillations over an arbitrary solid surface. In the first few sections the boundary-layer equation is employed to follow the historical development and simplify the presentation. A more general approach, starting with the full Navier–Stokes equation and employing inner and outer expansions, is then described.

Very little numerical results are available. Most of the analytical data were derived via asymptotic expansions. Such typical results are presented and their comparison with experimental data is discussed. A brief account of technical or physiological problems that fall in this category is given at the end of the chapter.

4.2 Oscillations of an Infinite Flat Plate

One of the simplest and most beautiful examples that indicate some of the characteristic features of oscillatory flow is the flow over an infinite flat plate that oscillates harmonically in its plane. The problem is very similar to

Rayleigh's problem and is often referred to in literature as Stokes flow (Stokes, 1851a). As in Rayleigh's case (see Chapter 3), the Navier–Stokes equations reduce to

$$\frac{\partial u}{\partial t} = \nu \frac{\partial^2 u}{\partial y^2} .$$ *(4.2.1)

The boundary conditions read

$$u \to 0, \quad \text{as } y \to \infty,$$ *(4.2.2a)

$$u = u_w(t), \quad \text{as } y = 0,$$ *(4.2.2b)

and the motion of the wall is assumed to be given by

$$u_w = u_o e^{-i\omega t},$$ *(4.2.3)

where u_0 is the constant amplitude of the wall.

We shall seek a solution in the form

$$u(y,t) = f(y)e^{-i\omega t}.$$ *(4.2.4)

The function f then satisfies the equation

$$f'' + \frac{i\omega}{\nu} f = 0.$$ *(4.2.5)

The characteristic equation becomes $\lambda^2 + i\omega/\nu = 0$ and the square root of the imaginary unit i introduces a real and an imaginary part:

$$\lambda = \pm(i-1)\sqrt{\frac{\omega}{2\nu}} .$$ *(4.2.6)

One of the two roots is excluded so that the function f dies out as $y \to \infty$ to satisfy the boundary condition (4.2.2a), and the solution that meets the condition at the wall finally reads

$$u = u_0 \exp\left[(i-1)\sqrt{\frac{\omega}{2\nu}}\; y\right] e^{-i\omega t}.$$ *(4.2.7)

Only the real part of the above expression is physically meaningful and thus,

$$u = u_0 \exp\left(-\sqrt{\frac{\omega}{2\nu}}\; y\right)\cos\left(\omega t - \sqrt{\frac{\omega}{2\nu}}\; y\right).$$ *(4.2.8)

Some characteristic features of unsteady boundary layers are already present in this simple example. First we notice that the fluctuations of the motion die out exponentially as we move away from the wall. In fact, a typical length that characterizes the amplitude decrease is the quantity $\sqrt{\nu/\omega}$. In other words, the length scale of variation of the amplitude in the normal direction is $\sqrt{\nu/\omega}$. Smaller values of viscosity or larger values of frequency imply smaller thickness of the extent to which the fluid is disturbed by the wall oscillation. Mathematically of course the fluid is disturbed

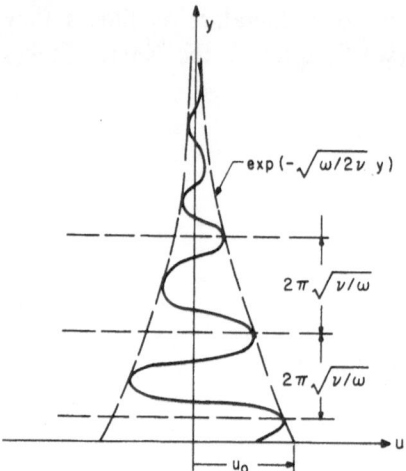

all the way to infinity, regardless of the value of ν or ω. The concept is identical to the concept of the thickness of the Blasius layer, that is the boundary layer generated in the case of steady flow over a semiinfinite flat plate.

Notice that ω can be considered as the inverse of a characteristic time t_0 and thus our characteristic length becomes $\sqrt{\nu t_0}$, which we encountered in the previous chapters. A second important observation is the fact that the oscillations have the same frequency ω throughout the flow, but there is a phase lag proportional to the distance from the wall. The solution is schematically depicted in Fig. 4.1, where one typical profile $u(y,t)$ is sketched for a fixed t and the envelope of all profiles, that is, the exponential function $\exp(-\sqrt{\omega/2\nu}\, y)$ is shown.

The oscillatory flow just described is very often referred to in literature as a "Stokes layer." It is remarkable that quite often Stokes layers are discovered embedded in other flow fields with properties almost independent of the other fields. An example will be given in the next chapter where we consider oscillations imposed on a developed laminar boundary layer. It is then found that for very large frequencies, a Stokes layer is generated, with profiles independent of the profiles of the steady layer.

4.3 Oscillations of a Blunt Body

Let us now consider the oscillatory motion of a body with finite dimensions immersed in a fluid at rest. The problem has interested theoreticians for nearly one hundred years (Rayleigh, 1883) and is considered as one of the typical cases whereby the nonlinear effects give rise to a peculiar and rather

Fig. 4.2 Schematic of the steady-streaming
streamline pattern about a circular cylinder
oscillating in the horizontal direction.

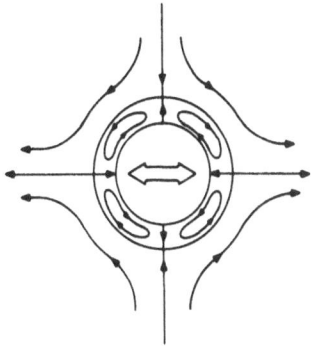

unexpected phenomenon, the phenomenon of streaming. The most common
experiment is that of a circular cylinder performing linear oscillations with its
axis being displaced perpendicular to itself as depicted schematically in Fig.
4.2. Experimental data indicate (Carrière, 1929; Andrade, 1931; Schlichting,
1932; Holtsmark et al., 1954) that fluid particles are carried along the
boundary of the cylinder up to points that may be identified as stagnation
points and then they are shot away in the form of jets. An analogous
phenomenon is observed at the top and the bottom of the cylinder where one
may identify attaching stagnation points. This motion is certainly not continuous but intermittent and the streamlines shown in Fig. 4.2 represent motion
only in the mean.

Before we proceed with the analysis of this problem, we should emphasize
some assumptions: The flow is assumed to be incompressible and the
wavelength of sound propagation is assumed to be very large compared with
a typical length of the problem. Finally, a word of caution is necessary with
regard to the system of reference. Traditionally a hydrodynamic system of
reference is attached on the immersed body. For steady flow this introduces
no ambiguities and a Galilean transformation will allow us to view the fluid
at rest and the body in motion. In the present case accelerations are involved
and such a transformation is not possible.

The Stokes layer

Let us assume that the boundary-layer approximation holds and that the
viscous flow thus generated is governed by Eqs. (1.2.23) and (1.2.24). For this
to be true we assume that the Stokes layer is thin compared to the size of the
body L. That is, we assume that the number $\omega L^2/\nu$ is large

$$\frac{\omega L^2}{\nu} \gg l. \tag{4.3.1}$$

If the frame of reference is attached to the body and the outer flow performs oscillations about a zero mean, the boundary conditions are

$$u^* = v^* = 0, \quad \text{at } y = 0, \tag{4.3.2a}$$

$$u^* \to U_e^* = \tfrac{1}{2} U_0^*(x)(e^{i\omega t^*} + e^{-i\omega t^*}), \quad \text{at } y \to \infty. \tag{4.3.2b}$$

The following derivation follows the analysis of Stuart (1966), but the notation is changed to emphasize the fact that no similarity variables are used and therefore x, y, and t can be employed to denote dimensionless and stretched variables instead of ξ, η, and τ. The stream function ψ^* is introduced according to

$$u^* = \frac{\partial \psi^*}{\partial y^*}, \qquad v^* = -\frac{\partial \psi^*}{\partial x^*}, \tag{4.3.3}$$

and the momentum equation Eq. (1.2.4) reads

$$\frac{\partial^2 \psi^*}{\partial y^* \partial t^*} + \frac{\partial \psi^*}{\partial y^*} \frac{\partial^2 \psi^*}{\partial x^* \partial y^*} - \frac{\partial \psi^*}{\partial x^*} \frac{\partial^2 \psi^*}{\partial y^{*2}} = \frac{\partial U_e^*}{\partial t^*} + U_e^* \frac{\partial U_e^*}{\partial x^*} + \nu \frac{\partial^3 \psi^*}{\partial y^{*3}}. \tag{4.3.4}$$

Anticipating a solution similar to Stokes' as described in the previous problem, we introduce the following dimensionless quantities that contain a familiar stretching of the normal coordinate

$$x = \frac{x^*}{L}, \qquad y = \frac{y^*}{\sqrt{2\nu/\omega}}, \qquad t = t^* \omega, \tag{4.3.5}$$

$$U_e(x,t) = \frac{U_e^*(x^*,t^*)}{U_\infty}, \qquad \psi = \psi^* \left(U_\infty \sqrt{\frac{2\nu}{\omega}} \right)^{-1}, \tag{4.3.6}$$

where U_∞ and L are a typical velocity and length of the problem, respectively.

Substitution of these expressions into Eq. (4.3.4) yields after some algebra:

$$\frac{\partial^2 \psi}{\partial y \partial t} - \frac{1}{2} \frac{\partial^3 \psi}{\partial y^3} - \frac{\partial U_e}{\partial t} = \frac{U_\infty}{L\omega} \left[-\frac{\partial \psi}{\partial y} \frac{\partial^2 \psi}{\partial y \partial x} + \frac{\partial \psi}{\partial x} \frac{\partial^2 \psi}{\partial y^2} + U_e \frac{\partial U_e}{\partial x} \right], \tag{4.3.7}$$

and the boundary conditions (4.3.2) become

$$\psi = \frac{\partial \psi}{\partial y} = 0, \quad \text{at } y = 0, \tag{4.3.8}$$

$$\frac{\partial \psi}{\partial y} \to U_e = \frac{U_0}{2} (e^{it} + e^{-it}), \quad \text{at } y \to \infty. \tag{4.3.9}$$

Experiments indicate that the most interesting phenomena are observed for large frequencies or, more accurately, for small values of the parameter $U_\infty/L\omega$. In this case we may disregard the right-hand side of the equation.

With the assumption that this parameter is small,

$$\varepsilon = \frac{U_\infty}{L\omega} \ll 1, \tag{4.3.10}$$

Eq. (4.3.7) can be approximated by

$$\frac{\partial^3 \psi}{\partial y^3} - 2\frac{\partial^2 \psi}{\partial y \partial t} = -2\frac{\partial U_e}{\partial t}. \tag{4.3.11}$$

If $\zeta = \partial\psi/\partial y$, Eq. (4.3.11) becomes

$$\frac{\partial^2 \zeta}{\partial y^2} - 2\frac{\partial \zeta}{\partial t} = -2\frac{\partial U_e}{\partial t}, \tag{4.3.12}$$

and a solution can be readily assumed in the form

$$\zeta(y,t) = \zeta_0(y)e^{it} + \overline{\zeta}_0(y)e^{-it}, \tag{4.3.13}$$

where the bar denotes the complex conjugate. This expression is real and equivalent to the harmonic function $A(y)\cos t + B(y)\sin t$, with $A(y)$ and $B(y)$ arbitrary real functions. It is easy to show the equivalence of the two expressions and the relation between the real and imaginary part of $\zeta_0(y)$ and $A(y)$, $B(y)$. The form of (4.3.13) is appropriate for demonstrating the nonlinear coupling and the generation of steady streaming. Equation (4.3.12) then reduces to an ordinary differential equation

$$\frac{d^2\zeta_0}{dy^2} - 2i\zeta_0 = -iU_0. \tag{4.3.14}$$

The characteristic equation becomes $\lambda^2 - 2i = 0$ and one of the two roots is excluded since ζ must remain finite as $y \to \infty$, according to Eq. (4.3.9). The complete solution of (4.3.12) is then

$$\frac{\partial\psi}{\partial y} = \zeta_0 e^{it} + CC = \left[\frac{U_0}{2} - Ce^{-(1+i)y}\right]e^{it} + CC, \tag{4.3.15}$$

where CC stands for the complex conjugate and the arbitrary constant C is nothing but $U_0/2$, so that $\partial\psi/\partial y$ vanishes at the wall as Eq. (4.3.8) requires. Thus finally

$$\psi = \left\{-\frac{U_0(x)}{2}(1-i)[1 - e^{(1+i)y}] + \frac{U_0 y}{2}\right\}e^{it} + CC, \tag{4.3.16}$$

which also satisfies the condition $\psi = 0$ at $y = 0$. This solution may be thought of as a generalization of Stokes solution and does not seem to introduce any new features other than the familiar phase shift and the exponential decay. Notice that the dependence on the axial coordinate x is introduced only through the factor $U_0(x)$ much like in the first approximation of the Blasius problem (see Section 2.4).

The phenomenon of streaming

The solution given by Eq. (4.3.16) is correct within an error of order ε, assuming of course that the derivatives of the right-hand side of Eq. (4.3.7) behave well and the perturbation analysis is nonsingular. A more accurate solution can be derived if we take into account the correction introduced by the right-hand side of Eq. (4.3.7), and the problem clearly indicates the need to expand in powers of ε. Let us therefore assume

$$\psi(x, y, t) = \frac{U_0(x)}{2} \left[\psi_0(y)e^{it} + \bar{\psi}_0(y)e^{-it} \right]$$

$$+ \varepsilon \left[\psi_1(x, y)e^{2it} + \bar{\psi}_1(x, y)e^{-2it} \right] + O(\varepsilon^2). \tag{4.3.17}$$

If we substitute the above expression in Eq. (4.3.7), we readily arrive at

$$\frac{U_0}{2} \left[i\frac{\partial \psi_0}{\partial y} e^{it} - i\frac{\partial \bar{\psi}_0}{\partial y} e^{-it} \right] + \varepsilon \left[i\frac{\partial \psi_1}{\partial y} e^{2it} - i\frac{\partial \bar{\psi}_1}{\partial y} e^{-2it} \right] + O(\varepsilon^2)$$

$$- \frac{U_0}{4} \left[\frac{\partial^3 \psi_0}{\partial y^3} e^{it} + \frac{\partial^3 \bar{\psi}_0}{\partial y^3} e^{-it} \right] - \frac{\varepsilon}{4} \left[\frac{\partial^3 \psi_1}{\partial y^3} e^{2it} + \frac{\partial^3 \bar{\psi}_1}{\partial y^3} e^{-2it} \right]$$

$$+ O(\varepsilon^2) - \frac{U_0}{2} (ie^{it} - ie^{-it})$$

$$= \varepsilon \left\{ - \left[\frac{U_0}{2} \left(\frac{\partial \psi_0}{\partial y} e^{it} + \frac{\partial \bar{\psi}_0}{\partial y} e^{-it} \right) + O(\varepsilon) \right] \right.$$

$$\times \left[\frac{1}{2} \frac{\partial U_0}{\partial x} \left(\frac{\partial \psi_0}{\partial y} e^{it} + \frac{\partial \bar{\psi}_0}{\partial y} e^{-it} \right) + O(\varepsilon) \right]$$

$$+ \left[\frac{1}{2} \frac{dU_0}{dx} (\psi_0 e^{it} + \bar{\psi}_0 e^{-it}) + O(\varepsilon) \right]$$

$$\times \left[\frac{U_0}{2} \left(\frac{\partial^2 \psi_0}{\partial y^2} e^{it} + \frac{\partial^2 \bar{\psi}_0}{\partial y^2} e^{-it} \right) + O(\varepsilon) \right]$$

$$+ \frac{U_0}{4} (e^{it} + e^{-it}) \frac{dU_0}{dx} (e^{it} + e^{-it}) \right\}. \tag{4.3.18}$$

Then we may collect terms of order ε^0 to get

$$\frac{\partial^3 \psi_0}{\partial y^3} - 2i\frac{\partial \psi_0}{\partial y} = -i \tag{4.3.19}$$

and its complex conjugate and thus identify that $\partial \psi_0 / \partial y$ is proportional to the function $\zeta_0(y)$ we solved for in the previous section

$$\zeta_0 = \frac{U_0}{2} \frac{d\psi_0}{dy},$$

(4.3.20)

therefore,

$$\psi_0 = -\frac{1}{2}(1-i)\left[1 - e^{-(1+i)y}\right] + y.$$

(4.3.21)

Collecting terms of order ε, we group them according to the factors e^{2it} and e^{-2it}, since they vary independently with time. Notice though that on the right-hand side, multiplying coefficients of e^{it} and e^{-it}, we form terms independent of time and these must be separately grouped. No such terms appear on the left-hand side and hence the steady terms in the equation give

$$0 = -U_0 \frac{dU_0}{dx} \frac{\partial\psi_0}{\partial y} \frac{\partial\bar{\psi}_0}{\partial y} + \frac{1}{2}U_0 \frac{dU_0}{dx}\left(\psi_0 \frac{\partial^2\bar{\psi}_0}{\partial y^2} + \bar{\psi}_0 \frac{\partial^2\psi_0}{\partial y^2}\right) + U_0 \frac{dU_0}{dx}.$$

(4.3.22)

At this point we may start questioning the validity of our method, since it would be most unlikely that the function ψ_0 we have already solved for would satisfy the above equation as well. Apparently the nonlinearity of the right-hand side of Eq. (4.3.7) has created a family of terms that do not vary with time. If we had anticipated that, we could have included a term in our original expansion. Let us replace (4.3.17) with

$$\psi = \frac{U_0}{2}\left(\psi_0 e^{it} + \bar{\psi}_0 e^{-it}\right) + \varepsilon\left\{\psi_s + \frac{1}{2}\left(\psi_1 e^{2it} + \bar{\psi}_2 e^{-2it}\right)\right\} + O(\varepsilon^2).$$

(4.3.23)

Indeed the real quantity $\psi_s(x, y)$ gives rise to a steady term and the equivalent of Eq. (4.3.22) becomes

$$-\frac{\partial^3\psi_s}{\partial y^3} = -U_0 \frac{dU_0}{dx} \frac{\partial\psi_0}{\partial y} \frac{\partial\bar{\psi}_0}{\partial y} + \frac{1}{2}U_0 \frac{dU_0}{dx}\left(\psi_0 \frac{\partial^2\bar{\psi}_0}{\partial y^2} + \bar{\psi}_0 \frac{\partial^2\psi_0}{\partial y^2}\right)$$

$$+ U_0 \frac{dU_0}{dx}.$$

(4.3.24)

The appropriate boundary conditions are

$$\psi_1 = \psi_s = 0, \quad \text{at } y = 0,$$

(4.3.25)

$$\frac{\partial\psi_1}{\partial y} = \frac{\partial\psi_s}{\partial y} = 0, \quad \text{at } y = 0,$$

(4.3.26)

$$\frac{\partial\psi_1}{\partial y} = \frac{\partial\psi_s}{\partial y} = 0, \quad \text{at } y \to \infty.$$

(4.3.27)

We can now solve for the functions ψ_1 and ψ_s to derive the oscillatory and steady corrections. This mathematical peculiarity has a very important

physical interpretation and provides the explanation of the phenomenon that
we have already described. It proves that forced outer-flow oscillations about
a vanishing mean produce in the Stokes layer an oscillatory motion which
averages to a nonvanishing mean, the flow ψ_s. That is, the response of fluid
particles within the boundary layer is oscillatory in character, but the
excursions in one direction are larger than in the opposite direction so that,
in the mean, the particles appear to be displaced continuously. This phenomenon is known in literature as "steady streaming" or "acoustic streaming" or
"nonlinear streaming."

4.4 The Streaming Layer

Equation (4.3.24) can be solved in closed form. Its complementary solution is
$A + By + Cy^2$ where A, B, and C are arbitrary constants and together with
the particular solution we have

$$\psi_s(x, y) = A + By + Cy^2 + U_0 \frac{dU_0}{dx} \left(-\frac{1}{8} e^{-2y} - \frac{3}{2} e^{-y} \cos y \right.$$

$$\left. - e^{-y} \sin y - \frac{1}{2} ye^{-y} \sin y \right). \tag{4.4.1}$$

The particular solution should die out at infinity and this requires that
both B and C are zero in order to meet the boundary condition (4.3.27).
However, once again we arrive at an impasse, because with $B = C = 0$ we
cannot satisfy both conditions (4.3.25) and (4.3.26). The least we can do is
allow $C = 0$, and calculate A and B from Eqs. (4.3.25) and (4.3.26). We thus
arrive at

$$\psi_s = U_0 \frac{dU_0}{dx} \left(\frac{13}{8} - \frac{3}{4} y - \frac{1}{8} e^{-2y} - \frac{3}{2} e^{-y} \cos y - e^{-y} \sin y \right.$$

$$\left. - \frac{1}{2} ye^{-y} \sin y \right). \tag{4.4.2}$$

As y goes to infinity, therefore, the x component of the streaming velocity
does not vanish, since $B \neq 0$, but tends asymptotically to

$$u_s^*(x, \infty) = \varepsilon \frac{\partial \psi_s^*}{\partial y^*} = U_\infty \varepsilon \frac{\partial \psi}{\partial y} = -\frac{3}{4\omega} U_0^* \frac{dU_0^*}{dx^*}. \tag{4.4.3}$$

This means that the streaming phenomenon is proportional to the pressure
gradient but inversely proportional to the frequency. Moreover it implies that
at the edge of the oscillating boundary layer, the Stokes layer, the steady part
of the nonlinear response is finite. This finite velocity may be in turn
considered as the inner boundary condition of a second boundary layer
equivalent to the layer that the body would generate if its skin were in
motion with a steady velocity $U_s = u_s(L, \infty)$. The thickness of such a layer

would then be inversely proportional to a Reynolds number based on U_s. If U_∞ is a typical velocity of the imposed oscillatory flow, then from Eq. (4.4.3), $U_s \sim (1/\omega)U_\infty^2/L$ and the streaming Reynolds number becomes $Re_s = U_s L/\nu = U_\infty^2/\omega\nu$. Thus, the streaming boundary-layer thickness, the oscillatory boundary-layer thickness, and their ratio can be written as

$$\delta_s = \frac{L}{\sqrt{Re_s}} = \frac{L\sqrt{\omega\nu}}{U_\infty} , \tag{4.4.4}$$

$$\delta_{in} = \sqrt{\frac{\nu}{\omega}} , \tag{4.4.5}$$

$$\frac{\delta_s}{\delta_{in}} = \frac{L\omega}{U_\infty} = \frac{1}{\varepsilon} . \tag{4.4.6}$$

This is schematically depicted in Fig. 4.3. The amplitude of the outer flow oscillation $U_0(x)$ is shown as a uniform profile since it is assumed that both δ_s and δ_{in} are still much smaller than the scale of the potential flow. Within the oscillatory part of the boundary layer, the amplitudes of the first harmonic, $u_0 = \partial\psi_0/\partial y$ is shown. The steady-streaming profile $u_s(x, y)$ appears to penetrate well within the potential flow region.

The basic theoretical arguments of Rayleigh (1883) that explained the phenomenon of streaming are practically a perturbation scheme. Much later, Schlichting (1932) recognized the fact that the streaming layer may extend further than the Stokes layer into the potential flow and that therefore condition (4.3.27) should be abandoned. Stuart's (1963, 1966) contribution to the problem was a clearer reformulation and the identification of the significance of the parameter Re_s, the streaming Reynolds number. A theory in terms of inner and outer expansions and a careful investigation of all possible situations for different values of the parameters involved is contained in Riley (1965, 1967). In the second paper the author presents a detailed critical review of both experimental and theoretical work to that date. The main points of Riley's theory are outlined in the next section.

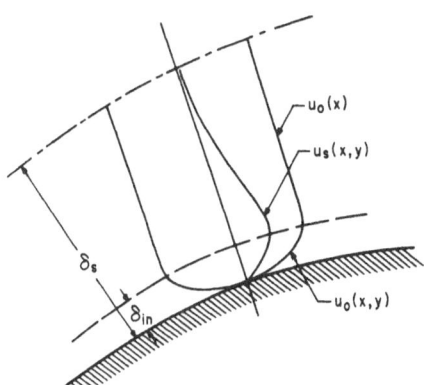

Fig. 4.3 Stokes and streaming layers and corresponding velocity profiles.

Investigation of the flow field outside the Stokes layer requires a very careful asymptotic analysis. To order ε^0 the flow outside the Stokes layer is inviscid. This is the pure oscillatory flow we shall get if we multiply the classical potential solution for the steady flow about a circular cylinder by $e^{i\omega t}$. It has been demonstrated that the Stokes layer contains the oscillatory part of the viscous response. However, terms of order ε introduce the steady-streaming layer, a purely viscous phenomenon, which extends beyond the Stokes layer. Assuming that the streaming flow away from the Stokes layer does not interact with the potential flow, we may analyze it by solving the steady-flow boundary-layer equations. This assumption has been discussed by Riley (1965) and Stuart (1966). In terms of a new, stretched variable, and a new stream function

$$Y = \frac{yU_\infty}{L(\omega\nu)^{1/2}}, \tag{4.4.7}$$

$$\phi_s(x, Y) = \frac{\psi_s}{U_\infty(\nu/\omega)^{1/2}}, \tag{4.4.8}$$

the governing equation, that is, the boundary-layer equation, becomes

$$\frac{\partial\phi}{\partial y}\frac{\partial^2\phi}{\partial x\partial Y} - \frac{\partial\phi}{\partial x}\frac{\partial^2\phi}{\partial Y^2} = \frac{\partial^2\phi}{\partial Y^3}. \tag{4.4.9}$$

The symbol ϕ should not be confused here with a potential function. As ε tends to infinity the ratio of the two thicknesses δ_{in}/δ_s tends to zero. The streaming layer therefore is essentially driven by a moving skin situated at the edge of the Stokes layer. Within the accuracy of our approximation, however, we impose the boundary condition at the wall

$$\frac{\partial\phi}{\partial Y} = -\frac{3}{4}U_0\frac{dU_0}{dx}, \qquad \phi = 0, \quad \text{at } Y = 0. \tag{4.4.10}$$

The streaming flow vanishes at the edge of the streaming layer

$$\frac{\partial\phi}{\partial Y} \to 0, \quad \text{as } Y \to \infty. \tag{4.4.11}$$

The problem can be solved numerically for any body shape. In fact, Haddon and Riley (1979) solve the corresponding full Navier–Stokes streaming equations numerically. Approximate solutions have been derived by Riley (1965) and Stuart (1966). The last author expands the solution in powers of an artificial small parameter, say σ:

$$\phi(x, Y) = \phi_0(x) + \sigma\phi_1(x, Y) + \sigma^2\phi_2(x, Y) + \cdots. \tag{4.4.12}$$

It turns out that the solution converges in practice as far as $\sigma = 1$. The functions $\phi_1, \phi_2, \ldots,$ satisfy the following differential equations:

$$\frac{\partial^3 \phi_1}{\partial Y^3} + \frac{\partial \phi_0}{\partial x}\frac{\partial^2 \phi_1}{\partial Y^2} = 0, \tag{4.4.13}$$

$$\frac{\partial^3 \phi_2}{\partial Y^3} + \frac{\partial \phi_0}{\partial x}\frac{\partial^2 \phi_2}{\partial Y^2} = \left(\frac{\phi_0''}{\phi_0'}\right)\left(\frac{3}{4} U_0 \frac{dU_0}{dx}\right)^2 e^{-2\phi_0' Y}, \tag{4.4.14}$$

with solutions expressed in terms of ϕ_0:

$$\phi_1(x, Y) = -\frac{3}{4} U_0 \frac{dU_0}{dx} e^{-\phi_0' Y}, \tag{4.4.15}$$

$$\phi_2(x, Y) = \frac{\phi_0''}{2\phi_1'^4}\left(\frac{3}{4} U_0 \frac{dU_0}{dx}\right)^2 \left(e^{-\phi_0' Y} - \frac{1}{2} e^{-2\phi_0' Y}\right). \tag{4.4.16}$$

However ϕ_0 satisfies a very complicated nonlinear ordinary differential equation

$$\frac{1}{2}(\phi_0^2)' + \frac{3}{4} U_0 \frac{dU_0}{dx} + \left(\frac{3}{4} U_0 \frac{dU_0}{dx}\right)^2 \frac{\phi_0''}{4\phi_0'^3}$$

$$-\left(\frac{3}{4} U_0 \frac{dU_0}{dx}\right)^3 \frac{1}{36\phi_0'^6}(\phi_0'\phi_0''' - 7\phi_0''^2) + E = 0, \tag{4.4.17}$$

where E stands for an infinite series of terms that contain fractions of powers of higher-order derivatives of ϕ_0. This equation can be solved in powers of x if U_0 can be approximated by a polynomial (Stuart 1966).

4.5 Inner and Outer Expansions

Considering the problem in a more general framework, Riley (1967) identifies two dimensionless parameters. The quantity ε defined in the previous section and a Reynolds number Re_ω based on a typical velocity constructed by the product of ω and L

$$\varepsilon = \frac{U_\infty}{\omega L}, \qquad Re_\omega = \frac{\omega L^2}{\nu}. \tag{4.5.1}$$

Three Reynolds numbers can thus be defined for such problems, depending on the choice of the velocity. The "oscillation" Reynolds number $Re_\omega = \omega L^2/\nu$, the "outer-flow" Reynolds number $Re = U_\infty L/\nu$ and the "streaming" Reynolds number $Re_s = U_\infty^2/\omega L$. The last two are related to ε and Re_ω via the expressions

$$Re = \varepsilon Re_\omega, \qquad Re_s = \varepsilon^2 Re_\omega. \tag{4.5.2}$$

The problem however is characterized by only two dimensionless parameters and we choose here the ones given by Eq. (4.5.1). The first is essentially the inverse of a Strouhal number and the second is a Reynolds number.

If U_∞, ω^{-1}, and L are used as typical velocity, time, and length, respectively, to nondimensionalize the full Navier–Stokes equations, then Eqs. (1.2.14) and (1.2.10) read

$$\frac{\partial u_i}{\partial t} + \varepsilon u_j \frac{\partial u_i}{\partial x_j} = -\frac{\partial p}{\partial x_i} + \frac{1}{Re_\omega} \frac{\partial^2 u_i}{\partial x_j \partial x_j}, \tag{4.5.3}$$

$$\frac{\partial}{\partial t}(\nabla^2 \psi) + \varepsilon \left[\frac{\partial \psi}{\partial y} \frac{\partial}{\partial x}(\nabla^2 \psi) - \frac{\partial \psi}{\partial x} \frac{\partial}{\partial y}(\nabla^2 \psi) \right] = \frac{1}{Re_\omega} \nabla^4 \psi, \tag{4.5.4}$$

where x and y are coordinates fixed on the body and $y = 0$ defines the contour of the body. The parameters ε and Re_ω appear as natural dimensionless parameters. In the earlier sections we assumed that $Re_\omega = \omega L^2 / \nu$ is large and invoked directly the boundary-layer approximation. This is a process that requires great caution, since it involves essentially a double expansion in powers of two small parameters. The ordering of the terms depends on the relative magnitude of the two parameters involved. In this section we essentially follow the work of Riley (1967) emphasizing the cases we feel that deserve attention.

Throughout this chapter we assume that the parameter ε is very small. Let us further assume that Re_ω is very large,

$$\varepsilon \ll 1, \qquad \frac{1}{Re_\omega} \ll 1, \tag{4.5.5}$$

but such that

$$\varepsilon \ll \frac{1}{Re_\omega}, \tag{4.5.6}$$

which results to

$$Re = \varepsilon Re_\omega \ll 1. \tag{4.5.7}$$

Two important problem categories satisfy the above conditions:

(a) $Re_s \ll 1$, i.e., $\varepsilon^2 \ll 1/Re_\omega$,   (4.5.8a)
(b) $Re_s = O(1)$, i.e., $\varepsilon \ll 1/Re_\omega$.   (4.5.8b)

The solution to the problem of harmonic excitation can be sought in the form of an expansion in powers of ε

$$\psi(x, y, t; \varepsilon, R_\omega) = \psi_0(x, y; Re_\omega)e^{it} + \varepsilon\psi_1(x, y; Re_\omega)e^{2it}$$
$$+ \varepsilon\psi_s(x, y; Re_\omega) + O(\varepsilon^2), \tag{4.5.9}$$

where the terms of order ε have been split into the fluctuating and the steady-streaming parts. Substituting this expansion in Eq. (4.5.4), collecting powers of ε, and separating the oscillating from the steady terms, we arrive at

$$i\nabla^2\psi_0 - \frac{1}{Re_\omega}\nabla^4\psi_0 = 0, \tag{4.5.10}$$

$$i\nabla^2\psi_1 - \frac{1}{Re_\omega}\nabla^4\psi_1 = -\frac{\partial\psi_0}{\partial y}\frac{\partial}{\partial x}(\nabla^2\psi_0) + \frac{\partial\psi_0}{\partial x}\frac{\partial}{\partial y}(\nabla^2\psi_0), \tag{4.5.11}$$

$$-\frac{1}{Re_\omega}\nabla^4\psi_s = -\frac{\partial\psi_0}{\partial y}\frac{\partial}{\partial x}(\nabla^2\bar{\psi}_0) + \frac{\partial\psi_0}{\partial x}\frac{\partial}{\partial y}(\nabla^2\bar{\psi}_0) - \frac{\partial\bar{\psi}_0}{\partial y}\frac{\partial}{\partial x}(\nabla^2\psi_0)$$

$$+ \frac{\partial\bar{\psi}_0}{\partial x}\frac{\partial}{\partial y}(\nabla^2\psi_0), \tag{4.5.12}$$

where an overbar denotes the complex conjugate. At infinity, ψ_0 merges into a uniform harmonic oscillation and all other terms should vanish. At the wall the no-slip and no-penetration condition should be satisfied

$$\psi_i = \frac{\partial\psi_i}{\partial y} = 0, \quad \text{for } i = 0, 1, \ldots, \quad \text{on } y = 0. \tag{4.5.13}$$

The next step is to approximate the solution to Eqs. (4.5.10)–(4.5.12) by dropping the terms of order $1/Re_\omega$, thus generating the inviscid solution, which in the terminology of inner–outer expansions can be the outer solution. However, such a process may easily lead to errors if it is not recognized that this is essentially a double expansion. This will become clear if the complete expansion is written as in Riley (1967):

$$\psi = \psi_{00}e^{it} + \frac{1}{Re_\omega}\psi_{01}e^{it} + \frac{1}{Re_\omega^2}\psi_{02}e^{it} + \cdots$$

$$+ \varepsilon\left[(\psi_{10}e^{2it} + \psi_{s0}) + \frac{1}{Re_\omega}(\psi_{11}e^{2it} + \psi_{s1}) + \cdots\right] + O(\varepsilon^2). \tag{4.5.14}$$

What could be misleading in this equation is the fact that the terms of the expansion are not written in order of decreasing magnitude. The parameter ε is smaller than $1/Re_\omega$ or even $(1/Re_\omega)^2$. Only then is it possible to obtain Eqs. (4.5.10)–(4.5.12). The term ψ_0 in (4.5.10) can be thus further expanded in powers of $1/Re_\omega$ and then again, one may proceed to the term of order Re_ω^{-n} only if

$$\varepsilon \ll Re_\omega^{-n}. \tag{4.5.15}$$

However, the limitations are not as severe as described, since the terms of the expansion and, hence, the terms of the equation are first grouped according to the frequency of oscillation. All terms $\psi_{0j}e^{it}$ for $j = 0, 1, \ldots$ are thus linearly independent of the terms $\psi_{1j}e^{2it}$, etc. With (4.5.8) valid, the equation that governs ψ_{00} is

$$\nabla^2\psi_{00} = 0, \tag{4.5.16}$$

and the well-known solution of inviscid flow that violates the no-slip condition at the wall can be readily derived. The classical procedure dictates the
introduction of inner dependent and independent variables

$$\Psi = \psi \sqrt{\frac{Re_\omega}{2}} , \qquad Y = y \sqrt{\frac{Re_\omega}{2}} , \qquad (4.5.17)$$

and an inner expansion

$$\Psi = \Psi_{00} e^{it} + \frac{1}{Re_\omega} \Psi_{01} e^{it} + \cdots$$

$$+ \varepsilon \left[(\Psi_{10} e^{it} + \Psi_{s0}) + \frac{1}{Re_\omega} (\Psi_{11} e^{it} + \Psi_{s1}) + \cdots \right] + O(\varepsilon^2). \qquad (4.5.18)$$

The terms ψ_{00} and Ψ_{00} in these expansions represent the potential flow and
the boundary-layer flow, respectively. Both oscillate with the basic frequency. The correction to these solutions due to the displacement effect is
represented by the terms ψ_{01} and Ψ_{01} and is familiar to fluid mechanicists as
the displacement effect on the outer flow and higher-order boundary-layer,
respectively.

Substituting Eq. (4.5.18) in the governing equation, we can follow the
classical procedure to rederive the boundary-layer equation for Ψ_{00} and the
solution obtained in Section 4.3.

$$\Psi_{00} = U_0 \left\{ Y - \frac{1}{2}(1 - i)[1 - e^{-(1+i)Y}] \right\}, \qquad (4.5.19)$$

which satisfies the conditions at $Y = 0$ and matches with the outer flow as
$Y \to \infty$. As in the previous section, $U_0(x)$ is the amplitude of the outer flow
oscillation evaluated at the wall.

To match with the outer flow, Riley expands this expression for large Y
and expresses it in terms of the outer variable Y

$$\Psi_{00} \underset{Y \to \infty}{\to} U_0(x) \left[y e^{it} - Re_\omega^{-1/2} e^{i(t - \pi/4)} \right]. \qquad (4.5.20)$$

It now becomes obvious that the matching process involves contributions of
one term of the inner expansion to more than one term of the outer
expansion. That is, Ψ_{00} feeds into ψ_{00} and ψ_{01}. Riley (1975) carries further
investigations of higher-order boundary-layer theory and provides numerical
results.

Returning now to the terms of order ε, we find that the equation for Ψ_{10}
with potential flow boundary conditions has the trivial solution

$$\Psi_{10} = 0, \qquad (4.5.21)$$

and the corresponding equation for Ψ_{10} is satisfied by

$$\Psi_{10} = U_0 \frac{dU_0}{dx} \left[\frac{1+i}{4\sqrt{2}} e^{-(1+i)Y\sqrt{2}} + \frac{iY}{2} e^{-(1+i)Y} - \frac{1+i}{4\sqrt{2}} \right], \qquad (4.5.22)$$

which meets the boundary-layer conditions

$$\Psi_{10} = \frac{\partial \Psi_{10}}{\partial Y} = 0, \quad \text{at } Y = 0, \tag{4.5.23}$$

$$\frac{\partial \Psi_{10}}{\partial Y} = 0, \quad \text{as } Y \to \infty. \tag{4.5.24}$$

Turning to the streaming terms, we identify two cases:

(a) $Re_s \ll 1$.
  With $\nabla^2 \Psi_0 = 0$, Eq. (4.5.13) reduces to

$$\nabla^4 \Psi_s = 0. \tag{4.5.25}$$

The outer flow therefore is essentially a creeping viscous flow. This flow vanishes far from the body, but on the skin of the body it matches with the inner streaming flow. The inner-layer equation is satisfied by the expression (4.4.2), which meets the no-slip and no-penetration condition on the wall. At the interface, we have

$$\frac{\partial \Psi}{\partial Y}\bigg|_{y=0} \to -\frac{3}{4} U_0 \frac{dU_0}{dx}. \tag{4.5.26}$$

(b) $Re_s = O(1)$.
  In this case the terms of order ε in the governing equation yield no information about the streaming term. One has to proceed to higher-order terms to arrive at

$$i\nabla^2 \Psi_3 e^{it} - \frac{\partial(\Psi_{00}, \nabla^2 \Psi_2)}{\partial(x, y)} e^{it} - \frac{1}{Re_s^{1/2}} \frac{\partial(\Psi_{01}, \nabla^2 \Psi_{s0})}{\partial(x, y)} e^{it}$$

$$- \frac{\partial(\Psi_{10}, \nabla^2 \Psi_{s0})}{\partial(x, y)} \tag{4.5.27}$$

$$= \frac{1}{Re_s} \nabla^4 \Psi_{s0}.$$

Separating again the steady from the unsteady terms, we arrive at

$$\frac{1}{Re_s} \nabla^4 \Psi_{s0} + \frac{\partial \Psi_{s0}}{\partial y} \frac{\partial}{\partial x} (\nabla^2 \Psi_{s0}) - \frac{\partial \Psi_{s0}}{\partial x} \frac{\partial}{\partial y} (\nabla^2 \Psi_{s0}) = 0. \tag{4.5.28}$$

The steady part of the outer flow is therefore governed by the full Navier–Stokes equations with Re_s the streaming Reynolds number. Note that it is the streaming Reynolds number that scales the effects of diffusion and convection. The outer streaming flow assumes the character of a boundary layer if Re_s is large. Two thin layers of different thicknesses can then be identified, the Stokes layer and the streaming layer as described in the previous sections.

4.6 Typical Results

Oscillatory flows about a zero mean are found in a variety of natural and physiological phenomena, but are not very common in engineering applications. The study of model problems defined in this chapter has been essential for the understanding of the physical mechanisms involved and the mathematical tools necessary for the analysis of more complex problems. Numerical results to the problems described are rather limited.

We first provide some results of the classical problem of the oscillating circular cylinder. We then proceed to describe two more interesting problems of oscillating external flows about a zero mean. The first is the oscillatory flow over an infinite wavy wall and the second is the pulsating flow over an infinite flat surface; both find immediate application in the study of the effects of water waves on fluid and sediment transport. Internal flow problems are only briefly mentioned at the end of the chapter.

The circular cylinder

One of the most interesting features of the oscillatory flow about a blunt body is the generation of closed recirculating streaming bubbles as shown schematically in Fig. 4.2. It should be noticed that the direction of the external flow is opposite to the flow next to the skin of the body. This phenomenon has given rise to conflicting experimental evidence in the early part of the century (Carrière, 1929; Andrade, 1931; Schlichting, 1932). It appeared that in some cases the streaming flow was jettisoned away from the cylinder in the direction of the oscillation, whereas in other cases the opposite was true. The phenomenon has now been adequately explained (Rott, 1964; Riley, 1967). It turns out that for a large oscillatory-flow Reynolds number $Re_\omega = \omega L^2/\nu$, the steady-streaming term changes sign within the Stokes layer. This is perhaps due to phase shifts within the Stokes layer that may lead to sign changes of convective nonlinear terms on the right-hand side of Eq. (4.3.24). As a result the streaming flow is in the opposite direction of the outer flow and closes in a loop to maintain continuity. For smaller values of Re_ω, that is, for thicker Stokes layers, the domain of observation becomes more narrow. In the extreme case, the flow pattern depicted schematically in Fig. 4.2 effectively stretches in the direction away from the body until only the inner layer can be observed. The streaming flow then appears directed toward the cylinder along the line of oscillation.

The change of sign of the streaming profiles has been predicted analytically by Riley (1965) and Stuart (1966). Bertelsen (1974) has calculated velocity profiles based on earlier theories in order to compare with his experimental results. He evaluated numerically solutions of the form of Eq.

(4.4.12) to calculate streaming velocity profiles at $x = 0.5$ and for two streaming Reynolds numbers. For the flow about a circular cylinder, Stuart (1966) assumes the pressure gradient term in the form of a polynomial:

$$-\frac{3}{4} U_0 \frac{dU_0}{dx} = 12x_1 - 32x_1^3 + \frac{128}{5} x_1^5 \ldots \,, \tag{4.6.1}$$

where

$$x_1 \equiv x - \pi/4. \tag{4.6.2}$$

The solution to Eq. (4.4.17) then becomes

$$\phi_0(x) = 2\sqrt{3}\left(x_1 - \frac{8}{17} x_1^3 + \frac{229056}{1116985} x_1^3 + \ldots\right). \tag{4.6.3}$$

In Fig. 4.4 we display the experimental data of Bertelsen contrasted with analytical predictions calculated from the solutions of Riley (1965) and Stuart (1966). In these experiments a brass tube was placed inside a larger cylindrical container filled with a viscous liquid. The inner cylinder was vibrated rectilinearly and velocity profiles were produced by photographic methods. Notice that the Stokes layer is very thin and experimental points are not found in the region where the profile turns down and the flow eventually reverses direction.

Discrepancies between theory and experiment, especially in the outer region of the field, were attributed at first to higher-order effects. To assess the effects of a finite Reynolds number Re_s, Riley (1975) carried out the boundary-layer analysis to second order and this resulted in some improvement as shown in Fig. 4.5.

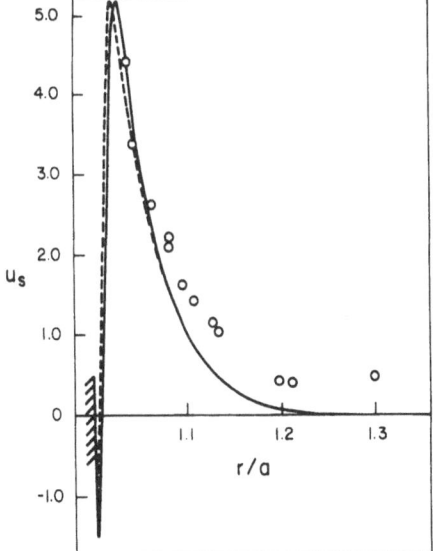

Fig. 4.4 Streaming velocity profiles for $\varepsilon = 0.05$, $Re_\omega = 20$, $Re_s = 400$. ———, from Riley (1965); – – –, from Stuart (1966); ○, Bertelsen's (1974) experimental points. The wall, $r/a = 1$, is denoted by the cross hatch.

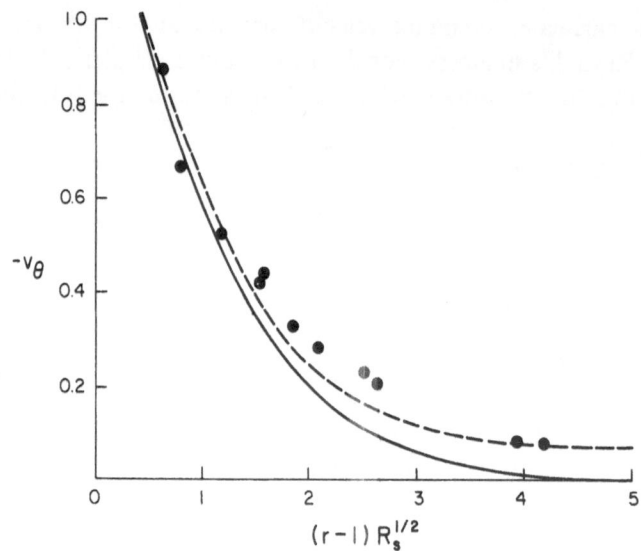

Fig. 4.5 Streaming velocity profiles for $\varepsilon = 0.05$, $Re_\omega = 20$, $Re_s = 400$. ———, Riley (1975); – – –, Haddon and Riley (1979); ●, Bertelsen (1974).

In a subsequent effort (Haddon and Riley, 1979), it was attempted to solve the problem in the closed domain between the two cylinders that Bertelsen considered in his experiments. To eliminate completely any errors due to boundary-layer type of approximations, Haddon and Riley solved the full Navier–Stokes form of Eq. (4.4.9), which in polar coordinates becomes

$$\frac{1}{r} \frac{\partial(\phi, \nabla^2\phi)}{\partial(r, \theta)} = \frac{1}{Re_s} \nabla^4\phi. \tag{4.6.4}$$

Note that the coordinate r is dimensionless but not stretched and therefore the Reynolds number is present on the right-hand side of the equation.

Numerical solution to the problem is sought in the form of a coupled unsteady system by introducing the vorticity Ω and the potential function Φ,

$$\frac{\partial \Omega}{\partial t} + \frac{1}{r} \frac{\partial \Phi}{\partial r} \frac{\partial \Omega}{\partial \theta} - \frac{1}{r} \frac{\partial \Phi}{\partial \theta} \frac{\partial \Omega}{\partial r} = \frac{1}{Re_s} \nabla^2\Omega, \tag{4.6.5}$$

$$\nabla^2\Phi = \Omega. \tag{4.6.6}$$

An artificial initial condition is then introduced and numerical integration proceeds until the flow tends to a steady-state form. For small Reynolds numbers, the leading term of an expansion is used as an initial condition. With a solution for Re_s at hand, further solutions for larger Re_s are generated using as a first guess the solution for Re_s. The finite-differencing scheme is formulated in an explicit form with respect to time and, therefore, it is subject to stability limitations. Typical results are shown in Fig. 4.5 together with the results of higher-order boundary-layer calculations. The theoretical predictions fall definitely within the domain of experimental uncertainty.

The wavy wall

Consider now the two-dimensional pure oscillatory flow about a wavy wall whose surface is given by

$$y^* = \alpha \cos \kappa x^*, \tag{4.6.7}$$

where α is the amplitude of the wavy shape and $2\pi/\kappa$ is the wavelength. The flow should satisfy the no-slip and no-penetration condition at the wall and merge smoothly to the potential oscillation at infinity

$$\psi^* = 0; \frac{\partial \psi^*}{\partial y^*} \cos \theta - \frac{\partial \psi^*}{\partial x^*} \sin \theta = 0, \quad \text{at } y^* = \alpha \cos \kappa x^*. \tag{4.6.8}$$

$$\frac{\partial \psi^*}{\partial y^*} = U_\infty \cos \omega t, \quad \frac{\partial \psi^*}{\partial x^*} = 0, \quad \text{as } y \to \infty. \tag{4.6.9}$$

where θ is the inclination of the wavy wall. Lyne (1971b) introduces a conformal transformation in order to transfer the boundary condition to a coordinate axis.

$$\zeta^* = \xi^* + i\eta^* = z^* - i\alpha e^{i\kappa z^*}, \quad \text{with } z^* = x^* + iy^*. \tag{4.6.10}$$

Assuming that α is small, the Jacobian of the transformation becomes

$$J = \left| \frac{d\zeta^*}{dz^*} \right|^2 = 1 + 2\alpha\kappa e^{-\kappa y^*} \cos \kappa x^* + O(\alpha^2) \tag{4.6.11}$$

and, in terms of the transformed coordinates,

$$J = 1 + 2\alpha\kappa e^{-\kappa \eta^*} \cos \kappa \xi^* + O(\alpha^2). \tag{4.6.12}$$

Neglecting terms of order α^2, we may impose the boundary conditions now at the wall

$$\psi^* = \frac{\partial \psi^*}{\partial \eta^*} = 0, \quad \text{at } \eta^* = 0. \tag{4.6.13}$$

A new stream function is defined as

$$u^* = J^{1/2} \frac{\partial \Psi^*}{\partial \eta^*}, \quad v^* = -J^{1/2} \frac{\partial \Psi^*}{\partial \xi^*}, \tag{4.6.14}$$

which satisfies the equation

$$\frac{\partial}{\partial t^*} D^{*2}\Psi^* - \frac{\partial(\Psi^*, JD^{*2}\Psi^*)}{\partial(\xi^*, \eta^*)} = \nu D^2(JD^{*2}\Psi^*), \tag{4.6.15}$$

where

$$D^{*2} \equiv \frac{\partial^2}{\partial \xi^{*2}} + \frac{\partial^2}{\partial \eta^{*2}}. \tag{4.6.16}$$

The problem has two characteristic lengths: the amplitude of the wavy wall α and the wavelength $2\pi/\kappa$. Based on these two lengths then, two oscillatory Reynolds numbers can be defined:

$$Re_{\omega 1} = \frac{\omega}{2\nu\kappa^2}, \qquad Re_{\omega 2} = \frac{\omega a^2}{2\nu}, \qquad (4.6.17)$$

whereas the streaming Reynolds number is

$$Re_s = \frac{2U_\infty}{\nu\omega}. \qquad (4.6.18)$$

Lyne (1971) further defines a small parameter

$$\varepsilon = a\sqrt{\frac{\omega}{2\nu}}, \qquad (4.6.19)$$

which is the ratio of the amplitude of the wall to the thickness of the Stokes layer. In terms of dimensionless stretched coordinates

$$\xi = \xi^*(2\nu/\omega)^{-1/2}, \qquad \eta = \eta^*(2\nu/\omega)^{-1/2}, \qquad (4.6.20)$$

$$t = \omega t^*, \qquad (4.6.21)$$

and a dimensionless stream function

$$\Psi = \Psi^*\left[U_\infty(2\nu/\omega)^{1/2}\right]^{-1}, \qquad (4.6.22)$$

Equation (4.6.15) now becomes

$$\frac{2}{Re_s^{1/2}}\frac{\partial}{\partial t}(D^2\Psi) - \frac{\partial(\Psi, JD^2\Psi)}{\partial(\xi, \eta)} = Re_s^{-1/2}D^2(JD^2\Psi), \qquad (4.6.23)$$

where

$$D^2 = \frac{\partial^2}{\partial\xi^2} + \frac{\partial^2}{\partial\eta^2}. \qquad (4.6.24)$$

This equation can be integrated in terms of finite differences subject to the boundary conditions (4.6.9) and (4.6.13). Expanding in powers of ε

$$\Psi = \Psi_0 + \varepsilon\Psi_1 + O(\varepsilon^2), \qquad (4.6.25)$$

one readily arrives at the classical Stokes layer for Ψ_0, which in terms of the new variables takes the form

$$\Psi_0 = \eta\cos t + 2^{-1/2}\left[e^{-\eta}\sin\left(t - \eta + \frac{\pi}{4}\right) - \sin\left(t + \frac{\pi}{4}\right)\right]. \qquad (4.6.26)$$

The equation of order ε now becomes

$$\frac{2}{\sqrt{Re_s}}\frac{\partial}{\partial t}D^2\Psi_1 - \frac{\partial(\Psi_0, D^2\Psi_1)}{\partial(\xi, \eta)} - \frac{\partial(\Psi_1, D^2\Psi_0)}{\partial(\xi, \eta)}$$

$$- \frac{\partial(\Psi_0, 2ke^{-k\eta}\cos k\xi D^2\Psi_0)}{\partial(\xi, \eta)}$$

$$= \frac{1}{\sqrt{Re_s}}\left[D^4\Psi_1 + D^2(2ke^{-k\eta}\cos k\xi D^2\Psi_0)\right],$$

where

$$k = Re_{\omega 1}^{-1/2} = \kappa(2\nu/\omega)^{1/2}. \tag{4.6.28}$$

The boundary conditions that the disturbance Ψ_1 should meet are homogeneous

$$\Psi_1 = \frac{\partial \Psi_1}{\partial \eta} = 0, \quad \text{at } \eta = 0, \tag{4.6.29}$$

$$\frac{\partial \Psi_1}{\partial \eta} = \frac{\partial \Psi_1}{\partial \xi} = 0, \quad \text{as } \eta \to \infty. \tag{4.6.30}$$

Following Benjamin (1959), a solution is sought in the form

$$\Psi_1 = \Re\left[F(\eta, t) + \frac{\partial \Psi_0}{\partial \eta} e^{-k\eta} e^{ik\xi} \right], \tag{4.6.31}$$

where $\Re$ denotes the "real part of" and $\partial \Psi_0/\partial \eta$ can be obtained from (4.6.26). Equation (4.6.27) and the boundary conditions on F take the form

$$\frac{2}{ik\sqrt{Re_s}} \frac{\partial}{\partial t} (F'' - k^2 F) + \frac{\partial \Psi_0}{\partial \eta} (F'' - k^2 F) - \frac{\partial^3 \Psi_0}{\partial y^3} F$$

$$= -\frac{1}{ik\sqrt{Re}} (F^{iv} - 2k^2 F'' + k^4 F), \tag{4.6.32}$$

$$F = 0, F' = -\partial^2 \Psi_0/\partial \eta^2, \quad \text{at } \eta = 0, \tag{4.6.33}$$

$$F' = 0, F = 0, \quad \text{as } \eta \to \infty, \tag{4.6.34}$$

where prime denotes differentiation with respect to η.

It is very intriguing that Eq. (4.6.32) is almost identical to the Orr–Sommerfeld equation that governs the stability characteristics of plane parallel flows over a flate plate. Moreover, it now becomes apparent that with the assumption that $\varepsilon = a\sqrt{\omega/2\nu}$ is small, the only parameter left in the problem is the quantity $k\sqrt{Re_s}$, which is nothing but the square root of the ratio ot two Reynolds numbers

$$k\sqrt{Re_s} = \sqrt{\frac{Re_s}{Re_{\omega 1}}}. \tag{4.6.35}$$

This ratio is proportional to the amplitude of oscillation of the fluid far from the wall to the wavelength of the wall.

Numerical solutions to Eq. (4.6.32) are obtained (Lyne, 1971b) and steady streaming is calculated by integration over one period of oscillation. Note that this is necessary since Ψ_1 contains both the oscillatory and the steady disturbance.

Three characteristic cases can be identified depending on the values of the parameters $Re_s, Re_{\omega 1}$ and $Re_{\omega 1}$. First let us consider the case $Re_s/Re_{\omega 1} \ll 1$, that is, the case when the wavelength of the wall $L = 2\pi/\kappa$ is much larger

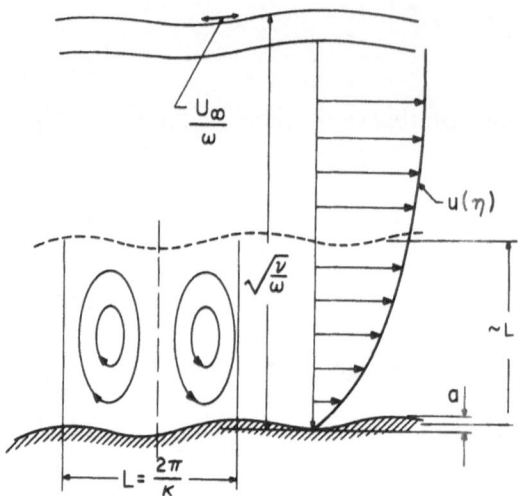

Fig. 4.6 Steady streaming
over a wavy wall for $Re_s/Re_\omega \ll 1$ and $Re_\omega \ll 1$.
The wavelength of the
wall L is larger than the
particle amplitude U_∞/ω,
but smaller than the
thickness of the Stokes
layer $(\nu/\omega)^{1/2}$.

than the amplitude of oscillation of the fluid far from the wall U_∞/ω. Two special subcases are considered: $Re_{\omega 1} \ll 1$ and $Re_{\omega 1} \gg 1$. For a very small oscillating Reynolds number, $Re_{\omega 1} \ll 1$, that is, if the wavelength of the wall is much smaller than the Stokes layer, Lyne finds that the steady streaming is confined to a boundary layer whose thickness is of the order of a wavelength. The situation is schematically depicted in Fig. 4.6.

Assuming that $Re_{\omega 1} \gg 1$, but the ratio $Re_s/Re_{\omega 1}$ is always small, we recover the classical case investigated by Schlichting (1932) and all the subsequent investigators. This is the case whereby the wavelength of the wall dwarfs the Stokes layer, in other words, the oscillatory viscous layer is a very thin layer compared to the geometry of the solid boundary. It is discovered in this case, in qualitative agreement with the results described in the previous subsection, that a thin steady-streaming bubble is contained in the Stokes layer but the flow eventually reverses its direction and streaming extends further into the potential flow. This is shown schematically in Fig. 4.7.

Finally the case $Re_s/Re_{\omega 1} \gg 1$, that is, the case where the amplitude of a fluid particle is much larger than the wavelength of the wall appears to generate a cascade of recirculating bubbles, stacked one on top of the other and alternating in direction. All of the streaming bubbles, however, are confined within the Stokes layer, driven from within very thin viscous layers that are formed next to the wall.

The same problem has been attacked independently by Sleath (1974, 1976) with a very specific application in mind, namely, the oscillatory flow generated by surface waves on a wavy bottom. Sleath expanded first in terms of the parameter

$$\varepsilon_1 = \frac{\alpha}{L} = \frac{\alpha\kappa}{2\pi}. \tag{4.6.36}$$

This is the ratio of the amplitude to the wavelength of the wavy wall. A

Fig. 4.7 Steady streaming over a wavy wall for $Re_s / Re_\omega \ll 1$ and $Re_\omega \gg 1$. The wavelength of the wall L is larger than the amplitude of the oscillation U_∞ / ω, as well as the thickness of the Stokes layer $\sqrt{\nu / \omega}$.

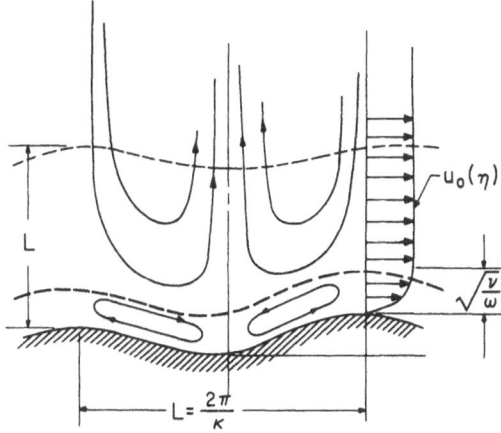

second expansion is then necessary in terms of the ratio

$$\lambda = \frac{U_\infty}{\omega L} = \frac{U_\infty \kappa}{2\pi\omega}, \qquad (4.6.37)$$

the ratio of the amplitude of the outer flow particle oscillations to the wavelength of the wall. If this parameter is small, then the two cases depicted in Figs. 4.6 and 4.7 are recovered. Sleath (1976) presents his numerical results in the form of curves similar to the schematics of Figs. 4.6 and 4.7 for $\alpha / L = 0.3$ and two values of $Re_{\omega 1}$, 0.051 and 0.688. This seems to be a rather unlikely case for ocean waves on shallow beaches since the lengths of the water waves and the mean particle oscillation near the surface are much larger than the wavelength of the ripples formulated on a sandy bed. Even in this case, an observation of the streamline patterns of Figs. 4.6 and 4.7 indicates that for both the cases of Re_ω, large and small, the direction of steady streaming on the wall is always away from the troughs and toward the peaks. This, of course, tends to reinforce the wave pattern of the bottom by sharpening the peaks. It should be noted here that in the terminology of investigators working with ocean waves, the term "mass transport" is more common than the term "steady streaming."

The case of large λ, a case much more important in ocean engineering, is considered further by Sleath. However the interest is concentrated on the formation of bottom ripples, a process involving the entrainment of sediment particles and their transport. Assumptions have to be made on the entrainment process, and the ripple wavelength is determined by virtue of a maximizing process. More details of this specific problem can be found referenced in the article by George and Sleath (1979).

Mass transport in water waves

A very interesting problem where oscillations of the external flow are controlled by the development of a viscous layer on a solid wall is the response of a constant depth body of water to standing or traveling surface waves. The inviscid part of the flow has been studied first by Stokes (1847) and Rayleigh (1876). Stokes demonstrated that in the irrotational part of the flow, except for the periodic motion, a secondary field is generated that has been known to investigators as mass-transport flow. It was later proved that Stokes' solution can be derived, in general, for irrotational flows and the smallness of the wave amplitude is not a necessary assumption. Stokes' theory predicts a strong mean velocity in the direction of wave propagation at the surface and weak velocity on the bottom but in the opposite direction. Experiments conducted much later actually indicated the opposite trend. Longuet-Higgins (1953) derived a theoretical solution that explained the experimental data available at the time. More recent experiments showed further discrepancies on the predictions of the flow velocity on the water bed. However, carrying the Longuet-Higgins' expansion to a higher order proved sufficient to account for all the discrepancies (Sleath, 1972).

The stream function ψ is expanded in powers of a small parameter ε such that the first term contains the purely periodic motion

$$\psi = \varepsilon\psi_1 + \varepsilon^2\psi_2 + \varepsilon^3\psi_3 + \cdots . \tag{4.6.38}$$

The no-slip and no-penetration are decreed at the wall, but at the edge of the boundary layer the velocity is required to merge to the inner limit of the flow due to surface waves (U.S. Beach Erosion Board, 1963)

$$u = u_0 + u_\infty \cos(\omega t - kx) + A\,\frac{u_\infty^2 k}{2\omega}\cos 2(\omega t - kx)$$
$$+ B\,\frac{u_\infty^3 k^2}{4\omega^2}\cos 3(\omega t - kx) + \cdots , \qquad\qquad *(4.6.39)$$

where k and ω are the wave number and the angular velocity of the surface wave, respectively, u_0 is the mean flow, u_∞ is the velocity amplitude, A and B depend on k, ω, and the depth of the water h. In particular,

$$A = \frac{3}{2\sinh^2 kh}\left\{1 + O\left[\left(\frac{u_\infty k}{\omega}\right)^2\right]\right\}. \qquad\qquad *(4.6.40)$$

In the expansion (4.6.38) the small parameter ε could be the quantity $U_\infty k/\omega$. However, Sleath (1980) suggests that if A were large, that is, if k were small, it may be more appropriate to define

$$\varepsilon = \frac{U_\infty k A^{1/2}}{\omega}. \tag{4.6.41}$$

Substituting in the full Navier–Stokes equations expressed in terms of ψ and collecting powers of ε, one arrives at differential equations for ψ_1, ψ_2, etc.

The solution to the first equation is

$$\varepsilon\psi_1 = \Re\left\{ \frac{u_\infty}{\beta}\left[\frac{\beta}{k}\sinh ky + \left(\frac{1-i}{2}\right)(e^{-\alpha y} - 1)\right]e^{i(\omega t - kx)}\right.$$
$$\left. + O\left[\frac{u_\infty k^2 \nu}{\beta\omega}\right]\right\},$$

*(4.6.42)

where $\Re$ denotes the real part of the expression that follows and the quantities β and α are given by

$$\beta = \left(\frac{\omega}{2\nu}\right)^{1/2}, \qquad \alpha = (1+i)\beta.$$

*(4.6.43)

The nonlinear terms of the momentum equations now couple to generate steady terms due to the dependence of ψ_1 on x, similar to the solutions described in earlier chapters. Sleath (1972) provides solutions for ψ_2, ψ_3, and ψ_4 and demonstrates that the inclusion of higher-order terms results in considerable improvement of the theoretical predictions. For kh less than 1.1, it is indicated that the second-order terms result in an increase of the mass-transport velocity, whereas at higher values of kh the opposite is true, in agreement with the experimental evidence.

A point of interest to the students of unsteady viscous flow is the approach introduced by Longuet-Higgins (1953), which is essentially a return to the Lagrangian coordinates. If ξ and η are the coordinates of a fluid particle at time t, then the velocity can be expressed in the form

$$\frac{\partial\xi}{\partial t} = u(\xi,\eta,t) = u(0,0,t) + \xi\frac{\partial u(0,0,t)}{\partial x} + \eta\frac{\partial u(0,0,t)}{\partial y}$$
$$+ \frac{\xi^2}{2}\frac{\partial^2 u}{\partial x^2} + \xi\eta\frac{\partial^2 u}{\partial x\partial y} + \frac{\eta^2}{2}\frac{\partial^2 u}{\partial y^2} + \cdots .$$

(4.6.44)

The Lagrangian coordinates ξ and η can then be also expanded in powers of ε. Sleath formulates the problem in terms of these coordinates and calculates steady streaming as the average velocity over a cycle of the oscillation

$$u_s = \frac{1}{T^2}\int_0^T\left[\xi(t+T) - \xi(t)\right]dt,$$

(4.6.45)

$$v_s = \frac{1}{T^2}\int_0^T\left[\eta(t+T) - \eta(t)\right]dt.$$

(4.6.46)

We shall also find a return to Lagrangian formulation in Chapter 7, where numerical integration proceeds in terms of Lagrangian coordinates through regions of reversed flow.

The problem may become more complex if one considers the effects of transition to turbulence. For quite some time, the inability of the theory to explain the experimental phenomenon was attributed to turbulence. The fact is that, as Sleath (1970) demonstrated experimentally, a set of secondary vortices is generated that has been for some time interpreted erroneously as turbulence. A large number of publications have appeared addressing the modeling of turbulent motion in purely oscillatory external and internal flows as described in Knight's (1978) review article. Modeling of unsteady turbulent flows is described in Chapter 6, but mostly for problems of external aerodynamics involving a nonvanishing mean flow. The material and the references of Knight (1978) seem to have followed a completely independent path.

Internal flow problems

Although the main interest in this monograph is in external aerodynamics, a few internal flow oscillations display all the characteristics of the flows described in this chapter and deserve a brief mention. Consider, for example, oscillatory flow in a curved pipe. Let r_0 be the radius of the pipe centerline and a the internal radius of the circular cross section, such that $a \ll r_0$ (Fig. 4.8). With purely harmonic oscillations of the pressure gradient, Lyne

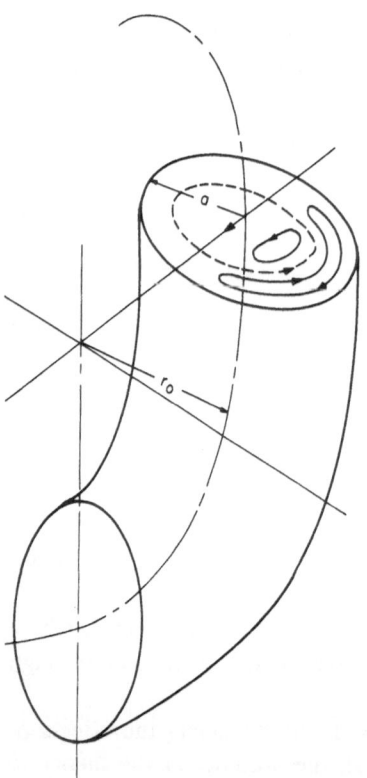

Fig. 4.8 Oscillatory flow through a curved pipe.

(1970a) identifies two dimensionless parameters that characterize the flow:

$$\varepsilon^2 = \frac{U_0}{r_0 a \omega^2}\,, \qquad Re_s = \frac{U_0^2 a}{r_0 \omega \nu} \tag{4.6.47}$$

where U_0 is a typical velocity along the pipe and ω is the frequency of the oscillation. The quantity U_0/ω is an average of a fluid particle amplitude of oscillation and ε is the ratio of this diplacement to a typical length of the problem, the quantity $\sqrt{r_0 a}$. The second parameter is a streaming Reynolds number. A third parameter is constructed based on ε and Re_s

$$Re_\omega = \frac{\omega a^2}{2\nu} = \frac{Re_s}{2\varepsilon^2}\,. \tag{4.6.48}$$

This is an oscillation Reynolds number representing the ratio of the radius of the pipe to the thickness of the Stokes layer $\sqrt{2\nu/\omega}$. Lyne (1970a) derives solutions to the problem in terms of matched asymptotic expansions based on the assumptions that Re_ω is very large, but allows the streaming Reynolds number to be either large or small.

It is well known that for steady fully developed viscous flow through a curved pipe, a secondary motion, "centrifuging," is generated that appears in the form of closed recirculating bubbles in planes perpendicular to the pipe axis. The situation here is different. The core of the pipe contains plug flow, that is, an inviscid and spatially uniform flow that oscillates with frequency

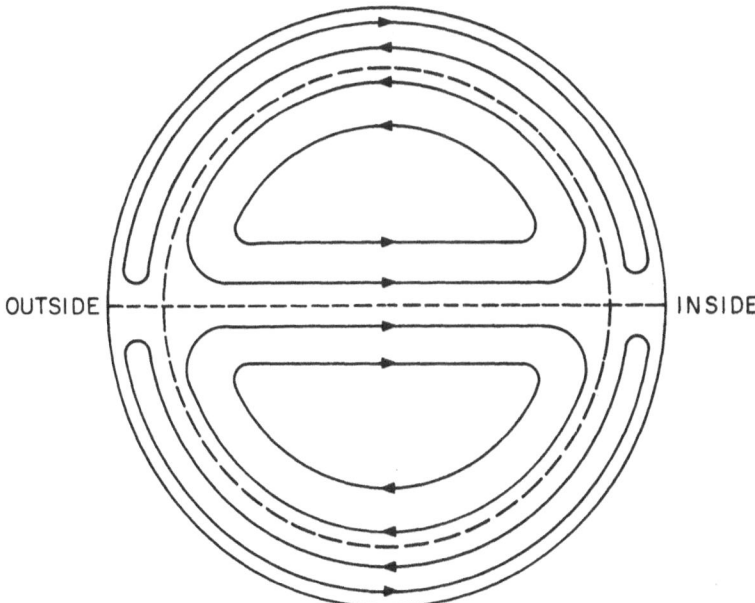

Fig. 4.9 Schematic representation of streaming flow in the cross section of a curved circular pipe.

ω. The Stokes layer is very thin. Steady streaming is now a secondary flow reminiscent of the centrifuging flow but actually opposite in direction. Some steady-streaming streamlines are shown in Fig. 4.9. There is again a pair of thin recirculating bubbles in the direction opposite to the main core streaming flow.

An impressive variety of fluid mechanics problems can be grouped under the category of oscillations with zero mean or a mean small compared to the driving oscillation. For example, the flow around an isolated island in an ocean in which there is a system of large-scale oscillating currents (Longuet-Higgins, 1970), the fluid flow in the cochlea of the mammal ear (Hallauer, 1974), the fluctuating flow through tubes of varying cross sections with immediate applications to blood flow (Hall, 1973; Schneck and Walburn, 1976). A more complete account of such contributions can be found in the review articles of Stuart (1972) and Riley (1975).

REFERENCES

Andrade, E. N., 1931. *Proc. R. Soc. London*, *A* **134**, 445.
Benjamin, T. B., 1959. *J. Fluid Mech.*, **6**, 101–205.
Bertelsen, A. F., 1974. *J. Fluid Mech.*, **64**, 589–597.
Carrière, Z., 1929. *J. Phys. Radium*, **10**, 198.
Dvorak, V., 1874. *Ann. Phys. Lpz.*, **151**, 634.
Faraday, M., 1831. *Trans. R. Soc. London*, **121**, 229.
George, C. B., and Sleath, J. F. A., 1979. *J. Hydraul. Res.*, 303–313.
Haddon, E. W., and Riley, N., 1979. *Q. J. Mech. Appl. Math.*, **32**, 263–282.
Hall, P., 1973. "Some Unsteady Viscous Flows and Their Stability," Ph.D. Thesis, London University.
Hallauer, W. L., 1974. "Nonlinear Mechanical Behavior of the Cochlea," Ph.D. Dissertation, Stanford University.
Holtsmark, J., Johnsen, I., Sikkeland, T., and Skavlem, S., 1954. *J. Acoust. Soc. Am.*, **26**, 26–39.
Knight, D. W., 1978. *J. Hydraul. Div.*, **104**, 839–855.
Lighthill, M. J., 1964. In *Laminar Boundary Layers*, ed. Rosenhead, L., Oxford Press, London and New York.
Longuet-Higgins, M. S., 1953. *Philos. Trans. R. Soc. A* **245**, 535–581.
Longuet-Higgins, M. S., 1970. *J. Fluid Mech.*, **42**, 701–720.
Lyne, W. H., 1971a. *J. Fluid Mech.*, **45**, 13–31.
Lyne, W. H., 1971b. *J. Fluid Mech.*, **50**, 33–48.
Rayleigh, Lord, 1876. *Philos. Mag.*, **5**, 257.
Rayleigh, Lord, 1883. *Philos. Trans. A* **175**, 1–21.
Rayleigh, Lord, 1894. *Theory of Sound*, second edition, Macmillan, New York.
Riley, N., 1965. *Mathematica*, **12**, 161–175.
Riley, N., 1967. *J. Inst. Math. Appl.* **3**, 419–434.
Riley, N., 1975. *SIAM Rev.*, **17**, 274–297.
Rott, N., 1964. In *Theory of Laminar Flows*, ed. Moore, F. K., Princeton Univ. Press, Princeton, New Jersey, pp. 412–438.
Schlichting, H., 1932. *Phys. Z.*, **33**, 327–335.
Schneck, D. J., and Walburn, F. J., 1976. *J. Fluids Eng.*, **98**, 707–714.
Sleath, J. F. A., 1970. *J. Fluid Mech.*, **42**, 111–123.
Sleath, J. F. A., 1972. *J. Marine Res.*, **30**, 295–304.
Sleath, J. F. A., 1974. *J. Marine Res.*, **32**, 13–24.
Sleath, J. F. A., 1976. *J. Hydraul. Res.*, **14**, 69–81.
Sleath, J. F. A., 1980. Private communication.

Stokes, G. G., 1847. *Trans. Cambridge Philos. Soc.*, **8**, 441–452.

Stokes, G. G., 1851a. *Trans. Cambridge Philos. Soc.*, **9**, Pt. II, 8–106; see also, *Math. Phys.*, **3**, 1–441.

Stokes, G. G., 1851b. *Trans. Cambridge Philos. Soc.*, **9**, 20–21.

Stuart, J. T., 1963. In *Laminar Boundary Layers*, Clarendon, Oxford, pp. 349–408.

Stuart, J. T., 1966. *J. Fluid Mech.*, **24**, 673–687.

Stuart, J. T., 1972. In *Recent Research of Unsteady Boundary Layers*, ed. Eichelbrenner, E. A., Laval University Press, Quebec, Canada **2**, 1–59.

CHAPTER 5

Oscillating Flows with Nonvanishing Mean

5.1 Introduction

This chapter may be considered a continuation of Chapter 4. It deals with an extension of the theory to flows with a nonvanishing mean. Most of the introductory remarks made in Section 4.1 of the previous chapter are therefore still valid. It is emphasized again that our interest is confined to the response of viscous flows to external disturbances, transient or periodic. No attempt is made to estimate possible interactions of the viscous flow with the external disturbing flow or with vibrations induced to solid surfaces.

The regions where viscous forces are important in external aerodynamics are essentially boundary layers and wakes. Their development influences the outer flow, and the term widely accepted for the calculation of such effects is "weak interaction" of the viscous field with the outer flow. One step further involves in turn the feedback from the readjusted outer flow to the viscous flow and this is known as "strong interaction."

Very little has been done up to now in the area of wakes and the large-scale turbulence generated by bluff bodies. There have been some interesting experimental studies, but our understanding of the phenomena involved and our analytical capabilities are minimal. Surprisingly enough, unsteady wakes appear easier to understand and analyze. This is because unsteadiness and, especially, periodicity give rise to well-ordered large-scale eddies. The vortices that for steady flow occupy the wake region and rearrange themselves in a totally erratic way exchanging momentum and energy with each other appear to organize themselves in periodically disturbed flows. The wake of an oscillating airfoil, for example, is made up of a few discrete vortices in a pattern that periodically repeats itself with the frequency of the mechanical oscillation.

Purely numerical solutions of the problem of unsteady laminar boundary layers have appeared in the last decade. However, great computational simplification can be accomplished by virtue of perturbation methods. Moreover, asymptotic expansions and perturbation methods provide a much clearer view into the physical aspects of a problem. In this chapter, therefore, we devote a few sections to such methods, starting with the work of Lighthill, which historically has opened the area.

Engineering applications of unsteady boundary layers are numerous and of great importance in external aerodynamics. In almost all the real-life situations, airfoils encounter an unsteady environment. Helicopter blades rotate in a plane inclined with respect to the direction of the helicopter motion and experience a mean flow that fluctuates in direction and magnitude with the frequency of the rotor. The incidence may vary from $1°$ to $15°$ in one cycle. In the cascades of turbomachinery, a blade row finds itself in the wake of the earlier row, which contains the periodic disturbance of the wakes of the previous stage of blades. Even for systems designed to operate in steady environments, unsteadiness enters due to unexpected effects. For example, wings are flown through gusts, turbine engine fans experience distorted inlet flows due to crosswinds, etc. It therefore appears that a thorough investigation of such problems will inevitably involve considerations of unsteady viscous effects.

5.2 Perturbation Methods

Expansion in powers of the amplitude

Let us assume once again that the boundary-layer equations govern the unsteady flow. That is, we assume that it is possible to study the response of the laminar boundary layer to fluctuations of the outer flow velocity using the equations

$$\frac{\partial u}{\partial x} + \frac{\partial v}{\partial y} = 0, \qquad\qquad *(5.2.1)$$

$$\frac{\partial u}{\partial t} + u\frac{\partial u}{\partial x} + v\frac{\partial u}{\partial y} = \frac{\partial U_e}{\partial t} + U_e\frac{\partial U_e}{\partial x} + v\frac{\partial^2 u}{\partial y^2}. \qquad *(5.2.2)$$

The validity of the boundary-layer approximation in this case will be investigated in the following sections.

We are seeking solutions to Eqs. (5.2.1) and (5.2.2) subject to the boundary conditions

$$u = v = 0, \quad \text{at } y = 0, \qquad\qquad *(5.2.3)$$

$$u \rightarrow U_e, \qquad \text{as } y \rightarrow \infty, \qquad\qquad *(5.2.4)$$

where the outer flow is given by

$$U_e(x,t) = U_0(x) + \frac{\varepsilon}{2} U_1(x)(e^{i\omega t} + e^{-i\omega t}), \qquad *(5.2.5)$$

with ε a small parameter representative of the ratio of the fluctuating to the steady part of the outer flow.

With the experience already gained from the previous chapter, we may expect that the presence of the wall and the forced oscillation will give rise again to a Stokes layer with a thickness $\delta_{in} \propto \sqrt{\nu/\omega}$. Since the steady boundary layer will have a thickness $\delta \propto \sqrt{\nu L/U_\infty}$, we readily conclude that the two layers will be of the same order of magnitude if the ratio $\delta_{in}/\delta = \sqrt{U_\infty/L\omega}$ is of order one. Moreover, for very large frequencies we expect that the Stokes layer will be much thinner than the steady boundary layer.

At this point a word of caution seems pertinent. As Stuart (1971) has emphasized, the boundary-layer equation is certainly an appropriate mathematical model for the steady part of this problem. However, the unsteady disturbances may violate the boundary-layer approximation. This would be the case if a wavelike perturbation were traveling along the outer stream, with a wavelength of order δ and a wave speed comparable to the outer flow velocity. This point will be reconsidered in a later section.

Following Lighthill (1954), let us assume that the solution exists and can be expanded in powers of ε as follows:

$$u(x, y, t) = u_0(x, y) + \frac{\varepsilon}{2}\left(u_1 e^{i\omega t} + \bar{u}_1 e^{-i\omega t}\right) + \cdots, \qquad *(5.2.6)$$

$$v(x, y, t) = v_0(x, y) + \frac{\varepsilon}{2}\left(v_1 e^{i\omega t} + \bar{v}_1 e^{-i\omega t}\right) + \cdots, \qquad *(5.2.7)$$

where the quantities u_1, v_1 are complex and an overbar denotes the complex conjugate. The boundary conditions (5.2.3) and (5.2.4) then become

$$u_i = v_i = 0, \qquad \text{at } y = 0, \text{ for } i = 0, 1, \qquad *(5.2.8)$$

$$u_0 \to U_0, u_1 \to U_1, \quad \text{at } y \to \infty. \qquad *(5.2.9)$$

Notice that the form of the expansions (5.2.6) and (5.2.7) permits a phase advance or delay in the response of the boundary layer. The phase angle will then be a function of x and y and will be given by $\phi = \tan^{-1} \{$Imaginary (u_1)/Real $(u_1)\}$. A schematic representation of the complex vectors is shown in Fig. 5.1. The vectors U_0 and u_0 are fixed and represent the mean flow. The vectors U_1 and u_1 rotate with the constant angular velocity ω having their toes at the tips of the vectors U_0 and u_0, respectively. The projections on the horizontal axis, as shown in the figure, represent the instantaneous velocities of the outer flow and the boundary-layer flow, respectively.

We may now substitute the above expressions in Eqs. (5.2.1) and (5.2.2)

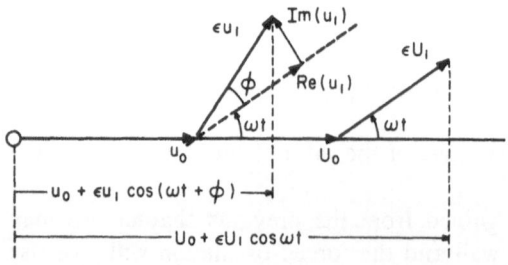

Fig. 5.1 Polar diagram of the complex velocities.

and collect terms of order ε^0, and ε^1. Notice that for terms of order ε we collect separately coefficients of $e^{i\omega t}$ and $e^{-i\omega t}$ since these quantities vary independently with time. We arrive then at two equations that are the complex conjugate of each other, and hence it will be necessary to work with only one of the two. After some algebra we finally arrive at

$$\frac{\partial u_0}{\partial x} + \frac{\partial v_0}{\partial y} = 0, \qquad\qquad *(5.2.10)$$

$$u_0 \frac{\partial u_0}{\partial x} + v_0 \frac{\partial u_0}{\partial y} = U_0 \frac{dU_0}{dx} + \nu \frac{\partial^2 u_0}{\partial y^2}, \qquad\qquad *(5.2.11)$$

$$\frac{\partial u_1}{\partial x} + \frac{\partial v_1}{\partial y} = 0, \qquad\qquad *(5.2.12)$$

$$\begin{aligned} i\omega u_1 + u_0 \frac{\partial u_1}{\partial x} + u_1 \frac{\partial u_0}{\partial x} + v_0 \frac{\partial u_1}{\partial y} + v_1 \frac{\partial u_0}{\partial y} \\ = i\omega U_1 + U_0 \frac{\partial U_1}{\partial x} + U_1 \frac{\partial U_0}{\partial x} + \nu \frac{\partial^2 u_1}{\partial y^2}. \end{aligned} \qquad *(5.2.13)$$

In the first two equations we recover as we expected the steady-state equations, which together with the boundary conditions (5.2.8) and (5.2.9) can be solved by any of the familiar methods. Equations (5.2.12) and (5.2.13) then are linear in character and can be solved by any of the conventional methods to generate the first-order correction. It is observed that according to this method and at this level of approximation, the averaged flow is nothing but the steady flow, which corresponds to the averaged boundary conditions. Nonlinear effects will influence the time-averaged flow only if terms of ε^2 are included in the expansion. Any of the steady-state solutions available can be used as coefficients of Eqs. (5.2.12) and (5.2.13). These equations can then be solved to generate the unsteady correction to the steady flow.

No exact closed form solutions to Eqs. (5.2.10)–(5.2.13) exist. However, these equations can be solved numerically for any body configuration. To investigate the physical significance of high and low frequencies and in an effort to present numerical results, Lighthill (1954) expanded his solutions in terms of the frequency parameter as well. One of the two cases is described in the following subsection.

The high–frequency approximation

An approximate solution to Eq. (5.2.12 and 13) can be derived in closed form if we assume that the frequency of oscillation is very large (Lighthill, 1954). The time derivatives of the velocities, that is, the terms containing $i\omega$ in Eq. (5.2.13) dominate over most of the other terms. The large accelerations will be balanced only by the viscous diffusion term and the equation reduces to

$$iwu_1 = i\omega U_1 + \nu \frac{\partial^2 u_1}{\partial y^2}.$$
*(5.2.14)

The reader should notice that the same terms are dominant in motions started impulsively from rest, where again accelerations are large (see Section 3.4) and the approximation is good for small times. A quick inspection of Eq. (5.2.14) indicates that with the high-frequency approximation, the correction due to unsteadiness, that is, the functions (u_1, v_1) are independent of the mean flow (u_0, v_0). The physical significance of this fact is obvious. It implies that the unsteady response of the boundary layer is approximately independent of the local as well as the upstream behavior of the mean viscous flow. The unsteady part of the flow is driven only by the unsteady component of the externally imposed fluctuations. It is further interesting to note that for larger frequencies, the thickness of the inner layer, $\delta_{in} = (\nu/\omega)^{1/2}$, decreases but the outer-flow unsteadiness penetrates across the boundary layer via the pressure term and drives the Stokes layer, unaffected by the presence of the mean boundary-layer flow. Equation (5.2.14) can be rewritten as follows:

$$\frac{\partial^2(u_1 - U_1)}{\partial y^2} - \frac{i\omega}{\nu}(u_1 - U_1) = 0,$$
*(5.2.15)

and its closed form solution is

$$u_1 - U_1 = C_1 \exp\left(y\sqrt{\frac{i\omega}{\nu}}\right) + C_2 \exp\left(-y\sqrt{\frac{i\omega}{\nu}}\right),$$
*(5.2.16)

where C_1 and C_2 are constants of integration. Conditions (5.2.8) and (5.2.9) now are met if $C_1 = 0$ and $C_2 = -U_1$, and thus

$$u_1(x, y) = U_1(x)\left[1 - \exp\left(-y\sqrt{\frac{i\omega}{\nu}}\right)\right].$$
*(5.2.17)

It is instructive to calculate the response of the skin friction within the approximation of large frequency

$$\tau = \mu \frac{\partial u}{\partial y} = \mu \frac{\partial u_0}{\partial y} + \frac{\varepsilon}{2} \mu \frac{\partial u_1}{\partial y} e^{i\omega t}$$

$$= \mu \frac{\partial u_0}{\partial y} + \frac{\varepsilon}{2} \mu U_1 \sqrt{\frac{\omega}{2\nu}} (1 + i)e^{i\omega t}.$$
*(5.2.18)

The reader is reminded that only the real part of this expression is physically meaningful. We therefore conclude that within the high-frequency approximation the wall shear is always leading the outer flow by a phase angle of 45°.

Lighthill (1954) proceeds further to expand in powers of $i\omega$ for the case of small frequencies. He then derives a closed-form solution using the integral method of Kármán and Polhausen. Integral methods are today considered a little outmoded since exact numerical solutions of boundary-layer equations can be generated in a few seconds with a modern computing machine. Nevertheless, Lighthill was able to derive essentially all the important physical information for this particular problem.

Most of the subsequent investigations are based on expansions in powers of the frequency ω or the quantity $\xi = x\omega/U_\infty$. Expansions for small or large values of these parameters have been carried out by Lam and Rott (1960), Rott and Rosenweig (1960), Illingworth (1958), Kestin et al. (1961), Gersten (1965), King (1966), Goshal and Goshal (1970), Ishigaki (1971), and others. Ackerberg and Phillips (1972) reconsidered the expansions for small and large values of the parameter ξ. They found that for large ξ, a double boundary layer is developed. The inner layer is a Stokes layer oscillating with zero-mean flow while a modified Blasius motion exists outside and convects the mean flow downstream. A very similar problem is the case of unsteady stagnation flow that was studied independently by Rott (1956) and Glauert (1956) and later by Tokuda and Yang (1966).

5.3 The Method of Averaging

An alternative method of analysis is due to Lin (1956) and is based on the method of averaging. The method is essentially developed to treat random oscillations of arbitrary amplitude, as, for example, turbulence. Its application to the problems of organized oscillations with discrete frequencies inherits the basic advantages and disadvantages of the original method. Large amplitudes of oscillations can be easily handled. However, the systems of equations are not closed and higher-order moments must be modeled in an arbitrary way.

Consider the decomposition of all unsteady quantities into their mean and fluctuating part

$$q(\mathbf{r}, t) = \bar{q}(\mathbf{r}) + q'(\mathbf{r}, t), \qquad *(5.3.1)$$

where[†]

$$\bar{q}(\mathbf{r}) = \lim_{T \to \infty} \frac{1}{T} \int_0^T q(\mathbf{r}, t)\,dt \qquad *(5.3.2)$$

and by definition $\overline{q'} = 0$. After decomposing the velocity field (u, v) and the

---

[†]An overbar in this section is used to denote time average whereas in the previous section it is used to denote complex conjugate. This is unfortunate but no other symbol was found that would be less confusing or less cumbersome.

outer flow velocity U_e in this fashion, we substitute in Eqs. (5.2.1) and (5.2.2) to arrive at

$$\frac{\partial \bar{u}}{\partial x} + \frac{\partial \bar{v}}{\partial y} + \frac{\partial u'}{\partial x} + \frac{\partial v'}{\partial y} = 0, \qquad *(5.3.3)$$

$$\frac{\partial u'}{\partial t} + \bar{u}\frac{\partial \bar{u}}{\partial x} + u'\frac{\partial u'}{\partial x} + \bar{u}\frac{\partial u'}{\partial x} + u'\frac{\partial \bar{u}}{\partial x} + \bar{v}\frac{\partial \bar{u}}{\partial y} + v'\frac{\partial u'}{\partial y} + \bar{v}\frac{\partial u'}{\partial y} + v'\frac{\partial \bar{u}}{\partial y}$$

$$= \frac{\partial U_e'}{\partial t} + \bar{U}_e\frac{\partial \bar{U}_e}{\partial x} + U_e'\frac{\partial U_e'}{\partial x} + \bar{U}_e\frac{\partial U_e'}{\partial x} + U_e'\frac{\partial \bar{U}_e}{\partial x} + \nu\frac{\partial^2 \bar{u}}{\partial y^2} + \nu\frac{\partial^2 u'}{\partial y^2} \, .$$

$$*(5.3.4)$$

Upon averaging Eqs. (5.3.3) and (5.3.4) and noting that the operation of averaging can be interchanged with the operation of differentiation, we derive equations for the mean flow:

$$\frac{\partial \bar{u}}{\partial x} + \frac{\partial \bar{v}}{\partial y} = 0, \qquad *(5.3.5)$$

$$\bar{u}\frac{\partial \bar{u}}{\partial x} + \bar{v}\frac{\partial \bar{u}}{\partial y} + \overline{u'\frac{\partial u'}{\partial x}} + \overline{v'\frac{\partial u'}{\partial y}} = \bar{U}_e\frac{d\bar{U}_e}{dx} + \overline{U_e'\frac{\partial U_e'}{\partial x}} + \nu\frac{\partial^2 \bar{u}}{\partial y^2} \, . \quad *(5.3.6)$$

The reader will notice that terms containing a fluctuating factor, as, for example, the term $u'\partial\bar{u}/\partial x$, average to zero but the average of the product $u'\partial u'/\partial x$ generates a nonvanishing contribution to the equation for the mean quantities. Such terms are very familiar to students of turbulence as Reynolds stresses. It is due to their presence that the solution deviates from the quasisteady solution if the amplitude of the oscillation is not very small. Equations (5.3.5) and (5.3.6) cannot be solved for the mean part of the flow without prior knowledge of the fluctuating part.

Equations that govern the fluctuating part of the flow can be derived by subtracting Eqs. (5.3.5) and (5.3.6) from Eqs. (5.3.3) and (5.3.4), respectively. We thus arrive at

$$\frac{\partial u'}{\partial x} + \frac{\partial v'}{\partial y} = 0, \qquad *(5.3.7)$$

$$\frac{\partial u'}{\partial t} + \left(\bar{u}\frac{\partial u'}{\partial x} + \bar{v}\frac{\partial u'}{\partial y}\right) + \left(u'\frac{\partial \bar{u}}{\partial x} + v'\frac{\partial \bar{u}}{\partial y}\right)$$

$$+ \left(u'\frac{\partial u'}{\partial x} + v'\frac{\partial u'}{\partial y}\right) - \left(\overline{u'\frac{\partial u'}{\partial x}} + \overline{v'\frac{\partial u'}{\partial y}}\right) \qquad *(5.3.8)$$

$$= \frac{\partial U_e'}{\partial t} + \bar{U}_e\frac{\partial U_e'}{\partial x} + U_e'\frac{\partial \bar{U}_e}{\partial x} + U_e'\frac{\partial U_e'}{\partial x} - \overline{U_e'\frac{\partial U_e'}{\partial x}} + \nu\frac{\partial^2 u'}{\partial y^2} \, .$$

The problem is clearly coupled and once again we have to resort to an approximation. Equations for the second-order moments can easily be derived. This, however, does not solve the problem, because the process introduces third-order moments. The hypothesis now (Lin, 1956) is that we deal with quantitites that fluctuate about a mean with a very large frequency, a case quite common, for example, in turbomachinery flows. We then assume that the time derivatives in Eq. (5.3.8) are large and balance directly with the viscous term. Thus, the nonlinear terms are again disregarded, even though the amplitude of the oscillation is not necessarily small. The system of Eqs. (5.3.7) and (5.3.8) then becomes

$$\frac{\partial u'}{\partial x} + \frac{\partial v'}{\partial y} = 0, \qquad\qquad *(5.3.9)$$

$$\frac{\partial u'}{\partial t} = \frac{\partial U'_e}{\partial t} + v\frac{\partial^2 u'}{\partial y^2}, \qquad\qquad *(5.3.10)$$

and a closed-form solution for specific flow configurations or an exact numerical solution for arbitrary body contours is possible. For example, if the outer flow fluctuates harmonically,

$$U'_e(x,t) = U_1(x)\sin \omega t \qquad\qquad *(5.3.11)$$

with $U_1(x)$ an arbitrary function of x, the solution to our problem is

$$u'(x,y,t) = U_1(x)\sin \omega t$$

$$- \exp\left(-y\sqrt{\frac{\omega}{2v}}\right)\sin\left[\omega t - y\sqrt{\frac{\omega}{2v}}\right], \qquad *(5.3.12)$$

$$v'(x,y,t) = -\int_0^y\left(\frac{\partial u'}{\partial x}\right)dy. \qquad\qquad *(5.3.13)$$

In this way we have essentially solved for the fluctuating part of the flow first and we are ready to turn back to the equations for the mean flow. Now the nonlinear terms $u'\partial u'/\partial y$ and $v'\partial u'/\partial y$ can be calculated and thus become a forcing function for the mean flow, Eqs. (5.3.5) and (5.3.6).

Recapitulating, we note that the method is valid for large frequencies only. It estimates by a crude assumption the oscillatory part of the motion. However, it is tailored to calculate immediately the steady-streaming effect on the mean flow. Not being restricted to small amplitudes of oscillation, it would be most appropriate for the calculation of nonlinear effects. Very few numerical results have been obtained with this method and since the 1950s no effort has been made to use it in practical engineering problems.

5.4 The Boundary-Layer Approximation

Consider again the full Navier–Stokes equations for two-dimensional incompressible flow and the energy equation [See Eqs. (1.2.4)–(1.2.6)] written in the traditional notation:

$$\frac{\partial u}{\partial x} + \frac{\partial v}{\partial y} = 0, \tag{5.4.1}$$

$$\frac{\partial u}{\partial t} + u\frac{\partial u}{\partial x} + v\frac{\partial u}{\partial y} = -\frac{\partial p}{\partial x} + \frac{1}{Re}\left(\frac{\partial^2 u}{\partial y^2} + \frac{\partial^2 u}{\partial x^2}\right), \tag{5.4.2}$$

$$\frac{\partial v}{\partial t} + u\frac{\partial v}{\partial x} + v\frac{\partial v}{\partial y} = -\frac{\partial p}{\partial y} + \frac{1}{Re}\left(\frac{\partial^2 v}{\partial y^2} + \frac{\partial^2 v}{\partial x^2}\right), \tag{5.4.3}$$

$$\frac{\partial T}{\partial t} + u\frac{\partial T}{\partial x} + v\frac{\partial T}{\partial y} = \frac{1}{Re}\frac{1}{Pr}\left[\frac{\partial^2 T}{\partial y^2} + \frac{\partial^2 T}{\partial x^2}\right]$$

$$+ Ec\left\{\left(\frac{\partial u}{\partial y}\right)^2 + \frac{2}{Re}\left[\left(\frac{\partial u}{\partial x}\right)^2 + \left(\frac{\partial v}{\partial y}\right)^2\right]\right. \tag{5.4.4}$$

$$\left. + \frac{\partial u}{\partial x}\frac{\partial v}{\partial y}\right]\right\},$$

where $Pr = \nu/\alpha$, $Ec = U_\infty^2/c(T_w - T_\infty)$ and $Re = U_\infty L/\nu$ are the Prandtl, Eckert and Reynolds numbers, respectively. All quantities are rendered dimensionless as described in Chapter 1. Usually at this point the boundary-layer approximation is introduced, that is, with the assumption of $Re \gg 1$, the quantities y and v are stretched by the factor $\sqrt{Re}$ and the terms of order $1/Re$ are neglected. This results in the elimination of Eq. (5.4.3) altogether. However, here all the terms of order $1/Re$ have been retained for reasons that will soon become apparent.

Consider outer flows given by:

$$Q_e(x,t) = Q_0(x) + \varepsilon Q_1(x)\cos\omega t, \tag{5.4.5}$$

where Q_e represents the instantaneous value of any of the velocity components or temperature. This corresponds to small amplitude harmonic oscillations of the flow parameters imposed on an otherwise time-independent free-stream value Q_0. Here ω is the angular frequency of the oscillations and ε is a small dimensionless parameter. Solutions to Eqs. (5.4.1)–(5.4.4) are then assumed in the form (Telionis and Romaniuk, 1977; Romaniuk, 1978)

$$q(x,y,t) = q_0(x,y) + \frac{\varepsilon}{2}\left[q_1(x,y)e^{i\omega t} + \overline{q_1}e^{-i\omega t}\right]$$

$$+ \varepsilon^2\left[q_s(x,y) + \tfrac{1}{2}\left(q_2 e^{2i\omega t} + \overline{q_2}e^{-2i\omega t}\right)\right] + \cdots, \tag{5.4.6}$$

where q represents any of the four unknown field quantities: u, v, T, and p. Subscript 0 denotes the steady solution, i.e., the $O(\varepsilon^0)$ approximation. Complex notation has been introduced and q_1 represents a complex amplitude of the oscillatory part of the solution. An overbar denotes again the complex conjugate. Streaming components q_s (representing u_s, v_s, T_s, and p_s) must be included in addition to the second harmonics in order to balance the nonlinear steady terms appearing in the equations of order ε^2.

The assumed form of the solution for the velocity components and temperature is now substituted into Eqs. (5.4.1)–(5.4.4). Collecting coefficients of ε^0 gives

$$\frac{\partial u_0}{\partial x} + \frac{\partial v_0}{\partial y} = 0, \tag{5.4.7}$$

$$u_0 \frac{\partial u_0}{\partial x} + v_0 \frac{\partial u_0}{\partial y} = -\frac{\partial p_0}{\partial x} + \frac{1}{Re}\left(\frac{\partial^2 u_0}{\partial y^2} + \frac{\partial^2 u_0}{\partial x^2}\right), \tag{5.4.8}$$

$$u_0 \frac{\partial v_0}{\partial x} + v_0 \frac{\partial v_0}{\partial y} = -\frac{\partial p_0}{\partial y} + \frac{1}{Re}\left(\frac{\partial^2 v_0}{\partial y^2} + \frac{\partial^2 v_0}{\partial x^2}\right), \tag{5.4.9}$$

$$u_0 \frac{\partial T_0}{\partial x} + v_0 \frac{\partial T_0}{\partial y} = \frac{1}{Re}\frac{1}{Pr}\left(\frac{\partial^2 T_0}{\partial y^2} + \frac{\partial^2 T_0}{\partial x^2}\right)$$

$$+ Ec\left\{\left(\frac{\partial u_0}{\partial y}\right)^2 + \frac{2}{Re}\left[\left(\frac{\partial u_0}{\partial x}\right)^2 + \left(\frac{\partial v_0}{\partial y}\right)^2\right.\right. \tag{5.4.10}$$

$$\left.\left. + \frac{\partial u_0}{\partial x}\frac{\partial v_0}{\partial y}\right]\right\}.$$

Except for the time derivatives, the above set of equations has the same form as Eqs. (5.4.1)–(5.4.4).

We may now proceed with the classical steps of deriving the boundary-layer equations. At the zeroth-order approximation, assuming that $Re \gg 1$, all the terms of order $1/Re$ are neglected. The quantities y and v_0 are stretched by the factor $\sqrt{Re}$

$$Y = y\sqrt{Re}, \qquad V_0 = v_0\sqrt{Re}. \tag{5.4.11}$$

This results in eliminating Eq. (5.4.9), since all its terms, except for $\partial p_0/\partial y$, are of order $1/\sqrt{Re}$, and pressure becomes a known function of x only, which can be calculated from Bernoulli's equation. The governing equations then assume the familiar boundary-layer form:

$$\frac{\partial u_0}{\partial x} + \frac{\partial v_0}{\partial y} = 0, \tag{5.4.12}$$

$$u_0 \frac{\partial u_0}{\partial x} + V_0 \frac{\partial u_0}{\partial Y} = U_0 \frac{dU_0}{\partial x} + \frac{\partial^2 u_0}{\partial Y^2}, \tag{5.4.13}$$

$$u_0 \frac{\partial T_0}{\partial x} + V_0 \frac{\partial T_0}{\partial Y} = \frac{1}{Pr}\left[\frac{\partial^2 T_0}{\partial Y^2}\right] + Ec\left(\frac{\partial u_0}{\partial Y}\right)^2, \tag{5.4.14}$$

with the boundary conditions:

$$u_0 = V_0 = 0,\ T_0 = 1, \quad \text{at } Y = 0; \tag{5.4.15}$$

$$u_0 = U_0,\ T_0 = 0, \quad \text{as } Y \to \infty. \tag{5.4.16}$$

The terms of order $1/Re$ neglected here may be comparable to the steady-streaming terms if $1/Re = O(\varepsilon^2)$. They are therefore formally included in the equations describing the second-order steady contribution and some numerical tests have been performed to determine their significance. The outcome of these tests will be discussed later together with other results.

Collection of coefficients of ε yields

$$\frac{\partial u_1}{\partial x} + \frac{\partial v_1}{\partial y} = 0, \tag{5.4.17}$$

$$i\omega u_1 + u_0 \frac{\partial u_1}{\partial x} + u_1 \frac{\partial u_0}{\partial x} + v_0 \frac{\partial u_1}{\partial y} + v_1 \frac{\partial u_0}{\partial y}$$
$$= i\omega U_1 + U_0 \frac{\partial U_1}{\partial x} + U_1 \frac{\partial U_0}{\partial x} + \frac{1}{Re} \left(\frac{\partial^2 u_1}{\partial y^2} + \frac{\partial^2 u_1}{\partial x^2} \right), \tag{5.4.18}$$

$$i\omega T_1 + u_0 \frac{\partial T_1}{\partial x} + u_1 \frac{\partial T_0}{\partial x} + v_0 \frac{\partial T_1}{\partial y} + v_1 \frac{\partial T_0}{\partial y}$$
$$= \frac{1}{RePr} \left(\frac{\partial^2 T_1}{\partial y^2} + \frac{\partial^2 T_1}{\partial x^2} \right) + 2Ec \frac{\partial u_0}{\partial y} \frac{\partial u_1}{\partial y}, \tag{5.4.19}$$

with the boundary conditions

$$u_1 = v_1 = 0, \qquad T_1 = T_{1w}, \quad \text{at } y = 0; \tag{5.4.20}$$

$$u_1 = U_1, \qquad T_1 = T_{1e}, \quad \text{as } y \to \infty. \tag{5.4.21}$$

The boundary conditions for T_1 allow in general the temperatures at the wall and at the edge of the layer to fluctuate with the amplitudes T_{1w} and T_{1e}, respectively. The above boundary-layer approximation is valid only if the function $U_1(x)$ is not changing sharply with x. Otherwise, the ratio U_∞/L is not a good estimate of the order of magnitude of the derivative $\partial u_1/\partial x$. This would be the case, for example, if the unsteadiness were induced in the outer flow by a control surface or if a wavelike disturbance with a wavelength of the order of the boundary-layer thickness were traveling downstream.

A question also arises whether the condition $Re \gg 1$ is sufficient for the boundary-layer approximation to hold for the oscillatory motion, and if not, what should the criterion be for the validity of the boundary-layer approximation at this level? In order for the solution to Eqs. (5.4.17)–(5.4.21) to represent a more significant correction to the steady part than the neglected terms of $O(1/Re)$, the inequality $1/Re \ll \varepsilon$ must be satisifed. In fact a stronger inequality $1/Re \ll \varepsilon^2$ must hold to make the second-order steady-streaming correction more important than the ellipticity. This is not a very restrictive condition: an $\varepsilon = 1/10$ would require $Re = O(10^3)$, which is still a low value for the external flows being discussed.

For the oscillatory motion, the characteristic time is $T_1 = 1/\omega$ and the characteristic length scale is $L_1 = U_\infty/\omega$. The typical velocity amplitude is assumed the same as for the mean flow, i.e., U_∞. The actual amplitude of the oscillatory outer flow is of course of the order εU_∞. However, ε is factored out when the equations that govern the oscillatory flow are derived and the proper outer-flow velocity amplitude becomes of the order U_∞. Second derivatives of u_1^* with respect to x^* and y^* may then be estimated as follows:

$$\frac{\partial^2 u_1^*}{\partial x^{*2}} = O\left[\frac{U_\infty}{(U_\infty/\omega)^2}\right], \tag{5.4.22}$$

and with the y^*-scale smaller than the x^*-scale by a factor of $Re_1^{1/2}$,

$$\frac{\partial^2 u_1^*}{\partial y^{*2}} = O\left[\frac{U_\infty}{(U_\infty/\omega)^2} Re_1\right], \tag{5.4.23}$$

where $Re_1 = U_\infty(U_\infty/\omega)/\nu = U_\infty^2/\omega\nu$ is the Reynolds number based on the length scale of the oscillatory motion. The condition allowing the use of boundary-layer approximation is expressed in terms of the second derivative

$$\frac{\partial^2 u_1^*}{\partial x^{*2}} \ll \frac{\partial^2 u_1^*}{\partial y^{*2}}, \tag{5.4.24}$$

which, via Eqs. (5.4.22) and (5.4.23), results in the inequality

$$Re_1 = \frac{U_\infty^2}{\omega\nu} \gg 1. \tag{5.4.25}$$

This is a necessary condition for the boundary-layer approximation to hold at this level.

On the other hand, as mentioned earlier, the Reynolds number must be large:

$$Re = \frac{U_\infty L}{\nu} \gg 1 \gg \frac{1}{\varepsilon^2} \tag{5.4.26}$$

so that the mean pressure gradient normal to the wall is negligible and the second-order correction due to unsteadiness is more significant than the neglected diffusion terms.

The boundary-layer thicknesses for the mean and oscillatory flows are determined by Re and Re_1, respectively. In general it might be necessary to introduce two separate scales to describe these two parts of motion, depending on the ratio δ_0/δ_1, where

boundary-layer thickness
of the mean flow:
$$\delta_0 \propto \frac{L}{\sqrt{Re}} = \sqrt{\frac{\nu L}{U_\infty}} \tag{*5.4.27}$$

and

boundary-layer thickness
of the oscillatory flow:
$$\delta_1 \propto \frac{L_1}{\sqrt{Re_1}} = \sqrt{\frac{\nu}{\omega}}. \tag{*5.4.28}$$

Then

$$\frac{\delta_0}{\delta_1} = \sqrt{\frac{\omega L}{U_\infty}} = \sqrt{St} \, , \qquad\qquad\qquad *(5.4.29)$$

where St is the Strouhal number.

The Strouhal number characterizes the relative importance of purely time-dependent inertia terms with respect to convection terms:

$$St = \frac{\omega L}{U_\infty} = \underbrace{(\omega U_\infty)}_{O\left[\frac{\partial u^*}{\partial t^*}\right]} \Bigg/ \underbrace{\left(U_\infty \frac{U_\infty}{L}\right)}_{O\left[u^* \frac{\partial u^*}{\partial x^*}\right]} . \qquad *(5.4.30)$$

$$\begin{array}{cc} \text{local} & \text{convective} \\ \text{acceleration} & \text{acceleration} \end{array}$$

Two cases, corresponding to the two extreme values of St have to be considered.

case 1:

$$\sqrt{St} \gg 1, \qquad \frac{\delta_0}{\delta_1} \gg 1. \qquad\qquad\qquad *(5.4.31)$$

This corresponds to Lighthill's high frequency, where changes in time are dominating and the convective terms can be neglected. The "oscillatory boundary layer" (often called Stokes sublayer) is very thin and a proper description of the situation requires separate treatment of the inner (oscillatory) and outer (steady) parts of the boundary layer.

case 2:

$$\sqrt{St} \ll 1, \qquad \frac{\delta_0}{\delta_1} \ll 1. \qquad\qquad\qquad *(5.4.32)$$

In this case temporal changes are negligibly small compared to convection terms. This corresponds to a quasi-steady situation. Here the periodic motion is so slow that the oscillatory boundary layer is much thicker than the "inner" layer corresponding to the mean flow. Such flows are not encountered very often in practice. For example, the flow with average velocity of 20 ft/sec over a body of characteristic length $L = 5$ ft should have to oscillate with frequency 0.5 Hz to be considered quasi-steady. Table 5.1 gives the smallest and the largest values of Strouhal number (St) and equivalent ratios δ_0/δ_1 for some of the experimental data available today.

The experimental information on laminar oscillating flows is rather limited. In addition to the work of Hill and Stenning (1960), work on laminar flows has also been reported by Despard and Miller (1971) and Morkovin et al. (1972). The first concentrates on separation and will be discussed again in

Table 5.1

	Laminar		Turbulent
	Hill and Stenning (1960)	Karlsson (1959)	Houdeville et. al (1976) Cousteix et al. (1977)
St_{min}	0.10	1.13	1.63
$(\delta_0/\delta_1)_{min}$	0.316	1.06	1.28
St_{max}	10.0	140.7	4.59
$(\delta_0/\delta_1)_{max}$	3.16	11.86	2.14

Chapter 7. The second contains a lucid description of the physical character-istics of oscillatory flows, reports on some spot-checks of Hill and Stenning's data, and proceeds with transition to turbulence.

Stretching the quantities y and v at the order ε^1 by a different factor than in the equations of order ε^0 introduces a very complex notation that requires transformations from the one system to the other. If we assume that the dimensionless numbers Re and Re_1 are not far from each other, then stretching the two fields by the same factor, say $\sqrt{Re}$, permits a straightfor-ward coupling between quantities u_0, v_0 and u_1, v_1 and Eqs. (5.4.17)–(5.4.19) are written as follows:

$$\frac{\partial u_1}{\partial x} + \frac{\partial V_1}{\partial Y} = 0, \tag{5.4.33}$$

$$i\omega u_1 + u_0 \frac{\partial u_1}{\partial x} + u_1 \frac{\partial u_0}{\partial x} + V_0 \frac{\partial u_1}{\partial Y} + V_1 \frac{\partial u_0}{\partial Y}$$
$$= i\omega U_1 + U_0 \frac{\partial U_1}{\partial x} + U_1 \frac{\partial U_0}{\partial x} + \frac{\partial^2 u_1}{\partial Y^2}, \tag{5.4.34}$$

$$i\omega T_1 + u_0 \frac{\partial T_1}{\partial x} + u_1 \frac{\partial T_0}{\partial x} + V_0 \frac{\partial T_1}{\partial Y} + V_1 \frac{\partial T_0}{\partial Y}$$
$$= \frac{1}{Pr} \frac{\partial^2 T_1}{\partial Y^2} + 2Ec \frac{\partial u_0}{\partial Y} \frac{\partial u_1}{\partial Y}. \tag{5.4.35}$$

5.5 Numerical Results and Their Physical Significance

It is well known, as described in the introduction of this and the earlier chapter, that any disturbances in the oncoming free stream velocity are transmitted instantly to the outer incompressible flow. This is true, provided the acoustic wavelength associated with the frequency of fluctuations is large compared with the length of the body. However, the response of the flow within the boundary layer is not instantaneous. The familiar effects of

momentum, vorticity, and energy convection and diffusion generate variations in the amplitudes and phases of the oscillations.

Some insight can be gained by mere observation of the equations. Consider first the case of a flat plate. Then, no pressure gradient exists if the flow
is steady. Once some unsteadiness sets in, the term $\partial U_e / \partial t$ generates the
pressure field necessary to accelerate the viscous flow. This disturbance is
instantly transmitted across the boundary layer and drives the velocity
fluctuations as shown in Eq. (5.4.34). In the absence of any other pressure
gradient effects, it gives rise to the term $i\omega U_1$ and therefore tends to generate
a phase lead in the flow. This effect is more clearly identified in terms of real
variables. Consider, for example, an outer-flow fluctuation given by

$$U_e(t) = U_0 + U_1 \sin \omega t, \tag{5.5.1}$$

where U_0 and U_1 are constants. The unsteady contribution to the pressure
gradient then becomes

$$\frac{\partial U_e}{\partial t} = U_1 \omega \sin(\omega t + \pi/2). \tag{5.5.2}$$

The tendency to create phase leads in the flow are counteracted by the
viscous stresses in the flow. The inertia of the fluid resisting the imposed
fluctuations is represented by the term $i\omega u_1$ in Eq. (5.4.34). If brought to the
right-hand side of the equation, it can be seen that it produces a phase lag.
The two effects are therefore counteracting each other. In the inner part of
the boundary layer inertia effects are negligible and the unsteady pressure
gradient drives the flow, generating large phase leads. In the outer part of the
boundary layer we encounter sizable inertia effects that counterbalance the
pressure gradient effects and in fact generate small phase lags.

For outer flows that contain a space variation of pressure, the situation
may change drastically. In terms of real variables, for example, an outer-
velocity distribution given by

$$U_e(x,t) = U_0(x) + U_1(x)\sin \omega t \tag{5.5.3}$$

will give rise to pressure gradient terms of the form

$$\frac{\partial U_e}{\partial t} + U_e \frac{\partial U_e}{\partial x} = U_1 \omega \sin(\omega t + \pi/2) + U_0 \frac{dU_0}{dx}$$
$$+ \left(U_0 \frac{dU_1}{dx} + U_1 \frac{dU_0}{dx} \right)\sin \omega t. \tag{5.5.4}$$

Depending therefore on the sign of dU_0/dx and dU_1/dx, some of the terms
may have a phase difference of 90° with respect to the outer flow. This in
fact will be the case for adverse mean pressure gradients, that is, flows over
the leeward side of a body, which usually lead to separation.

The problem of harmonically oscillating boundary layers is widely accepted as a convenient test case for calculation methods. Most of the
computer codes developed in the late 1960s and early 1970s were tested

against the classical perturbation results of Lighthill (1954) and the experimental data of Hill and Stenning (1960). A detailed description of numerical procedures is contained in Chapter 3. In this section we briefly consider the steps required for the calculation of periodic flows.

In purely numerical methods all the derivatives in the differential equations are expressed in finite difference form and the system is solved numerically. As a result, boundary and initial conditions are needed. The most convenient way to start such calculations is to derive the solution for steady flow and use the results as the initial conditions for the calculations of the unsteady flow. This has been essentially done by Farn and Arpaci (1966), Tsahalis and Telionis (1974), and McCroskey and Philippe (1975), who integrated numerically the laminar boundary-layer equations for periodic flows. The numerical methods employed for laminar flow are identical with the calculations of unsteady turbulent boundary layers, provided the mathematical models are parabolic. More details on this topic and references are included in the next chapter.

As an example consider here a periodic Howarth flow. That is, a linearly retarded outer flow (Howarth, 1938)

$$U_0 = (1 - bx)/(1 - b) \tag{5.5.5}$$

on which we impose a periodic disturbance in the form of Eq. (5.2.5)

$$U_1 = \varepsilon \frac{1 - bx}{1 - b} , \tag{5.5.6}$$

where ω is the dimensionless frequency. A steady flow field is generated (Tsahalis and Telionis, 1974) by solving numerically the steady-state boundary-layer equations for an outer-flow distribution given by

$$U_e = \frac{1 - bx}{1 - b} (1 + \varepsilon). \tag{5.5.7}$$

This information is stored in the (x, y) plane to serve as an initial condition. Unsteadiness is then introduced by incrementing time and considering a change of the outer-flow distribution to

$$U_e = \frac{1 - bx}{1 - b} (1 + \varepsilon \cos \omega \Delta t). \tag{5.5.8}$$

Notice that time derivatives of the outer flow at the first step vanish because of the specific choice of the point of starting the calculations. Physically the mathematical conditions described correspond to a steady decelerating flow, which smoothly goes through some dynamic disturbance leading to a periodic flow. The first steps of the calculations are not part of the periodic flow. In fact, it proved necessary to go through two or three oscillations before a purely periodic flow sets in. Typical numerical results showing the transient behavior of the flow (Tsahalis and Telionis, 1974) are shown in Fig. 5.2.

It is emphasized again that numerical methods integrating directly the

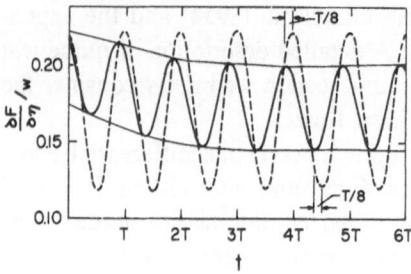

Fig. 5.2 Dimensionless wall shear
for $\xi_s = \omega x_s/U_\infty = 3.16$ and U_e
$= 10/9\ (1 - 9x/10) + x\cos\omega t/90$.
- - -, quasi-steady; ——, unsteady. Here x_s is the distance to
mean separation.

differential equations require storage of information in all mesh points of the
two-dimensional physical space (x, y). As time is incremented and a new
(x, y) plane is swept, the new information replaces the old and the process is
repeated. To generate a periodic solution, it is necessary to sweep the domain
of integration $T/\Delta t$ times for one period, where T is the period and Δt the
time increment. On the other hand, methods based on perturbation techniques essentially eliminate one variable, with a drastic reduction in computer space and time requirements. Such are the methods described in
Sections 5.2 and 5.3. No initial conditions are then needed and a periodic
flow field can be generated by a single sweep of the space plane x, y. Typical
results of both methods are discussed in this section.

A property of the flow, traditionally selected for comparison is the phase
angle of the skin friction or, equivalently, the phase angle of the flow in the
innermost layer of the boundary layer. Figure 5.3 shows the skin friction
phase angle as a function of the frequency parameter $\xi = \omega x/U_\infty$ for
oscillations imposed on a flat plate flow. Lighthill's asymptotic solutions for

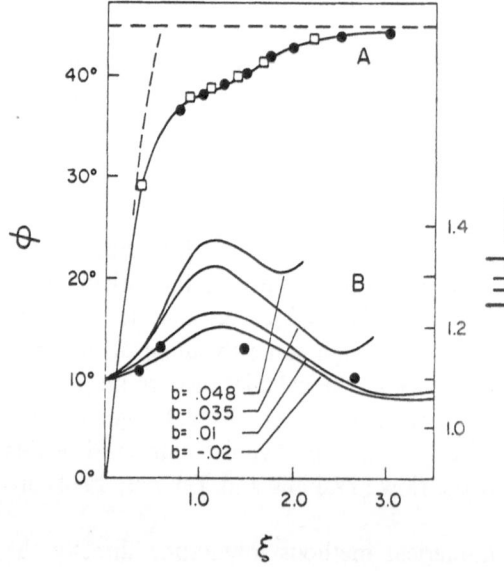

Fig. 5.3 The skin friction phase shift
(curves A, scale on left) - - -,
Lighthill (1954); □, Ackerberg and
Phillips (1972); •, Cebeci (1977);
——, Telionis and Romaniuk
(1977). Maxima of the amplitude of
velocity oscillations (curves B, scale
on right) for different values of the
pressure gradient parameter b; ——,
Telionis and Romaniuk (1977); •,
Hill and Stenning (1960) for $b = 0$.

Fig. 5.4 Profiles of the amplitude of velocity oscillations at $\xi = 2.844$ for $b \cong 0.035$; ——, Telionis and Romaniuk (1978); O, experimental and –·–, numerical results of Hill and Stenning (1960); – – –, asymptotic solution according to Lighthill (1954).

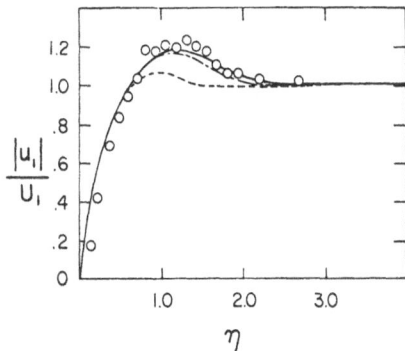

small and large frequencies are represented by a sharp, nearly linear increase and a flat horizontal curve, respectively, indicating that skin friction quickly reaches a periodic oscillation, leading the outer flow by 45°. It is very interesting that all the subsequent improvements are in full agreement, as Fig. 5.3 shows, and that all of them indicate an unexpected behavior in the neighborhood of $\xi = \omega x / U_\infty = 1$. It appears, indeed, that bridging of Lighthill's two asymptotic branches is not done in a smooth fashion. Instead, the curve has two points of inflection and changes curvature in a small region that extends until approximately $\xi = 2.0$.

The amplitude of the velocity fluctuation has a characteristic overshoot as shown in Fig. 5.4. This implies that the oscillation is in a sense amplified as one enters the boundary layer and the amplitude ratio $|u_1/U_1|$ exceeds the value 1. The picture may be a little misleading as McCroskey (1978) pointed out to the author. McCroskey maintains that in the calculations reported in McCroskey and Philippe (1975) no overshoot is observed if the instantaneous velocity is made dimensionless by the instantaneous outer-flow velocity and the equations are solved in terms of similarity coordinates

$$\tilde{\xi} = x/x_0, \qquad \tilde{\eta} = (U_e/2\tilde{\xi}\nu)^{1/2} y. \qquad (5.5.9)$$

In the physical space, the thickness of the boundary-layer fluctuates with the frequency of the outer-flow disturbance. As a result the steady part of the momentum of the outer stream is brought momentarily closer to the wall and appears as an overshoot if plotted in terms of the quantities of Fig. 5.3.

As the quantity $\xi = \omega x / U_\infty$ increases, the peak of the profile $|u_1/U_1|$ is moved closer to the wall and the region of $u_1/U_1 \cong 1$ extends deeper in the boundary layer. Physically this implies that a larger portion of the boundary layer oscillates in phase and with the same amplitude as the outer flow. For very large values of the distance x or the frequency ω, the unsteadiness in the largest part of the boundary layer resembles an oscillating plug flow. This behavior is clearly represented by the value of η at which the fluctuation amplitude peaks. Figure 5.5 shows this function for different values of the pressure gradient.

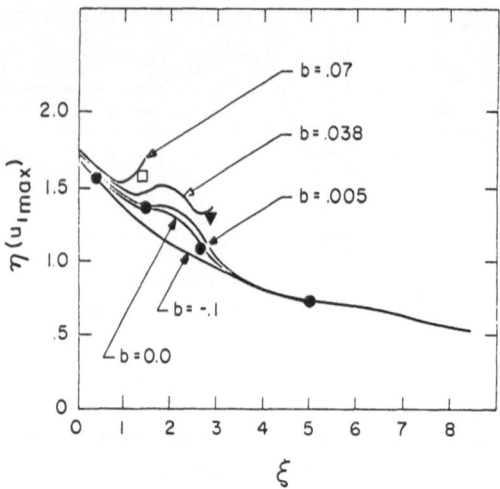

Fig. 5.5 Position of the maximum velocity overshoot as function of frequency parameter for different values of pressure gradient parameter b.

In Fig. 5.6 the effect of the frequency parameter again becomes apparent. It can be seen that the amplitude of the oscillation increases mildly with the pressure gradient parameter. However, it appears that the behavior of the maximum overshoot is not monotonic with the frequency parameter ξ.

Another characteristic feature of the flow is the phase angle whose wall values we plotted in Fig. 5.3. Figure 5.7 now shows the profile of this quantity. The competing effects of the unsteady pressure term $\partial U_e / \partial t$ and the boundary-layer inertia $\partial u / \partial t$ result in positive and negative values of this function as described earlier in this section. The boundary-layer flow in the neighborhood of the wall leads the outer flow by a phase angle of 45°. The trend reverses itself in the outer part of the boundary layer and the largest part of the boundary layer lags by phase angles of the order of 10°.

Turning now to the temperature field, we find that the numerical results (Telionis and Romaniuk, 1978) are in excellent agreement with the asymptotic solution of Lighthill for low frequencies. Figure 5.8 gives this comparison for a mild adverse pressure gradient and no dissipation. For intermediate

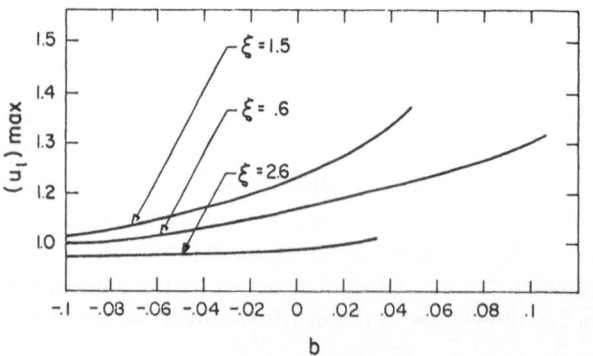

Fig. 5.6 Magnitude of the velocity overshoot as function of pressure gradient for different values of the frequency parameter.

Fig. 5.7 Phase angles for oscillatory flow. – – –, $\xi = 2.897$, Lighthill (1954), high-frequency approximation; – · – · –, theoretical, O, experimental, $\xi = 2.897$, Hill and Stenning (1960); ———, Tsahalis and Telionis (1974).

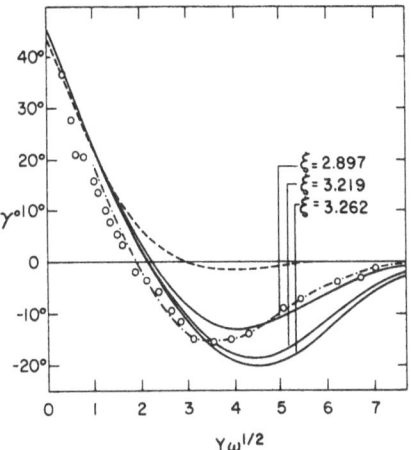

frequencies, no comparison is possible, since Lighthill's approximation holds for large or small frequencies only. There is no experimental data known to the author on temperature distribution in oscillating laminar boundary layers. Therefore, only the asymptotic solutions obtained by Lighthill's method for extreme values of frequency can be used for comparison.

A comparison of temperature profiles generated via Lighthill's asymptotic analysis and the semi-numerical method of Telionis and Romaniuk (1978), for high frequencies, indicate some discrepancies. The asymptotic profiles of Lighthill are well behaved and have only one point of inflection. The numerical profiles, however, are wiggly and may possess three points of inflection. This behavior is probably inherited from the velocity field, through coupling of the equations. A better agreement of the numerical solution with the high-frequency asymptotic approximation would obviously

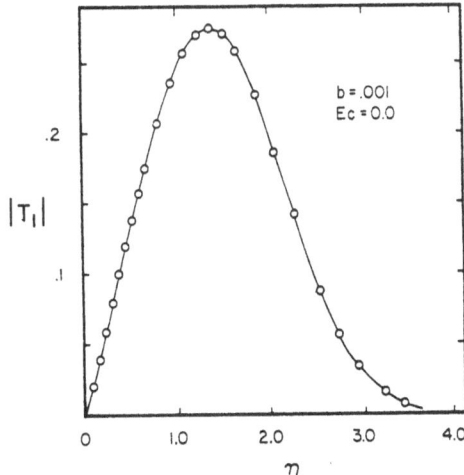

Fig. 5.8 Amplitude of oscillatory temperature profile, low frequency, no dissipation case: ———, Telionis and Romaniuk (1978); O, asymptotic solution Lighthill (1954).

be expected for larger ξ. However, for frequency parameters larger than 10, separation is approached even for the mild adverse pressure gradients considered, and the disagreement is bound again to grow. For larger adverse pressure gradients, separation occurs for even smaller ξ's, before the high-frequency region is reached. For larger frequencies, the T_1 amplitude quickly decreases, as we move away from the wall, thus generating a Stokes thermal wave.

For moderately large frequencies, the convergence of the numerical solution is rather slow and, in fact, the numerical results oscillate around the asymptotic solution as the parameter ξ grows. Such fluctuations are reminiscent of numerical instabilities but in this case they are a feature of the solution. Similar undulations have been indeed observed in the velocity profiles (Cebeci, 1977; Telionis and Romaniuk, 1978). Ackerberg and Phillips (1972) have indicated that the asymptotic results oscillate around the numerical results if plotted against the frequency parameter.

Figure 5.9 shows the temperature amplitude profiles for three different values of the Eckert number disclosing the effects of dissipation (Telionis and Romaniuk, 1978). The temperature at the wall and the outer boundary is assumed constant and its oscillations in the boundary layer are generated by the velocity field fluctuations only. It appears that for larger values of Ec, the T_1 amplitude, as well as the fluctuating part of the wall heat transfer, grow appreciably.

The next two figures show the effect of dissipation and frequency on the temperature amplitude for flows with outer temperatures oscillating in phase

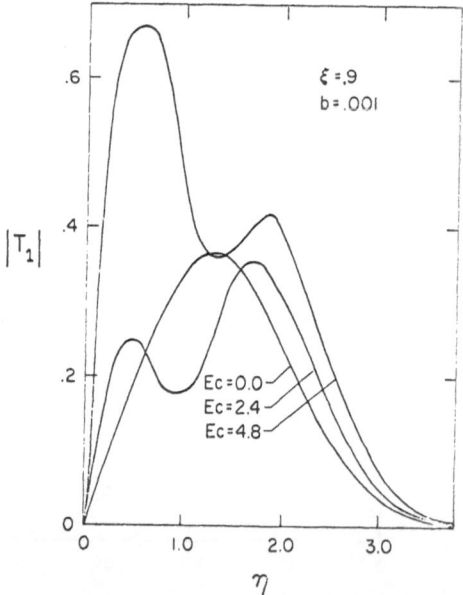

Fig. 5.9 Profiles of the amplitude of temperature oscillations for $T_1(\xi, 0)$ = 0.

Fig. 5.10 Profiles of the amplitudes of temperature oscillations for $T_1(\xi, 0) = 1$; ———, adiabatic wall; ---, cold wall.

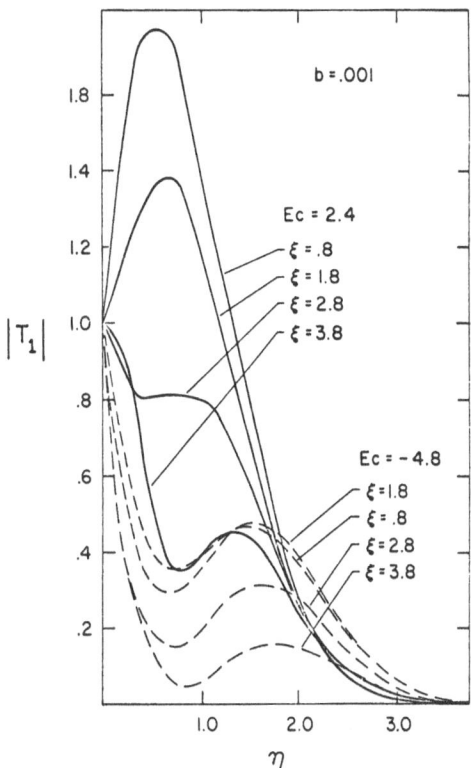

$b = .001$

$Ec = 2.4$

$\xi = .8$
$\xi = 1.8$
$\xi = 2.8$
$\xi = 3.8$

$|T_1|$

$Ec = -4.8$

$\xi = 1.8$
$\xi = .8$
$\xi = 2.8$
$\xi = 3.8$

η

with the outer velocity. For the case of an adiabatic wall, Fig. 5.10 shows that the temperature amplitudes decrease very sharply next to the wall and reach their minima and local maxima at distances from the wall that do not depend on frequency. Similar behavior of the T_1 amplitude is exhibited in this figure for a cold wall case. This means that for a given Eckert number, oscillatory heat transfer changes sign at fixed distances from the wall. Therefore the thicknesses of the layers being locally heated or cooled depend only on the Eckert number and not on the frequency of the oscillations. Figure 5.10 also shows that for $Ec = -4.8$, the amplitude of the temperature oscillation has an overshoot for low- and intermediate-frequency values. With the increase of ξ, the T_1 amplitude decreases very sharply with η regardless of the value of the Eckert number. The thickness of the oscillatory thermal boundary layer seems to increase for larger Ec.

In Fig. 5.11 profiles of the in-phase part of the η component of T_1 gradient $(\partial T_1/\partial \eta)$ are plotted for the case of a mild adverse pressure gradient and relatively low frequency, with no outer temperature oscillations imposed. Recall that this data could not have been obtained with expansion in powers of ω or its inverse. Several values of the Eckert number are considered, corresponding to the following cases: cold wall $(Ec = -2.4)$; no dissipation $(Ec = 0.0)$; adiabatic wall $(Ec = 2.4)$; and hot wall $(Ec = 4.8)$.

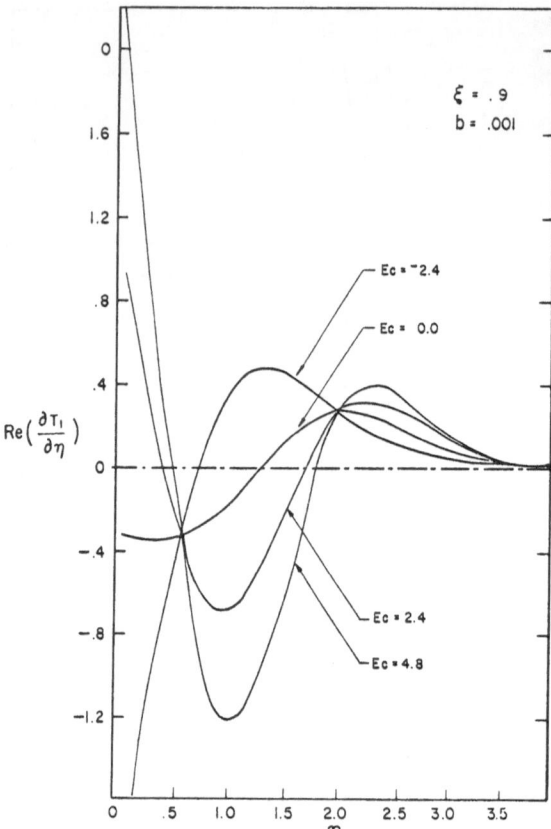

Fig. 5.11 Profiles of the in-phase part of the temperature amplitude gradient for a moderate frequency and different Eckert numbers.

In most cases the temperature gradient changes sign twice, indicating that there are regions where the excessive heat inflow generates local maxima or minima in the oscillating temperature profile. This is of course in agreement with general trends of the T_1 profiles, exhibited in the previous figure. An interesting feature of the profiles in Fig. 5.11 is that they intersect each other at fixed distances from the wall. For low values of the frequency parameter, there are two such intersections, but when $\xi \gtrsim 3.5$, a third may appear (Romaniuk, 1978). Physically, each intersection corresponds to a narrow region where the direction and magnitude of the oscillatory heat transfer does not depend on the dissipation and is determined entirely by other factors, most probably by the velocity field variations.

The quantity $|\partial T_1/\partial \eta|/(\partial T_0/\partial \eta)$ estimated at the wall will be denoted by ∇T_{1w} and referred to as the relative oscillatory wall heat transfer. Curve A in Fig. 5.12 shows the dependence of ∇T_{1w} on the frequency parameter, for $Ec = 0.0$ (no dissipation) and fixed outer temperature. The magnitude of ∇T_{1w} depends very strongly on ξ, especially for small and intermediate values of the frequency parameter and seems to approach the small positive value of 0.04 through damped oscillations. This is a very interesting charac-

Fig. 5.12 Wall values of oscillating temperature gradient (curve A) and streaming temperature gradient (curve B) as functions of frequency for $Ec = 0.0$.

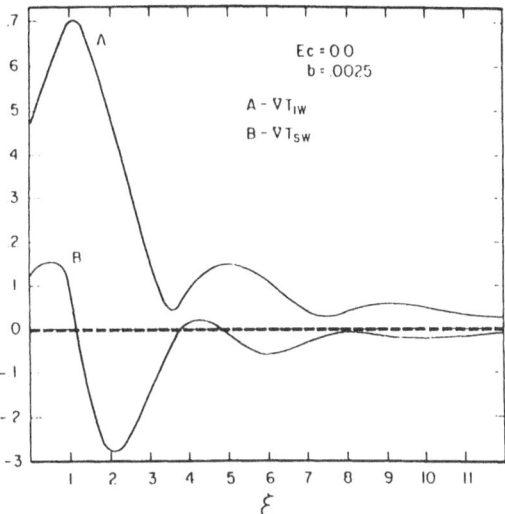

teristic: variations of the velocity field, oscillatory in time, generate a phenomenon periodic in ξ! The period seems to be $\Delta\xi \cong 4.0$. Other quantities, like $(u_1)_{\max}$, $(u_s)_{\max}$, $(\partial T_s/\partial\eta)_{\text{wall}}$, etc., also exhibit similar periodic dependence on ξ as Ackerberg and Phillips (1972) reported.

When dissipation effects are included, the character of ∇T_{1w} drastically changes (Romaniuk, 1978). Except for a very mild maximum at $\xi = 1.14$, the value of ∇T_{1w} is almost independent of ξ. This is consistent with the conclusion derived from Fig. 5.10, that is, the presence of dissipation has a damping effect on the amplitude of the temperature oscillations. One can add that large Ec has a similar damping effect on the oscillatory wall heat transfer.

5.6 Nonlinear Effects

The nonlinear character of the governing equations gives rise to steady contributions departing from the averaged flow. The phenomenon deserves a clearer physical description. Consider the steady flow $u(x, y)$ that corresponds to a steady outer-flow distribution $U_0(x)$. Consider further the periodic response of the boundary layer $u_p(x, y, t)$ due to a periodic disturbance imposed on the outer flow $U_0(x) + \varepsilon U_1(x)e^{i\omega t}$. By definition, the time average of the outer unsteady flow yields the steady part $U_0(x)$. However, due to nonlinear effects, the time average of the unsteady response of the boundary layer is not equal to its steady counterpart, $\overline{u_p(x, y, t)} \neq u(x, y)$.

Such nonlinear effects have been extensively studied for oscillating outer flows with a vanishing mean. The established terminology is "nonlinear" or "steady" or "acoustic streaming" and this has been the topic of the earlier chapter. Very little has been done to extend the concept to flows with nonvanishing mean. No special provision is of course necessary if the complete set of equations is numerically solved. In fact, such methods would be more appropriate for the investigation of nonlinear effects because they are not limited to small amplitudes, a domain in which the nonlinearities are weak. However, we gain a much better insight into the mechanics of nonlinear effects through an extension of the perturbation method described in Section 5.2.

Collecting terms of order ε^2 in Eqs. (5.2.1) and (5.2.2) gives two independent sets of equations: one for the second harmonic amplitudes u_2, v_2, T_2 and another for the "nonlinear streaming" contributions u_s, v_s, T_s. The first set of equations can be used to estimate the second-order correction to the unsteady part of the motion, representing a distortion of the harmonic response due to the nonlinearity of the boundary-layer equations. Of interest here is the second set, concerning the steady-streaming part of the solution. Assuming that the elliptic terms of the zeroth-order equations $(\partial^2 u_0 / \partial x^2)/ Re$ and $(\partial^2 T_0/\partial x^2)/ Re$ contribute at this level of the expansion, the streaming equations in their boundary-layer form read (Telionis and Romaniuk, 1978):

$$\frac{\partial u_s}{\partial x} + \frac{\partial v_s}{\partial y} = 0, \tag{5.6.1}$$

$$u_0 \frac{\partial u_s}{\partial x} + v_0 \frac{\partial u_s}{\partial y} + u_s \frac{\partial u_0}{\partial x} + v_s \frac{\partial u_0}{\partial y}$$

$$= -2\Re\left[\bar{u}_1 \frac{\partial u_1}{\partial x} + \bar{v}_1 \frac{\partial u_1}{\partial y} \right] + U_1 \frac{dU_1}{dx} + \frac{\partial^2 u_s}{\partial y^2} + \frac{1}{\varepsilon^2 Re} \frac{\partial^2 u_0}{\partial x^2}, \tag{5.6.2}$$

$$u_0 \frac{\partial T_s}{\partial x} + v_0 \frac{\partial T_s}{\partial y} + u_s \frac{\partial T_0}{\partial x} + v_s \frac{\partial T_0}{\partial y}$$

$$= \frac{1}{Pr}\left[\frac{\partial^2 T_s}{\partial y^2} + \frac{1}{\varepsilon^2 Re} \frac{\partial^2 T_0}{\partial x^2} \right] - 2\Re\left[\bar{u}_1 \frac{\partial T_1}{\partial x} + \bar{v}_1 \frac{\partial T_1}{\partial y} \right] \tag{5.6.3}$$

$$+ 2Ec\left[\frac{\partial u_0}{\partial y} \frac{\partial u_s}{\partial y} + \frac{1}{\varepsilon^2 Re} \left(\frac{\partial u_0}{\partial x} \right)^2 \right],$$

with homogeneous boundary conditions

$$u_s = v_s = T_s = 0, \quad \text{at } y = 0 \text{ and } y \to \infty. \tag{5.6.4}$$

Here $\Re$ denotes the real part of the expression that follows it.

The validity of the boundary-layer approximation at this level may again be questioned. The characteristic length scale for the steady-streaming part of the flow is the same as for the mean flow, i.e., $L_s = L$ since they are both steady. The characteristic velocity, however, is different and, as shown by

Stuart (1966), it can be estimated by

$$U_s = \frac{U_\infty^2}{\omega L} .$$ (5.6.5)

Second derivatives of u_s^* with respect to x^* and y^* can then be estimated

$$\frac{\partial^2 u_s^*}{\partial x^{*2}} = O\left(\frac{U_s}{L_s^2}\right) = O\left(\frac{U_\infty^2}{\omega L^3}\right),$$ (5.6.6)

$$\frac{\partial^2 u_s^*}{\partial y^{*2}} = O\left[\frac{U_s}{L_s^2\left(\frac{\nu}{U_s L_s}\right)}\right] = O\left[\frac{U_\infty^4}{\omega^2 L^3 \nu}\right].$$ (5.6.7)

The condition

$$\frac{\partial^2 u_s^*}{\partial x^{*2}} \ll \frac{\partial^2 u_s^*}{\partial y^{*2}}$$ (5.6.8)

leads to

$$\frac{U_\infty^2}{\omega L^3} \ll \frac{U_\infty^4}{\omega^2 L^3 \nu}$$ (5.6.9)

or

$$\frac{U_\infty^2}{\omega \nu} \gg 1,$$ (5.6.10)

which is the same requirement as Eq. (5.4.26). We may therefore conclude that if the boundary-layer approximation is valid for the oscillatory motion, it is also valid for the steady-streaming part of the solution.

As mentioned before, the terms of order $1/Re$ neglected at the $O(1)$ level may be significant compared to $O(\varepsilon^2)$ terms. It has to be determined now under what condition the second-order correction to the solution is more important than the neglected "elliptic" terms of the mean part of the flow. This is expressed by the relationship

$$\frac{1}{Re}\frac{\partial^2 u_0}{\partial x^2} \ll \varepsilon^2 \frac{\partial^2 u_s}{\partial y^2},$$ (5.6.11)

and since both x and y are of order 1, this is equivalent to

$$\frac{1}{\varepsilon^2 Re} \ll \frac{u_s}{u_0} .$$ (5.6.12)

However,

$$\frac{u_s}{u_0} = \frac{u_s^*}{u_0^*} = O\left(\frac{U^2}{\omega L}\frac{1}{U_\infty}\right) = O\left(\frac{U_\infty}{\omega L}\right);$$ (5.6.13)

therefore,

$$\varepsilon^2 \gg \frac{\nu}{U_\infty L} \frac{\omega L}{U_\infty} = \frac{1}{Re} \frac{\omega L}{U_\infty} \tag{5.6.14}$$

and finally

$$\frac{1}{Re_1} = \frac{\omega \nu}{U_\infty^2} \ll \varepsilon^2 \ll 1 \tag{5.6.15}$$

or

$$\frac{U_\infty^2}{\omega \nu} \gg \frac{1}{\varepsilon^2} \gg 1. \tag{5.6.16}$$

This condition imposes a particular relationship between the oscillatory Reynolds number Re_1 and the small parameter ε. For given Re_1 the parameter ε cannot be chosen arbitrarily small, even though in principle it would improve the accuracy of the solution, because then the ellipticity of the original equations would become more important than the steady-streaming correction.

The condition (5.6.16) is usually satisfied and holds for most flows of practical interest. For example, in the case of air flow at room temperature, with a characteristic velocity of 7 m/sec, the oscillatory motion with amplitude equal to 10% of the mean velocity ($\varepsilon = 0.1$) would satisfy condition (5.6.16) as long as the frequency of oscillations does not exceed 10^5 Hz. In general, inequality (5.6.16) holds for flows with $Re_1 \gg O(\varepsilon^{-3})$.

In deriving the governing equations in this chapter, the velocity and temperature fields were decomposed into mean, oscillatory, and steady-streaming components, each of them having a rather precise physical interpretation. The mean part corresponds to the solution of a boundary-layer flow with the same mean outer boundary conditions as the unsteady flow but in the absence of time-dependent oscillations. The oscillatory component represents the time-dependent response of the boundary-layer velocity and temperature fields to imposed oscillations. Mathematically it is simply the difference between the actual "total" solution and its time average. Since the outer oscillatory motion and the boundary-layer response are harmonic, the explicit time dependence is eliminated and the unknown variables are the amplitude and phase of the physical property considered. They are combined here, for mathematical convenience, in the form of the complex amplitude of oscillations.

Steady-streaming components of the solution constitute the time-independent response of the boundary layer to external oscillations. The term "steady streaming" has a clear physical meaning for the velocity field and is self-explanatory. Mathematically the streaming part represents the difference between the time average of the unsteady solution and the steady solution for a given geometry of the flow, that is, the quantity $\overline{u_p(x, y, t)} -$

Fig. 5.13 Steady-streaming
profiles.

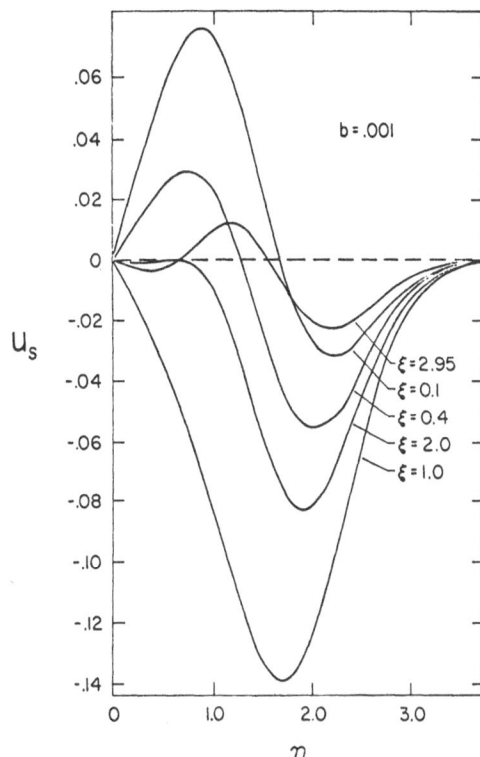

U_s

$\xi = 2.95$
$\xi = 0.1$
$\xi = 0.4$
$\xi = 2.0$
$\xi = 1.0$

$b = .001$

η

$u(x, y)$. For the temperature field, the term "streaming" may be misleading. It was adopted here by analogy to the velocity field.

The streaming contribution to the velocity field as calculated by Telionis and Romaniuk (1978) for a very mild adverse pressure gradient is shown in Fig. 5.13, where the u_s profile is plotted for different values of the frequency parameter ξ. Positive and negative values of a profile indicate that the secondary streaming flow follows some kind of a recirculating pattern, similar to the ones observed by Schlichting (1932) for the oscillations of a cylinder with zero-mean flow, and Schneck and Walburn (1976) for internal pulsating flows. The magnitude of u_s at a given distance from the wall depends strongly on the value of ξ and for $\xi \approx 1$, u_s reaches its largest value within the boundary layer. This phenomenon will be discussed later together with the streaming effects on temperature.

Profiles of streaming for the temperature field, for the case of no dissipation, are shown in Fig. 5.14. It should be emphasized here that this is a time-independent correction to the mean profile and the corresponding heat transfer represents a net gain or loss of heat. Depending on the value of the frequency parameter, the correction to the temperature field at a fixed distance from the wall may vary significantly and even change sign. The

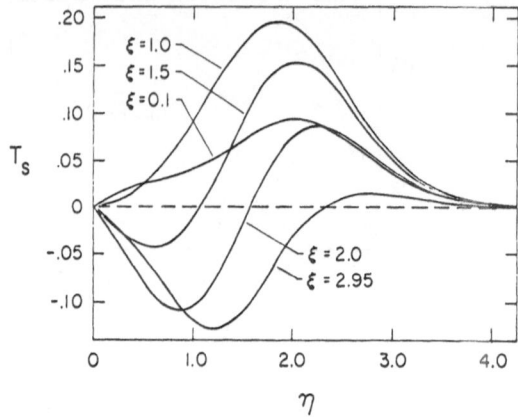

Fig. 5.14 Steady-stream-ing temperature profiles for $Ec = 0.0$.

magnitude of T_s varies with ξ in a manner similar to the u_s variations—namely, it reaches the largest value for $\xi \approx 1$.

For the case of no dissipation, the effect of "temperature streaming" on the wall heat transfer is very small, but it becomes quite significant for larger Eckert numbers. The effect of dissipation shifts the streaming temperature to larger and always positive values. The slopes at the wall appear also dramatically increased, especially for larger ξ. The magnitude of ∇T_{sw} increases monotonically with ξ and reaches values 5 to 6 times larger than ∇T_{1w}. Even though the magnitude of the steady-streaming contribution to the temperature field is very small, its effect on the wall heat transfer may be quite important. For example, in the case of outer-velocity oscillations with amplitude equal to about 10% of the mean flow velocity (i.e., $\varepsilon = 0.1$), and large dissipation ($Ec = 4.8$), the streaming contribution to the wall heat transfer may reach 30% of the mean wall heat transfer. Unfortunately, there are no experimental data known to the author confirming these findings as well as most of the other conclusions reached in this section.

In Fig. 5.15 the extrema of streaming velocity and temperature profiles, defined simply as the largest positive or negative deviations from zero at a given ξ station, are plotted versus the frequency parameter. These quantities reach their own extrema in the neighborhood of $\xi = 1.15$. This means that the "strongest" velocity and temperature streaming takes place for this particular value of ξ. It is interesting to note that minimum velocity streaming occurs where the amplitude of the oscillatory velocity is the highest, and then the maximum (positive) values of u_s are reached in the ξ region where $(u_1)_{max}$ is the lowest.

As mentioned before, the terms of order $1/Re$ neglected in the $O(\varepsilon^0)$ equations may become important in comparison with quantities $O(\varepsilon^2)$, depending on the value of the parameter $k = 1/\varepsilon^2 R$. Numerical results indicate that for $k = 10^{-5}$, the eventual correction is of the order of the truncation error and, for $k = 10^{-4}$, it becomes quite significant. Romaniuk

Fig. 5.15 (A) Relative oscillatory wall heat transfer $(\partial T_1/\partial\eta)/(\partial T_0/\partial\eta)_{\text{wall}}$. (B) The maxima of temperature streaming $(T_s)_{\max}$. (C) The maxima of velocity amplitude $|u_1|_{\max}$. (D) Relative "streaming wall heat transfer" $(\partial T_s/\partial\eta)/(\partial T_0/\partial\eta)_{\text{wall}}$. (E) The extrema of the velocity streaming $(u_s)_{\max}$ or $(u_s)_{\min}$.

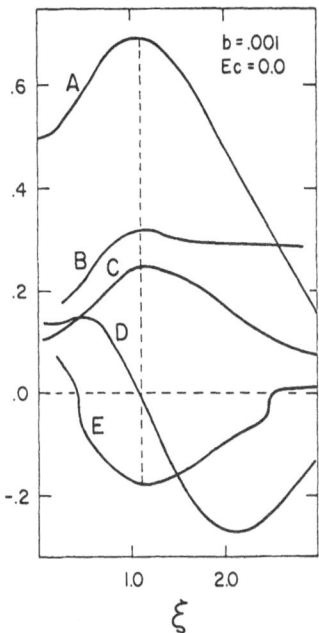

(1978) attempted the comparison of the profiles of velocity and temperature streaming with and without the correction for this particular k at fixed ξ. In the region far from the wall, the influence of the correction is negligible but it is quite large for intermediate and small values of η. The correction would become larger for larger k until eventually it overpowers the "pure streaming" itself. Then, of course, in order to obtain more accurate solutions, one should solve the full Navier–Stokes equations rather than calculate second-order corrections of the streaming type.

5.7 Traveling Waves

Long waves

Lighthill's theory (Lighthill, 1954) can be easily extended to include an outer-flow disturbance, which is traveling along the boundary, provided the wavelength of the disturbance is much larger than the thickness of the boundary layer. The problem has been tackled both experimentally and numerically by Patel (1975), and a brief description is included in this section. If the wavelength is comparable to the size of the boundary-layer thickness, then the full Navier–Stokes equations must be invoked and the character of the equations drastically changes. This is not a very common case in unsteady aerodynamics, unless one would be interested in the convection of disturbances. The problem then falls in the category of hydrodynamic stability, a discipline about which a lot has been written. The common points of stability analysis with the theory of oscillating boundary layers are brought out in the latter part of this section.

Consider an outer-flow disturbance in the form

$$U_e(x,t) = U_0(x) + \frac{\varepsilon}{2} U_1(x) e^{i(\omega t - \kappa x)} + CC, \qquad *(5.7.1)$$

where ω and κ are the frequency and wavenumber of the outer-flow oscillation and CC stands for the complex conjugate of the preceding expression. The solution is assumed in a similar form as in Eqs. (5.2.6) and (5.2.7).

$$u(x, y, t) = u_0(x, y) + \frac{\varepsilon}{2} u_1(x, y) e^{i(\omega t - \kappa x)} + CC, \qquad *(5.7.2)$$

$$v(x, y, t) = v_0(x, y) + \frac{\varepsilon}{2} v_1(x, y) e^{i(\omega t - \kappa x)} + CC, \qquad *(5.7.3)$$

and the boundary conditions are (5.2.8) and (5.2.9).

Substituting in the boundary-layer equation and collecting terms of order 1, we obtain again Eqs. (5.2.10) and (5.2.11), which govern the mean part of the flow. The terms of order ε, the counterpart of Eqs. (5.2.12) and (5.2.13), are (Patel, 1975)

$$\frac{\partial u_1}{\partial x} + \frac{\partial v_1}{\partial y} - i\kappa u_1 = 0, \qquad *(5.7.4)$$

$$i\omega u_1 + u_1 \frac{\partial u_0}{\partial x} + u_0 \frac{\partial u_1}{\partial x} - i\kappa u_0 u_1 + v_1 \frac{\partial u_0}{\partial y} + v_0 \frac{\partial u_1}{\partial y}$$

$$= i\omega U_1 - i\kappa U_0 U_1 + U_0 \frac{\partial U_1}{\partial x} + U_1 \frac{\partial U_0}{\partial x} + \nu \frac{\partial^2 u_1}{\partial y}. \qquad *(5.7.5)$$

A closed-form solution to this equation is not possible, but numerical solutions can be easily derived using any code for two-dimensional steady boundary-layer flow. High- and low-frequency approximations are also possible and have been derived by Patel. For high frequencies, an approximate form of the above equations, following Lighthill's arguments, are

$$\frac{\partial u_1}{\partial x} + \frac{\partial v_1}{\partial y} - i\kappa u_1 = 0, \qquad *(5.7.6)$$

$$i\omega u_1 - i\kappa u_0 u_1 = i\omega U_1 - i\kappa U_0 U_1 + \nu \frac{\partial^2 u_1}{\partial y^2}. \qquad *(5.7.7)$$

Note that according to our assumptions, the wavenumber is the same order of magnitude as ω. This can be easily brought out by expressing κ in terms of ω:

$$\kappa = \frac{\omega}{Q}. \qquad *(5.7.8)$$

The quantity Q then has the physical meaning of the wave velocity propagation. In Patel's experiments this is equal to 77% of the outer flow. The reduced amplitude ratio u_1/U_1 therefore satisfies the equation

$$\frac{d^2}{dy^2}\left(\frac{u_1}{U_1}\right) + \frac{i\omega}{\nu}\left(\frac{U_0}{Q}\frac{u_0}{U_0} - 1\right)\frac{U_0}{Q} = \frac{i\omega}{\nu}\frac{U_0}{Q} . \qquad *(5.7.9)$$

This can be solved numerically for any mean flow profile.

Patel (1975) describes solutions for both low and high frequencies and compares his analytical results with his experimental data. It is demonstrated that the profiles are qualitatively similar to those corresponding to standing waves, but the trends can be exaggerated; for example, overshoots of the fluctuating part of the velocity more than 100% over the outer-flow fluctuation are possible. Typical results are shown in Fig. 5.16, where the skin friction phase angle is plotted against the reduced frequency ξ. The asymptotic value of the phase is approaching 150° compared to 45° for standing waves. In the same figure turbulent flow results are also plotted for comparison, but a more detailed discussion will be found in the next chapter.

For both standing and traveling waves, the numerical results indicate that the steady part of the flow deviates very little from its steady counterpart. In other words, for amplitudes of up to 30% of the outer flow, the averaged field is essentially quasi-steady. Moreover, the behavior of the fluctuating part of the motion is practically independent of the amplitude of the external oscillations. Apparently nonlinearities become important only for amplitudes beyond the 50% range or perhaps in the neighborhood of separation where the induced amplitudes substantially grow.

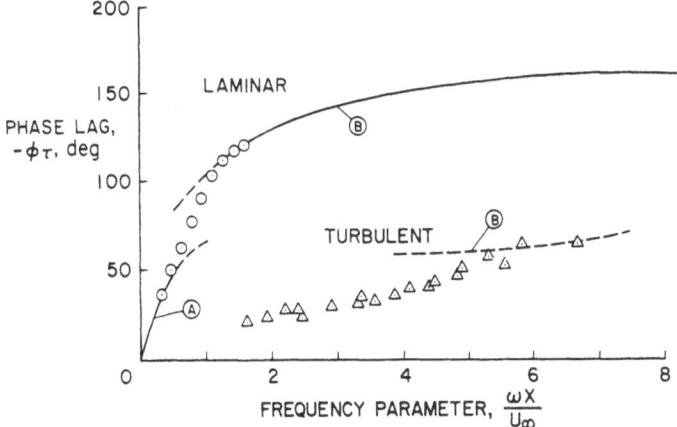

Fig. 5.16 The phase lag as a function of the frequency parameter for traveling waves. O, laminar experiments; △, turbulent experiments. (A) Low-frequency approximation; (B) high-frequency approximation.

Short waves

The study of short waves induced by the outer flow requires a careful reexamination of the governing equations. If the disturbance changes with x at the same rate it changes with y, then the boundary-layer approximation is not valid and the full Navier–Stokes equations must be invoked.

It is very interesting and a well-known fact in stability analysis that an elliptic disturbance, that is, a disturbance governed by diffusion in all space directions can be superimposed on a boundary layer. In real life, this type of disturbance is very common. Acoustic disturbances or free stream turbulence may fall in this category, but their amplitude is very small and their effects are relatively insignificant for the unsteady aerodynamic characteristics of a body. The possibility that such disturbances may affect macroscopic properties, as, for example, transition to a turbulent boundary layer or separation, is being investigated recently but this type of analysis is beyond the scope of this volume.

To demonstrate the physical aspects of the problem, we chose here to rephrase the analyses of Nayfeh, *et al.* (1975) and Saric and Nayfeh (1975). Both are developed in terms of multiple scales (Nayfeh, 1973; Chapter 6), but neither addresses the present problem of interest. The first is a study of inviscid disturbances, namely, the propagation of sound through a duct with variable cross section; the second is a study of the growth or decay of hydrodynamic instabilities imposed on a nonparallel flow.

In anticipation of a boundary-layer type of flow, let us assume a slow axial scale and the corresponding derivative

$$x_1 = \varepsilon x, \qquad \frac{\partial}{\partial x} = \varepsilon \frac{\partial}{\partial x_1}. \tag{5.7.10}$$

The solution is then assumed in the form

$$u(x, y, t) = u_0(x_1, y) + \varepsilon u_{01}(x_1, y) + \cdots + \varepsilon_1 u_1(x, y, t)$$
$$+ \varepsilon_1^2 u_{11}(x, y, t) + \cdots, \tag{5.7.11}$$

$$v(x, y, t) = \varepsilon v_0(x_1, y) + \varepsilon^2 v_{01}(x_1, y) + \cdots + \varepsilon_1 v_1(x, y, t)$$
$$+ \varepsilon_1^2 v_{11}(x, y, t) + \cdots. \tag{5.7.12}$$

The parameter ε represents the ratio of the normal to the parallel velocity component, while the parameter ε_1 represents the ratio of the magnitude of the disturbance to the mean part of the flow. Both are assumed small and for the analyses referenced above, ε_1 is much smaller than ε. This is not necessary for the present development.

The governing equations are the full Navier–Stokes equations, namely, Eqs. (1.2.16)–(1.2.18). However, using a typical boundary-layer thickness δ instead of a length typical of the size of the body is more desirable here. The Reynolds number Re_δ based on the thickness δ then appears in the equation rather than Re. The boundary conditions of the problem under consideration require that the flow merges smoothly to the potential flow.

$$u(x,\delta,t) = U_0(x_1) + \varepsilon U_{01}(x_1) + \cdots + \varepsilon_1 U_1(x)e^{i\theta} + \cdots, \tag{5.7.13}$$

$$v(x,\delta,t) = \varepsilon V_0(x_1) + \varepsilon_1 V_1(x)e^{i\theta} + \cdots, \tag{5.7.14}$$

where θ depends on x and t according to

$$\frac{\partial\theta}{\partial x} = \kappa(x_1), \qquad \frac{\partial\theta}{\partial t} = -\omega. \tag{5.7.15}$$

The quantity κ is the wavelength that may vary slowly with the axial distance x, and ω is a constant frequency.

The expressions (5.7.11)–(5.7.12) are substituted in the governing equations and for $\varepsilon_1 \ll \varepsilon$, the terms of $O(\varepsilon)$ are collected to yield

$$\frac{\partial u_0}{\partial x_1} + \frac{\partial v_0}{\partial y} = 0, \tag{5.7.16}$$

$$u_0\frac{\partial u_0}{\partial x_1} + v_0\frac{\partial u_0}{\partial y} = -\frac{\partial p_0}{\partial x} + \frac{\varepsilon}{Re_\delta}\left(\varepsilon\frac{\partial^2 u_0}{\partial x_1^2} + \frac{\partial^2 u_0}{\partial y^2}\right). \tag{5.7.17}$$

For the diffusion term to balance the convection, it is necessary that Re_δ is $O(\varepsilon^{-1})$, which is a well-known fact in boundary-layer theory. Notice that then, the more traditional Reynolds number based on the length of the body is $Re = (L/\delta)R_\delta = O(\varepsilon^{-2})$. Moreover, in the derivation of the above equations the introduction of the slow scale anticipates the fact that the x derivatives are much smaller than the y derivatives, but at the end the derivatives $\partial/\partial x_1$ and $\partial/\partial y$ appearing in Eqs. (5.7.16) and (5.7.17) are of order 1. The second derivative $\partial^2/\partial x_1^2$ on the right-hand side can be dropped now and the system reduces to Prandtl's equations.

The case $\varepsilon_1 = O(\varepsilon)$ should not be discarded here since our analysis is then still valid. The terms u_0 and v_0 can still be separated because they are steady, whereas all higher-order terms are oscillating harmonically and can therefore be grouped together as linearly independent terms. Collecting terms of order ε_1 or, in the case of $\varepsilon_1 = O(\varepsilon)$, terms $O(\varepsilon)$ that fluctuate with frequency ω, we obtain

$$\frac{\partial u_1}{\partial x} + \frac{\partial v_1}{\partial y} = 0, \tag{5.7.18}$$

$$\frac{\partial u_1}{\partial t} + u_0\frac{\partial u_1}{\partial x} + v_1\frac{\partial u_0}{\partial y} = -\frac{\partial p_1}{\partial x} + \frac{1}{Re_\delta}\left(\frac{\partial^2 u_1}{\partial x^2} + \frac{\partial^2 u_1}{\partial y^2}\right), \tag{5.7.19}$$

$$\frac{\partial v_1}{\partial t} + u_0\frac{\partial v_1}{\partial x} = -\frac{\partial p_1}{\partial y} + \frac{1}{Re_\delta}\left(\frac{\partial^2 v_1}{\partial x^2} + \frac{\partial^2 v_1}{\partial y^2}\right). \tag{5.7.20}$$

With $Re_\delta = O(\varepsilon^{-1})$ the diffusion terms on the right-hand side of the equation belong to a higher-order equation and should not affect the harmonic fluctuations. In other words, the response of the boundary layer to such disturbances should be inviscid.

In stability theory one introduces a disturbance stream function $\psi(x, y, t)$:

$$\psi(x, y, t) = \left[\psi_1(x_1, y) + \varepsilon\psi_{11}(x_1, y) + \ldots \right]e^{i\theta}, \tag{5.7.21}$$

in terms of which Eqs. (5.7.19) and (5.7.20) reduce to the classical Orr-Sommerfeld equation

$$\left(\frac{\partial^2}{\partial y^2} - \kappa^2 \right)^2 \psi_1 - i\kappa Re_\delta\left[\left(u_0 - \frac{\omega}{\kappa} \right)\left(\frac{\partial^2\psi_1}{\partial y^2} - \kappa^2\psi_1 \right) - \frac{\partial^2 u_o}{\partial y^2}\psi_1 \right] = 0,$$

$$\tag{5.7.22}$$

The significance of retaining the term multiplying Re_δ is now obvious and a basic element of stability theory. The quantity $u_0 - \omega/\kappa$ changes sign within the boundary layer and therefore the factor of R_δ becomes very small in a narrow layer, the critical layer, which contains a singularity. The critical layer coincides with the wall for a standing wave and this would generate difficulties in the handling of the boundary conditions. Upstream propagating waves would be free of singularities, but such waves would be rather unlikely in practical applications.

Saric and Nayfeh (1975) proceed further to derive the equation of ψ_{11}, which contains the effects of a growing boundary layer on the stability of disturbances and their subsequent amplifications.

REFERENCES

Ackerberg, R. C., and Phillips, J. H., 1972. *J. Fluid Mech.* **51**, 137–157.
Cousteix, J., Desopper, A., and Houdeville, R., 1977. "Structure and Development of a Turbulent Boundary Layer in an Oscillatory External Flow," ONERA TP 14; also presented at the Symposium on Turbulent Shear Flows, Penn. State University, Pennsylvania, Apr.
Cebeci, T., 1977. *Proc. R. Soc. London A* **355**, 225–238.
Despard, R. A., and Miller, J. A., 1971. *J. Fluid Mech.*, **47**, 21–31.
Farn, C. L. S., and Arpaci, V. S., 1966. *AIAA J.* **4**, 730–732.
Gersten, K., 1965. "Heat Transfer in Laminar Boundary Layers with Oscillating Outer Flow," AGARDograph, No. 97, 423–475.
Glauert, M. B., 1956. *J. Fluid Mech.*, **1**, 97–110.
Goshal, S., and Goshal, A., 1970. *J. Fluid Mech.*, **43**, 465–476.
Houdeville, R., Desopper, A., and Cousteix, J., 1976. "Experimental Analysis of Average and Turbulent Boundary Layer," ONERA TP, No. 30, also Rech. Aerosp. No. 1976–4.
Howarth, L., 1938. *Proc. R. Soc. London, A* **164**, 547–549.
Hill, P. G., and Stenning, A. H., 1960. *J. Basic Eng.*, **82**, 593–608.
Illingworth, C. R., 1958. *J. Fluid Mech.*, **3**, 471–493.
Ishigaki, H., 1971. *J. Fluid Mech.*, **47**, 537–546.
Karlsson, S. K. F., 1959. *J. Fluid Mech.*, **5**, 622–636.
Kestin, J., Maeder, P., and Wang, H. E., 1961. *Appl. Sci. Res.*, **A10**, 1–22.
King, W. S., 1966. *AIAA J.*, **4**, 994–1001.
Lam, S. H., and Rott, N., 1960. *Theory of Linearized Time-Dependent Boundary Layers*, Cornell Univ. GSAE Rep. AFOSR TN-60-1100.
Lighthill, M. J., 1954. *Proc. R. Soc. London*, **224A**, 1–23.
Lin, C. C., 1956. *Proc. 9th Int. Congr. Appl. Mech.*, Brussels, **4**, 155–169.
McCroskey, W. J., 1978. Private communication.

McCroskey, W. J., and Philippe, J. J., 1975. *AIAA J.*, **13**, 71–79.

Morkovin, M. V., Loehrke, R. I., and Fejer, A. A., 1972. In *Recent Research of Unsteady Boundary Layers*, ed. Eichelbrenner, E. A., **1**, 60–128.

Nayfeh, A. H., 1973. *Perturbation Methods*, Wiley-Interscience, New York.

Nayfeh, A. H., Telionis, D. P., and Lekoudis, S. G., 1975. *Progress Astro Aero.*, **37**, 333–351.

Patel, M. H., 1975. *Proc. R. Soc. London, A* **347**, 99–123.

Romaniuk, M. S., 1978. Ph.D. Thesis, Virginia Polytechnic Institute and State University.

Rott, N., 1956. *J. Fluid Mech.*, **1**, 97–110.

Rott, N., and Rosenweig, M. L., 1960. *J. Aero. Sci.*, **27**, 741–747.

Saric, W. S., and Nayfeh, A. H., 1975. "Nonparallel Stability of Boundary Layers with Pressure Gradients and Suction," AGARD Conf. Proc. No. 224.

Schlichting, H., 1932. *Phys. Z.*, **33**, 327–335.

Schneck, D. J., and Walburn, F. J., 1976. *J. Fluids Eng.*, **98**, 707–714.

Stuart, J. T., 1966. *J. Fluids Mech.*, **24**, 673–687.

Stuart, J. T., 1971. In *Recent Research of Unsteady Boundary Layers*, ed. Eichelbrenner, E. A., **1**, 1–46.

Telionis, D. P., and Romaniuk, M. S., 1977. *Soc. Eng. Sci. Proc. 12th Annual Meeting*, 1169–1180.

Telionis, D. P., and Romaniuk, M. S., 1978. *AIAA J.*, **16**, 488–495.

Tokuda, N., and Yang, W. J., 1966. *Proc. Third Int. Heat Trans. Conf.*, **2**, 223–232.

Tsahalis, D. Th., and Telionis, D. P., 1974. *AIAA J.*, **12**, 1469–1476.

CHAPTER 6

Unsteady Turbulent Flows

6.1 Introduction

The problem of turbulence is certainly one of the well-known "unsolved" problems in mechanics. It has challenged some of the most respected scientists of our times. New physical concepts have thus been invented, and the efforts to understand the problems stimulated the development of elegant stochastic theories. However, such methods are far from producing meaningful engineering results for practical problems. The designer still relies on heuristic phenomenological models that may appear almost arbitrary to the rigorous analyst. It is within this framework that we introduce here one more dimension in the problem that further increases its complexity: time dependence.

The classification of turbulence methods according to the number of differential equations involved and descriptions of the methods for calculations of steady turbulent boundary layers can be found in a recent review article by Reynolds (1976). The value of methods based on closure assumptions rests upon the fact that they predict with reasonable accuracy the mean features of the flow, which has been sufficient, at least up to now, for engineering applications. This, of course, was accomplished with painful reconsiderations and readjustments based on a large number of experimental data.

A reader who follows up the developments of such models is left with a strong feeling that theories are tailored to fit particular experimental data. In fact, very often such theories fail to predict the flow if the boundary conditions depart substantially from the model experimental cases. Yet, it appears that for many years to come such methods will be the only available tools for predicting turbulent boundary layers for engineering applications. Detailed information on existing steady turbulent-flow models appears in recent publications (Reynolds, 1970; Mellor and Herring, 1973; Launder and Spalding, 1974).

The definition of unsteady turbulent flow is in itself challenging. The basic characteristic of turbulent flow is that it is unsteady. This time dependence, however, is random and due to hydrodynamic instabilities rather than externally imposed disturbances. How turbulence is generated and sustained is a topic about which a lot has been written and we shall not even attempt to outline it here. The problem we want to address ourselves to is the response of viscous turbulent flow to boundary conditions that change with time, and mainly the response of turbulent boundary layers to externally imposed transient or periodic changes. Two forms of unsteadiness are therefore involved: the erratic unsteady motion due to hydrodynamic instability and the organized motion due to the external disturbances; the first is characterized by a wide range of frequencies, whereas the second is usually driven by a discrete frequency. The two motions coexist and under certain conditions they may interact with each other.

The problem can be formulated in terms of the instantaneous velocity and pressure, but this method would require computing facilities generations ahead of the present machines. Averaging techniques are necessary and this is where a basic conceptual difficulty arises. An ensemble average can be introduced again, but only subject to a well-defined condition. For periodic flows, that is, turbulent flows on which a periodic disturbance with known frequency is imposed, samples at a fixed phase are taken and ensemble-averaged. For transient flows the phenomenon is assumed to be repeated for a large number of times, and each time, samples are received at a fixed time after the initiation of the process; these samples are then again ensemble-averaged. The ensemble averages so defined are now functions of phase or time, respectively. One could probably arrive at an alternative definition of a time-dependent averaged quantity using the time averaging method with a finite time interval of integration:

$$\tilde{f}(t) = \frac{1}{T} \int_t^{t+T} f\, dt. \tag{6.1.1}$$

In principle then the function $\tilde{f}$ is a varying averaged quantity. However such a definition would be meaningful only if the time scale of the random fluctuation were much smaller than the scale of the organized disturbances and this is not always the case. It is within the framework of the above ideas that we shall be able to discuss and analyze steady and unsteady turbulent boundary layers, even though both problems are inherently unsteady.

Problems of oscillating turbulent flows are very often encountered in practice in a variety of applications as described in the introduction to this monograph. In fact it seems to be the rule rather than the exception that flows in engineering applications are unsteady and turbulent rather than steady and laminar. Typical examples are the flows over helicopter blades or through turbomachinery cascades. For all practical purposes the analysis of such problems is based today on quasi-steady models. However, it has been experimentally demonstrated that the unsteady effects are very significant. The use of quasi-steady models is therefore not well justified. In the last few

years investigators attempted to calculate unsteady turbulent fields by integrating the complete unsteady flow equations and using quasi-steady
models for their closure. Although such methods represent a considerable
improvement, their success is limited. Proper and direct modeling of the
specific unsteady problem seems to be necessary.

In this chapter a brief description of numerical methods based essentially
on quasi-steady models is given. The need for a more complete theory is
demonstrated and some new ideas are described, which may lead to correct
modeling of the phenomenon.

Experimental evidence on an oscillating turbulent jet indicated that it is
possible for energy to be transferred from the organized to the random
motion. This problem will be the next step in the investigation of oscillating
turbulent flows; namely, the study of the possible interaction between the
organized and the random fluctuation. To the knowledge of the author no
analytical work has been reported in this area up to now. The problem
however is challenging and of great engineering significance.

6.2 The Triple Decomposition

Let the time average and the phase average of a field quantity be defined by

$$\bar{f}(\mathbf{x}) = \lim \frac{1}{T} \int_{t}^{t+T} f(\mathbf{x}, t)\, dt, \tag{6.2.1}$$

$$\langle f(\mathbf{x}, t) \rangle = \lim_{N \to \infty} \frac{1}{N} \sum_{n=0}^{N} f(\mathbf{x}, t + n\tau), \tag{6.2.2}$$

respectively, where τ is the period of an externally imposed fluctuation.
Norris and Reynolds (1975) proposed an equivalent formula for the phase
averaging (see also Acharya and Reynolds, 1975).

$$\langle f(\mathbf{x}, \phi_0) \rangle = \lim_{N \to \infty} \frac{1}{N} \sum_{n=0}^{N} f(\mathbf{x}, \phi_0 + n\omega\tau), \qquad 0 < \phi_0 < 2\pi, \tag{6.2.3}$$

where ϕ is the phase of the periodic motion. At every instant that the driving
outer disturbance is at a particular phase ϕ_0, a sample of the field quantity is
taken and stored. This is done repeatedly for a large number of samples and
their ensemble average is defined as the phase average of the quantity. The
organized oscillations of the field quantity can be defined as

$$f''(\mathbf{x}, t) = \langle f(\mathbf{x}, t) \rangle - \bar{f}(\mathbf{x}), \tag{6.2.4}$$

and the random fluctuations are

$$f'(\mathbf{x}, t) = f(\mathbf{x}, t) - \langle f(\mathbf{x}, t) \rangle. \tag{6.2.5}$$

We can then decompose the quantity into three parts:

$$f(\mathbf{x}, t) = \bar{f}(\mathbf{x}) + f''(\mathbf{x}, t) + f'(\mathbf{x}, t). \tag{6.2.6}$$

Primes here denote the fluctuating part and not derivatives.

In order to decompose a random quantity in this fashion experimentally, one needs to know the frequency of the organized oscillations. In all practical applications of turbulent flows this is possible, since the disturbance is externally imposed. Simple examples are the frequency of rotation of a helicopter blade or a turbomachinery rotor, etc. In the laboratory the externally imposed disturbance provides a convenient reference for the phase-averaging process. In many other applications, however, the organized fluctuations may be buried and sophisticated techniques are required to decode the signal and calculate the discrete frequency of the organized motion.

All the above, of course, holds only if the higher harmonics are negligible. Otherwise one may have to resort to a multiple decomposition, a process that would become very cumbersome for both the experiment and the analysis.

A set of simple identities may appear useful in deriving the equations of the following sections. Their derivation (Acharya and Reynolds, 1975) is straightforward and hence not shown here.

$$\langle f' \rangle = 0, \qquad \overline{f''} = 0, \qquad \overline{f'} = 0, \tag{6.2.7}$$

$$\overline{\langle f \rangle} = \bar{f}, \qquad \langle \bar{f} \rangle = \bar{f}, \tag{6.2.8}$$

$$\overline{\bar{f}g} = \bar{f}\bar{g}, \qquad \langle f''g \rangle = f'' \langle g \rangle, \qquad \langle \bar{f}g \rangle = \bar{f} \langle g \rangle, \tag{6.2.9}$$

$$\overline{f''g'} = \overline{\langle f''g' \rangle} = 0. \tag{6.2.10}$$

The three components, that is, the time-averaged or mean, the organized fluctuation, and the random fluctuation are shown schematically in Fig. 6.1.

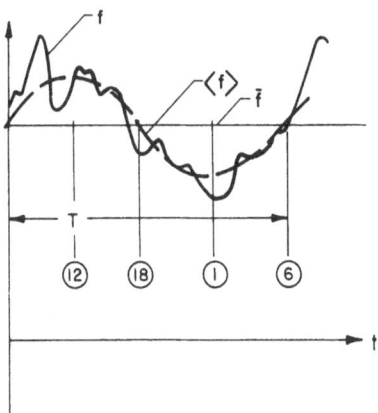

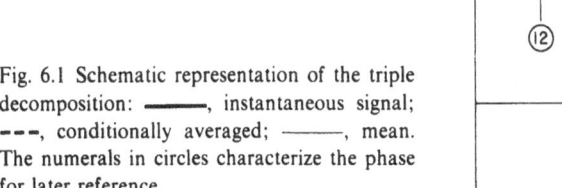

Fig. 6.1 Schematic representation of the triple decomposition: ——, instantaneous signal; - - -, conditionally averaged; ———, mean. The numerals in circles characterize the phase for later reference.

The equations for the mean and the organized motion

The equations that govern two-dimensional incompressible flow [Eqs. (1.2.10) and (1.2.11)] are repeated here for convenience:

$$\frac{\partial u_i}{\partial x_i} = 0, \tag{6.2.11}$$

$$\frac{\partial u_i}{\partial t} + u_j \frac{\partial u_i}{\partial x_j} = -\frac{\partial p}{\partial x_i} + \frac{1}{Re} \frac{\partial^2 u_i}{\partial x_j \partial x_j}, \tag{6.2.12}$$

where Re is the Reynolds number and all the quantities are dimensionless. Velocity and pressure are decomposed according to Eq. (6.2.6),

$$u_i = \bar{u}_i + u_i'' + u_i', \tag{6.2.13}$$

$$p = \bar{p} + p'' + p'. \tag{6.2.14}$$

Substituting Eq. (6.2.13) in Eq. (6.2.11) and time-averaging, we get

$$\frac{\partial \bar{u}_i}{\partial x_i} = 0. \tag{6.2.15}$$

Phase averaging (6.2.11) and subtracting (6.2.15) yields

$$\frac{\partial u_i''}{\partial x_i} = 0, \qquad \frac{\partial u_i'}{\partial x_i} = 0. \tag{6.2.16}$$

Norris and Reynolds (1975) have derived the most general form of the equations that govern the time-averaged and the organized part of the flow. Substitution of the triple decompositions in Eq. (6.2.12), time-averaging, and using Eqs. (6.2.15) and (6.2.16) give for the mean flow:

$$\bar{u}_j \frac{\partial \bar{u}_i}{\partial x_j} = -\frac{\partial \bar{p}}{\partial x_i} + \frac{1}{Re} \frac{\partial^2 \bar{u}_i}{\partial x_j \partial x_j} - \frac{\partial}{\partial x_j} \left(\overline{u_i' u_j'} \right) - \frac{\partial}{\partial x_j} \left(\overline{u_i'' u_j''} \right). \tag{6.2.17}$$

This is the familiar equation of momentum for the mean turbulent flow. Notice, however, that now there are two Reynolds stress terms on the right-hand side of the equation. One is the classical nonlinear contribution due to the random fluctuations, and the other is very similar in form but represents the nonlinear contribution of the organized fluctuations.

Taking the phase average of the momentum equation and using Eqs. (6.2.9), (6.2.15), and (6.2.16) yields

$$\frac{\partial u_i''}{\partial t} + \bar{u}_j \frac{\partial \bar{u}_i}{\partial x_j} + \bar{u}_j \frac{\partial u_i''}{\partial x_j} + u_j'' \frac{\partial \bar{u}_i}{\partial x_j} + u_j'' \frac{\partial u_i''}{\partial x_j} + \frac{\partial}{\partial x_j} \langle u_i' u_j' \rangle$$

$$= -\frac{\partial \bar{p}}{\partial x_i} - \frac{\partial p''}{\partial x_i} + \frac{1}{Re} \left(\frac{\partial^2 \bar{u}_i}{\partial x_j \partial x_j} + \frac{\partial^2 u_i''}{\partial x_j \partial x_j} \right). \tag{6.2.18}$$

Subtracting Eq. (6.2.17) from Eq. (6.2.18) gives the equation that governs the organized fluctuations

$$\frac{\partial u_i''}{\partial t} + \bar{u}_j \frac{\partial u_i''}{\partial x_j} + u_j'' \frac{\partial \bar{u}_i}{\partial x_j} + u_j'' \frac{\partial u_i''}{\partial x_j} = -\frac{\partial p''}{\partial x_i} + \frac{1}{Re} \frac{\partial u_i''}{\partial x_j \partial x_j} + \frac{\partial}{\partial x_j} \left(\overline{u_i'' u_j''} \right)$$

$$+ \frac{\partial}{\partial x_j} \left(\overline{u_i' u_j'} \right) - \frac{\partial}{\partial x_j} \left(\langle u_i' u_j' \rangle \right).$$

$$(6.2.19)$$

This equation is written here in a slightly different form than the one found in Norris and Reynolds (1975). The term $u_j'' \partial u_i''/\partial x_j$ is retained on the left-hand side and, together with the term $\partial u_i''/\partial t$, it represents the Euler operator, that is, the convective part of the organized fluctuations. Notice, however, that momentum is convected by the organized motion as well and that the operator is nonlinear. In practice and for small disturbances, this term may not be significant. Only then can we invoke linearity to assert that only the fundamental component is of interest and that the harmonics and their interaction can be neglected.

On the right-hand side of Eq. (6.2.19) we have again Reynolds stresses. Two of these terms are identical to the Reynolds stresses that appear in the equation for the mean flow, Eq. (6.2.17). They are the terms $\partial(\overline{u_i'' u_j''})/\partial x_j +$ $\partial(\overline{u_i'' u_j''})/\partial x_j$, they drive the mean motion and have been inherited directly from Eq. (6.2.17). The remaining Reynolds stress can be identified as the stress of the organized motion: $\partial(\langle u_i' u_j' \rangle)/\partial x_j$. In an attempt to close the system, we can derive equations that govern these quantities. This, however, will introduce higher moments, as described in all classical texts on turbulence. To explore the possibility of higher order closure schemes one will have to start with the equations that govern the fluctuations, u'', v'', that is, Eq. (6.2.19) and the momentum equation for the random disturbances (Hussain and Reynolds, 1970)

$$\frac{\partial u_i'}{\partial t} + \bar{u}_j \frac{\partial u_i'}{\partial x_j} + u_j'' \frac{\partial u_i'}{\partial x_j} + u_j' \frac{\partial \bar{u}_i}{\partial x_j} + u_j' \frac{\partial u_i''}{\partial x_j}$$

$$= -\frac{\partial p'}{\partial x_i} + \frac{1}{Re} \frac{\partial^2 u'}{\partial x_j \partial x_j} + \frac{\partial}{\partial x_j} \left(\langle u_i' u_j' \rangle - u_i' u_j' \right).$$

$$(6.2.20)$$

It is interesting to note that the organized fluctuations appear here only in the convective terms on the left-hand side of the equation. To derive equations that govern autocorrelations of u_i', autocorrelations of u_i'', and mixed correlations, we have to multiply Eqs. (6.2.20) and/or (6.2.19) with u_j' and/or u_j'' and average. The problem of course is very complicated, especially because two different methods of averaging, the time-averaging and the conditional ensemble averaging, are possible and both types of corresponding correlations are necessary. In the sections that follow we shall describe various efforts to model the Reynolds stresses of the mean and the organized motion, that is, the motion described by Eqs. (6.2.17) and (6.2.19).

The equivalent system of equations for two-dimensional boundary-layer flow in the traditional notation, and assuming that the normal stresses $\partial \overline{u'^2}/\partial x$ and $\partial \overline{v'^2}/\partial x$ are negligible, are the following:

For the mean flow

$$\frac{\partial \bar{u}}{\partial x} + \frac{\partial \bar{v}}{\partial y} = 0, \tag{6.2.21}$$

$$\bar{u}\frac{\partial \bar{u}}{\partial x} + \bar{v}\frac{\partial \bar{u}}{\partial y} = -\frac{\partial \bar{p}}{\partial x} + \frac{1}{Re}\frac{\partial^2 \bar{u}}{\partial y^2} - \frac{\partial}{\partial y}(\overline{u'v'} + \overline{u''v''}). \tag{6.2.22}$$

For the organized fluctuations

$$\frac{\partial u''}{\partial x} + \frac{\partial v''}{\partial y} = 0, \tag{6.2.23}$$

$$\frac{\partial u''}{\partial t} + u''\frac{\partial u''}{\partial x} + v''\frac{\partial u''}{\partial y} + \frac{\partial}{\partial x}(\bar{u}u'') + v''\frac{\partial \bar{u}}{\partial y} + \bar{v}\frac{\partial u''}{\partial y}$$

$$= -\frac{\partial p''}{\partial x} + \frac{1}{Re}\frac{\partial^2 u''}{\partial y^2} + \frac{\partial}{\partial y}(\overline{u''v''}) - \frac{\partial}{\partial y}(\langle u'v'\rangle - \overline{u'v'}). \tag{6.2.24}$$

Small perturbations

An alternative approach to the problem is based on the assumption that the disturbance is small throughout the turbulent boundary layer. The development of the earlier section is not restricted by such an assumption. However, in practice, a similar hypothesis would be necessary, in order to discard the higher harmonics of the organized motion and their mutual interaction. The approach presented in this section (Telionis, 1977; Romaniuk and Telionis, 1979) provides a simple mechanism for calculation of the effects of higher harmonics, but always within the small amplitude approximation.

Consider an outer-flow velocity distribution of the form

$$U_e(x,t) = U_0(x) + \frac{\varepsilon}{2}U_1(x)[e^{i\omega t} + CC], \tag{6.2.25}$$

where ε is a small dimensionless parameter and CC stands for the complex conjugate of the preceding quantity. We assume the response of the viscous layer in the form

$$u(x, y, t) = u_0(x, y) + \frac{\varepsilon}{2}[u_1(x, y)e^{i\omega t} + CC] + O(\varepsilon^2), \tag{6.2.26}$$

$$v(x, y, t) = v_0(x, y) + \frac{\varepsilon}{2}[v_1(x, y)e^{i\omega t} + CC] + O(\varepsilon^2), \tag{6.2.27}$$

where u and v are the velocity components parallel and perpendicular to the wall, respectively. The quantities u_i and v_i, $i = 1, 2, \ldots$, are in general random complex quantities. We therefore assume here that the boundary layer will respond to the outer flow with a deterministic fluctuation of the same frequency and an amplitude that contains random turbulent fluctuations. We further assume that the organized fluctuations are small compared to the mean flow. The difference with the analysis of the previous chapter lies in the fact that here all the terms in the expansion contain a random part. This essentially implies that up to order ε, there are two different kinds of random fluctuations: the fluctuations of the steady part of the flow and the fluctuations of the organized motion. It should be also emphasized that the random fluctuations are not small compared to the ensemble-averaged quantities. Our assumption here pertains to the relative magnitude of the steady part of the flow and its unsteady response to the outer-flow oscillations.

The steps that follow will be briefly described here. We first substitute the above expressions in the full Navier–Stokes equations, eliminate a few terms on the basis of the boundary-layer approximation, and collect powers of the small parameter ε. The process of making the boundary-layer assumption and collecting the powers of ε may be interchanged under certain conditions and comments on the proper procedure are included in the previous chapter. At this stage we have two sets of differential equations: the first is nonlinear and governs the mean flow (u_0, v_0). The second is linear and governs the amplitude of the fluctuations (u_1, v_1). Both sets of equations are stochastic and describe exactly the boundary-layer phenomena within an error of order ε^2.

We now further decompose the velocity components as follows:

$$u_i = \bar{u}_i + u_i', \tag{6.2.28}$$

$$v_i = \bar{v}_i + v_i', \tag{6.2.29}$$

where bars and primes denote the ensemble average and the random fluctuation, respectively, $i = 0, 1, \ldots$, and, by definition, $\overline{u_i'} = \overline{v_i'} = 0$. It is easy to show that both the average and the fluctuating components satisfy the continuity equation:

$$\frac{\partial \bar{u}_i}{\partial x} + \frac{\partial \bar{v}_i}{\partial y} = 0, \tag{6.2.30}$$

$$\frac{\partial u_i'}{\partial x} + \frac{\partial v_i'}{\partial y} = 0. \tag{6.2.31}$$

Substitution of the expressions (6.2.28) and (6.2.29) into the momentum equations of order 1 and ε, ensemble-averaging, and using Eqs. (6.2.30) and

(6.2.31) yield

$$\bar{u}_0 \frac{\partial \bar{u}_0}{\partial x} + \bar{v}_0 \frac{\partial \bar{u}_0}{\partial y} = U_0 \frac{dU_0}{dx} + \frac{1}{Re} \frac{\partial^2 u_0}{\partial y^2} - \frac{\partial}{\partial y} \left(\overline{u_0' v_0'} \right) - \frac{\partial}{\partial x} \left(\overline{u_0'^2} \right),$$

$$(6.2.32)$$

$$i\omega \bar{u}_1 + \bar{u}_0 \frac{\partial \bar{u}_1}{\partial x} + \bar{u}_1 \frac{\partial \bar{u}_0}{\partial x} + \bar{v}_0 \frac{\partial \bar{u}_1}{\partial y} + \bar{v}_1 \frac{\partial \bar{u}_0}{\partial y}$$

$$= i\omega U_1 + U_0 \frac{dU_1}{dx} + U_1 \frac{dU_0}{dx} + \frac{1}{Re} \frac{\partial^2 \bar{u}_1}{\partial y^2} \qquad (6.2.33)$$

$$- \frac{\partial}{\partial y} \left(\overline{u_0' v_1'} + \overline{u_1' v_0'} \right) - 2 \frac{\partial}{\partial x} \left(\overline{u_0' u_1'} \right).$$

Equation (6.2.32) is identical to the steady-state turbulent boundary-layer equation. If the x derivative of the random fluctuations is assumed negligible and the Reynolds stress modeled according to one of the phenomenological theories widely accepted (Reynolds, 1976), then the theory shows that within an error of order ε the mean flow is identical to the quasi-steady flow.

The nonlinear contributions to the organized fluctuations $(\bar{u}_1, \bar{v}_1)$ will require special attention. The Reynolds stresses in Eqs. (6.2.32) and (6.2.33) are of entirely different character. The first represents an ensemble average of random fluctuations of the mean flow. The second is a correlation of random fluctuations of the mean flow and the organized fluctuations.

It should be noted that Eq. (6.2.32) is the equation for steady turbulent flow. No nonlinear effects are present at this level of approximation. To capture the nonlinear feedback due to the organized motion, we shall have to include in our analysis terms of order ε^2. The reader may note that the approaches described in the previous and the present sections can be considered as the extensions of the methods of Lin and Lighthill to turbulent flow, (Sections 5.2 and 5.3) respectively.

Very little has been done up to now to model properly the phenomenon and close the equations. The methods suggested in literature are extensions of established phenomenological models. It is believed however that the form of the Reynolds stresses in Eqs. (6.2.22), (6.2.24), and (6.2.32), (6.2.33) may guide investigators to develop physically correct models that will permit accurate predictions of the flow.

6.3 Algebraic Models

In an effort to derive quickly engineering information badly needed in practical applications, some investigators employed straightforward extensions of existing programs to unsteady turbulent boundary-layer flows. For two-dimensional incompressible boundary-layer flow, the continuity and momentum equations can be written as

$$\frac{\partial U}{\partial x} + \frac{\partial V}{\partial y} = 0,$$ *(6.3.1)

$$\frac{\partial U}{\partial t} + U \frac{\partial U}{\partial x} + V \frac{\partial U}{\partial y} = \frac{\partial U_e}{\partial t} + U_e \frac{\partial U_e}{\partial x} + \nu \frac{\partial^2 U}{\partial y^2} + \frac{1}{\rho} \frac{\partial \tau}{\partial y},$$ *(6.3.2)

where U and V are the averaged velocity components that contain the mean and the organized oscillation of the flow. We introduce here a notation seldom used by the authors who contributed in this area, to distinguish from quantities defined in earlier sections. Quantities U and V represent the summation of mean and organized fluctuations of Section 6.1,

$$U = \bar{u} + u'', \qquad V = \bar{v} + v''.$$ *(6.3.3)

The functions U and V are therefore assumed to depend on time in a deterministic way while all the nonlinear effects are contained in the Reynolds stress τ. Some authors specify that U and V are ensemble averages, implying that a technique analogous to phase-averaging should be used. Others define U and V as the time average over a time interval large enough to average out the random fluctuations but also small enough to capture the organized fluctuations. This, in turn, implies that the externally imposed oscillation has a discrete frequency that lies outside the spectrum of turbulence.

Some methods are based on zero-equation models, that is, algebraic relations for closing the equations, whereas others are based on one- or two-equation models, employing one or more differential equations and expressing the Reynolds stress in terms of the turbulent energy. The straightforward methods described in the subsection that follows have a basic common characteristic: they are all based on steady models. Unsteadiness in the closure assumption enters only parametrically, via the time dependence of functions like the outer-flow velocity distribution or through acceleration terms in the expressions for the pressure gradient or turbulence convection. In a sense, such methods can therefore be termed quasi-steady methods because they are based on the steady turbulence models and the arbitrary constants and functions determined by extensive comparison with steady turbulent boundary-layer flow.

The initial efforts to decompose the Reynolds stress and define a truly unsteady model are described in the second subsection that follows.

Quasi-steady models

Models based on the mixing-length concept have proved to be very successful up to now despite their simplicity. This has been verified very recently by Burggraf (1975) who undertook a comparative study of representative methods. Cebeci and Keller (1972) and Abbott and Cebeci (1971) suggested an extension of their well-tested steady-flow model to unsteady flows. They proposed to retain the expression for the inner and outer eddy viscosity models. Dynamic effects are then introduced in the turbulence models through the unsteady term in the pressure gradient as well as parametrically, via the instantaneous values of the displacement thickness $\delta_1(x, t)$ and the outer flow velocity $U_e(x, t)$. The Reynolds stress in Eq. (6.3.2) is assumed proportional to the mean velocity gradient

$$\tau = \epsilon \frac{\partial U}{\partial y} .$$

*(6.3.4)

A two-layer eddy viscosity model is then introduced. In the inner layer the eddy viscosity is proportional to the velocity gradient

$$\epsilon_i = \rho l^2 \left| \frac{\partial U}{\partial y} \right| ,$$

*(6.3.5)

where l is the mixing length given by

$$l = k_1 y \left[1 - \exp\left(\frac{-y}{A} \right) \right],$$

*(6.3.6)

with $k_1 = 0.41$ and A the Van Driest damping factor (Van Driest, 1956). The latter quantity is traditionally assumed to be inversely proportional to the friction velocity $u_\tau = (\tau_w / \rho)^{1/2}$, where τ_w is the skin friction at the wall

$$A = A^+ \nu / u_\tau = A^+ \nu (\tau_w / \rho)^{-1/2}$$

*(6.3.7)

and $A^+ = 26$.

To account for flows with pressure gradients as well as flows with heat transfer, Cebeci (1970) assumed that a characteristic velocity in the Stokes flow that models the inner part of the turbulent boundary layer is the friction velocity at the edge of the viscous sublayer. The damping factor then becomes

$$A = A^+ \nu \left(\frac{\tau_s}{\rho} \right)^{-1/2} .$$

*(6.3.8)

This assumption permits a straightforward extension to unsteady flows. An estimate of the shear stress τ_s can be derived by solving the approximate form of the momentum equation in the viscous sublayer

$$0 = -\frac{dp}{dx} + \frac{d\tau}{dy} .$$

*(6.3.9)

Integrating across the viscous sublayer from the wall to $y_s^+ = y_s u_\tau / \nu = 11.8$ and substituting in (6.3.8), yields

$$A = \frac{\nu A^+}{u_\tau} \left[1 - 11.8 \frac{\nu}{\rho u_\tau^3} \frac{\partial p}{\partial x} \right].$$

*(6.3.10)

The acceleration effects of the outer flow now enter via the pressure gradient

$$-\frac{1}{\rho}\frac{\partial p}{\partial x} = \frac{\partial U_e}{\partial t} + U_e\frac{\partial U_e}{\partial x}.$$
*(6.3.11)

Low Reynolds number effects are taken into account by readjusting the values of k_1 and A^+ according to the equations

$$k_1 = 0.40 + \frac{0.19}{1 + 0.49Z^2},$$
*(6.3.12)

$$A^+ = 26 + \frac{14}{1 + Z^2}$$
*(6.3.13)

for $Z \geqslant 0.3$, where $Z = Re_\theta \times 10^{-3}$ and Re_θ is the Reynolds number based on the momentum thickness θ,

$$Re_\theta = U_e\theta/\nu.$$
*(6.3.14)

Cebeci and his associates propose for the outer layer

$$\epsilon_0 = \frac{0.0168}{1 + 5.5(y/\delta)^6}\left|\int_0^\infty (U_e - U)\,dy\right|.$$
*(6.3.15)

Telionis and Tsahalis (1975) adopted the same inner model for the eddy viscosity but used in the outer region a Clauser model (Clauser, 1956):

$$\epsilon_0 = \frac{k_2 U_e}{\nu}\delta_1\gamma,$$
*(6.3.16)

where $k_2 = 0.0168$, δ_1 is the displacement thickness, and γ is the intermittency factor

$$2\gamma = 1 - \text{erf}\left[5\left(\frac{y}{\delta} - 0.78\right)\right].$$
*(6.3.17)

In a later publication, Telionis (1976) attempted a comparative study of algebraic models. In this paper the models of Kays (1971) and Alber (1971) were also extended to account for unsteadiness. These models involve expressions of the pressure gradient and the generalization proposed is essentially based on Eq. (6.3.11). It is also noted that the approximate form of the momentum equation given by Eq. (6.3.9) may not be very accurate. This is based essentially on the assumption that the nonlinear terms of the convective derivative are negligible next to the wall. However, there is no justification for disregarding the local acceleration $\partial U/\partial t$ especially in the case of large frequency oscillations.

A better approximation for Eq. (6.3.9) should be

$$\rho\frac{\partial U}{\partial t} = -\frac{\partial p}{\partial x} + \frac{\partial \tau}{\partial y},$$
*(6.3.18)

but its integration across the viscous sublayer is impossible, since the function U is unknown. However, a good approximation of the variations of

U can be given in terms of Lighthill's theory, described in the previous chapter, since the flow in this layer is nearly laminar. For external oscillations with frequency ω and α a small amplitude parameter, it may be assumed that in the viscous sublayer

$$U = U_0 + \alpha U_1 \cos\left(\omega t + \frac{\pi}{4}\right),$$

*(6.3.19)

and further assuming that

$$U_0(x, y) = U_1(x, y) = \frac{\tau_w y}{\mu},$$

*(6.3.20)

with τ_w the average skin friction, Eq. (6.3.18) can be integrated to yield the instantaneous skin friction at $y = y_s$

$$\tau_s = \frac{\partial p}{\partial x} y_s - \rho \alpha \omega \tau_w \frac{y^2}{2\mu} \sin\left(\omega t + \frac{\pi}{4}\right) + \tau_w.$$

*(6.3.21)

Cebeci and Keller (1972) limited their calculations to spatially one-dimensional flows and compared their results to those of Bradshaw (1969). Dwyer *et al.* (1970) also developed a technique based on the quasi-steady model of the mixing-length type and integrated the boundary-layer equations by a finite difference method. McCroskey and Philippe (1975) later used this method, checked it against previous theoretical and experimental results, and calculated the flow fields about airfoils. The quasi-steady mixing-length model was also used by Gupta and Trimpi (1974), who computed the development of a compressible turbulent boundary layer on a semiinfinite flat plate after the passage of a shock wave and a trailing driver-gas, driven-gas interface. More recently, Cebeci (1977) employed this model again to calculate oscillatory flows over a flat plate and compare with experimental data.

It is well known that the laminar shear stress is very small compared to the Reynolds stress and outside the viscous sublayer, the term $(1/Re)\partial^2 U/\partial y^2$ in Eq. (6.3.2) is one order of magnitude smaller than the term $\partial \tau/\partial y$. With all algebraic models however, the Reynolds stress is proportional to the square of the velocity gradient. As a result, the dominant derivative on the right-hand side of Eq. (6.3.2) remains the second derivative with respect to y. The differential equation is parabolic and its numerical integration can be implemented by any of the methods described in the previous chapter. The investigations referenced in this section are based on straightforward numerical integrations in the three-coordinate space: x, y, and t.

The models described here may be adequate for crude engineering estimates. However, comparisons with experimental data indicate large discrepancies even for the simplest possible configurations as, for example, a flat plate. The relative performance of various algebraic models was attempted by Telionis (1976) and some typical results will be included in a later section.

Unsteady models

The perturbation method described in Section 6.2 can be used in numerical experiments that test the performance of quasi-steady and unsteady models. As described before, the first step in the efforts to calculate turbulent boundary layers with organized oscillations has been to assume that steady models are still valid. Unsteadiness was taken into account either implicitly, via the time dependence of the mean field or through the $\partial/\partial t$ terms in the model equation for turbulent energy and dissipation.

To compare the relative performance of quasi-steady and unsteady models, Romaniuk and Telionis (1979) have chosen algebraic expressions for the modeling of the Reynolds stresses. Consider Prandtl's classical eddy viscosity model

$$\tau = \epsilon \frac{\partial u}{\partial y}, \qquad \epsilon = \rho l^2 \left| \frac{\partial u}{\partial y} \right|, \qquad \text{*(6.3.22)}$$

where l is the mixing length. A quasi-steady version of this model yields the following expression for the Reynolds stresses of Eqs. (6.2.32) and (6.2.33):

$$\tau_0 = \overline{u_0' v_0'} = \rho l^2 \left(\frac{\partial \bar{u}_0}{\partial y} \right)^2, \qquad \text{*(6.3.23)}$$

$$\tau_1 = \overline{u_0' v_1'} + \overline{u_1' v_0'} = 2\rho l^2 \frac{\partial \bar{u}_0}{\partial y} \frac{\partial \bar{u}_1}{\partial y}. \qquad \text{*(6.3.24)}$$

Any quasi-steady model therefore is equivalent to the assumption that the total Reynolds stress is decomposed into a mean part τ_0 and an oscillatory part τ_1 in a way determined by the governing equation. Physically, it implies that the oscillatory Reynolds stress τ_1 is proportional to the gradient of the organized oscillations $\partial u_1/\partial y$, but the eddy viscosity of the oscillatory motion is proportional to the gradient of the mean flow. Clearly, all these assumptions may be an optimistic initial attempt to solve the problem but bear no physical justification and yield very poor results.

The simplest possible model that represents decoupling of the Reynolds stresses τ_0 and τ_1 is the quasi-laminar model that assumes that $\tau_1 = 0$. This assumption is essentially equivalent to a laminar oscillatory correction on a steady turbulent boundary layer. It is well known that the boundary layer, laminar or turbulent, responds to local disturbances in an almost inviscid manner. The hypothesis here is that the outer-flow pressure fluctuations are instantly carried across the turbulent boundary layer, without interaction with the random fluctuations. Or, equivalently, that the turbulent eddies undergo an oscillatory deformation that does not affect their entity and the process of their mutual interaction. The present author and his associates were experimenting with this idea when the work of Acharya and Reynolds (1975), developed for internal flow, was brought to their attention. Reynolds and his associates describe a quasi-laminar model and report that their analytical results are not satisfactory for internal flows.

Let us now assume that the oscillatory part of the Reynolds stress is given in terms of an eddy viscosity

$$\tau_1 = \epsilon_1 \frac{\partial \bar{u}_1}{\partial y},$$
*(6.3.25)

$$\epsilon_1 = \begin{cases} \epsilon_{1i}, & \text{if } \epsilon_{1i} < \epsilon_{1o}, \\ \epsilon_{1o}, & \text{if } \epsilon_{1i} \geqslant \epsilon_{1o}, \end{cases}$$
*(6.3.26)

where ϵ_{1i} is the eddy viscosity in the inner region:

$$\epsilon_{1i} = \frac{l^2}{\nu} \left| \frac{\partial \bar{u}_1}{\partial y} \right|$$
*(6.3.27)

and ϵ_{1o} is valid in the outer region of the boundary layer

$$\epsilon_{1o} = k_2 \frac{U_1 \delta_1}{\nu} \gamma.$$
*(6.3.28)

The difference between this model and the quasi-steady approach is that the eddy viscosity in the inner layer is assumed to be proportional to the gradient of the oscillatory velocity profile rather than the mean profile. The empirical constants k_1 and k_2 were originally estimated for steady flows. Therefore one may expect that they would require readjustment for oscillatory flow. As a first step in the investigation, these constants were assigned the same values, i.e., $k_1 = 0.41$ and $k_2 = 0.0168$. The mixing length was assumed again proportional to the distance from the wall as indicated by Eq. (6.3.6).

Our assumption is that the random field of the organized oscillation is not directly dependent on the mean motion. Purely oscillatory turbulent flow could be generated, if the mean part of the outer flow were zero, i.e., if in Eq. (6.2.25), $U_0 = 0$. Neglecting the streaming effects then, we would arrive at a purely oscillatory motion, that is, a velocity field with $\bar{u}_0 = 0$. The Reynolds stress of such a field would have to depend only on $\bar{u}_1$ and its derivatives. Such flows have been investigated experimentally by Jonsson and Carlsen (1976). In this paper an eddy viscosity formulation was employed. However, eddy viscosity values were estimated by comparison to experimental data and no mixing length concepts were introduced.

6.4 One- and Two-Equation Models

One of the most elegant methods and the first to be extended to unsteady flow is the method of the turbulent kinetic energy (Townsend, 1956; Bradshaw et al., 1967). Bradshaw (1969) solved the problem of unsteady flow over an infinite flat plate, thus eliminating the x coordinate, while Patel and Nash (1972) considered a truly unsteady case whereby the equations were integrated numerically with respect to the three independent variables t, x, and y. The method is based on the equation for the turbulent intensity:

$$\overline{q^2} = \overline{u'^2} + \overline{v'^2} + \overline{w'^2} . \qquad \qquad *(6.4.1)$$

The turbulent kinetic energy is then $\rho \overline{q^2}/2$ and the equation that governs this quantity is

$$\frac{D}{Dt}\left(\frac{\overline{q^2}}{2}\right) - \frac{\tau}{\rho}\frac{\partial U}{\partial y} + \frac{\partial}{\partial y}\left[\overline{v'\left(\frac{p'}{\rho} + \frac{q^2}{2}\right)}\right] + e = 0, \qquad *(6.4.2)$$

where the D/Dt stands for the convective derivative

$$\frac{D}{Dt} = \frac{\partial}{\partial t} + U\frac{\partial}{\partial x} + V\frac{\partial}{\partial y} , \qquad \qquad *(6.4.3)$$

e is the turbulent dissipation, and the normal Reynolds stress terms have been neglected.

Hyperbolic models

The Reynolds stress τ is related to the quantity $\overline{q^2}$ and the terms in (6.4.3) are modeled according to the equations (Patel and Nash, 1972)

$$\tau = a_1\rho\overline{q^2} , \qquad \qquad *(6.4.4)$$

$$\overline{v'\left(\frac{p'}{\rho} + \frac{q^2}{2}\right)} = \frac{1}{U_e}\left(\overline{q^2}\right)_{max}\overline{q^2}\,a_2, \qquad *(6.4.5)$$

$$e = \frac{\left(\overline{q^2}\right)^{3/2}}{L} , \qquad \qquad *(6.4.6)$$

where a_1 is assumed to be a universal constant equal to 0.15, and the quantities a_2 and L are universal empirical functions of y/δ, shown in Fig. 6.2. Here the boundary-layer thickness δ is defined by the relationship $\delta = y(u = 0.975\,U_e)$.

Bradshaw and his associates note that in most practical applications, the convection and diffusion terms are negligible and Eq. (6.4.2) reduces to

$$-\frac{\tau}{\rho}\frac{\partial U}{\partial y} + e = 0. \qquad \qquad *(6.4.7)$$

It can be readily seen then that the assumption (6.4.6) is equivalent to

$$\tau = a_1\rho L^2\left(\frac{\partial U}{\partial y}\right)^2, \qquad \qquad *(6.4.8)$$

which is essentially Prandtl's mixing length formulation. The present author

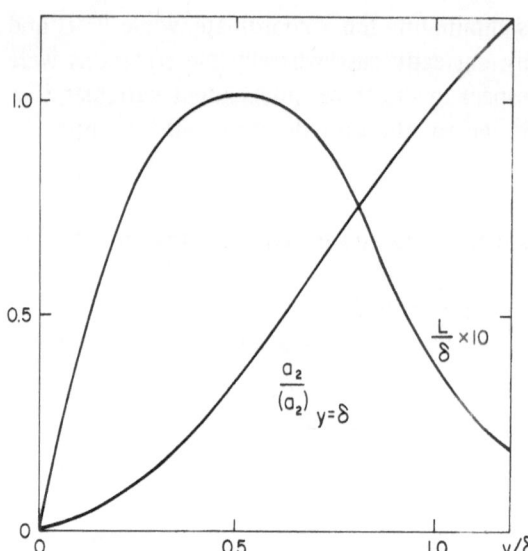

Fig. 6.2 Empirical functions for the turbulent energy equation.

feels that a straightforward extension of Eq. (6.4.7) to unsteady flow should include the local acceleration of the turbulent kinetic energy, which may not be small if large frequency oscillations are imposed. The equivalent simplified but unsteady version of Eq. (6.4.2), which for steady flow reduces to Prandtl's mixing length formula, is

$$\frac{1}{2}\frac{\partial \overline{q^2}}{\partial t} - \frac{\tau}{\rho}\frac{\partial U}{\partial y} + e = 0. \qquad\qquad *(6.4.9)$$

One may thus conclude that for unsteady flow the functional dependence of τ on L and $\partial U/\partial y$ given by (6.4.8) is not sufficient. It is very interesting to solve this equation in order to arrive at the generalization of the mixing length formula for unsteady flow, as dictated by the turbulent energy equation and the closure assumption of Bradshaw et al (1967), that is, Eqs. (6.4.4) and (6.4.6). Assume that the mean flow oscillates harmonically

$$U = \bar{u}_0 + \bar{u}_1 e^{i\omega t}. \qquad\qquad *(6.4.10)$$

Assuming further that the Reynolds stress also fluctuates about a mean and in phase with the velocity

$$\frac{\tau}{\rho} = \frac{\tau_0}{\rho} + \frac{\tau_1}{\rho}e^{i\omega t}, \qquad\qquad *(6.4.11)$$

we can bring Eq. (6.4.9) to the form

$$\frac{i\omega}{2a_1}\frac{\tau_1}{\rho}e^{i\omega t} - \left(\frac{\tau_0}{\rho} + \frac{\tau_1}{\rho}e^{i\omega t}\right)\left(\frac{\partial \bar{u}_0}{\partial y} + \frac{\partial \bar{u}_1}{\partial y}e^{i\omega t}\right)$$
$$+ \left(\frac{\tau_0}{\rho a_1} + \frac{\tau_1}{\rho a_1}e^{i\omega t}\right)^{3/2}\frac{1}{L} = 0, \qquad\qquad *(6.4.12)$$

No perturbation principles are invoked here and the expansion in (6.4.10) and (6.4.11) is simply decomposition to mean and oscillatory parts. The steady terms in Eq. (6.4.12) reduce to Eq. (6.4.8) expressed in terms of mean quantities

$$\tau_0 = \rho a_1 L^2 \left(\frac{\partial \bar{u}_0}{\partial y} \right)^2. \qquad \text{*(6.4.13)}$$

This may be considered as yet another indication that the mean part of an oscillating turbulent boundary layer is not influenced by nonlinear effects. Physically, this is due to the fact that the nonlinear terms in the turbulent energy equation, that is, the convective terms usually represent a negligible contribution and can be omitted.

Neglecting the higher harmonics in Eq. (6.4.12), we can solve for the fluctuating part of the Reynolds stress

$$\frac{\tau_1}{\rho} \left[-\frac{\partial \bar{u}_0}{\partial y} + \frac{3}{2a_1 L} \left(\frac{\tau_0}{\rho a_1} \right)^{1/2} + \frac{i\omega}{2a_1} \right] = \frac{\tau_0}{\rho} \frac{\partial \bar{u}_1}{\partial y}. \qquad \text{*(6.4.14)}$$

Some qualitative characteristics can now be identified. Dividing through by $(\tau_0/\rho a_1)^{1/2}$ renders the factor within the brackets dimensionless and indicates that τ_1 is proportional to the product of the gradients of the mean and the oscillatory velocity components

$$\tau_1 \propto \left(\frac{\partial \bar{u}_0}{\partial y} \right) \left(\frac{\partial \bar{u}_1}{\partial y} \right). \qquad \text{*(6.4.15)}$$

It is very interesting to note that a straightforward expansion of the algebraic models described in the previous section resulted in the same equation (Eq. 6.3.24). Such a closure model therefore has the potential of further development. The coefficient of proportionality, however, is complex and this implies that the Reynolds stress does not oscillate in phase with the velocity field. The departure from the in-phase variation grows with frequency.

Characteristics

According to the refined model we describe in this section, the Reynolds stress is considered unknown and governed by Eq. (6.4.2). In this way one more equation is added to the system of Eqs. (6.3.1) and (6.3.2) and the model becomes a "one-equation model." If the laminar stresses are omitted from Eq. (6.3.2), the complete system of equations becomes

$$\frac{\partial U}{\partial x} + \frac{\partial V}{\partial y} = 0, \qquad \text{*(6.4.16)}$$

$$\frac{DU}{Dt} + \frac{1}{\rho} \frac{\partial p}{\partial x} - \frac{1}{\rho} \frac{\partial \tau}{\partial y} = 0, \qquad \text{*(6.4.17)}$$

$$\frac{D\tau}{Dt} - 2a_1\tau \frac{\partial U}{\partial y} + 2a_1 \frac{\tau_{max}}{\rho U_e} \frac{\partial(a_2\tau)}{\partial y} + 2a_1 \frac{\tau^{3/2}}{L\rho^{1/2}} = 0. \qquad \text{*(6.4.18)}$$

This is a system of three equations to be solved for the three unknowns U, V, and τ.

There is a striking difference between the above system and the systems of boundary-layer equations described up to now. Dropping the second derivative with respect to the coordinate y changes the character of the system from parabolic to hyperbolic. The exact form of the characteristic directions has not been reported in literature for unsteady flow. This is a simple application of the theory presented in the first chapter. If the equation that describes the characteristic surface is

$$\phi(x, y, t) = 0, \qquad\qquad\qquad\qquad *(6.4.19)$$

then a straightforward application of Eq. (1.3.3) leads to the following equation:

$$\begin{vmatrix} \dfrac{\partial \phi}{\partial x} & \dfrac{\partial \phi}{\partial y} & 0 \\[2ex] \dfrac{\partial \phi}{\partial t} + U \dfrac{\partial \phi}{\partial x} + V \dfrac{\partial \phi}{\partial y} & 0 & -\dfrac{1}{\rho}\dfrac{\partial \phi}{\partial y} \\[2ex] -2a_1\tau\dfrac{\partial \phi}{\partial y} & 0 & \dfrac{\partial \phi}{\partial t} + U \dfrac{\partial \phi}{\partial x} + (V + 2B)\dfrac{\partial \phi}{\partial y} \end{vmatrix} = 0.$$

$$*(6.4.20)$$

In the above determinant the columns again correspond to the independent variables, here U, V, and τ, respectively; and the quantity B combines the following parameters

$$B = a_1 a_2 \frac{\tau_{\max}}{\rho U_e} \qquad\qquad\qquad\qquad *(6.4.21)$$

One solution can be readily identified: $\partial\phi/\partial y = 0$, which implies again that the normal direction to the wall is a characteristic. This is essentially due to the boundary-layer approximation and the elimination of the second momentum equation. Its physical counterpart is the fact that within the boundary-layer approximation, pressure disturbances propagate instantly across the boundary layer, even in supersonic flow. Factoring $\partial\phi/\partial y$ out gives

$$\left(\frac{\partial \phi}{\partial t} + U \frac{\partial \phi}{\partial x} + V \frac{\partial \phi}{\partial y} \right)\left(\frac{\partial \phi}{\partial t} + U \frac{\partial \phi}{\partial x} + V \frac{\partial \phi}{\partial y} + 2B \frac{\partial \phi}{\partial y} \right)$$
$$- \frac{2a_1\tau}{\rho}\left(\frac{\partial \phi}{\partial y} \right)^2 = 0. \qquad\qquad *(6.4.22)$$

There are two alternative interpretations to this equation. Consider first the variable s along the instantaneous streamline. Then Eq. (6.4.22) becomes

$$\frac{\partial \phi}{\partial s}\left(\frac{\partial \phi}{\partial s} + 2B \frac{\partial \phi}{\partial y} \right) - \frac{2a_1\tau}{\rho}\left(\frac{\partial \phi}{\partial y} \right)^2 = 0. \qquad *(6.4.23)$$

Fig. 6.3 Characteristics in a turbulent boundary layer.

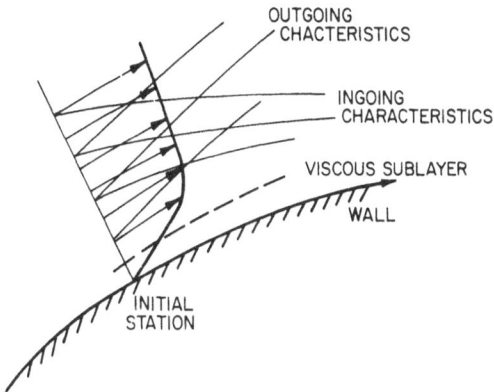

If α_t, α_x, α_y, and α_s are the directional cosines of the characteristic directions with respect to the axes t, x, y, and s, respectively, Eqs. (6.4.22) and (6.4.23) become

$$(\alpha_t + U\alpha_x + V\alpha_y)(\alpha_t + U\alpha_x + V\alpha_y + 2B\alpha_y) - \frac{2a_1\tau}{\rho}\alpha_y^2 = 0, \qquad *(6.4.24)$$

$$\alpha_s(\alpha_s + 2B\alpha_y) - \frac{2a_1\tau}{\rho}\alpha_y^2 = 0. \qquad *(6.4.25)$$

The second equation can be solved in terms of the ratio α_y/α_s, which represents the slope of the characteristic curves in the (s, y) plane

$$\frac{\alpha_y}{\alpha_s} = \left(\frac{2a_1\tau}{\rho}\right)^{-1}\left[B \pm \left(B^2 + \frac{2a_1\tau}{\rho}\right)^{1/2}\right]. \qquad *(6.4.26)$$

The other two characteristics are therefore inclined with respect to the streamlines as shown schematically in Fig. 6.3.

In the three-coordinate space, we can calculate the slopes of the characteristic lines that are defined by the intersection of the characteristic surface and the planes (t, x), (t, y), and (x, y) by setting α_y, α_x, and α_t equal to zero, respectively. Setting $\alpha_t = 0$, we recover the steady two-dimensional case (Bradshaw et al., 1967):

$$\frac{\alpha_y}{\alpha_x} = \left[\left(\frac{2a_1\tau}{\rho} - V^2 - VB\right)\right]^{-1} U\left\{(V + B) \pm \left(B^2 + \frac{2a_1\tau}{\rho}\right)^{1/2}\right\}. \qquad *(6.4.27)$$

Setting $\alpha_y = 0$, we find that the characteristic equation reduces to

$$(\alpha_t + U\alpha_x)^2 = 0, \qquad *(6.4.28)$$

and its slope is just U.

Numerical solution to the hyperbolic equations

Hyperbolic systems of equations are conveniently integrated by the method of characteristics. The mesh is aligned with the characteristics and the system reduces to ordinary differential equations. Singleton and Nash (1973) propose instead to march downstream, essentially following the traditional method of solution of the boundary-layer equation. Singleton and Nash (1973) introduce a transformation to account for the thickening of the boundary layer

$$y = S(x,t)\eta^{\beta},$$
(6.4.29)

where S is taken to be equal to 1.25δ and β is a constant larger than 1. They then introduce three-dimensionality effects by considering the flow about an infinite yawed cylinder, in the sense proposed by Sears (1947). A third velocity and Reynolds stress component is thus introduced, but the problem can still be solved in two space dimensions and time. The equations are rewritten in the form

$$\frac{\partial F}{\partial t} = A \frac{\partial F}{\partial \eta} + B \frac{\partial F}{\partial x} + C,$$
(6.4.30)

where F represents a dependent variable, A, B, and C are functions of the dependent variables and, in addition, C contains derivatives of U and W with respect to η. Following the work of Nash (1972), a mesh pattern is introduced with primary and secondary points as shown in Fig. 6.4. Fourth-order differences in the η direction are written

$$\left(\frac{\partial F}{\partial \eta}\right)^n_{k,m} = \frac{1}{\Delta \eta}\left[\frac{1}{12} F^n_{k-2,m} - \frac{2}{3} F^n_{k-1,m} + \frac{2}{3} F^n_{k+1,m} \right.$$

$$\left. - \frac{1}{12} F^n_{k+2,m} \right] + O(\Delta \eta)^4$$
(6.4.31)

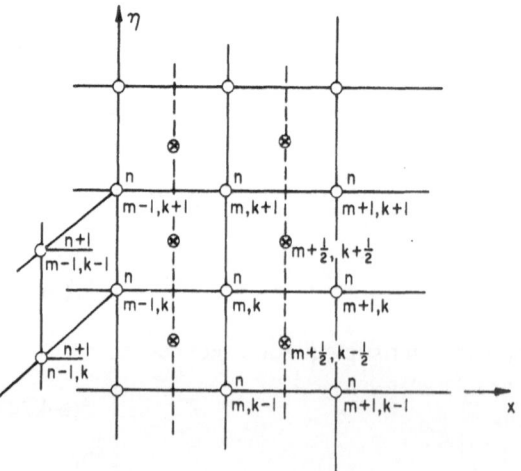

Fig. 6.4 Uniform grid for the scheme of Singleton and Nash (1973); O, primary points; ⊗, secondary points.

except for the point $k = K - 1$ at the edge of the boundary layer, where

$$\left(\frac{\partial F}{\partial \eta} \right)^n_{k,m} = \frac{1}{\Delta \eta} \left[\frac{1}{6} F^n_{K-3,m} - F^n_{K-2,m} + \frac{1}{2} F^n_{K-1,m} + \frac{1}{3} F^n_{K,m} \right] + O(\Delta \eta)^3.$$

(6.4.32)

A two-step scheme is employed whereby the values at secondary points are estimated according to the formula

$$F^n_{m+1,k+1/2} = \frac{1}{2} (F^n_{m,k+1} + F^n_{m,k}) + \frac{1}{8} \Delta \eta \left[\left(\frac{\partial F}{\partial \eta} \right)^n_{m,k+1} + \left(\frac{\partial F}{\partial \eta} \right)^n_{m,k} \right]$$
$$+ \frac{1}{4} \Delta x \left[\left(\frac{\partial F}{\partial x} \right)^n_{m,k+1} + \left(\frac{\partial F}{\partial x} \right)^n_{m,k} \right]$$

(6.4.33)

$$+ O\left[(\Delta \eta)^2 \right] + O\left[(\Delta x)^2 \right] + O\left[\Delta x (\Delta \eta)^2 \right].$$

For the calculations of unsteady flow, Singleton and Nash use an explicit second-order accurate, finite-difference scheme. New time points are calculated at secondary points in the new time plane according to the formula

$$F^{n+1}_{m,k+1/2} = \frac{1}{2} (F^n_{m,k} + F^n_{m,k+1}) + \frac{1}{8} \Delta \eta \left[\left(\frac{\partial F}{\partial \eta} \right)^n_{m,k} + \left(\frac{\partial F}{\partial \eta} \right)^n_{m,k+1} \right]$$
$$+ \frac{1}{2} \Delta t \left[\left(\frac{\partial F}{\partial t} \right)^n_{m,k} + \left(\frac{\partial F}{\partial \eta} \right)^n_{m,k+1} \right]$$

(6.4.34)

$$+ O\left[(\Delta \eta)^2 \right] + O\left[(\Delta x)^2 \right] + O\left[\Delta x (\Delta \eta)^2 \right].$$

These expressions are then used to derive more accurate values at the primary mesh points. With appropriate boundary and initial conditions, information is derived in the two-dimensional space for time t. Time then is incremented to $t + \Delta t$ and information on a new time plane is obtained. Marching, therefore, proceeds first in the x direction sweeping the two-dimensional physical space and then, in time, progressing in the future.

There is no theory available for the investigation of the stability of nonlinear equations. The stability characteristics of such a scheme, Nash (1972) determined by trial and error. The criterion of stability was based on inspection of calculated profiles of shear stress gradients.

The step sizes Δt, Δx, and $\Delta \eta$, are constrained by the characteristic direction. In the numerical calculations of Singleton and Nash (1973), the following inequalities constrain the mesh sizes:

$$\Delta t \leqslant \frac{\eta^{\beta-1}}{\sqrt{a_1 \tau_x}} \beta S \Delta \eta,$$

(6.4.35)

$$\Delta t \leqslant \frac{\Delta x}{U}.$$

(6.4.36)

If the above conditions are met, then the calculations do not step outside of the region of influence of the point of interest and the Courant–Friedrichs–Lewy condition is satisfied.

Nash *et al* (1975) employ the method described above to calculate the response of a turbulent boundary layer to oscillatory fluctuations of the outer stream. They investigate flow fields with adverse pressure gradients and proceed until the point of flow reversal. In this effort and due to the neglect of the laminar shear stress, the calculations cannot be extended to the wall. Instead the solution is matched at $y/\delta = 0.05$ to an approximate solution (Townsend, 1961).

The same group (Nash and Patel, 1975; Patel and Nash, 1975; Nash, 1976) later extended their work by introducing a refinement in the neighborhood of the wall, to meet the inner boundary condition. This is essentially the law of the wall with appropriate modifications in order to handle regions of partially reversed flow. Patel and Nash (1975) investigate a flow that progressively goes through a minimum of the outer-flow velocity. This generates a small but growing region of recirculating flow approximately in the middle of the domain of integration. More details on the physical significance of their findings are included in the next section.

6.5 Numerical Results and Experimental Data

There is no doubt that the problem of unsteady turbulent boundary layers is far from being solved. Extensions of classical theories to unsteady flows have been introduced as described in earlier sections, but little effort has been directed toward careful comparisons between the results of different methods or between numerical results and experimental data. The emphasis in this monograph is on the computational aspects of the problem and the physical conclusions that one may draw from them. However, it is believed that the most recent and most exciting experimental findings deserve some attention here. Certainly, any theoretician who seeks to model effectively such complex flows should be fully aware of the physical aspects of the problem as disclosed by experiments.

Relative performance of analytic methods

Quantities usually calculated in turbulent boundary-layer calculations are the displacement thickness δ_1 and the wall shear τ. A comparison of the relative performance of different methods (Singleton and Nash, 1973; Nash, *et al.*, 1975; Kuhn and Nielsen, 1973; Telionis and Tsahalis, 1975) has been attempted by Cousteix, *et al.* (1976). Oscillating flows over a flat plate and a configuration that imposes an adverse pressure gradient were examined. In particular, outer flows were chosen according to the formulas

$$U_e = U_0(1 + \varepsilon \sin \omega t), \qquad *(6.5.1)$$

$$U_e = U_0\left[1 + (\varepsilon_0 + \varepsilon_1 \sin \tilde{\omega} t)x\right], \qquad *(6.5.2)$$

where ε_i are dimensionless amplitudes, x and t are dimensionless downstream distance and time, respectively, and $\tilde{\omega}$ is the reduced frequency

$$\tilde{\omega} = \frac{\omega^* L}{U_\infty}. \qquad (6.5.3)$$

All quantities were calculated at the point $x = 1.0$ for two different frequencies $\tilde{\omega} = 1.57$ and 15.7, and different values of the amplitude parameters.

Figures 6.5 and 6.6 show the periodic variation of the displacement thickness and the skin friction for a small amplitude ($\varepsilon = 0.125$) and two different frequencies ($\tilde{\omega} = 1.57$ and 15.7). Unexpected discrepancies appear in the displacement thickness for low frequency and amplitude. Similar calculations for higher amplitudes indicate only some departure from the harmonic response of both quantities. The maxima of all curves appear to be more pointed, but very little changes in the phase angles can be observed. All methods indicate that the skin friction phase lead does not exceed $10°$, although the displacement thickness phase lead may reach values of $30°$ or $40°$. Moreover, large departures from the quasi-steady values are indicated in the plots of displacement thickness.

In Fig. 6.7 calculated and measured values of the skin friction phase lead are plotted versus the frequency parameter $\tilde{\omega}$. The case of laminar flow is also indicated for comparison. Unfortunately, the experimental data of Karlsson (1959) are widely dispersed in the neighborhood of the wall. Karlsson actually measured the fluctuating components of the velocity. An

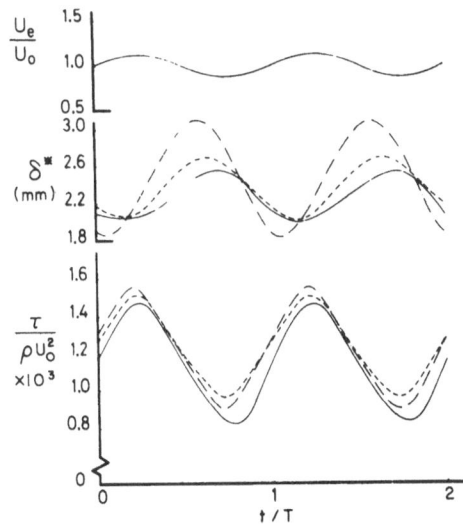

Fig. 6.5 Displacement thickness and wall shear for $\varepsilon = 0.125$, $\tilde{\omega} = 1.57$, $x = 1$, $Re = 10^7$, $U_e = U_0 (1 + \varepsilon \sin \tilde{\omega} t)$. ———, Cousteix, et al. (1976); ———, Nash et al. (1975); —·—, Kuhn and Nielsen (1973); ·····, Telionis and Tsahalis (1975).

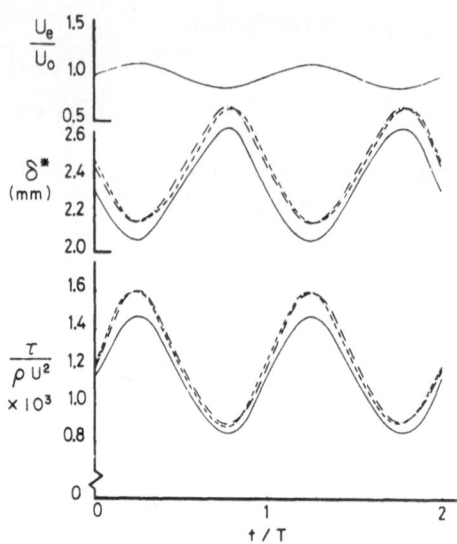

Fig. 6.6 Displacement thickness and wall shear for $\varepsilon = 0.125$, $\tilde{\omega} = 15.7$, $Re = 10^7$, $\tilde{x} = 1$, $U_e = U_0 (1 + \varepsilon \sin \omega t)$. ———, Cousteix et al. (1976); – – –, Nash et al. (1975); –·–, Kuhn and Nielsen (1973).

estimate of the phase angle from these data can be derived by extrapolation followed by calculation of the ratio of the imaginary to the real part of the quantities involved. The dispersion of the analytical results shown in this figure is equally disheartening.

Quasi-steady calculations with outer-flow distributions of the type given by Eq. (6.2) indicate that the wall shear may vanish during a portion of the period. In fact, it is probable that separation is in the neighborhood of this point as indicated by the large values of the displacement thickness. Calculated results with the quasi-steady models and the corresponding unsteady models are shown in Figs. 6.8 and 6.9 for small and large frequencies, respectively. It is most interesting to note that the unsteady boundary layer

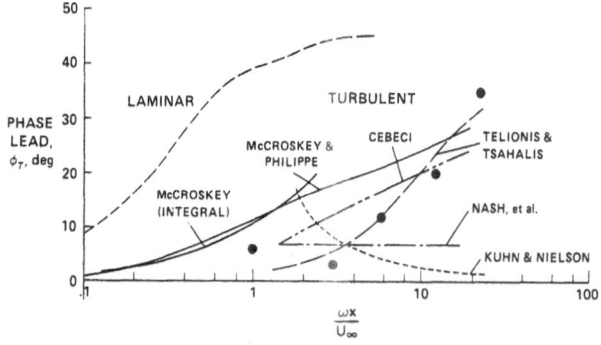

Fig. 6.7 The skin-friction phase angle for turbulent oscillations as a function of the frequency parameter $\omega x / U_0$. Theoretical results; – – –, Kuhn and Nielsen (1973); –·–·–, Nash et al. (1975), $Re = 10^7$; ———, Telionis and Tsahalis (1975); –··–··, Cebeci (1959); – – – –, McCroskey and Philippe (1975). Experiment: ●, Karlsson (1959).

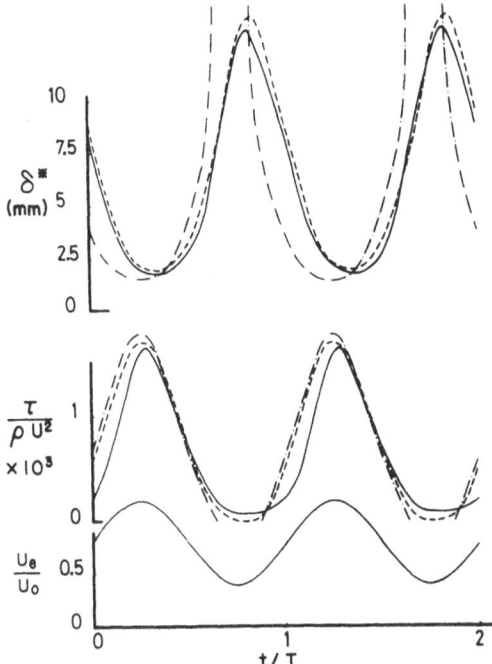

Fig. 6.8 Displacement thickness and wall shear for $\varepsilon_0 = -0.2$, $\varepsilon_1 = 0.4$, $\tilde{\omega} = 1.57$, $\tilde{x} = 1$, $Re = 10^7$, $U_e = U_0$ $[1 + (\varepsilon_0 + \varepsilon_1 \sin \tilde{\omega} t)x]$. ———, Cousteix et al. (1976); – – –, Nash et al. (1975); –·–, Kuhn and Nielsen (1973).

remains attached and well behaved for a pressure gradient, which, if averaged, would induce separation. It is noted further that the larger frequencies and, therefore, the much larger pressure gradients that the boundary layer experiences periodically do not seem to affect the phenomenon. In other words, the strong but periodic pressure gradients due to the term

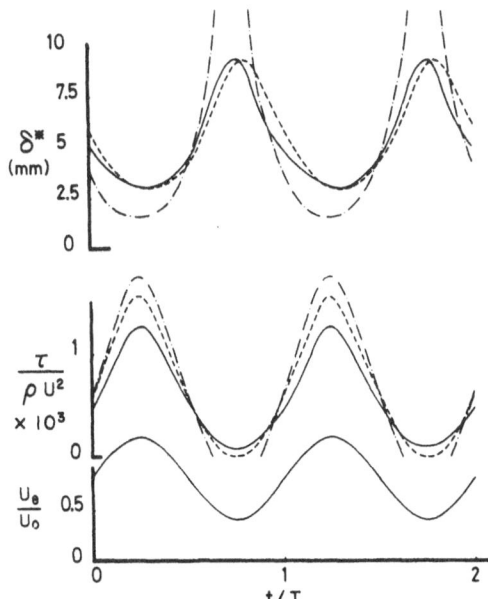

Fig. 6.9 Displacement thickness and wall shear for $\varepsilon_0 = -0.2$, $\varepsilon_1 = 0.4$, $\tilde{\omega} = 15.7$, $\tilde{x} = 1$, $Re = 10^7$, $U_e = U_0$ $[1 + (\varepsilon_0 + \varepsilon_1 \sin \tilde{\omega} t)\tilde{x}]$. ———, Cousteix et al. (1976); –·–·–, Nash et al. (1975); –·–··–, quasi-steady calculation.

$\partial U_e/\partial t$, which in fact grow larger with frequency, do not induce separation. The results indicate that the point of the flow reversal is displaced downstream, at least for the range of frequencies investigated.

Nash, et al. (1975) calculated unsteady boundary layers and the upstream propagation of flow reversal but terminated their calculations at the point of zero wall shear. Turbulent boundary layers that remain attached, even over a thin recirculating region, have been calculated by Telionis and Tsahalis (1975), Patel and Nash (1975), and others. In these references transient flows with ever-steepening adverse pressure gradients are investigated. Telionis and Tsahalis (1975) chose a linearly decelerating flow, which eventually separates. In such calculations, it is necessary that flow reversal propagates upstream faster than other disturbances as described in detail in the earlier chapter with regard to laminar flows. It is then necessary to drop a few of the mesh points at the downstream end of the domain. To avoid this, Patel and Nash (1975) and later Nash (1976) and Nash and Scruggs (1978) chose an outer flow with a linear decrease followed by a linear increase of the velocity:

$$U_e(x,t) = U_0, \qquad\qquad \text{for } t \leqslant 0, \text{ all } x_1, \qquad (6.5.4)$$

$$U_e/U_0 = 1 - \frac{x}{x_1}\left[1 - f(t)\right], \qquad \text{for } t \geqslant 0, 0 \leqslant x \leqslant x_1, \qquad (6.5.5)$$

$$U_e/U_0 = 1 - \frac{2x_1 - x}{x_1}\left[1 - f(t)\right], \quad \text{for } t \geqslant 0, x_1 \leqslant x < x_2, \qquad (6.5.6)$$

where x_2 is the downstream extent of integration, x_1 is a prescribed value of x, and f is an arbitrary function of time. Nash and Scruggs proposed

$$f = 1 - \frac{(1 - f_f)t}{t_f}, \quad \text{for } 0 \leqslant t \leqslant t_f, \qquad (6.5.7)$$

$$f = f_f, \quad \text{for } t_f \leqslant t, \qquad (6.5.8)$$

with f_f and t_f some prescribed constants. Such flows and their equivalent oscillatory variations were essentially proposed for investigation of the relationship between unsteady flow reversal and separation. Of interest in this chapter is the modeling of turbulent boundary layers with partially reversed velocity profiles.

To integrate through such regions, the basic assumption made by all the investigators is that for negative $\partial U/\partial y$ the eddy viscosity also changes sign. However, it was further assumed that the basic character of the closure models is not necessarily different in the reversing flow region. In other words, the Reynolds stress effect is equivalent to shear stresses opposing the motion and pushing toward deceleration of the local flow, regardless of the direction of the outer flow. For eddy viscosity models, Telionis and Tsahalis (1975) simply use in the inner layer an eddy viscosity given by

$$\epsilon_i = \rho l^2 \left| \frac{\partial U}{\partial y} \right|. \qquad *(6.5.9)$$

The damping effect can be modeled exactly in the same way, since A may then vanish with τ_w or τ_s, as indicated by Eqs. (6.3.7) and (6.3.6). For models based on the turbulent energy method, Patel and Nash (1975) propose a modification of Eq. (6.4.18):

$$\frac{D\tau}{Dt} + 2a_1|\tau|\frac{\partial U}{\partial y} + 2a_1\phi + 2a_1\frac{\partial}{\partial y}(a_2\tau) + 2a_1\frac{\tau}{L}|\tau|^{1/2} = 0, \qquad *(6.5.10)$$

where

$$\phi = \Gamma\left(|\tau|\frac{\partial U}{\partial y} + \tau\left|\frac{\partial U}{\partial y}\right|\right) \qquad *(6.5.11)$$

with Γ some large number. The inclusion of the last term does not have any effect on the resultant shear stress and essentially serves to maintain the proper direction of the shear stress vector according to the equation

$$\tau = a_1\overline{q^2}\,\mathrm{sgn}\left(\frac{\partial U}{\partial y}\right). \qquad *(6.5.12)$$

In a later publication, Nash and Scruggs (1978) propose an alternative assumption for closing the turbulent energy equation

$$\tau = -k\left(\overline{q^2}\right)^{1/2}L\frac{\partial U}{\partial y}. \qquad *(6.5.13)$$

This assumption results in a drastic change of the character of the differential equations. A second derivative of the mean velocity U reappears and the system becomes again parabolic.

It should be emphasized that according to the models described, the Reynolds stress in the neighborhood of a vanishing $\partial U/\partial y$ is zero. This occurs at the point of zero skin friction and, from then on, there may exist further downstream a point where $\partial U/\partial y$ vanishes within the flow.

Typical velocity profiles for partially reversed laminar and turbulent flows (Nash and Scruggs, 1978) are shown in Figs. 6.10 and 6.11 for $Re = 10^4$ and $Re = 10^7$, respectively.

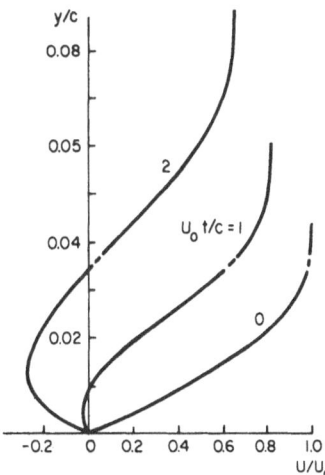

Fig. 6.10 Laminar boundary layer penetrating a reversed-flow region from Nash and Scruggs (1977). (Laminar frozen flow: $Re = 10^4$, $f_f = 0.5$, $t_f = 2.0$, $x/c = 0.5$.)

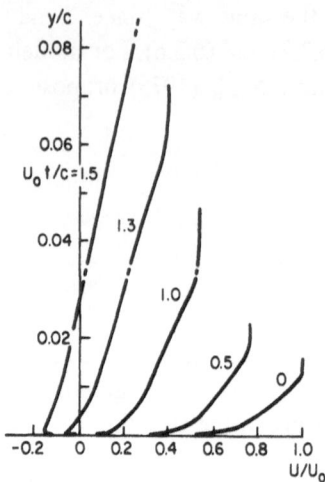

Fig. 6.11 Turbulent boundary layer penetrating a reversed-flow region from Nash and Scruggs (1978). (Turbulent frozen flow: $Re = 10^7$, $f_f = 0.25$, $t_f = 1.5$, $x/c = 0.66$.)

Comparison with experiments

Very little effort has been made to compare the analytical findings with existing experimental data. This is inexcusable, because the experimental information supplied by Karlsson (1959) indicates some clearly identifiable trends. Telionis (1976) collected all the available eddy viscosity models, some of which have been extended to unsteady flows (Alber, 1971; Cebeci and Smith, 1968; Kays, 1971), and ran calculations in an effort to estimate their relative performance in comparison to Karlsson's experiments. The overall performance of quasi-steady, zero-equation models is not satisfactory.

Some typical results calculated by methods referenced earlier are shown in Fig. 6.12. In this figure we plot the "in-phase" and "out-of-phase" velocity components. These are the parts of the velocity fluctuation in phase and at a phase of 90° with respect to the outer-flow velocity, respectively. Karlsson estimated these velocities by assuming that the response is harmonic:

$$u_1(x, y, t) = F(x, y)\cos(\omega t + \phi), \tag{6.5.14}$$

where F is the amplitude function and ϕ is the phase angle. The "in-phase" and "out-of-phase" velocity components can then be obtained from $u_1(x, y, t)$ as follows:

$$u_{in}(x, y) = F(x, y)\cos\phi = \sqrt{2} \; \frac{\overline{u_1 \cos \omega t}}{\left(\overline{\cos^2 \omega t}\right)^{1/2}} \tag{6.5.15}$$

$$u_{out}(x, y) = F(x, y)\sin\phi = \sqrt{2} \; \frac{\overline{u_1 \cos(\omega t + \pi/2)}}{\left(\overline{\cos^2 \omega t}\right)^{1/2}} \tag{6.5.16}$$

Fig. 6.12 Profiles of the amplitude of velocity oscillations for an intermediate frequency, $\omega/2\pi = 1.0$. ——, Romaniuk and Telionis (1979), quasi-steady and Shamroth and Kreskovsky (1974); – – –, Cebeci (1977); –·–·–, Telionis and Tsahalis (1975); $\triangledown$, $\triangle$, $\bullet$ experiment, Karlsson (1959).

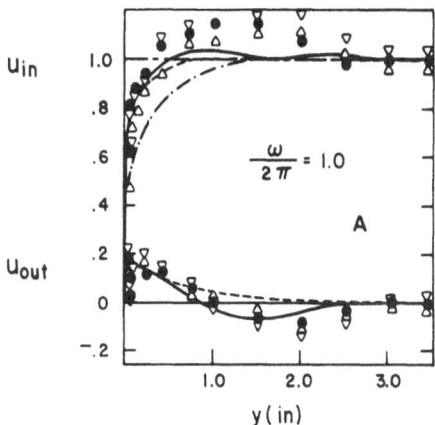

where an overbar denotes a time average. In the notation of Section 6.2, the definition of these quantities is a lot simpler.

$$u_{\text{in}}(x, y) = \mathfrak{R}(u_1), \tag{6.5.17}$$

$$u_{\text{out}}(x, y) = \mathfrak{I}(u_1), \tag{6.5.18}$$

where $\mathfrak{R}$ and $\mathfrak{I}$ are operators designating the real and the imaginary part of the quantity that follows.

Figure 6.12 compiles the analytical results of quasi-steady models for the frequency $\tilde{\omega} = 1$, which appears to be the threshold between low- and high-frequency regimes and analytically the most difficult to model. We have attempted here to retain the symbols used by Karlsson to denote data for different amplitudes. It appears that Karlsson was unable to control accurately the amplitude of the oscillations, because his amplitude ratios are not the same for any two of the frequencies he investigated. In all the figures that follow, the symbols $\bullet$, $\triangle$, and $\triangledown$ represent the highest, intermediate, and lowest amplitude, respectively. Perturbation methods are valid for small amplitudes of oscillations and, to a first approximation, results are independent of the amplitude parameter. Therefore, agreement should be expected with the experimental data that correspond to the lowest amplitude of Karlsson (1959).

More experimental data are available today, which offer valuable new information for the understanding of the phenomena involved. However, such data are confined to a single frequency of oscillation (Houdeville, et al., 1976; Cousteix, et al., 1977a, b; Houdeville and Cousteix, 1978) or to traveling waves (Patel, 1977). Most recent reports contain descriptions of investigations of pressure gradients (Schachenmann and Rockwell, 1976) and separation (Kenison, 1977). Karlsson's data remain the most complete experimental information for a flat plate with a wide range of frequencies and amplitudes.

Calculations carried out for turbulent boundary layers with quasi-laminar

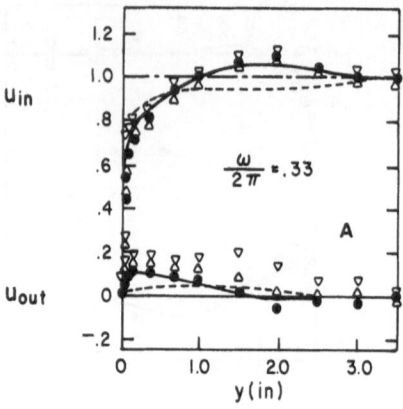

Fig. 6.13 Velocity amplitude profiles, $\omega/2\pi$ = 0.33. ———, Romaniuk and Telionis (1979); – – –, Cebeci (1977), ∇, $\triangle$, •, Karlsson (1959).

models, that is, $\tau_1 = 0$, as described in Section 6.3, indicate some interesting results (Romaniuk and Telionis, 1979). Velocity overshoots of the correct order of magnitude are predicted, that is, 15%–30% of the outer-flow amplitude of oscillation. It should be noted that quasi-steady models usually predict overshoots of the order of 1% or 2% at most. However the quasi-laminar model grossly overestimates the "out-of-phase" components and the phase angle.

In Fig. 6.13 experimental data and analytical results are shown for the lowest frequency for which data are available, that is, $\omega/2\pi = 0.33$. In this figure, the unsteady model is based on the assumptions described by Eqs. (6.3.25)–(6.3.28). It is clear that in this case, decoupling of the Reynolds stress improves considerably the model. However the same model is not as satisfactory for $\omega/2\pi = 1.0$. Some improvement may be achieved by assuming a constant value for ϵ_{1o}. Results based on this assumption are shown in Figs. 6.14–6.16.

It appears that for high frequencies most models are acceptable. All evidence indicates that with increasing frequency the overshoot is approach-

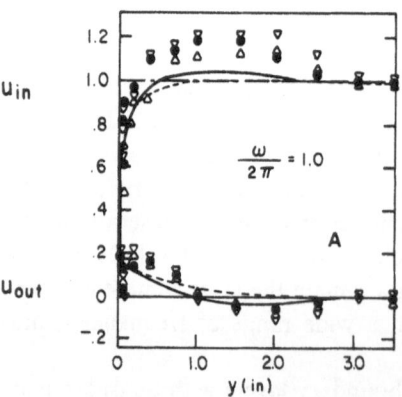

Fig. 6.14 Profiles of amplitude of velocity oscillations, $\omega/2\pi = 1.0$. ———, Romaniuk and Telionis (1979); – – –, Cebeci (1977); ∇, $\triangle$, •, Karlsson (1959).

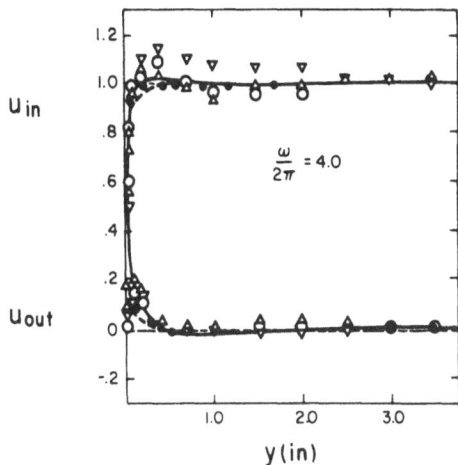

Fig. 6.15 Amplitude of velocity oscillations for moderately high frequency, $\omega/2\pi = 4.0$. ———, Romaniuk and Telionis (1979) (no damping): ●, Telionis and Romaniuk (1979) (with damping); ---, quasi-steady; △, ▽, ○ experiment, Karlsson (1959).

ing the wall. This means that for high frequencies, almost the entire boundary layer oscillates in phase with the outer flow, except for a thin region next to the wall. This is also the case for oscillating laminar boundary layers.

The analogy between the behavior of laminar and turbulent boundary layers goes further than that. In both cases, the dependence of the maximum overshoot on the frequency for laminar flows is very strong and, for a certain value of the frequency parameter $\tilde{\omega}$, the velocity overshoot indicates a clearly pronounced maximum (Romaniuk and Telionis, 1979). The experimental points determined from Karlsson's data show considerably large scattering due to the amplitude differences. However, some characteristic features of the distributions can be noticed. The low-frequency limit of the overshoot seems to be $\Delta u_{max} = 0.1$, regardless of the value of the amplitude. This is the same limit as for laminar boundary layers (Telionis and Romaniuk, 1978). The experimental points for large amplitude oscillations do not follow any regular pattern. However, the data for intermediate and small amplitudes

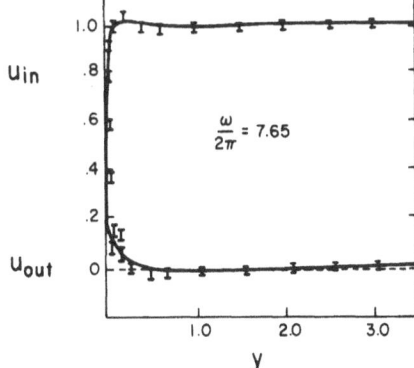

Fig. 6.16 Amplitude of velocity oscillations for high frequency, $\omega/2\pi = 7.65$. ———, Romaniuk and Telionis (1979) (no damping); I, experimental data of Karlsson (1959).

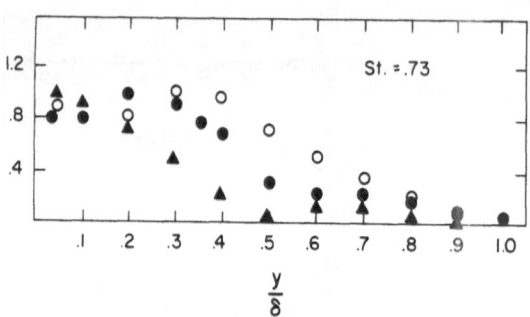

Fig. 6.17 Distribution of the normalized amplitude of the oscillatory Reynolds stress, $|\tau_1|/|\tau_{1max}|$, across the boundary layer for $\tilde{\omega} = 0.73$. O, rough estimate derived from Cousteix et al. (1977a); ●, Romaniuk and Telionis (1979) (exponential damping); ▲, quasi-steady model.

indicate that Δu_{max} is the largest in the vicinity of $\omega/2\pi = 2.0$. For the velocity and distance from the leading edge corresponding to Karlsson's experimental conditions, this frequency is equivalent to $\tilde{\omega} = 8.5$, which would be very high for laminar flow.

The best theoretical prediction of the velocity profiles was obtained using a model with exponential damping of the outer-layer eddy viscosity (Telionis and Romaniuk, 1978). The amplitude of the oscillatory Reynolds stress corresponding to this model was calculated numerically and compared with the experimental profile derived from the data of Cousteix, et al. (1977a). These data represent an ensemble average of the turbulence level and the Reynolds stresses and disclose their organized periodic character. As Fig. 6.17 shows, there is a qualitative agreement between the two profiles. Unfortunately, these results are valid only for one specific value of the reduced frequency $\tilde{\omega} = 0.73$. Reynolds stress measurements covering a wider range of frequencies are needed to give a physical base for model assumptions. In Fig. 6.17 the normalized τ_1 profile for the quasi-steady model is also given for comparison. Due to the normalization process, all the profiles start approximately at the same point, that is, at about $\tau_1/\tau_{max} \simeq 1.0$.

Recent experimental information

Some very interesting features of oscillating turbulent boundary layers were disclosed in the last few years by the experimental efforts of a group of investigators from Toulouse (Houdeville et al., 1976; Cousteix et al., 1976, 1977a, b). All of these experiments are characterized by not so small amplitudes of oscillation, of the order 30–40% of the mean outer flow. As a result some clear nonlinear effects become obvious. Perhaps their most exciting discovery, in the opinion of the present author, is that the turbulence level at a point fluctuates with the same frequency as the outer flow. That is, the random disturbances have an amplitude that follows the outer-flow fluctuations. Figure 6.18 is taken from Houdeville et al. (1976) and corresponds to a reduced frequency of oscillation $\omega\delta_0/U_0 = 3.7 \times 10^{-3}$, where δ_0 and U_0 are

Fig. 6.18 The time history of the
level of turbulence, $(u'^2)^{1/2}$ at different distances from the wall from
Houdeville *et al.* (1976).

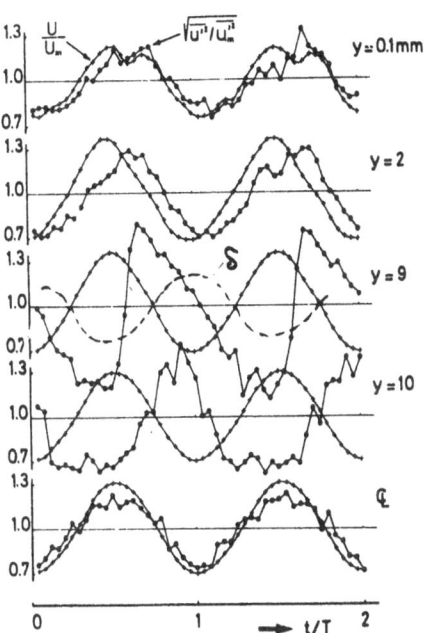

the mean boundary-layer thickness and outer-flow velocity, respectively. It
shows the reduced turbulence level variation $(\overline{u'^2}/(\overline{u'^2})_m)^{1/2}$ and the reduced
average flow U/U_m. In the notation of Houdeville *et al.* (1976), the subscript
m denotes a time average. In our notation the quantity U/U_m should be
$\langle u \rangle / \bar{u}$. We observe that the turbulence level follows with some delay the
fluctuations of the ensemble-averaged motion, but as one proceeds away
from the wall and approaches the edge of the boundary layer, larger phase
shifts and a periodic intermittency phenomenon appear. More surprising,
near the edge of the boundary layer, at a certain instant and while the
averaged velocity and boundary-layer thickness are at a maximum and
minimum, respectively, the turbulence level violently increases. In other
words when the boundary layer goes through its smallest thickness, the
turbulence suddenly bursts out into the free flow. Inspecting the ensemble-
averaged profiles, which are given only for four points of the period, we also
observe a clear asymmetry of the periodic variation, that is, a deviation from
the sinusoidal behavior.

In a later publication (Cousteix *et al.*, 1977a), the same authors give
instantaneous profiles for a higher-frequency parameter $\omega\delta_0/U_0 = 12.7 \times$
10^{-3} for the quantities $(\overline{u'^2})^{1/2}$ and $\overline{(u'v')}$. In Fig. 6.19, we have carried over
the experimental data from Cousteix *et al.* (1977a), as well as from Patel
(1977), who undertook a similar investigation. Patel has used a time constant
larger than the period of oscillation and essentially obtained the average of
the instantaneous turbulence level profiles.

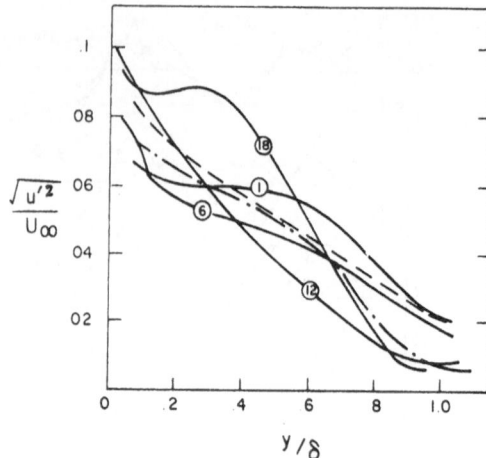

Fig. 6.19 The level of turbulence profiles: ———, four characteristic instances of a period $\omega t = 0, \pi/2, \pi,$ and $3\pi/2$ (curves 1, 6, 12, and 18, respectively as shown in Fig. 6.1), Cousteix et al. (1977a); – – –, averaged through a period and steady flow, Patel (1977); –·–·–, steady flow, Klebanoff (1954).

Fluctuating turbulent boundary layers with an adverse pressure gradient were investigated experimentally in a conical diffuser by Schachenmann and Rockwell (1976). In this study the frequency parameter was varied between 0.63 and 7.33, but the amplitude of the oscillations was only $\varepsilon = 0.069$. However, the pressure gradient effect appears to have a strong influence on the fluctuating part of the motion. First, the overshoot of the fluctuating component u'' increases with downstream distance. Overshoot values reach magnitudes 60% higher than the value of the outer stream. The same effects were encountered in a numerical study for laminar oscillations by Telionis and Tsahalis (1974). Second, the profile of the fluctuating velocity component for large frequencies of oscillations indicates two peaks, and away from the wall it undershoots the value of the outer flow. Such a behavior was also encountered by Telionis and Romaniuk (1978) and was found to carry over to temperature profiles as well. Third, for distances further downstream, the phase angle shows some negative values and eventually it becomes negative across the entire thickness of the boundary layer. The velocity fluctuations follow the fluctuations of the outer flow. Unfortunately the statistical information is not extensive. The root-mean-square of the total velocity is given and it is shown to increase with downstream distance. However, if this quantity contains the RMS of the organized fluctuations, then the turbulence level is buried in it, and very little can be deduced about the statistics of the turbulent boundary layer.

Some very interesting problems that probably belong to the same family, although not strictly dealing with external turbulent boundary layers were investigated most recently. Soutif et al. (1979) studied the response of a turbulent jet to oscillations generated by two fluidic devices that fluctuate symmetrically or antisymmetrically. The authors presented instantaneous profiles and the downstream evolution of periodic and turbulent intensities. Most important of all, they indicated that the externally imposed fluctuation may transfer energy to the turbulent motion. Indeed 20 jet thicknesses downstream of the disturbance, the periodic fluctuations die out but the turbulent intensity grows to a value 76% larger than the corresponding undisturbed jet intensity. Thomas and Shukla (1976) have looked into the wall region of fully developed fluctuating turbulent pipe flow. They report on the interaction between the bursting effect and the imposed fluctuations and compare their experimental results with the theoretical model based on the concept of surface renewal. The work of Binder and Didelle (1975), Mainardi and Panday (1979), and Mainardi et al. (1979) may also be found useful to the investigators of unsteady turbulent boundary layers.

Recapitulating, we may say that a clear picture has started to emerge from the available experimental data. There are of course many parameters involved and a lot more work is necessary to provide a complete description of the phenomenon. The frequency parameter is not easy to vary in all the experimental layouts. However, a relatively wide range has been explored by different investigators. The amplitude parameter is fixed or very difficult to adjust for some of the experimental setups. Very little information is available for flows with pressure gradients that have been proved to have a strong influence on the fluctuating part of the motion. One of the most surprising conclusions that has emerged from many experimental studies is that the mean profile is very little affected even for large amplitudes of oscillation.

Comparing the results of different experiments will be very difficult and it has been attempted by Telionis (1979) but with no enthusiasm. This is due to the fact that the parameter $\omega x / U_0$ is no longer a similarity variable as is the case for laminar flows. In the laminar flow equations the independent variables x and t appear always in the combination $\partial/\partial t + \partial u/\partial x$. However, in the turbulent flow equations, x is also introduced in the Reynolds stress. The parameters $\omega \delta_1 / U_0$ or $\omega \theta / U_0$ have been suggested for comparison, but these too cannot be used for a global presentation of the experimental data. It appears that the only method of comparing experimental data can be through analytical extrapolations based on a successful theoretical method and, to the knowledge of the author, no such method is yet fully satisfactory. The theoreticians at this point must develop new models and the experimentalists will be requested to provide more information on the statistics of the turbulent boundary layers and not only a set of mean velocity profiles.

6.6 A Computer Program

We chose here to include a computer program due to Cebeci and Carr (1978) for the calculation of turbulent boundary layers based on the two-layer eddy viscosity model. This is the most simple model and up to now has proved to be almost equally successful with one- and two-equation models. The program listed in this section is based on the box method described in Chapter 2. In fact one of the reasons for choosing this particular program is the need to include more detailed information about this important technique for integrating nonlinear parabolic equations.

Various elements and details of the method have been scattered in Chapters 2, 4, 7, and the present. This is because different aspects of computational methods are discussed in different parts of the book. If one is particularly interested in all the details of the program that follows, he may need to trace back the original source.

The Cebeci–Carr program can be used to calculate unsteady two-dimensional laminar or turbulent boundary layers. The calculations are performed along the t-direction for a fixed x-station. At each t-station the equations are solved across the layer and are iterated until some prescribed convergence criterion is satisfied. An empirical formula is used to calculate the transition location when needed and the closure conditions are satisfied by using the two-layer eddy viscosity formulas described earlier in this chapter. The grid across the layer is a nonuniform one in which the step size increases away from the wall according to a geometric progession. A flow chart for the computer program is shown in Fig. 6.20.

Input

There are 5 READ statements at the beginning of the subroutine INITIAL. Corresponding to these statements, 5 data cards are necessary. The data should be read in the following order and format:

1st READ Statement
Punch the title in an 80-column alphanumeric field.

2nd READ Statement
Read the following data in 10I3 format and in the following order:

NXT	Number of grid points in the t-direction
NZT	Number of grid points in the x-direction
NTR	The x-station where transition begins. For all laminar flows, or flows for which transition is to be calculated, NTR > NZT
IBDY	Flag controlling the initial steps at $x = 0$ = 1, sharp edge flow = 2, stagnation flow
ISD	Surface distance flag. If surface distance is calculated the pressure gradient parameter P denoted by P2 is also calculated from the external velocity distribution specified in card 5. = 1, surface distance calculated = 2, surface distance input
IP2	P_2 flag: This option allows the pressure gradient parameter to be either input or calculated from the given external velocity distribution. If ISD = 1, IP2 must also be equal to 1.

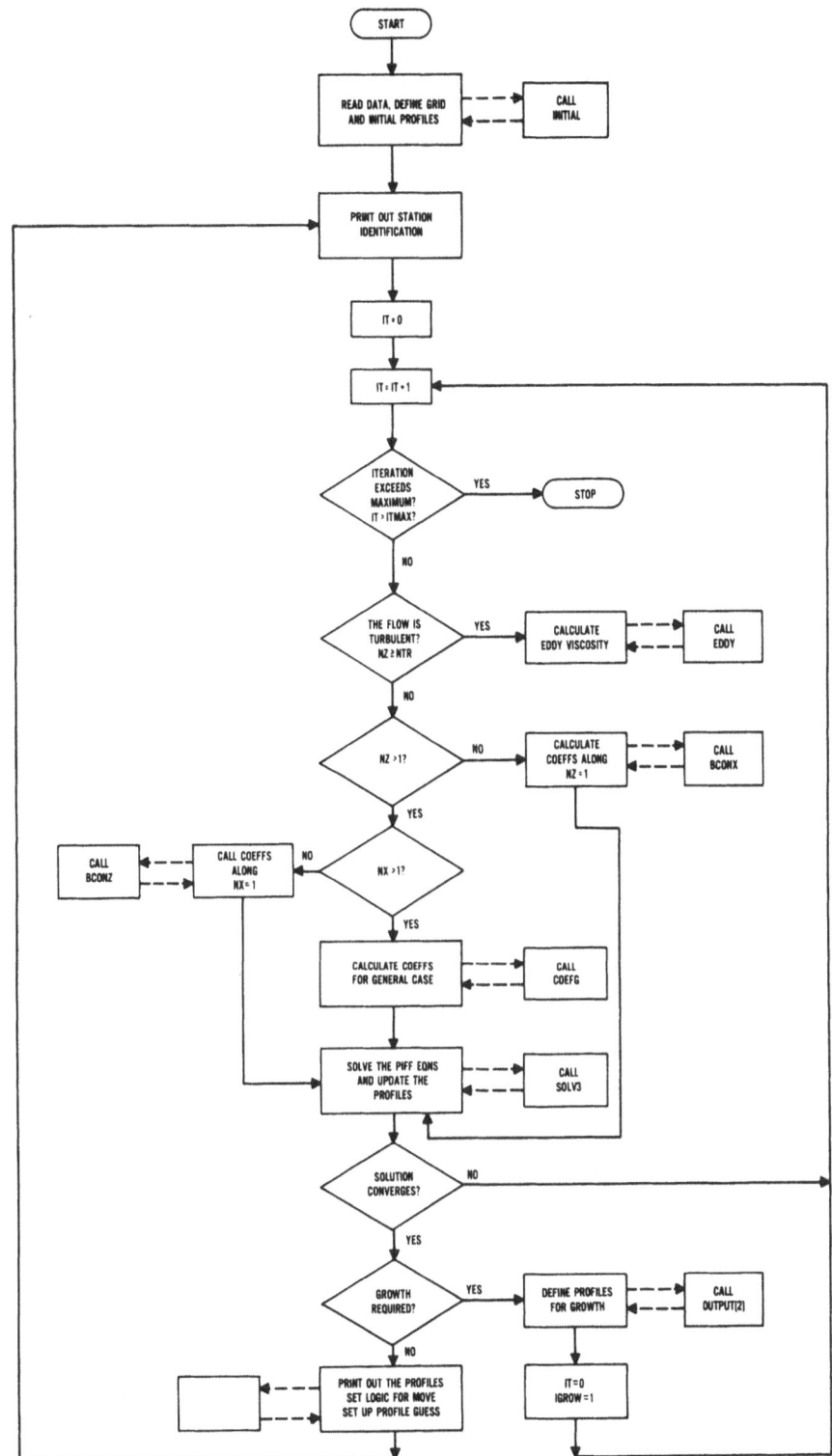

Fig. 6.20 Flowchart for the computer program of Cebeci & Carr (1978).

	= 1, P_2 calculated
	= 2, P_2 input
INTR	Transition flag
	= 1, transition calculated
	= 2, transition input
KPHA	Phase angle flag. This flag provides the calculation of phase angle between wall shear and external velocity and phase angle between displacement thickness and external velocity for both laminar and turbulent flows.
	= 0, phase angle is not calculated
	= 1, phase angle is calculated
IPOP	Flag for in-phase and out-of-phase components of an oscillating turbulent flow on a flat plate.
	= 0, not computed
	= 1, computed
N	Number of t-stations in one cycle to perform the phase angles or in-phase angles or out-of-phase components.

3rd READ Statement

Read data in a 7F10.0 format

ETAE	Transformed boundary-layer thickness, η_∞
DETA(1)	Initial $\Delta\eta$ spacing, $\Delta\eta_1$
VGP	Variable grid parameter
UINF	Reference freestream velocity, u_∞ (ft/sec)
CB	Amplitude of the fluctuating external velocity
OMEGA	Radial frequency ω (rad/sec)
CA	Slope of velocity at the stagnation point of an airfoil, $(du_0/dx)_{x=0}$

4th READ Statement

This contains the input t-stations read in F10.0 format.

5th READ Statement

Card 5 contains the geometry of the two-dimensional body and the steady-state external velocity distribution $u_0(x)$, or the dimensionless pressure gradient parameter $P(x)$ in 3F10.0 format. The geometry of the body is either read in by specifying the surface distance of the body or is computed from the (x/c) and (y/c) coordinates of the body. Here c is a reference length of the body, which for an airfoil is usually taken as the chord. The computer names for the usual (x/c), (y/c) coordinates are denoted by ZC and YC, respectively. Note that when surface distance denoted by Z in field 1 is to be calculated, then the third field must contain the external velocity distribution from which the pressure gradient P2 must be calculated. If surface distance Z is input, and P2 is to be calculated, the third field may be blank. There will be NZT cards of this type.

Read in the following order, NZT cards in 3F10.0 format.

AC or Z	dimensionless distance along the wall or surface distance

YC or UO   dimensionless distance perpendicular to the wall, or the dimensionless edge velocity

UO or P2   dimensionless edge velocity or pressure gradient parameter

Glossary and output

At each station the following information is provided

NX   number of t-grid point
NZ   number of x-grid point
X t, time coordinate
Z x, space coordinate

Moreover, the iteration point is given

V(WALL)   the derivative f'' at $y = 0$
DELVN   the difference of the current and previous value of f''.

If the flag INTR is given the value 1, then close to transition the following values are printed.

RTHETA   the Reynolds number based on the momentum thickness
RTHT   the same quantity as computed by the formula
$$= 1.174[1 + (22,400/\text{Re}_x)]\text{Re}_x^{0.46}$$
for a given Reynolds number based on x in the range $10^5 < \text{Re}_x < 40 \times 10^7$

If transition occurs and a station is interpolated, new values of P2, U and Z are printed which correspond to P, UO and x.

The output boundary-layer parameters include profiles f, f', f'' and b as a function of the similarity variable η and grid parameter j. Here

ETA η
F f
U f'
V f''
B $b \, (= 1 + \epsilon_m)$; equals 1.0 for laminar flows

They also include displacement thickness δ_1, momentum thickness θ, local skin-friction coefficient C_f, Reynolds numbers based on δ_1, θ and x, that is, Re_{δ_1}, Re_θ and Re_x respectively, and shape factor H. The definition of these parameters and their computer notation is

DELSTR $\delta_1 = \displaystyle\int_0^\infty (1 - u/U_e)\,dy$

THETA $\theta = \displaystyle\int_0^\infty u/U_e(1 - u/U_e)\,dy$

CF $C_f = 2\tau_w/\rho u_0^2$
RDELST $\text{Re}_{\delta_1} = \delta_1 U_0/\nu$
RTHETA $\text{Re}_\theta = \theta U_0/\nu$
RZ $\text{Re}_z = z U_0/\nu$
H δ_1/θ

```
      COMMON/BLC0/ NXT,NZT,NX,NZ,NTR,NP,ITMAX,INTR,IBDY,KPHA,IPOP,N,
     1             NPT,CNU,ETAE,VGP,A(61),ETA(61),DETA(61)
      COMMON/BLC1/ CA,CB,OMEGA,OMX,REDFR,X(41),Z(30),UO(30),RZ(30),
     1             P1(30),P2(30),P3(41,30),UE(41,30)
      COMMON/PROF/ DELV(61),F(61,41,2),U(61,41,2),V(61,41,2),B(61,41,2)
C - - - - - - - - - - - - - - - - - - - - - - - - - - - - - - - - - - -
      CALL INTIAL
      WRITE(6,9000)
   25 WRITE(6,9100) NX,NZ,X(NX),Z(NZ)
      OMX   = OMEGA*X(NX)
   30 IT    = 0
      IGROW = 0
   60 IT    = IT+1
      IF(IT .LE. ITMAX) GO TO 65
      WRITE(6,2500)
      STOP
   65 IF(NZ .GE. NTR) CALL EDDY
      IF(NZ .GT. 1) GO TO 70
      CALL BCONX
      GO TO 95
   70 IF(NX .GT. 1) GO TO 90
      CALL BCONZ
      GO TO 95
   90 CALL COEFG
   95 CALL SOLV3
      IF(V(1,NX,2) .GE. 0.0) GO TO 61
      IF(NZ .GE. NTR) GO TO 61
      WRITE(6,9300)
      STOP
C  CHECK FOR CONVERGENCE
   61 IF(NZ .GE. NTR) GO TO 62
C - LAMINAR FLOW
      IF(ABS(DELV(1)) .GT. 0.001) GO TO 60
      GO TO 100
C - TURBULENT FLOW
   62 IF(ABS(DELV(1)/(V(1,NX,2)+0.5*DELV(1))) .GT. 0.02) GO TO 60
  100 IF(NP .EQ. NPT) GO TO 200
      IF(ABS(V(NP,NX,2)) .LE. 0.001) GO TO 200
      IF(IGROW .EQ. 1) GO TO 200
      IGROW = 1
      LL    = 2
      CALL OUTPUT(LL)
      GO TO 30
  200 LL    = 1
      CALL OUTPUT(LL)
      GO TO 25
C - - - - - - - - - - - - - - - - - - - - - - - - - - - - - - - - - - -
 2500 FORMAT(1H0,16X,25HITERATIONS EXCEEDED ITMAX)
 9000 FORMAT(1H1,30H** BOUNDARY LAYER CALCULATIONS//)
 9100 FORMAT(1H0,4HNX =,I3,5X,4HNZ =,I3,5X,3HX =,F10.5,5X,3HZ =,F10.5)
 9300 FORMAT(1H0,33H** LAMINAR SEPARATION OCCURRED **)
      END

      SUBROUTINE INTIAL
      COMMON/BLC0/ NXT,NZT,NX,NZ,NTR,NP,ITMAX,INTR,IBDY,KPHA,IPOP,N,
     1             NPT,CNU,ETAE,VGP,A(61),ETA(61),DETA(61)
      COMMON/BLC1/ CA,CB,OMEGA,OMX,REDFR,X(41),Z(30),UO(30),RZ(30),
     1             P1(30),P2(30),P3(41,30),UE(41,30)
      COMMON/PROF/ DELV(61),F(61,41,2),U(61,41,2),V(61,41,2),B(61,41,2)
      DIMENSION TITLE(20),ZC(30),YC(30)
C - - - - - - - - - - - - - - - - - - - - - - - - - - - - - - - - - - -
C     IBDY  = 1       FLAT PLATE
C     IBDY  = 2       AIRFOIL
C     ISD   = 1       CALCULATED SURFACE DISTANCE
C     ISD   = 2       INPUT SURFACE DISTANCE
C     IP2   = 1       P2 CALCULATED
C     IP2   = 2       P2 READ IN
C     INTR  = 1       CALCULATED TRANSITION
```

```
C      INTR  = 2        INPUT TRANSITION
C      KPHA  = 0        DO NOT CALCULATE PHASE ANGLES
C      KPHA  = 1        CALCULATE PHASE ANGLES
C      IPOP  = 0        DO NOT CALCULATE PHASE COMPONENTS
C      IPOP  = 1        CALCULATE PHASE COMPONENTS
C      N                NUMBER OF X-STATIONS IN ONE CYCLE
C - - - - - - - - - - - - - - - - - - - - - - - - - - - - - - - - - - - -
       NPT   = 61
       CNU   = 1.6E-04
       ITMAX = 6
       READ(5,8001) TITLE
       WRITE(6,9011) TITLE
       READ(5,8000) NXT,NZT,NTR,IBDY,ISD,IP2,INTR,KPHA,IPOP,N
       READ(5,8100) ETAE,DETA(1),VGP,UINF,CB,OMEGA,CA
       READ(5,8300) (X(I),I=1,NXT)
       WRITE(6,9200) NXT,NZT,NTR,ETAE,DETA(1),VGP,CB,OMEGA,UINF,CA
       GO TO (10,30), ISD
    10 READ(5,8400) (ZC(I),YC(I),UO(I),I=1,NZT)
       WRITE(6,9400) (I,ZC(I),YC(I),I=1,NZT)
C  CALCULATE Z
       Z(1)  = 0.0
       UO(1) = UO(1)*UINF
       IF(NZT .EQ. 1) GO TO 60
       SUM1  = 0.C
       DO 20 I=2,NZT
       UO(I) = UO(I)*UINF
       SUM1  = SUM1+SQRT((ZC(I)-ZC(I-1))**2+(YC(I)-YC(I-1))**2)
    20 Z(I)  = SUM1
       GO TO 60
    30 READ(5,8400) (Z(I),UO(I),P2(I),I=1,NZT)
C
    50 IF(IP2 .EQ. 2) GO TO 55
    60 IF(IBDY .EQ. 1) P2(1) = 0.0
       IF(IBDY .EQ. 2) P2(1) = 1.0
    55 DO 90 I=1,NZT
       IF(I .EQ. 1) GO TO 82
       IF(IP2 .EQ. 2) GO TO 81
       IF(I .EQ. NZT) GO TO 70
       A1    = (Z(I)-Z(I-1))*(Z(I+1)-Z(I-1))
       A2    = (Z(I)-Z(I-1))*(Z(I+1)-Z(I))

       A3    = (Z(I+1)-Z(I))*(Z(I+1)-Z(I-1))
       DUDS  = -(Z(I+1)-Z(I))/A1*UO(I-1)+(Z(I+1)-2.0*Z(I)+
      1         Z(I-1))/A2*UO(I)+(Z(I)-Z(I-1))/A3*UO(I+1)
       GO TO 80
    70 A1    = (Z(I-1)-Z(I-2))*(Z(I)-Z(I-2))
       A2    = (Z(I-1)-Z(I-2))*(Z(I)-Z(I-1))
       A3    = (Z(I)-Z(I-1))*(Z(I)-Z(I-2))
       DUDS  = (Z(I)-Z(I-1))/A1*UO(I-2)-(Z(I)-Z(I-2))/A2*UO(I-1)+
      1         (2.0*Z(I)-Z(I-2)-Z(I-1))/A3*UO(I)
    80 P2(I) = Z(I)/UO(I)*DUDS
    81 REDFR = OMEGA*Z(I)/UO(I)
       GO TO 84
    82 IF(IBDY .EQ. 1) GO TO 81
       REDFR = OMEGA/CA
    84 RZ(I) = UO(I)*Z(I)/CNU
       DO 85 K=1,NXT
       OMX   = OMEGA*X(K)
       UE(K,I)=UO(I)*(1.0+CB*COS(OMX))
       IF(K .GT. 1 .AND. I .GT. 1) GO TO 83
       IF(K .EQ. 1 .AND. I .GT. 1) GO TO 73
       IF(K .GT. 1) GO TO 71
       IF(IBDY .EQ. 1) P3(K,I) = 0.0
       IF(IBDY .EQ. 2) P3(K,I) = (1.0+CB*COS(OMX))**2
       GO TO 85
    71 IF(IBDY .EQ. 1) P3(K,I) = 0.0
       IF(IBDY .EQ. 2) P3(K,I) = (1.0+CB*COS(OMX))**2-REDFR*CB*SIN(OMX)
       GO TO 85
```

```
  73 P3(K,I)=P2(I)*(1.0+CB*COS(OMX))**2
     GO TO 85
  83 P3(K,I)=P2(I)*(UE(K,I)/UO(I))**2-CB*REDFR*SIN(OMX)
  85 CONTINUE
  90 CONTINUE
  95 WRITE(6,9000) (Z(I),UO(I),P2(I),I=1,NZT)
 100 DO 110 I=1,NZT
 110 P1(I) = 0.5*(P2(I)+1.0)
     NX   = 1
     NZ   = 1
C
     IF((VGP-1.0) .LE. 0.001) GO TO 105
     NP   = ALOG((ETAE/DETA(1))*(VGP-1.0)+1.0)/ALOG(VGP) + 1.0
     GO TO 112
 105 NP   = ETAE/DETA(1) + 1.0
 112 IF(NP .LE. NPT) GO TO 115
     WRITE(6,9300)
     STOP
 115 ETA(1)= 0.0
     DO 120 J=2,NPT
     DETA(J)=VGP*DETA(J-1)
     A(J)  = 0.5*DETA(J-1)
 120 ETA(J)= ETA(J-1)+DETA(J-1)
     ETANPQ= 0.25*ETA(NP)
     ETAU15= 1.5/ETA(NP)
     DO 130 J=1,NP
     ETAB  = ETA(J)/ETA(NP)
     ETAB2 = ETAB**2
     F(J,NX,2)= ETANPQ*ETAB2*(3.0-0.5*ETAB2)
     U(J,NX,2)= 0.5*ETAB*(3.0-ETAB2)
     V(J,NX,2)= ETAU15*(1.0-ETAB2)
     B(J,NX,2)= 1.0
 130 CONTINUE
     RETURN
C - - - - - - - - - - - - - - - - - - - - - - - - - - - - - - -
8000 FORMAT(10I3)
8001 FORMAT(20A4)
8100 FORMAT(8F10.0)
8300 FORMAT(F10.0)
8400 FORMAT(3F10.0)
9000 FORMAT(///1H0,46H** EXTERNAL STEADY STATE VELOCITY DISTRIBUTION/
    1        1H ,24H** AND PRESSURE GRADIENT/1H0,4X,1HZ,9X,2HUO,8X,2HP2/
    2        (1H ,3F10.5))
9011 FORMAT(1H0,20A4)
9200 FORMAT(///1H0,12H** CASE DATA/1H0,3X,6HNXT =,I3,14X,6HNZT =,
    1        I3,14X,6HNTR =,I3/1H ,3X,6HETAE =,E14.6,3X,6HDETA1=,
    2        E14.6,3X,6HVGP =,E14.6/1H ,3X,6HCB   =,E14.6,3X,
    3        6HOMEGA=,E14.6,3X,6HUINF =,E14.6/1H ,3X,6HCA   =,E14.6)
9300 FORMAT(1H0,37HNP EXCEEDED NPT -- PROGRAM TERMINATED)
9400 FORMAT(//1H0,22H** INPUT BODY GEOMETRY/1H0,3H NZ,6X,3HZ/C,11X,
    1        3HY/C/(1H ,I3,2E14.6))
     END

     SUBROUTINE EDDY
     COMMON/BLC0/ NXT,NZT,NX,NZ,NTR,NP,ITMAX,INTR,IBDY,KPHA,IPOP,N,
    1            NPT,CNU,ETAE,VGP,A(61),ETA(61),DETA(61)
     COMMON/BLC1/ CA,CB,OMEGA,OMX,REDFR,X(41),Z(30),UO(30),RZ(30),
    1            P1(30),P2(30),P3(41,30),UE(41,30)
     COMMON/PROF/ DELV(61),F(61,41,2),U(61,41,2),V(61,41,2),B(61,41,2)
C - - - - - - - - - - - - - - - - - - - - - - - - - - - - - - -
     IFLG  = 0
     RZ2   = SQRT(RZ(NZ))
     RZ216 = RZ2*0.16
     RZ4   = SQRT(RZ2)
     CN    = 1.0
     CRSQV = CN*RZ4*SQRT(ABS(V(1,NX,2)))/26.0
     SUM   = 0.0
     F1    = 0.0
```

```
      DO 30 J=2,NP
      F2      = U(J,NX,2)*(1.0-U(J,NX,2))
      SUM     = SUM+(F1+F2)*A(J)
   30 F1      = F2
      RT      = RZ2*SUM
      IF(RT .LE. 425.) GO TO 35
      IF(RT .GT. 6000.) GO TO 38
      XPI     = RT/425.-1.0
      PI      = .55*(1.0-EXP(-.243*SQRT(XPI)-2.98*XPI))
      GO TO 40
   35 PI      = 0.0
      GO TO 40
   38 PI      = .55
   40 EDVO    = .0168*(1.55/(1.0+PI))*RZ2*(U(NP,NX,2)*ETA(NP)-F(NP,NX,2))
      J       = 1
   50 IF(IFLG .EQ. 1) GO TO 100
      YOA     = CRSQV*ETA(J)
      EL      = 1.0
      IF(YOA .LT. 4.0) EL = (1.0-EXP(-YOA))**2
      EDVI    = RZ216*ETA(J)**2*V(J,NX,2)*EL
      IF(EDVI .LT. EDVO) GO TO 200
      IFLG    = 1
  100 EDV     = EDVO
      GO TO 300
  200 EDV     = EDVI
  300 B(J,NX,2) = 1.0+EDV
      J       = J+1
      IF(J .LE. NP) GO TO 50
      RETURN
      END

      SUBROUTINE BCONX
      COMMON/BLC0/ NXT,NZT,NX,NZ,NTR,NP,ITMAX,INTR,IBDY,KPHA,IPOP,N,
     1             NPT,CNU,ETAE,VGP,A(61),ETA(61),DETA(61)
      COMMON/BLC1/ CA,CB,OMEGA,OMX,REDFR,X(41),Z(30),UO(30),RZ(30),
     1             P1(30),P2(30),P3(41,30),UE(41,30)
      COMMON/PROF/ DELV(61),F(61,41,2),U(61,41,2),V(61,41,2),B(61,41,2)
      COMMON/BLCC/ S1(61),S2(61),S3(61),S4(61),S5(61),S6(61),
     1             R1(61),R2(61),R3(61)
C - - - - - - - - - - - - - - - - - - - - - - - - - - - - - - - - - -
      CEL     = 0.0
      IF(NX .EQ. 1) GO TO 300
      IF(IBDY .EQ. 1) GO TO 300
      DELX    = X(NX)-X(NX-1)
      CEL     = 2.0/(CA*DELX)
  300 U(NP,NX,2) = 1.0+CB*COS(OMX)
      P22     = -2.0*P2(NZ)
      DO 500 J=2,NP
      UB      = 0.5*(U(J,NX,2)+U(J-1,NX,2))
      VB      = 0.5*(V(J,NX,2)+V(J-1,NX,2))
      FVB     = 0.5*(F(J,NX,2)*V(J,NX,2)+F(J-1,NX,2)*V(J-1,NX,2))
      USB     = 0.5*(U(J,NX,2)**2+U(J-1,NX,2)**2)
      DERBV   = (B(J,NX,2)*V(J,NX,2)-B(J-1,NX,2)*V(J-1,NX,2))/DETA(J-1)
      IF(NX .GT. 1) GO TO 400
      R2B     = 0.0
      GO TO 450
  400 CUB     = 0.5*(U(J,NX-1,2)+U(J-1,NX-1,2))
      CFVB    = 0.5*(F(J,NX-1,2)*V(J,NX-1,2)+F(J-1,NX-1,2)*V(J-1,NX-1,2))
      CUSB    = 0.5*(U(J,NX-1,2)**2+U(J-1,NX-1,2)**2)
      CDERBV= (B(J,NX-1,2)*V(J,NX-1,2)-B(J-1,NX-1,2)*V(J-1,NX-1,2))/
     1          DETA(J-1)
      CR2B    = CDERBV+P1(NZ)*CFVB-P2(NZ)*CUSB+P3(NX-1,NZ)
      R2B     = -CR2B-CEL*CUB
C
  450 P1A     = A(J)*P1(NZ)
      S1(J)   = B(J,NX,2)+P1A*F(J,NX,2)
      S2(J)   = -B(J-1,NX,2)+P1A*F(J-1,NX,2)
      S3(J)   = P1A*V(J,NX,2)
```

```
         S4(J) = P1A*V(J-1,NX,2)
         S5(J) = A(J)*(P22*U(J,NX,2)-CEL)
         S6(J) = A(J)*(P22*U(J-1,NX,2)-CEL)
C
         R1(J) = F(J-1,NX,2)-F(J,NX,2)+DETA(J-1)*UB
         R3(J-1)=U(J-1,NX,2)-U(J,NX,2)+DETA(J-1)*VB
         R2(J) = DETA(J-1)*(R2B-P3(NX,NZ)-(DERBV+P1(NZ)*FVB-
        1         P2(NZ)*USB-CEL*UB))
   500 CONTINUE
         R1(1) = 0.0
         R2(1) = 0.0
         R3(NP)= 0.0
         RETURN
         END

         SUBROUTINE BCONZ
         COMMON/BLC0/ NXT,NZT,NX,NZ,NTR,NP,ITMAX,INTR,IBDY,KPHA,IPOP,N,
        1             NPT,CNU,ETAE,VGP,A(61),ETA(61),DETA(61)
         COMMON/BLC1/ CA,CB,OMEGA,OMX,REDFR,X(41),Z(30),UO(30),RZ(30),
        1             P1(30),P2(30),P3(41,30),UE(41,3C)
         COMMON/PROF/ DELV(61),F(61,41,2),U(61,41,2),V(61,41,2),B(61,41,2)
         COMMON/BLCC/ S1(61),S2(61),S3(61),S4(61),S5(61),S6(61),
        1             R1(61),R2(61),R3(61)
C - - - - - - - - - - - - - - - - - - - - - - - - - - - - - - - - - - - -
         U(NP,NX,2) = 1.0+CB*COS(OMX)
         BEL   = 0.0
         IF(NZ .GT. 1) BEL = 0.5*(Z(NZ)+Z(NZ-1))/(Z(NZ)-Z(NZ-1))
         P1P   = P1(NZ)+BEL
         P2P   = P2(NZ)+BEL
         DO 100 J=2,NP
C DEFINITION OF AVERAGED QUANTITIES
         FB    = 0.5*(F(J,NX,2)+F(J-1,NX,2))
         UB    = 0.5*(U(J,NX,2)+U(J-1,NX,2))
         VB    = 0.5*(V(J,NX,2)+V(J-1,NX,2))
         FVB   = 0.5*(F(J,NX,2)*V(J,NX,2)+F(J-1,NX,2)*V(J-1,NX,2))
         USB   = 0.5*(U(J,NX,2)**2+U(J-1,NX,2)**2)
         DERBV = (B(J,NX,2)*V(J,NX,2)-B(J-1,NX,2)*V(J-1,NX,2))/DETA(J-1)
         IF(NZ .GT. 1) GO TO 10
         CFB   = 0.0
         CUB   = 0.0
         CVB   = 0.0
         GO TO 20
    10 CFB   = 0.5*(F(J,NX,1)+F(J-1,NX,1))
         CUB   = 0.5*(U(J,NX,1)+U(J-1,NX,1))
         CVB   = 0.5*(V(J,NX,1)+V(J-1,NX,1))
         CFVB  = 0.5*(F(J,NX,1)*V(J,NX,1)+F(J-1,NX,1)*V(J-1,NX,1))
         CUSB  = 0.5*(U(J,NX,1)**2+U(J-1,NX,1)**2)
         CDERBV= (B(J,NX,1)*V(J,NX,1)-B(J-1,NX,1)*V(J-1,NX,1))/DETA(J-1)
C
C COEFFICIENTS OF THE DIFFERENCED MOMENTUM EQUATION
    20 S1(J) = B(J,NX,2)/DETA(J-1)+(P1P*F(J,NX,2)-BEL*CFB)*0.5
         S2(J) = -B(J-1,NX,2)/DETA(J-1)+(P1P*F(J-1,NX,2)-BEL*CFB)*0.5
         S3(J) = 0.5*(P1P*V(J,NX,2)+BEL*CVB)
         S4(J) = 0.5*(P1P*V(J-1,NX,2)+BEL*CVB)
         S5(J) = -P2P*U(J,NX,2)
         S6(J) = -P2P*U(J-1,NX,2)
C
C DEFINITIONS OF RJ
         R1(J) = F(J-1,NX,2)-F(J,NX,2)+DETA(J-1)*UB
         R3(J-1)=U(J-1,NX,2)-U(J,NX,2)+DETA(J-1)*VB
         IF(NZ .EQ. 1) GO TO 30
         CLB   = CDERBV+P1(NZ-1)*CFVB-P2(NZ-1)*CUSB+P3(NX,NZ-1)
         CRB   = -P3(NX,NZ)+BEL*(CFVB-CUSB)-CLB
         GO TO 40
    30 CRB   = -P2(NZ)
    40 R2(J) = CRB-(DERBV+P1P*FVB-P2P*USB-BEL*(CFB*VB-CVB*FB))
   100 CONTINUE
         R1(1) = 0.0
```

```
      R2(1) = 0.0
      R3(NP)= 0.0
      RETURN
      END
      SUBROUTINE COEFG
      COMMON/BLC0/ NXT,NZT,NX,NZ,NTR,NP,ITMAX,INTR,IBDY,KPHA,IPOP,N,
     1            NPT,CNU,ETAE,VGP,A(61),ETA(61),DETA(61)
      COMMON/BLC1/ CA,CB,OMEGA,OMX,REDFR,X(41),Z(30),UO(30),RZ(30),
     1            P1(30),P2(30),P3(41,30),UE(41,30)
      COMMON/PROF/ DELV(61),F(61,41,2),U(61,41,2),V(61,41,2),B(61,41,2)
      COMMON/BLCC/ S1(61),S2(61),S3(61),S4(61),S5(61),S6(61),
     1            R1(61),R2(61),R3(61)
C - - - - - - - - - - - - - - - - - - - - - - - - - - - - - - - -
      U(NP,NX,2) = 1.0+CB*COS(OMX)
      DELX   = X(NX)-X(NX-1)
      DELZ   = Z(NZ)-Z(NZ-1)
      ZB     = 0.5*(Z(NZ)+Z(NZ-1))
      UOB    = 0.5*(UO(NZ)+UO(NZ-1))
      CEL    = ZB/DELZ
      BEL    = ZB/(DELX*UOB)
      CEL2   = 0.5*CEL
      P1B    = 0.5*(P1(NZ)+P1(NZ-1))
      P2B    = 0.5*(P2(NZ)+P2(NZ-1))
      P3B    = 0.25*(P3(NX,NZ)+P3(NX-1,NZ)+P3(NX,NZ-1)+P3(NX-1,NZ-1))
      P2B2   = 2.0*P2B
      P3B4   = 4.0*P3B
      BEL2   = 2.0*BEL
      DO 500 J=2,NP
      FB     = 0.5*(F(J,NX,2)+F(J-1,NX,2))
      FVB    = 0.5*(F(J,NX,2)*V(J,NX,2)+F(J-1,NX,2)*V(J-1,NX,2))
      FB4    = 0.5*(F(J,NX-1,2)+F(J-1,NX-1,2))
      FVJ2   = F(J,NX,1)*V(J,NX,1)+F(J,NX-1,1)*V(J,NX-1,1)+
     1         F(J,NX-1,2)*V(J,NX-1,2)
      FVJ1   = F(J-1,NX,1)*V(J-1,NX,1)+F(J-1,NX-1,1)*V(J-1,NX-1,1)+
     1         F(J-1,NX-1,2)*V(J-1,NX-1,2)
      FBI1   = 0.25*(F(J,NX-1,1)+F(J-1,NX-1,1)+F(J,NX,1)+F(J-1,NX,1))
      FVB234 = 0.5*(FVJ2+FVJ1)
      UB     = 0.5*(U(J,NX,2)+U(J-1,NX,2))
      USB    = 0.5*(U(J,NX,2)**2+U(J-1,NX,2)**2)
      UB2    = 0.5*(U(J,NX,1)+U(J-1,NX,1))
      UB4    = 0.5*(U(J,NX-1,2)+U(J-1,NX-1,2))
      UJ1    = U(J-1,NX,1)+U(J-1,NX-1,1)+U(J-1,NX-1,2)
      UJ2    = U(J,NX,1)+U(J,NX-1,1)+U(J,NX-1,2)
      UBI1   = 0.25*(U(J,NX-1,1)+U(J-1,NX-1,1)+U(J,NX,1)+U(J-1,NX,1))
      UBK1   = 0.25*(U(J,NX-1,1)+U(J-1,NX-1,1)+U(J,NX-1,2)+U(J-1,NX-1,2))
      USJ2   = U(J,NX,1)**2+U(J,NX-1,1)**2+U(J,NX-1,2)**2
      USJ1   = U(J-1,NX,1)**2+U(J-1,NX-1,1)**2+U(J-1,NX-1,2)**2
      UB234  = 0.5*(UJ2+UJ1)
      USB234 = 0.5*(USJ2+USJ1)
      VB     = 0.5*(V(J,NX,2)+V(J-1,NX,2))
      VJ1    = V(J-1,NX,1)+V(J-1,NX-1,1)+V(J-1,NX-1,2)
      VJ2    = V(J,NX,1)+V(J,NX-1,1)+V(J,NX-1,2)
      VB234  = 0.5*(VJ2+VJ1)
      BVJ1   = B(J-1,NX,1)*V(J-1,NX,1)+B(J-1,NX-1,1)*V(J-1,NX-1,1)+
     1         B(J-1,NX-1,2)*V(J-1,NX-1,2)
      BVJ2   = B(J,NX,1)*V(J,NX,1)+B(J,NX-1,1)*V(J,NX-1,1)+
     1         B(J,NX-1,2)*V(J,NX-1,2)
      DERBV  = (B(J,NX,2)*V(J,NX,2)-B(J-1,NX,2)*V(J-1,NX,2))/DETA(J-1)
C
      CM1    = UB234
      CM2    = UB2-2.0*UBK1
      CM3    = VB234
      CM4    = FB4-2.0*FBI1
      CM5    = FVB234
      CM6    = UB4-2.0*UBI1
      CM7    = CM1*CM6-CM3*CM4
      CM8    = BVJ2-BVJ1
```

```
      CM9   = CM1+CM6
      CM10  = -CM8+DETA(J-1)*(-P1B*CM5+USB234*P2B-P3B4+CEL2*CM7+BEL2*
     1        CM2)
C
      S1(J) = B(J,NX,2)+A(J)*(P1B*F(J,NX,2)+CEL2*(FB+CM4))
      S2(J) = -B(J-1,NX,2)+A(J)*(P1B*F(J-1,NX,2)+CEL2*(FB+CM4))
      S3(J) = A(J)*(P1B*V(J,NX,2)+CEL2*(VB+CM3))
      S4(J) = A(J)*(P1B*V(J-1,NX,2)+CEL2*(VB+CM3))
      S5(J) = A(J)*(-P2B2*U(J,NX,2)-CEL2*(2.0*UB+CM9)-BEL2)
      S6(J) = A(J)*(-P2B2*U(J-1,NX,2)-CEL2*(2.0*UB+CM9)-BEL2)
C
      R1(J) = F(J-1,NX,2)-F(J,NX,2)+DETA(J-1)*UB
      R2(J) = CM10-DETA(J-1)*(DERBV+P1B*FVB-P2B*USB-CEL2*(UB*UB+CM9*
     1        UB-VB*FB-CM4*VB-CM3*FB)-BEL2*UB)
      R3(J-1)=U(J-1,NX,2)-U(J,NX,2)+DETA(J-1)*VB
  500 CONTINUE
      R3(NP)= 0.0
      R1(1) = 0.0
      R2(1) = 0.0
      RETURN
      END

      SUBROUTINE SOLV3
      COMMON/BLC0/ NXT,NZT,NX,NZ,NTR,NP,ITMAX,INTR,IBDY,KPHA,IPOP,N,
     1             NPT,CNU,ETAE,VGP,A(61),ETA(61),DETA(61)
      COMMON/PROF/ DELV(61),F(61,41,2),U(61,41,2),V(61,41,2),B(61,41,2)
      COMMON/BLCC/ S1(61),S2(61),S3(61),S4(61),S5(61),S6(61),
     1             R1(61),R2(61),R3(61)
      DIMENSION A11(61),A12(61),A13(61),A21(61),A22(61),A23(61),
     1          G11(61),G12(61),G13(61),G21(61),G22(61),G23(61),
     2          W1(61),W2(61),W3(61),DELF(61),DELU(61)
C - - - - - - - - - - - - - - - - - - - - - - - - - - - - - - - - - -
      W1(1) = R1(1)
      W2(1) = R2(1)
      W3(1) = R3(1)
      A11(1)= 1.0
      A12(1)= 0.0
      A13(1)= 0.0
      A21(1)= 0.0
      A22(1)= 1.0
      A23(1)= 0.0
      G11(2)=-1.0
      G12(2)=-0.5*DETA(1)
      G13(2)= 0.0
      G21(2)= S4(2)
      G23(2)=-2.0*S2(2)/DETA(1)
      G22(2)= G23(2)+S6(2)
      DO 500 J=2,NP
      IF(J .EQ. 2) GO TO 100
      DEN   = (A13(J-1)*A21(J-1)-A23(J-1)*A11(J-1)-A(J)*
     1        (A12(J-1)*A21(J-1)-A22(J-1)*A11(J-1)))
      G11(J)=(A23(J-1)+A(J)*(A(J)*A21(J-1)-A22(J-1)))/DEN
      G12(J)=-(1.0+G11(J)*A11(J-1))/A21(J-1)
      G13(J)= (G11(J)*A13(J-1)+G12(J)*A23(J-1))/A(J)
      G21(J)= (S2(J)*A21(J-1)-S4(J)*A23(J-1)+A(J)*(S4(J)*
     1        A22(J-1)-S6(J)*A21(J-1)))/DEN
      G22(J)= (S4(J)-G21(J)*A11(J-1))/A21(J-1)
      G23(J)= (G21(J)*A12(J-1)+G22(J)*A22(J-1)-S6(J))
  100 A11(J)= 1.0
      A12(J)=-A(J)-G13(J)
      A13(J)= A(J)*G13(J)
      A21(J)= S3(J)
      A22(J)= S5(J)-G23(J)
      A23(J)= S1(J)+A(J)*G23(J)
      W1(J) = R1(J)-G11(J)*W1(J-1)-G12(J)*W2(J-1)-G13(J)*W3(J-1)
      W2(J) = R2(J)-G21(J)*W1(J-1)-G22(J)*W2(J-1)-G23(J)*W3(J-1)
      W3(J) = R3(J)
  500 CONTINUE
```

```
   DELU(NP)  = W3(NP)
   E1        = W1(NP)-A12(NP)*DELU(NP)
   E2        = W2(NP)-A22(NP)*DELU(NP)
   DELV(NP)  = (E2*A11(NP)-E1*A21(NP))/(A23(NP)*A11(NP)-A13(NP)*
  1           A21(NP))
   DELF(NP)  = (E1-A13(NP)*DELV(NP))/A11(NP)
   J         = NP
600 J        = J-1
   E3        = W3(J)-DELU(J+1)+A(J+1)*DELV(J+1)
   DELV(J)   = (A11(J)*(W2(J)+E3*A22(J))-A21(J)*W1(J)-E3*A21(J)*A12(J)
  1           )/(A21(J)*A12(J)*A(J+1)-A21(J)*A13(J)-A(J+1)*
  2           A22(J)*A11(J)+A23(J)*A11(J))
   DELU(J)   =-A(J+1)*DELV(J)-E3
   DELF(J)   = (W1(J)-A12(J)*DELU(J)-A13(J)*DELV(J))/A11(J)
   IF(J .GT. 1) GO TO 600
   WRITE(6,9100) V(1,NX,2),DELV(1)
   DO 700 J=1,NP
   F(J,NX,2)= F(J,NX,2)+DELF(J)
   U(J,NX,2)= U(J,NX,2)+DELU(J)
700 V(J,NX,2)= V(J,NX,2)+DELV(J)
   U(1,NX,2)= 0.0
   RETURN
C - - - - - - - - - - - - - - - - - - - - - - - - - - - - - -
9100 FORMAT(1H ,5X,8HV(WALL)=,E14.6,5X,6HDELVW=,E14.6)
   END
   SUBROUTINE OUTPUT(LL)
   COMMON/BLC0/ NXT,NZT,NX,NZ,NTR,NP,ITMAX,INTR,IBDY,KPHA,IPOP,N,
  1             NPT,CNU,ETAE,VGP,A(61),ETA(61),DETA(61)
   COMMON/BLC1/ CA,CB,OMEGA,OMX,REDFR,X(41),Z(30),UO(30),RZ(30),
  1             P1(30),P2(30),P3(41,30),UE(41,30)
   COMMON/PROF/ DELV(61),F(61,41,2),U(61,41,2),V(61,41,2),B(61,41,2)
   DIMENSION NPK(41),RTHT(30),RTHETA(30),ALFAA(61,41),ALFAB(61,41),
  1             SUMA(61),SUMB(61)
C - - - - - - - - - - - - - - - - - - - - - - - - - - - - - -
   IF(LL .EQ. 2) GO TO 150
   NPK(NX)=NP
   IF(RZ(NZ) .EQ. 0.0) GO TO 140
   ZRZ    = Z(NZ)/SQRT(RZ(NZ))
   DELSTR= ZRZ*(ETA(NP)-F(NP,NX,2)/U(NP,NX,2))
   CF     = 2.0*V(1,NX,2)/SQRT(RZ(NZ))
   RDELST= UO(NZ)*DELSTR/CNU
   SUM1   = 0.0
   F1     = U(1,NX,2)/U(NP,NX,2)*(1.0-U(1,NX,2)/U(NP,NX,2))
   DO 50 J=2,NP
   F2     = U(J,NX,2)/U(NP,NX,2)*(1.0-U(J,NX,2)/U(NP,NX,2))
   SUM1   = SUM1+(F1+F2)*A(J)
50 F1     = F2
   THETA = ZRZ*SUM1
   RTHTA = UO(NZ)*THETA/CNU
   H      = DELSTR/THETA
C
C  CHECK FOR TRANSITION IF IT IS TO BE CALCULATED
   IF(INTR .EQ. 2) GO TO 150
   IF(P2(NZ) .GE. 0.0) GO TO 150
   IF(NZ .GE. NTR) GO TO 150
   IF(NX .GT. 1) GO TO 150
   IF(NZ .EQ. 1) GO TO 150
   IF(NZ .EQ. NZT) GO TO 150
   RZTR   = RZ(NZ)
   RTHETA(NZ)=RTHTA
   RTHT(NZ) = 1.174*(1.+22400./RZTR)*RZTR**0.46
   WRITE(6,9500) RTHETA(NZ),RTHT(NZ)
   IF(RTHETA(NZ)-RTHT(NZ)) 150,110,120
110 ZTR    = Z(NZ)
   WRITE(6,9600) ZTR
   GO TO 150
120 ZTR1   = Z(NZ-1)
```

```
          ZTR2   = Z(NZ)
          DRTH1  = RTHT(NZ-1)-RTHETA(NZ-1)
          DRTH2  = RTHT(NZ)-RTHETA(NZ)
          ZTR    = ZTR1+(DRTH1*(ZTR2-ZTR1))/(DRTH1-DRTH2)
          UOTR   = UO(NZ-1)+((ZTR-ZTR1)/(ZTR2-ZTR1))*(UO(NZ)-UO(NZ-1))
          P2TR   = P2(NZ-1)+((ZTR-ZTR1)/(ZTR2-ZTR1))*(P2(NZ)-P2(NZ-1))
          I      = NZT+2
          IF(NZT .EQ. 30) I = NZT+1
  100 I      = I-1
          Z(I)   = Z(I-1)
          RZ(I)  = RZ(I-1)
          UO(I)  = UO(I-1)
          P1(I)  = P1(I-1)
          P2(I)  = P2(I-1)
          DO 90 K=1,NXT
          UE(K,I)=UE(K,I-1)
   90 P3(K,I)=P3(K,I-1)
          IF(I .GT. (NZ+1)) GO TO 100
          P2(NZ)= P2TR
          P1(NZ)= 0.5*(1.0+P2(NZ))
          Z(NZ) = ZTR
          UO(NZ)= UOTR
          RZ(NZ)= Z(NZ)*UO(NZ)/CNU
          UE(1,NZ) = UO(NZ)*(1.0+CB*COS(OMEGA*X(1)))
          P3(1,NZ) = P2(NZ)*(1.0+CB*COS(OMEGA*X(1)))**2
          DO 125 K=2,NXT
          UE(K,NZ) = UO(NZ)*(1.0+CB*COS(OMEGA*X(K)))
          REDFR = OMEGA*Z(NZ)/UO(NZ)
          P3(K,NZ) = P2(NZ)*(UE(K,NZ)/UO(NZ))**2-CB*REDFR*SIN(OMEGA*X(K))
  125 CONTINUE
          NTR    = NZ
          IF(NZT .LT. 30) NZT = NZT+1
          DO 130 J=1,NP
          F(J,NX,2)= F(J,NX,1)
          U(J,NX,2)= U(J,NX,1)
          V(J,NX,2)= V(J,NX,1)
  130 B(J,NX,2)= B(J,NX,1)
          WRITE(6,9900) P2TR,UOTR,ZTR
          IF(NTR .EQ. NZ) RETURN
  140 RTHTA = 0.0
          RTHETA(NZ)=0.0
C
  150 NPO    = NP
          NP1    = NP+1
          IF(LL .EQ. 2) NP = NP+2
          IF(NP .GT. NPT) NP = NPT
          DO 160 J=NP1,NPT
          F(J,NX,2) = U(NPO,NX,2)*(ETA(J)-ETA(NPO))+F(NPO,NX,2)
          U(J,NX,2) = U(NPO,NX,2)
          V(J,NX,2) = V(NPO,NX,2)
  160 B(J,NX,2) = B(NPO,NX,2)
          IF(LL .EQ. 2) RETURN
C
  180 WRITE(6,9010)
          NPM1   = NP-1
          WRITE(6,9000)  (J,ETA(J),F(J,NX,2),U(J,NX,2),V(J,NX,2),B(J,NX,2),
         1                J=1,NPM1,3)
          WRITE(6,9000)  NP,ETA(NP),F(NP,NX,2),U(NP,NX,2),V(NP,NX,2),
         1                B(NP,NX,2)
          IF(NZ .EQ. 1) GO TO 10
          WRITE(6,9200) DELSTR,THETA,CF,RDELST,RTHTA,RZ(NZ),H,REDFR
          IF(NXT .EQ. 1) GO TO 10
          DELX   = X(2)-X(1)
          IF(IPOP .EQ. 0) GO TO 196
          IF(NZ .LT. NTR) GO TO 196
C  CALCULATE IN-PHASE AND OUT-OF-PHASE COMPONENTS OF AN OSCILLATING
C  TURBULENT FLOW
```

```
      COMX   = COS(OMX)
      SOMX   = SIN(OMX)
      DO 190 J=1,NPT
      ALFAA(J,NX) = U(J,NX,2)*COMX
  190 ALFAB(J,NX) = U(J,NX,2)*SOMX
      IF(NX .LT. NXT) GO TO 10
      I1     = 2
      I2     = N-1
  191 COEFF = 2.0/(CB*FLOAT(N-1))
      DO 195 J=1,NP
      SUMA(J)=0.0
      SUMB(J)=C.C
      DO 198 I=I1,I2
      SUMA(J)=SUMA(J)+ALFAA(J,I)
  198 SUMB(J)=SUMB(J)+ALFAB(J,I)
      SUMA(J)=(0.5*(ALFAA(J,I1-1)+ALFAA(J,I2+1))+SUMA(J))*COEFF
  195 SUMB(J)=-(0.5*(ALFAB(J,I1-1)+ALFAB(J,I2+1))+SUMB(J))*COEFF
      WRITE(6,9300) X(I2+1),(J,SUMA(J),SUMB(J),J=1,NP)
      IF(I2 .EQ. (NXT-1)) GO TO 196
      I1     = I1+(N-1)
      I2     = I2+(N-1)
      IF(I2 .LE. (NXT-1)) GO TO 191
      WRITE(6,9700)
C
  196 IF(KPHA .EQ. 0) GO TO 10
      IF(NX .LT. NXT) GO TO 10
C  CALCULATE PHASE ANGLES
      I1     = 2
      I2     = N-1
  211 AV1    = 0.0
      ADLSTR= 0.0
      DO 210 I=I1,I2
      ADLSTR= ADLSTR+(ZRZ*(ETA(NP)-F(NP,I,2)/U(NP,I,2)))
  210 AV1    = AV1+V(1,I,2)
      D1     = ZRZ*(ETA(NP)-F(NP,I1-1,2)/U(NP,I1-1,2))
      D2     = ZRZ*(ETA(NP)-F(NP,I2+1,2)/U(NP,I2+1,2))
      ADLSTR= OMEGA/6.2832*(0.5*(D1+D2)+ADLSTR)*DELX
      AV1    = OMEGA/6.2832*(0.5*(V(1,I1-1,2)+V(1,I2+1,2))+AV1)*DELX
      ALF2   = 0.0
      BTA2   = 0.0
      ALFBTA= 0.0
      ALFASQ= 0.0
      BETASQ= 0.0
      DO 220 I=I1,I2
      DLS    = ZRZ*(ETA(NP)-F(NP,I,2)/U(NP,I,2))
      ALF2   = ALF2+(UE(I,NZ)-UO(NZ))*(DLS-ADLSTR)
      BTA2   = BTA2+(DLS-ADLSTR)**2
      ALFBTA= ALFBTA+(UE(I,NZ)-UO(NZ))*(V(1,I,2)-AV1)
      ALFASQ= ALFASQ+(UE(I,NZ)-UO(NZ))**2
  220 BETASQ= BETASQ+(V(1,I,2)-AV1)**2
C  CALCULATE PHASE ANGLE BETWEEN WALL SHEAR AND UE
      ALFBTA= (0.5*((UE(I1-1,NZ)-UO(NZ))*(V(1,I1-1,2)-AV1)+
     1        (UE(I2+1,NZ)-UO(NZ))*(V(1,I2+1,2)-AV1))+ALFBTA)*DELX
      ALFASQ= (0.5*((UE(I1-1,NZ)-UO(NZ))**2+(UE(I2+1,NZ)-UO(NZ))**2)+
     1        ALFASQ)*DELX
      BETASQ= (0.5*((V(1,I1-1,2)-AV1)**2+(V(1,I2+1,2)-AV1)**2)+
     1        BETASQ)*DELX
      PHI    = ARCOS(ALFBTA/SQRT(ALFASQ*BETASQ))*57.29578
C  CALCULATE PHASE ANGLE BETWEEN DISPLACEMENT THICKNESS AND UE
      ALF2   = (0.5*((UE(I1-1,NZ)-UO(NZ))*(D1-ADLSTR)+(UE(I2+1,NZ)-
     1        UO(NZ))*(D2-ADLSTR))+ALF2)*DELX
      BTA2   = (0.5*((D1-ADLSTR)**2+(D2-ADLSTR)**2)+BTA2)*DELX
      PHI2   = ARCOS(ALF2/SQRT(ALFASQ*BTA2))*57.29578
      WRITE(6,9400) X(I2+1),PHI,PHI2
C
      IF(I2 .EQ. (NXT-1)) GO TO 10
      I1     = I1+(N-1)
```

```
         I2     = I2+(N-1)
         IF(I2 .LE. (NXT-1)) GO TO 211
         WRITE(6,9700)
C
C
   10 IF(NX .EQ. NXT) GO TO 200
      NX     = NX+1
  300 IF(NZ .GT. 1) GO TO 350
C INITIAL GUESS IN NX DIRECTION (NZ=1)
  310 DO 400 J=1,NPT
         F(J,NX,2)= F(J,NX-1,2)
         U(J,NX,2)= U(J,NX-1,2)
         V(J,NX,2)= V(J,NX-1,2)
         B(J,NX,2)= B(J,NX-1,2)
  400 CONTINUE
      GO TO 370
  350 IF(NX .EQ. 1) GO TO 500
C INITIAL GUESS IN NX DIRECTION (NZ .GT. 1)
      GO TO 310
  370 IF(NZ .EQ. 1) RETURN
      NP     = NPK(NX)
      IF(NX .EQ. 1) RETURN
      IF(NP .LT. NPK(NX-1)) NP = NPK(NX-1)
      RETURN
  200 NX     = 1
      WRITE(6,9800)
      IF(NZ .EQ. NZT) STOP
      NZ     = NZ+1
C SHIFT ALL NX PROFILES IN THE NZ DIRECTION
  500 DO 550 K=1,NXT
      DO 520 J=1,NPT
         F(J,K ,1)= F(J,K ,2)
         U(J,K ,1)= U(J,K ,2)
         V(J,K ,1)= V(J,K ,2)
         B(J,K ,1)= B(J,K ,2)
  520 CONTINUE
  550 CONTINUE
      GO TO 370
C - - - - - - - - - - - - - - - - - - - - - - - - - - - - - - - - - - - -
 9010 FORMAT(1H0,2X,1HJ,4X,3HETA,10X,1HF,13X,1HU,13X,1HV,13X,1HB)
 9000 FORMAT(1H ,I3,F10.6,4E14.6)
 9200 FORMAT(1H0,7HDELSTR=,E14.6,3X,7HTHETA =,E14.6,3X,7HCF   =,
     1         E14.6/1H ,7HRDELST=,E14.6,3X,7HRTHTA =,E14.6,3X,7HRZ   =,
     2         E14.6/1H ,7HH    =,E14.6,3X,7HREDFR =,E14.6)
 9300 FORMAT(/1H0,4X,22H** PHASE COMPONENTS **/1H ,18HCYCLE ENDS WITH X=
     1,       E14.6/1H0,2X,1HJ,3X,8HIN-PHASE,4X,12HOUT-OF-PHASE/
     2        (1H ,I3,2E14.6))


   *** TEST CASE 1 - HOWARTH'S LAMINAR FLOW


   ** CASE DATA
      NXT  = 1                  NZT  = 21              NTR  = 99
      ETAE = 0.800000E+01       DETA1= 0.200000E+00    VGP  = 0.100000E+01
      CB   = 0.0                OMEGA= 0.100000E+01    UINF = 0.100000E+01
      CA   = 0.200000E+03


   ** EXTERNAL STEADY STATE VELOCITY DISTRIBUTION
   ** AND PRESSURE GRADIENT
       Z          UO         P2
      0.0        1.00000     0.0
      0.05000    0.99375    -0.00629
      0.10000    0.98750    -0.01266
```

```
        0.15000      0.98125     -0.01911
        0.20000      0.97500     -0.02564
        0.25000      0.96875     -0.03226
        0.30000      0.96250     -0.03896
        0.35000      0.95625     -0.04575
        0.40000      0.95000     -0.05263
        0.45000      0.94375     -0.05960
        0.50000      0.93750     -0.06667
        0.55000      0.93125     -0.07369
        0.60000      0.92500     -0.08102
        0.65000      0.91875     -0.08825
        0.70000      0.91250     -0.09579
        0.75000      0.90625     -0.10345
        0.80000      0.90000     -0.11092
        0.85000      0.89375     -0.11880
        0.90000      0.88750     -0.12651
        0.95000      0.88125     -0.13461
        1.00000      0.87500     -0.14294
    ** BOUNDARY LAYER CALCULATIONS


NX =  1       NZ =  1      X =    0.0           Z =    0.0
        V(WALL)= 0.187500E+00      DELVW=  0.198620E+00
        V(WALL)= 0.386120E+00      DELVW= -0.486990E-01
        V(WALL)= 0.337421E+00      DELVW= -0.536625E-02
        V(WALL)= 0.332055E+00      DELVW= -0.669347E-04
     J     ETA            F              U              V                B
     1   0.0          0.0            0.0            0.331988E+00   0.100000E+01
     4   0.600000     0.597017E-01   0.198830E+00   0.329903E+00   0.100000E+01
     7   1.200000     0.237744E+00   0.393448E+00   0.316348E+00   0.100000E+01
    10   1.799999     0.528925E+00   0.574169E+00   0.282737E+00   0.100000E+01
    13   2.399999     0.921103E+00   0.728224E+00   0.228076E+00   0.100000E+01
    16   2.999998     0.139494E+01   0.845316E+00   0.161549E+00   0.100000E+01
    19   3.599998     0.192707E+01   0.922807E+00   0.983551E-01   0.100000E+01
    22   4.199997     0.249519E+01   0.966684E+00   0.507159E-01   0.100000E+01
    25   4.799996     0.308226E+01   0.987694E+00   0.219445E-01   0.100000E+01
    28   5.399996     0.367779E+01   0.996142E+00   0.792393E-02   0.100000E+01
    31   5.999995     0.427648E+01   0.998981E+00   0.237917E-02   0.100000E+01
    34   6.599995     0.487618E+01   0.999776E+00   0.592262E-03   0.100000E+01
    37   7.199994     0.547611E+01   0.999961E+00   0.121874E-03   0.100000E+01
    40   7.799994     0.607631E+01   0.999997E+00   0.206583E-04   0.100000E+01
    41   7.999993     0.627636E+01   0.100000E+01   0.109428E-04   0.100000E+01
    *-*-*-*-*-*-*-*-*-*-*-*-*-*-*-*-*-*-*-*-*-*-*-*-*-*-*-*-*-*-*-*-


NX =  1       NZ =  2      X =    0.0           Z =    0.05000
        V(WALL)= 0.331988E+00      DELVW= -0.972310E-02
        V(WALL)= 0.322265E+00      DELVW= -0.450226E-04
     J     ETA            F              U              V                B
     1   0.0          0.0            0.0            0.322220E+00   0.100000E+01
     4   0.600000     0.581813E-01   0.194094E+00   0.323867E+00   0.100000E+01
     7   1.200000     0.232522E+00   0.386156E+00   0.313812E+00   0.100000E+01
    10   1.799999     0.519068E+00   0.566277E+00   0.283169E+00   0.100000E+01
    13   2.399999     0.906733E+00   0.721254E+00   0.230550E+00   0.100000E+01
    16   2.999998     0.137691E+01   0.840133E+00   0.164844E+00   0.100000E+01
    19   3.599998     0.190651E+01   0.919551E+00   0.101347E+00   0.100000E+01
    22   4.199997     0.247316E+01   0.964962E+00   0.527956E-01   0.100000E+01
    25   4.799996     0.305951E+01   0.986931E+00   0.230901E-01   0.100000E+01
    28   5.399996     0.365474E+01   0.995860E+00   0.843160E-02   0.100000E+01
    31   5.999995     0.425332E+01   0.998894E+00   0.256168E-02   0.100000E+01
    34   6.599995     0.485297E+01   0.999754E+00   0.645788E-03   0.100000E+01
    37   7.199994     0.545289E+01   0.999957E+00   0.134743E-03   0.100000E+01
    40   7.799994     0.605288E+01   0.999996E+00   0.232829E-04   0.100000E+01
    41   7.999993     0.625288E+01   0.100000E+01   0.124532E-04   0.100000E+01
    DELSTR= 0.495709E-02    THETA =  0.189808E-02    CF  =  0.365695E-01
    RDELST= 0.307882E+02    RTHTA =  0.117888E+02    RZ  =  0.310547E+03
    H    =  0.261164E+01    REDFR =  0.114286E+01
    *-*-*-*-*-*-*-*-*-*-*-*-*-*-*-*-*-*-*-*-*-*-*-*-*-*-*-*-*-*-*-*-
```

```
NX =   1       NZ =   3      X =     0.0          Z =    0.10000
        V(WALL)=  0.322220E+00      DELVW= -0.998403E-02
        V(WALL)=  0.312236E+00      DELVW= -0.657526E-04
   J     ETA           F                U                V               B
   1    0.0          0.0              0.0            0.312170E+00    0.100000E+01
   4    0.600000     0.566133E-01     0.189204E+00   0.317602E+00    0.100000E+01
   7    1.200000     0.227123E+00     0.378592E+00   0.311117E+00    0.100000E+01
  10    1.799999     0.508849E+00     0.558047E+00   0.283535E+00    0.100000E+01
  13    2.399999     0.891792E+00     0.713937E+00   0.233062E+00    0.100000E+01
  16    2.999998     0.135810E+01     0.834651E+00   0.168257E+00    0.100000E+01
  19    3.599998     0.188502E+01     0.916079E+00   0.104488E+00    0.100000E+01
  22    4.199997     0.245010E+01     0.963110E+00   0.550052E-01    0.100000E+01
  25    4.799996     0.303568E+01     0.986103E+00   0.243212E-01    0.100000E+01
  28    5.399996     0.363058E+01     0.995550E+00   0.898349E-02    0.100000E+01
  31    5.999995     0.422905E+01     0.998798E+00   0.276242E-02    0.100000E+01
  34    6.599995     0.482866E+01     0.999729E+00   0.705297E-03    0.100000E+01
  37    7.199994     0.542858E+01     0.999952E+00   0.149013E-03    0.100000E+01
  40    7.799994     0.602857E+01     0.999996E+00   0.261163E-04    0.100000E+01
  41    7.999993     0.622857E+01     0.100000E+01   0.141165E-04    0.100000E+01
DELSTR=   0.713040E-02     THETA =   0.271337E-02     CF     =  0.251312E-01
RDELST=   0.440079E+02     RTHTA =   0.167466E+02     RZ     =  0.617187E+03
H     =   0.262788E+01     REDFR =   0.114286E+01
*-*-*-*-*-*-*-*-*-*-*-*-*-*-*-*-*-*-*-*-*-*-*-*-*-*-*-*-*-*-*-*-*-*-*-
NX =   1       NZ =   4      X =     0.0          Z =    0.15000
        V(WALL)=  0.312170E+00      DELVW= -0.102826E-01
        V(WALL)=  0.301887E+00      DELVW= -0.817980E-04
   J     ETA           F                U                V               B
   1    0.0          0.0              0.0            0.301805E+00    0.100000E+01
   4    0.600000     0.549919E-01     0.184141E+00   0.311074E+00    0.100000E+01
   7    1.200000     0.221523E+00     0.370720E+00   0.308233E+00    0.100000E+01
  10    1.799999     0.498217E+00     0.549423E+00   0.283818E+00    0.100000E+01
  13    2.399999     0.876195E+00     0.706213E+00   0.235609E+00    0.100000E+01
  16    2.999998     0.133841E+01     0.828816E+00   0.171804E+00    0.100000E+01
  19    3.599998     0.186247E+01     0.912348E+00   0.107804E+00    0.100000E+01
  22    4.199997     0.242585E+01     0.961100E+00   0.573694E-01    0.100000E+01
  25    4.799996     0.301058E+01     0.985194E+00   0.256553E-01    0.100000E+01
  28    5.399996     0.360513E+01     0.995206E+00   0.958888E-02    0.100000E+01
  31    5.999995     0.420347E+01     0.998690E+00   0.298533E-02    0.100000E+01
  34    6.599995     0.480305E+01     0.999701E+00   0.772255E-03    0.100000E+01
  37    7.199994     0.540296E+01     0.999946E+00   0.165874E-03    0.100000E+01
  40    7.799994     0.600294E+01     0.999995E+00   0.298419E-04    0.100000E+01
  41    7.999993     0.620294E+01     0.100000E+01   0.162542E-04    0.100000E+01
DELSTR=   0.888743E-02     THETA =   0.335972E-02     CF     =  0.199013E-01
RDELST=   0.545049E+02     RTHTA =   0.206045E+02     RZ     =  0.919922E+03
H     =   0.264529E+01     REDFR =   0.114286E+01
*-*-*-*-*-*-*-*-*-*-*-*-*-*-*-*-*-*-*-*-*-*-*-*-*-*-*-*-*-*-*-*-*-*-*-
NX =   1       NZ =   5      X =     0.0          Z =    0.20000
        V(WALL)=  0.301805E+00      DELVW= -0.106050E-01
        V(WALL)=  0.291200E+00      DELVW= -0.978548E-04
   J     ETA           F                U                V               B
   1    0.0          0.0              0.0            0.291103E+00    0.100000E+01
   4    0.600000     0.533129E-01     0.178890E+00   0.304263E+00    0.100000E+01
   7    1.200000     0.215707E+00     0.362511E+00   0.305143E+00    0.100000E+01
  10    1.799999     0.487134E+00     0.540372E+00   0.284004E+00    0.100000E+01
  13    2.399999     0.859882E+00     0.698041E+00   0.238190E+00    0.100000E+01
  16    2.999998     0.131775E+01     0.822587E+00   0.175493E+00    0.100000E+01
  19    3.599998     0.183873E+01     0.908327E+00   0.111310E+00    0.100000E+01
  22    4.199997     0.240028E+01     0.958911E+00   0.599049E-01    0.100000E+01
  25    4.799996     0.298408E+01     0.984194E+00   0.271057E-01    0.100000E+01
  28    5.399996     0.357824E+01     0.994824E+00   0.102561E-01    0.100000E+01
  31    5.999995     0.417644E+01     0.998568E+00   0.323399E-02    0.100000E+01
  34    6.599995     0.477597E+01     0.999669E+00   0.848130E-03    0.100000E+01
  37    7.199994     0.537587E+01     0.999939E+00   0.184847E-03    0.100000E+01
  40    7.799994     0.597586E+01     0.999995E+00   0.337430E-04    0.100000E+01
  41    7.999993     0.617586E+01     0.100000E+01   0.186796E-04    0.100000E+01
DELSTR=   0.104503E-01     THETA =   0.392278E-02     CF     =  0.166770E-01
RDELST=   0.636816E+02     RTHTA =   0.239044E+02     RZ     =  0.121875E+04
H     =   0.266401E+01     REDFR =   0.114286E+01
*-*-*-*-*-*-*-*-*-*-*-*-*-*-*-*-*-*-*-*-*-*-*-*-*-*-*-*-*-*-*-*-*-*-*-
```

```
NX =   1      NZ =  6      X =    0.0          Z =    0.25000
       V(WALL)=  0.291103E+00      DELVW= -0.109539E-C1
       V(WALL)=  0.280149E+00      DELVW= -0.115092E-03
   J     ETA         F              U              V              B
   1   0.C          0.0            0.0           0.280033E+00   0.100000E+01
   4   0.600000    0.515713E-01   0.173435E+00   0.297141E+00   0.100000E+01
   7   1.200000    0.209655E+00   0.353935E+00   0.301824E+C0   0.100000E+01
  10   1.799999    0.475562E+00   0.530849E+00   0.284080E+00   0.100000E+01
  13   2.399999    0.842785E+C0   0.689375E+00   0.240801E+00   0.100000E+01
  16   2.999998    0.129602E+01   0.815920E+00   0.179335E+00   0.100000E+01
  19   3.599998    0.181370E+01   0.903981E+00   0.115025E+00   0.100000E+01
  22   4.199997    0.237325E+01   0.956519E+00   0.626322E-01   0.100000E+01
  25   4.799996    0.295604E+01   0.983088E+00   0.286873E-01   0.100000E+01
  28   5.399996    0.354976E+01   0.994396E+00   0.109933E-C1   0.100000E+01
  31   5.999995    0.414780E+01   0.998430E+00   0.351333E-02   0.100000E+01
  34   6.599995    0.474729E+01   0.999632E+00   0.934214E-03   0.100000E+01
  37   7.199994    C.534718E+01   0.999932E+00   0.206550E-03   0.100000E+01
  40   7.799994    0.594716E+01   0.999994E+00   0.387246E-04   0.100000E+01
  41   7.999993    0.614716E+01   0.100000E+01   0.214701E-04   0.100000E+01
DELSTR=  0.119059E-01    THETA =  0.443553E-02    CF   =  0.143954E-01
RDELST=  0.720863E+02    RTHTA =  0.268557E+02    RZ   =  0.151367E+04
H     =  0.268420E+01    REDFR =  0.114286E+01
*-*-*-*-*-*-*-*-*-*-*-*-*-*-*-*-*-*-*-*-*-*-*-*-*-*-*-*-*-*-*-*-*-*-*-
NX =   1      NZ =  7      X =    0.0          Z =    0.30000
       V(WALL)=  0.280033E+00      DELVW= -0.113329E-01
       V(WALL)=  0.268701E+00      DELVW= -0.133248E-03
   J     ETA         F              U              V              B
   1   0.0          0.0            C.0           0.268567E+00   0.100000E+01
   4   0.600000    0.497615E-01   0.167759E+00   0.289678E+00   C.100000E+01
   7   1.200000    0.203345E+00   0.344956E+00   C.298249E+00   C.100000E+C1
  10   1.799999    0.463452E+00   0.520807E+00   0.284027E+00   0.100000E+01
  13   2.399999    0.824825E+00   0.680158E+00   0.243439E+00   0.100000E+01
  16   2.999998    0.127311E+01   0.808762E+00   0.18334CE+00   0.100000E+01
  19   3.599998    0.178724E+01   0.899264E+00   0.118970E+00   0.100000E+C1
  22   4.199997    0.234461E+01   0.953895E+00   0.655742E-01   0.100000E+01
  25   4.799996    0.292627E+01   0.981861E+00   0.304194E-01   0.100000E+01
  28   5.399996    0.351950E+01   0.993946E+00   0.118126E-01   0.100000E+01
  31   5.999995    0.411737E+01   0.998273E+00   0.382744E-02   0.100000E+01
  34   6.599995    0.471680E+01   0.999590E+00   0.103287E-02   0.100000E+01
  37   7.199994    0.531667E+01   0.999922E+00   0.232568E-03   0.100000E+01
  40   7.799994    0.591665E+01   0.999993E+00   C.444706E-04   0.100000E+01
  41   7.999993    0.611665E+01   0.100000E+01   0.252692E-04   C.100000E+01
DELSTR=  0.132999E-01    THETA =  0.491485E-02    CF   =  0.126439E-01
RDELST=  0.800073E+02    RTHTA =  0.295659E+02    RZ   =  0.180469E+C4
H     =  0.270607E+01    REDFR =  0.114286E+01
*-*-*-*-*-*-*-*-*-*-*-*-*-*-*-*-*-*-*-*-*-*-*-*-*-*-*-*-*-*-*-*-*-*-*-
NX =   1      NZ =  8      X =    0.0          Z -    0.35000
       V(WALL)=  0.268567E+00      DELVW= -0.117447E-01
       V(WALL)=  0.256823E+00      DELVW= -0.155162E-03
   J     ETA         F              U              V              B
   1   0.0          0.0            0.0           0.256667E+00   0.100000E+C1
   4   0.600000    0.478772E-01   0.161839E+00   0.281840E+00   C.100000E+01
   7   1.200000    0.196752E+00   0.335533E+00   0.294388E+00   0.100000E+01
  10   1.799999    0.450749E+00   0.510189E+00   0.283825E+00   0.100000E+01
  13   2.399999    0.805910E+00   0.670295E+00   0.246099E+00   0.100000E+01
  16   2.999998    0.124889E+01   0.801049E+00   0.187519E+00   0.100000E+01
  19   3.599998    0.175916E+01   0.894127E+00   0.123172E+00   0.100000E+01
  22   4.199997    0.231415E+01   0.951002E+00   0.687604E-01   0.100000E+01
  25   4.799996    0.289457E+01   0.980491E+00   0.323235E-01   0.100000E+01
  28   5.399996    0.348725E+01   0.993372E+00   0.127264E-01   0.100000E+01
  31   5.999995    0.408492E+01   0.998093E+00   0.418363E-02   0.100000E+01
  34   6.599995    0.468429E+01   0.999540E+00   0.114610E-C2   0.100000E+01
  37   7.199994    0.528414E+01   0.999912E+00   0.262093E-03   0.100000E+01
  40   7.799994    0.588412E+01   0.999992E+00   0.514313E-04   0.100000E+01
  41   7.999993    0.608412E+01   0.100000E+01   0.297837E-04   0.100000E+01
DELSTR=  0.146614E-01    THETA =  0.537079E-02    CF   =  0.112238E-01
RDELST=  0.876247E+02    RTHTA =  0.320989E+02    RZ   =  0.209180E+C4
H     =  0.272984E+01    REDFR =  0.114286E+01
*-*-*-*-*-*-*-*-*-*-*-*-*-*-*-*-*-*-*-*-*-*-*-*-*-*-*-*-*-*-*-*-*-*-*-
```

```
NX =   1      NZ =  9      X =      0.0           Z =      0.40000
         V(WALL)=  0.256667E+00        DELVW= -0.121982E-01
         V(WALL)=  0.244469E+00        DELVW= -0.181634E-03
   J     ETA          F                   U                V              B
   1    0.0          0.0              0.0            0.244287E+00    0.100000E+01
   4    0.600000    0.459102E-01     0.155649E+00    0.273584E+00    0.100000E+01
   7    1.200000    0.189844E+00     0.325616E+00    0.290204E+00    0.100000E+01
  10    1.799999    0.437384E+00     0.498923E+00    0.283448E+00    0.100000E+01
  13    2.399999    0.785926E+00     0.659796E+00    0.248771E+00    0.100000E+01
  16    2.999998    0.122320E+01     0.792702E+00    0.191889E+00    0.100000E+01
  19    3.599998    0.172927E+01     0.888503E+00    0.127658E+00    0.100000E+01
  22    4.199997    0.228164E+01     0.947797E+00    0.722225E-01    0.100000E+01
  25    4.799996    0.286068E+01     0.978954E+00    0.344275E-01    0.100000E+01
  28    5.399996    0.345274E+01     0.992754E+00    0.137529E-01    0.100000E+01
  31    5.999995    0.405017E+01     0.997885E+00    0.458946E-02    0.100000E+01
  34    6.599995    0.464947E+01     0.999483E+00    0.127798E-02    0.100000E+01
  37    7.199994    0.524931E+01     0.999899E+00    0.297499E-03    0.100000E+01
  40    7.799994    0.584928E+01     0.999991E+00    0.596816E-04    0.100000E+01
  41    7.999993    0.604928E+01     0.100000E+01    0.350289E-04    0.100000E+01
DELSTR=  0.160111E-01     THETA =  0.580995E-02     CF    =  0.100253E-01
RDELST=  0.950658E+02     RTHTA =  0.344966E+02     RZ    =  0.237500E+04
H     =  0.275581E+01     REDFR =  0.114286E+01
*-*-*-*-*-*-*-*-*-*-*-*-*-*-*-*-*-*-*-*-*-*-*-*-*-*-*-*-*-*-*-*-*-*-*-*-
```

```
NX =   1      NZ = 10      X =      0.0           Z =      0.45000
         V(WALL)=  0.244287E+00        DELVW= -0.126964E-01
         V(WALL)=  0.231591E+00        DELVW= -0.212776E-03
   J     ETA          F                   U                V              B
   1    0.0          0.0              0.0            0.231378E+00    0.100000E+01
   4    0.600000    0.438514E-01     0.149158E+00    0.264859E+00    0.100000E+01
   7    1.200000    0.182585E+00     0.315144E+00    0.285651E+00    0.100000E+01
  10    1.799999    0.423279E+00     0.486928E+00    0.282863E+00    0.100000E+01
  13    2.399999    0.764739E+00     0.648475E+00    0.251448E+00    0.100000E+01
  16    2.999998    0.119584E+01     0.783629E+00    0.196463E+00    0.100000E+01
  19    3.599998    0.169733E+01     0.882315E+00    0.132466E+00    0.100000E+01
  22    4.199997    0.224680E+01     0.944224E+00    0.760038E-01    0.100000E+01
  25    4.799996    0.282428E+01     0.977216E+00    0.367653E-01    0.100000E+01
  28    5.399996    0.341564E+01     0.992045E+00    0.149122E-01    0.100000E+01
  31    5.999995    0.401281E+01     0.997643E+00    0.505615E-02    0.100000E+01
  34    6.599995    0.461202E+01     0.999414E+00    0.143137E-02    0.100000E+01
  37    7.199994    0.521184E+01     0.999883E+00    0.338957E-03    0.100000E+01
  40    7.799994    0.581181E+01     0.999989E+00    0.702648E-04    0.100000E+01
  41    7.999993    0.601181E+01     0.100000E+01    0.411659E-04    0.100000E+01
DELSTR=  0.173658E-01     THETA =  0.623697E-02     CF    =  0.898210E-02
RDELST=  0.102431E+03     RTHTA =  0.367883E+02     RZ    =  0.265430E+04
H     =  0.278434E+01     REDFR =  0.114286E+01
*-*-*-*-*-*-*-*-*-*-*-*-*-*-*-*-*-*-*-*-*-*-*-*-*-*-*-*-*-*-*-*-*-*-*-*-
```

References

Abbott, D. E., and Cebeci, T. 1971. In *Fluid Dynamics of Unsteady Three-Dimensional and Separated Flows*, ed. Marshall, F. D., 202–222.

Acharya, M., and Reynolds, W. C. 1975. "Measurements and Predictions of a Fully Developed Turbulent Channel Flow with Imposed Controlled Oscillations," Stanford University Technical Report Number TF-8.

Alber, I. E. 1971. AIAA Paper No. 71–203.

Binder, G., and Didelle, H. 1975. "Improvement of Ejector Thrust Augmentation by Pulsating of Flapping Jets," 2nd Symposium on Jets Pumps and Ejectors and Gas Lift Techniques, Paper No. E2.

Bradshaw, P. 1969. "Calculation of Boundary Layer Development Using the Turbulent Energy Equation, VI. Unsteady Flow," NPL AERO Rept. 1288.

Bradshaw, P., Ferris, D. H., and Atwell, N. P. 1967. *J. Fluid Mech.*, **28**, 593–616.

Burggraf, O. R. 1973. "Comparative Study of Turbulence Models for Boundary Layers and Wakes," Aerospace Research Labs Report, ARL TR 74-0031.

Cebeci, T. 1970. *AIAA J.*, **8**, 2152–2156.

Cebeci, T. 1977. *Proc. R. Soc. London A***355**, 225–238.

Cebeci, T. and Carr, L. W. 1978. "A Computer Program for Calculating Laminar and Turbulent Boundary Layers for Two-Dimensional Time-Dependent Flows," NASA TM 78470.

Cebeci, T., and Keller, H. B. 1972. In *Recent Research on Unsteady Boundary Layers*, ed. Eichelbrenner, E. A., **II**, 1072–1105.

Cebeci, T., and Smith, A. M. O. 1968. In *Computation of Turbulent Boundary Layers*, AFOSR-IFP-Stanford Conference, **1**, 346–355.

Charnay, G., and Mathieu, J. 1976. *J. Fluids Eng.*, **98**, 278–283.

Clauser, F. H. 1956. In *Advances in Applied Mechanics*, Academic Press, Vol. IV, New York.

Cousteix, J., Desopper, A., and Houdeville, R. 1976. "Recherches sur les Couches Limites Turbulentes Instationnaires," ONERA TP No. 147.

Cousteix, J., Desopper, A., and Houdeville, R. 1977a. "Structure and Development of a Turbulent Boundary Layer in an Oscillatory External Flow," ONERA TP 14.

Cousteix, J., Desopper, A., and Houdeville, R. 1977b. "Structure and Development of a Turbulent Boundary Layer in an Oscillatory External Flow," Proceedings of Symposium on Turbulent Shear Flows, Penn State University, University Park, Philadelphia.

Dwyer, H. A., Doss, E. D., and Goldman, A. L. 1970. "A Computer Program for the Calculation of Laminar and Turbulent Boundary Layer Flows," NACA CR 114366.

Gupta, R. N., and Trimpi, R. L. 1974. " An Eddy-Viscosity Treatment of the Boundary Layer on a Flat Plate in an Expansion Tube," *Heat Transfer 1974, Jpn. Soc. Mech. Eng. Soc. Chem. Eng. Jpn.*, **2**, 339–343.

Houdeville, R., Desopper, A., and Cousteix, J. 1976. "Experimental Analysis of Average and Turbulent Boundary Layer," ONERA TP, No. 30, also Rech. Aerosp, No. 1976–4.

Houdeville, R., and Cousteix, J. 1978. "Premiers Résultats a'une Etude sur les Couches Limites Turbulentes en Ecoulement Pulsé avec Gradient de Pression Moyen Défavorable," 15th Colloquium on Applied Aerodynamics, Marseille.

Hussain, A. KM. F., and Reynolds, W. C. 1970. *J. Fluid Mech.* **41**, 241–258.

Jonsson, I. G., and Carlsen, N. A. 1976. *J. Hydraul. Res.*, **14**, 45–60.

Karlsson, S. K. F. 1959. *J. Fluid Mech.*, **5**, 622–636.

Kays, W. M. 1971. ASME Paper No. 71-HF-44.

Kenison, R. C. 1977. "An Experimental Study of the Effect of Oscillatory Flow on Separation Region in a Turbulent Boundary Layer," AGARD Symposium on Unsteady Aerodynamics, Ottawa.

Klebanoff, P. S. 1954. "Characteristics of Turbulence in a Boundary Layer with Zero Pressure Gradient," NACA TN 3178.

Kuhn, G. D., and Nielsen, J. N. 1973. "Studies of an Integral Method for Calculating Time-Dependent Turbulent Boundary Layers," Nielsen Engineering and Research, Inc., Rept. NEAR TR 57.

Launder, B. E. and Spalding, D. B. 1974. *Appl. Mech. Eng.*, **3**, 269–389.

McCroskey, W. J., and Phillipe, J. J. 1975. *AIAA J.*, **3**, 71–79.

Mainardi, H., and Panday, P. K. 1979. "A Study of Turbulent Pulsating Flow in a Circular Pipe," to appear.

Mainardi, H., Barriol, R., and Panday, P. K. 1979. "Characteristics of an Orifice Plate in Pulsating Flow," *Int. J. Mass Trans.*, to appear.

Mellor, G., and Herring, H. J. 1973. *AIAA J.* **11**, 590–599.

Nash, J. F. 1972. *J. Basic Eng.*, 94D, 131–141.

Nash, J. F. 1976. "Further Studies of Unsteady Boundary Layers with Flow Reversal," NASA CR-2767.

Nash, J. F., and Patel, V. C. 1975. "Calculations of Unsteady Turbulent Boundary Layers with Flow Reversal," NASA CR-2546.

Nash, J. F., and Scruggs, R. M. 1978. "Unsteady Boundary Layers with Reversal and Separation," AGARD Symposium on Unsteady Aerodynamics, Ottawa.

Nash, J. F., Carr, L. W., and Singleton, R. E. 1975. *AIAA J.*, **13**, 167–172.

Norris, H. L., III, and Reynolds, W. C. 1975. "Turbulent Channel Flow with a Moving Wavy Boundary," Stanford University Technical Report Number TF-7.

Patel, M. H. 1977. *Proc. R. Soc. London, A***353**, 121–144.

Patel, V. C., and Nash, J. F. 1972. In *Recent Research on Unsteady Boundary Layers*, ed. Eichelbrenner, E. A. **I**, 1106–1164.

Patel, V. C., and Nash, J. F. 1975. In *Unsteady Aerodynamics*, ed., Kinney, R. B., Vol. 1, 1975.

Reynolds, W. C. 1970. "Computation of Turbulent Flows—State-of-the-Art," Stanford University, Rept. MD-27.

Reynolds, W. C. 1976. In *Ann. Rev. Fluid Mech.*, **8**, 183–208.

Romaniuk, M. S., and Telionis, D. P. 1979. "Turbulence Models for Oscillating Boundary Layers," AIAA Paper No. 79-0069.

Schachenmann, A. A., and Rockwell, D. A. 1976. *J. Fluids Eng.* **98**, 695–702.

Shamroth, S. J., and Kreskovsky, J. P. 1974. "A Weak Interaction Study of the Viscous Flow about Oscillating Airfoils," NASA CR-132425.

Singleton, R. E., and Nash, J. F. 1973. *Proc. AIAA Comp. Fluid Dyn. Conf.*, 84–91.

Soutif, M., Favre-Marinet, M., and Binder, G. 1979. "Diffusion and Periodic Structure of Flapping Jets," to appear.

Telionis, D. P. 1976. *Arch. Mech.* **28**, 997–1010.

Telionis, D. P. 1977. "Unsteady Boundary Layers, Separated and Attached," AGARD Conference Proceedings No. 227, Paper No. 16.

Telionis, D. P. 1979. *J. Fluids Eng.*, **101** 29–43.

Telionis, D. P., and Romaniuk, M. S. 1978. *AIAA J.*, **16**, 488–495.

Telionis, D. P., and Tsahalis, D. Th. 1974. *AIAA J.*, **12**, 1469–1476.

Telionis, D. P., and Tsahalis, D. Th. 1975. *AIAA J.*, **14**, 468–474.

Thomas, L. C., and Shukla, R. K. 1976. *J. Fluids Eng.*, **98**, 27–32.

Townsend, A. A. 1956. *The Structure of Turbulent Shear Flow*, Cambridge University Press, London and New York.

Townsend, A. A. 1961. *J. Fluid Mech.*, **11**, 97–120.

Van Driest, E. R. 1956. *J. Aero. Sci.*, **23**, 1007–1011.

CHAPTER 7

Unsteady Separation

7.1 Introduction

The study of viscous phenomena and, in particular, boundary layers has interested investigators for a variety of reasons. It is often necessary to know the distribution of skin friction and heat transfer across the interface of fluids and solids. In internal fluid mechanics this information and the properties of viscous regions per se, for example, velocity profiles, flow rates, and perhaps the effects of viscosity on mixing and chemical reactions, is a final goal in itself. However, in external fluid dynamics it is necessary to consider the interaction between the viscous layer and the outer inviscid flow. This interaction is most violent if separation occurs.

Viscous flow theories must be employed in order to determine the location of separation that controls basic characteristics of lifting surfaces and perhaps the properties of small separated regions or turbulent wakes. Boundary-layer theory has been successful in predicting with reasonable accuracy the location of separation in steady fields (see review articles by Brown and Stewartson, 1969; Williams, 1977). In the last few decades it has been actively pursued to demonstrate that the classical boundary-layer theory can be used to determine the properties of unsteady flow fields as well. A breakthrough in this effort has been the identification of the fact that the classical criterion of separation for steady flow, that is, the vanishing of skin friction is no longer valid for unsteady flows. This idea was first presented by Sears (1956), Rott (1956), and Moore (1957). Numerical evidence appeared much later (Sears and Telionis, 1972; Telionis and Werle, 1973; see also review articles of Sears and Telionis, 1975; Williams, 1977; and Shen, 1979). This consists essentially of numerical integrations of laminar and turbulent boundary layers carried through the point of zero skin friction and into a region of partially reversed flow without any evidence of the separation singularity. Almost at the same time Despard and Miller (1971) published their experimental results and argued that separation in oscillatory flow is displaced with respect to its steady-state location but remains unaffected by oscillations of the outer stream. The point of zero skin friction oscillates back and forth, thus generating a thin layer of reversed flow that shoots periodically upstream from the location of separation.

Interest in this area was stimulated again and a variety of contributions, analytical and experimental, appeared in the 1970s. In all these contributions, the focus of the investigation is the location of separation. It is believed that the position of separation and perhaps the flow properties in its neighborhood must be known before one can proceed to study and calculate the wake. The mechanism by which the outer flow feeds energy and momentum into the wake is intimately connected with the phenomenon of separation. At the point of separation the vorticity produced in the boundary layer is shed into the flow. If the amount of vorticity produced is known, then even inviscid theories could model successfully an unsteady wake. This was most emphatically demonstrated by Crimi and Reeves (1972), who allowed a number of discrete vortices to be shed from a leading edge in order to study unsteady stall over an oscillating airfoil. However, up to now, very little has been done to study analytically the properties of an unsteady wake. Most of the work in the area is concentrated on the phenomenon of separation per se, as described in recent review articles (Sears and Telionis, 1975; Riley, 1975; Williams, 1977; Shen, 1978). In fact, the present chapter draws heavily on the article of Shen (1978).

Definition

There has been some disagreement in literature over terminology. Unfortunately, basic ideas have been obscured and considerable confusion has been generated due to poor nomenclature. Most investigators define separation as the point where "the flow ceases to follow the contour of the body and breaks away from the wall." There are many justified objections and the author feels that such a definition would be unambiguous and most general only in the limit of $Re \to \infty$, where Re is the Reynolds number. This definition would be equivalent to Prandtl's (1904) original description. Prandtl used the term "ablösung" to describe the phenomenon wherein "a fluid sheet projects itself into the free flow and effects a complete alteration of the motion." Some authors prefer to use the term "breakaway" or "catastrophic separation" for the phenomenon we defined as separation and reserve the term separation for the point of zero skin friction which we shall call "detachment."

For steady flow both theory and experiment indicate that separation and detachment coincide. For unsteady flow and large Reynolds numbers, this is not true. Detachment and a thin layer of reversed flow may appear at the bottom of a boundary layer that otherwise stays attached. In the limit of $Re \to \infty$, the viscous layer which contains the partially reversed flow collapses to the wall, at least until the point of separation. This has been recognized by Rott (1956), Lin (1956), Sears (1956), and Moore (1957). Experimental evidence was reported quite later (Vidal, 1959; Ludwig, 1964; Despard and Miller, 1971; Nash et al., 1975; Koromilas and Telionis, 1980).

In this limit, the thickness of the wall layers in both the fore and aft regions goes to zero, and the point of separation is the point at which the separation streamline meets the wall.

The main idea behind the above definition and the basic concern of aerodynamicists is the initiation of the wake. Separation, as defined above, should be its upstream limit. In principle, wakes are regions with very small variations of pressure containing large-scale turbulence. Unfortunately, what often appears to be unambiguous and clearly defined in theory may be obscured in practice. For very mild adverse pressure gradients and even for steady fields, the flow may be led through thin regions of partially reversed flow and into regions of random motion, while the outer flow deviates very little from the potential unseparated flow. In this case it would be hard to distinguish between separation and detachment. Finally, exact Navier–Stokes solutions (Mehta and Lavan, 1975; Mehta, 1977; O'Brien, 1975) indicate that for low Reynolds numbers, of the order of 10^3–10^4, separation occurs at the point of detachment.

The difficulties described up to now concern the correct interpretation of the behavior of analytical models. The next step, of course, is to compare the theoretical predictions with experimental data and here new difficulties arise. All the characteristic properties attributed to separation may appear gradually and over a considerable distance along the wall. For example, two of the most characteristic features of separation are the abrupt thickening of the boundary layer and the generation of a slow-moving turbulent wake made up of large-scale vortices. In practice both properties may appear so gradually, especially in the case of weak adverse pressure gradients, that it is very difficult to define a single position for separation. Even in the case of steady flow, which is considered well understood, the flow remains nearly parallel to the wall for up to 4 or 5 boundary-layer thicknesses beyond the point of zero skin friction and the eddies appear at first to be well ordered, even if downstream, the wake breaks into a totally random vortical motion (Telionis and Koromilas, 1978; Koromilas and Telionis, 1980; Mezaris and Telionis, 1980). This explains the poor comparison between theoretical results and experimental data.

Criteria for separation

With a specific definition in mind, theoreticians and experimentalists sought a criterion that would signal the location of the phenomenon. For low Reynolds numbers and in the case of steady flow, for all Reynolds numbers, the vanishing of the wall shear has been proved to be a reliable and convenient criterion of separation shown schematically in Fig. 7.1 (Prandtl, 1904):

$$\frac{\partial u}{\partial y} = 0, \quad \text{at } y = 0. \tag{7.1.1}$$

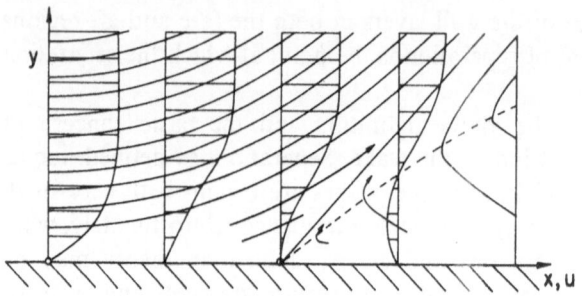

Fig. 7.1 Prandtl's sketch
of the streamline pattern
and velocity profiles in
the neighborhood of sepa-
ration.

A simple example that clearly demonstrates the inadequacy of this criterion
for higher Reynolds numbers is steady flow over moving walls. In this case
separation can be signaled conveniently, again in both theory and experiment, by the MRS criterion (Moore, 1957; Rott, 1956; Sears, 1956). Analytical evidence in favor of this criterion was provided most recently as described in a subsequent section. According to this criterion, separation
occurs at the station where the shear vanishes at a stagnation point within
the flow as shown schematically in Figs. 7.2a, b, and c:

$$\frac{\partial u}{\partial y} = 0, \quad \text{at } u = 0. \tag{7.1.2}$$

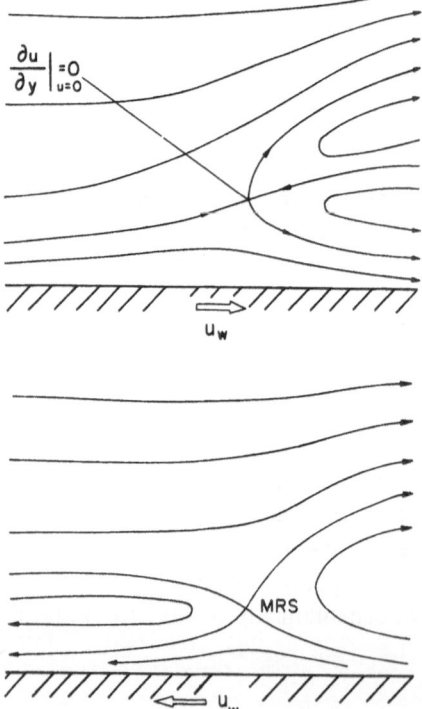

Fig. 7.2a Streamline pattern at a
point of separation over a downstream moving wall.

Fig. 7.2b Streamline pattern at a
point of separation over an upstream moving wall.

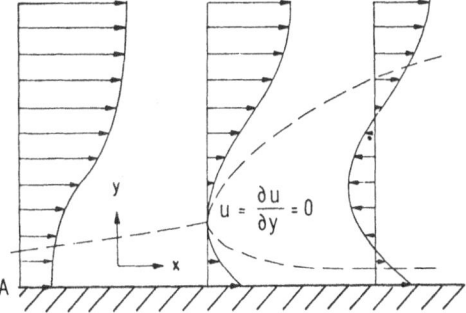

Fig. 7.2c Velocity profiles corresponding to the pattern of Fig. 7.2a.
The saddle point in Fig. 7.2a corresponds to the point $u = \partial u/\partial y = 0$
according to the MRS criterion.

For oscillatory flow, Despard and Miller (1971) uncovered evidence indicating that separation is displaced with respect to its position corresponding
to steady flow but remains unaffected during the cycle of oscillation.
However, they found that the point of zero skin friction oscillates along the
wall, thus generating a thin layer of reversed flow that shoots upstream from
separation and then moves downstream until it meets again the point of
separation. Based on their measurements and their observations, they propose as a criterion of separation, the station at which the wall shear
fluctuates between zero and a negative value or, equivalently, the furthest
upstream station along the wall at which the wall shear remains negative
throughout the entire cycle of oscillation. In Fig. 7.3 we have shown
schematically at each station the envelopes of the velocity profiles. In this
figure point A is the farthest upstream point at which the shear passes
through zero during a cycle of oscillation. According to the criterion of
Despard and Miller separation occurs at point S. During a cycle of the

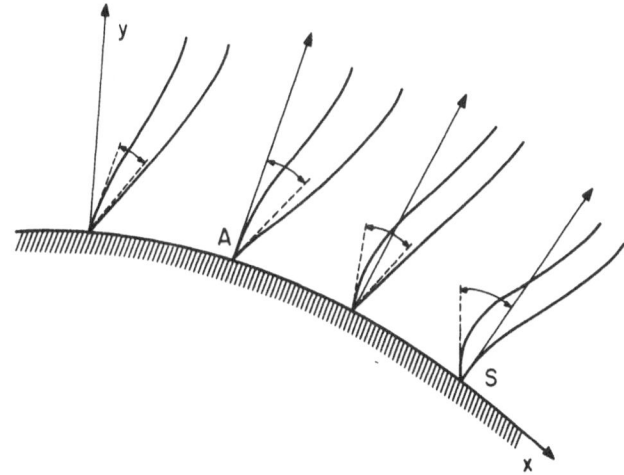

Fig. 7.3 Envelopes of velocity profiles for periodic oscillations. The dashed lines and the arrows
indicate the extreme positions of the tangents to the profile at the wall.

oscillation the point of zero skin friction oscillates between the points S and A. However, according to the observations of Despard and Miller, the outer flow remains attached until S.

More recent experimental results (Mezaris and Telionis, 1980) provide some evidence in favor of the Despard and Miller criterion and indicate that an alternative criterion may be used, based on the amplitude of the laminar oscillations. It appears that the contours of the constant amplitudes of oscillation close around a point away from the wall and peak at the station of separation. Actual experimental data from Mezaris and Telionis (1980) indicating this behavior are shown in Fig. 7.4. It is interesting to note that experiments with turbulent boundary layers indicate a similar behavior for the amplitude of the random oscillations (Kenison, 1977).

Very little has been done experimentally in the area of transient separation. Most of the investigations concern the entire flow field and, particularly, the shape of the wake rather than the details of the flow in the immediate neighborhood of separation. Such contributions have been described and referenced by this author (Telionis, 1979, 1980) and are beyond the scope of this monograph.

The proper criterion that signals separation in a theoretical investigation

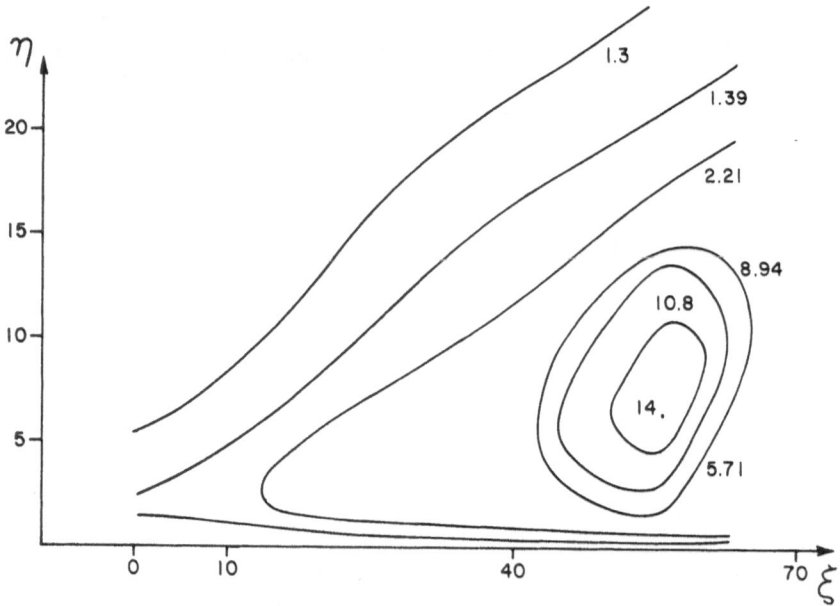

Fig. 7.4 Curves of constant amplitude of oscillation in the neighborhood of separation for $f = 0.2$ Hz obtained experimentally by Mezaris and Telionis (1980). The numbers in the figure denote the ratio of the velocity amplitude over the corresponding amplitude at the edge of the viscous region.

depends on the particular analytical model of the flow. If the full Navier–Stokes equations are adopted as a model, no specific criterion is available but then presumably the entire flow field can be predicted and there is no special need for identifying the specific location of separation. In fact, in this case the concept of separation itself becomes academic. If, however, the boundary-layer equations are used, laminar or turbulent, then the point of separation should be specified, in order to allow the continuation of the solution in terms of an appropriate wake model.

The boundary-layer approximation breaks down in the neighborhood of separation and therefore the boundary-layer equations are not an appropriate model to use in this neighborhood. If the solution of these equations is carried beyond their domain of validity and towards separation, it indicates a behavior very similar to the actual behavior of the flow, namely, for steady flow, vanishing of the wall shear. Moreover, the point of vanishing wall shear is approached in a singular manner, reminiscent of the behavior of the actual flow that starts turning away from the wall, indicating an abrupt growth of the normal component of the velocity. It is most remarkable that the approximate model of the boundary-layer equations predicts quite accurately the position of steady separation as well as some of its characteristic properties even though it is not valid in the neighborhood of separation.

Based on these observations, Moore (1957), Rott (1956), and Sears (1956) argued that the boundary-layer equations can be used to predict unsteady separation as well. Sears (1956) suggested as a criterion for unsteady separation the simultaneous vanishing of the shear and the velocity at a point within the boundary layer, that is, what today we call the MRS point, but in a frame of reference moving with separation.

Sears and Telionis (1971) proposed as a criterion for unsteady separation the appearance of singular behavior. It is of course well known that the actual flow and the solutions of the full Navier–Stokes equations are free of singularities (Dean, 1950). However, it was suggested that the singular response of the boundary-layer equations in the neighborhood of separation is an exaggerated simulation of the actual turning of the streamlines and the thickening of the boundary layer.

Following a parallel line of thought, Shen and Nenni (1975) proposed to associate separation directly with the condition that the boundary layer should become "unmatchable" with the outer flow in an inner-and-outer-expansion sense. This occurs, for example, when the induced normal velocity at the outer edge of the boundary layer attains such a magnitude as to invalidate the basic assumption of $v(Re)^{1/2} \sim O(1)$. Shen (1978) points out that by virtue of the continuity equation, the unmatchability condition is equivalent to a singular growth rate of the displacement thickness

$$\frac{\partial}{\partial x}(U_e \delta_1) > O(1). \tag{7.1.3}$$

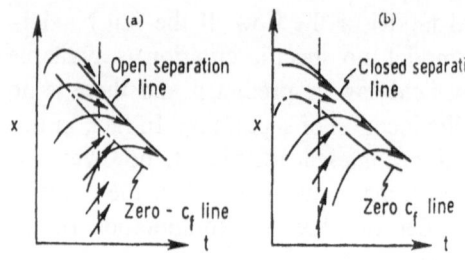

Fig. 7.5 Limiting streamlines in a $x-t$ plane showing the envelope of separation for (a) open separation and (b) closed separation.

An alternative point of view was presented by Wang (1979). Inspired by the mathematical similarities of unsteady two-dimensional and steady three-dimensional flow, Wang proposed to study in the $x-t$ plane the topography of the curves

$$\frac{dx}{dt} = \frac{\partial u}{\partial y}\bigg|_{y=0} \tag{7.1.4}$$

and search for envelopes. By analogy to their three-dimensional counterparts, he calls these lines *skin friction* lines. Wang then defines as unsteady separation the envelope of the skin friction lines in the $x-t$ plane. He proceeds further to distinguish between "open" and "closed" separation again in analogy to steady three-dimensional flow. If the separation line separates the $x-t$ plane into two unconnected regions, then separation is termed "closed." If the skin friction lines approach the separation line from both sides, then separation is termed "open." Schematic representations of the two situations are shown in Fig. 7.5.

The criteria described above are, to a reasonable extent, in agreement with each other. Their relative performance and the actual numerical methods for their implementation will be discussed in the sections that follow. Unfortunately, very little experimental information is available, and today, it is not possible to determine which of these criteria are more accurate. Credibility may be lent from one model to another by comparing analytical results.

7.2 Separation over Moving Walls

In the late 1950s Sears (1956) and Moore (1957) argued independently that if an observer were to move with the speed of separation he would see, at least in the immediate neighborhood of separation, a steady flow over a moving wall. Although this was not proved at that time, it was considered intuitively obvious to warrant subsequent theoretical and experimental investigation of the problem of steady separation over moving walls. Moore (1958) considered a flow involving a slowly moving point of separation and suggested a mathematical transformation that reduces part of the problem to the flow as

seen by an observer moving with the speed of separation. This problem
however was not completely free of unsteady terms. Telionis (1970) later
attempted to develop a transformation that would map the flow around
unsteady separation into steady separation over a moving wall.

In a similar and more successful effort, Williams and Johnson (1974b)
translated into mathematical formulation the argument of Moore, Rott, and
Sears. Consider again the two dimensional form of Eqs. (1.2.12)–(1.2.13) in
more traditional notation:

$$\frac{\partial u}{\partial x} + \frac{\partial v}{\partial y} = 0, \tag{7.2.1}$$

$$\frac{\partial u}{\partial t} + u\frac{\partial u}{\partial x} + v\frac{\partial u}{\partial y} = \frac{\partial U_e}{\partial t} + U_e\frac{\partial U_e}{\partial x} + \frac{1}{Re}\frac{\partial^2 u}{\partial y^2}, \tag{7.2.2}$$

and the classical boundary and initial conditions

$$u = v = 0, \qquad\qquad\qquad \text{at } y = 0, \tag{7.2.3}$$

$$u \to U_e(x,t), \qquad\qquad\quad \text{as } y \to \infty, \tag{7.2.4}$$

$$u = u_0(x, y), v = v_0(x, y), \quad \text{at } t = t_0. \tag{7.2.5}$$

A coordinate system that moves with the constant speed U_s is introduced by

$$x = \bar{x} + U_s t, \qquad y = \bar{y}, \qquad t = \bar{t}. \tag{7.2.6}$$

The observer who rides the system $(\bar{x}, \bar{y}, \bar{t})$ will see all velocity components
parallel to the wall decreased by the quantity U_s. To complete the physical
analogy, therefore, we introduce a new set of dependent variables

$$u = \bar{u} + U_s, \qquad v = \bar{v}, \qquad U_e = \bar{U}_e + U_s. \tag{7.2.7}$$

In terms of the new dependent and independent variables, Eqs. (7.2.1)–
(7.2.5) become

$$\frac{\partial \bar{u}}{\partial \bar{x}} + \frac{\partial \bar{v}}{\partial \bar{y}} = 0, \tag{7.2.8}$$

$$\frac{\partial \bar{u}}{\partial \bar{t}} + \bar{u}\frac{\partial \bar{u}}{\partial \bar{x}} + \bar{v}\frac{\partial \bar{u}}{\partial \bar{y}} = \frac{\partial \bar{U}_e}{\partial \bar{t}} + \bar{U}_e\frac{\partial \bar{U}_e}{\partial \bar{x}} + \frac{1}{Re}\frac{\partial^2 \bar{u}}{\partial \bar{y}^2}, \tag{7.2.9}$$

$$\bar{u} = -U_s, \bar{v} = 0, \qquad\qquad \text{at } \bar{y} = 0, \tag{7.2.10}$$

$$\bar{u} \to U_e - U_s = \bar{U}_e, \qquad\quad \text{as } \bar{y} \to \infty, \tag{7.2.11}$$

$$\bar{u} = u_0(x, y) - U_s, \bar{v} = v_0(x, y), \quad \text{at } \bar{t} = t_0. \tag{7.2.12}$$

We note that unsteadiness is introduced here only by the temporal
variations of $\bar{U}_e(x,t)$. The transformation represented by Eqs. (7.2.6) and

(7.2.7) is actually the same as the one proposed by Moore (1958) as described in the following section. However, it was Williams and Johnson (1974b) who identified a special case of unsteady flow that maps through this transformation on a steady flow. Consider an outer-flow distribution $U_e(x,t)$ that depends on a linear combination of x and t:

$$U_e(x,t) = U_e(\xi), \quad \text{with } \xi = Ax + Bt. \tag{7.2.13}$$

If U_s is set equal to $-B/A$, the initial and boundary conditions (7.2.10)–(7.2.12) become

$$\bar{u} = B/A, \bar{v} = 0, \qquad\qquad \text{at } \bar{y} = 0, \tag{7.2.14}$$

$$\bar{u} \to U_e(A\bar{x}) + \frac{B}{A}, \qquad\qquad \text{as } \bar{y} \to \infty, \tag{7.2.15}$$

$$\bar{u} = u_0(x, y) + \frac{B}{A}, \bar{v} = v_0(x, y), \qquad \text{at } \bar{t} = t_0. \tag{7.2.16}$$

We finally assume that the initial conditions $u_0(x, y)$ and $v_0(x, y)$ satisfy the steady-state version of the governing equations with boundary conditions

$$u_0 = v_0 = 0, \quad \text{at } y = 0, \tag{7.2.17}$$

$$u_0 \to U_e(A\bar{x}), \quad \text{as } y \to \infty. \tag{7.2.18}$$

In the transformed plane the boundary and initial conditions are independent of time, and therefore at least one solution to the problem given by Eqs. (7.2.8), (7.2.9) and subject to conditions (7.2.14)–(7.2.16) exists and is nothing but the functions $u_0(x, y) + B/A$ and $v_0(x, y)$. This is the steady flow over a wall with skin moving downstream with speed B/A.

The distribution $U_e(A\bar{x})$ is arbitrary. We can therefore choose a variation of the outer-flow velocity that induces separation. In the case of separation over a moving wall, the flow separates at the MRS point [Eq. (7.1.2)]. The inverse transformation therefore maps the neighborhood of steady separation over a moving wall to unsteady separation over a fixed wall. Hence, the point of separation is moving upstream with speed B/A. Sears' criterion is met since the MRS profile of the steady flow is mapped to an MRS profile moving with separation.

It now becomes obvious that the case of steady flow over moving walls is intimately connected with unsteady separating flow, since there exists one special case of the second that maps on the first. At least the qualitative features of the phenomenon can thus be studied by considering steady flow over moving walls.

Self-similar solutions

The problem of steady separating flow over moving walls can be approached in the Falkner–Skan variables that convert the system to an ordinary differential equation. Such solutions are rather unrealistic, especially for separating flows, since they predict that the flow is about to separate at all points over the wall but never does. There is indeed a specific value of the pressure gradient parameter for which the similarity profile has a vanishing slope at all points on the wall. For an outer-flow distribution given by

$$U_e^*(x) = Kx^{*m}, \tag{7.2.19}$$

with K and m constants, and in terms of a similarity variable and a modified stream function, f,

$$\eta = y^* \sqrt{\frac{m+1}{2} \frac{U_e^*(x)}{\nu x^*}}, \tag{7.2.20}$$

$$u^*(x, y) = U_e^*(x) f'(\eta), \tag{7.2.21}$$

the problem reduces to the third-order ordinary differential equation

$$f''' + ff'' + \beta(1 - f'^2) = 0, \tag{7.2.22}$$

where β is the pressure gradient parameter

$$\beta = \frac{2m}{1+m}. \tag{7.2.23}$$

This is a classical problem investigated extensively in the past. Solutions of this equation with moving wall-boundary conditions

$$f = 0, f' = B, \quad \text{at } \eta = 0, \tag{7.2.24}$$

$$f' = 1, \qquad \text{as } \eta \to \infty, \tag{7.2.25}$$

were considered first by Moore (1958), who was followed by many other investigators. Telionis and Werle (1972) obtained the solution for a few values of B as a test case of an existing code. Cebeci and Wilson (1972) repeated the calculations for a wider range of parameters. Inger and Swean (1973) considered a more general wall condition that provides for slip as well as blowing, including the effects of compressibility and heat transfer. Danberg and Fansler (1974) repeated the calculation extending the ratio of outer-flow velocities to $B = 3.5$ and reported results for negative wall velocities as well.

What seemed encouraging for such calculations is the fact that progressively larger values of the pressure gradient parameter β yield velocity profiles with smaller skin friction values, thus simulating the approach to separation. In fact modern integral methods very successfully employ such profiles for the calculations of developing boundary layers. Even more encouraging is the fact that the profile of zero skin friction, that is, the profile of separation is reached for $B = 0$, at $\beta = \beta_0 \equiv -0.19884$ but with a square root singularity in terms of the quantity $(\beta - \beta_0)$. Unfortunately the extension to flows with moving walls is not free of ambiguities. The method yields for every B an MRS profile at different values of β. However, such profiles do not correspond to a singular point of the skin friction curve. In

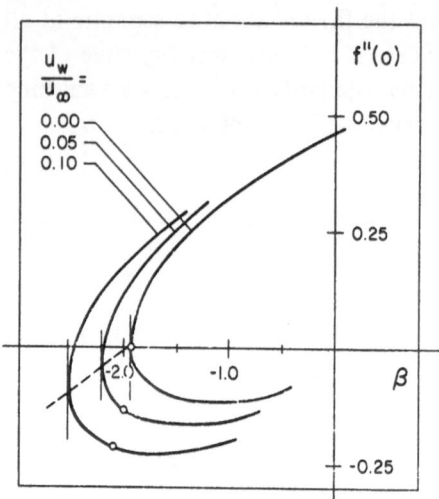

Fig. 7.6 The wall shear function for different wall velocities. ---, locus of the singularity; o, MRS point.

Fig. 7.6 we display the distribution of the wall shear $f''(0)$ as a function of the spacelike parameter β. The locus of all infinite slopes that correspond to singular points of the curves is marked by dashed lines. If a singularity were always present at the station of separation, then the two curves would have coincided. This finding is rather disappointing in view of the success of the analogy between self-similar and nonsimilar flows for $B = 0$.

Exact numerical solutions

To investigate the behavior of the steady boundary-layer equations in the neighborhood of zero skin friction, Werle and Davis (1972) chose a parabola at an angle of attack. In terms of new independent variables

$$\xi = \int_0^x U_e\, dx, \qquad \eta = \frac{U_e}{\sqrt{2\xi}}\, Y \tag{7.2.26}$$

and new dependent variables U, V

$$U = u/U_e, \qquad v = \frac{U_e}{\sqrt{2\xi}}\, V - \frac{\partial \eta}{\partial x}\, \sqrt{2\xi}\, U, \tag{7.2.27}$$

the continuity and momentum reduce to

$$2\xi \frac{\partial U}{\partial \xi} + U + \frac{\partial V}{\partial \eta} = 0, \tag{7.2.28}$$

$$2\xi U \frac{\partial U}{\partial \xi} + V \frac{\partial U}{\partial \eta} + \beta(U^2 - 1) = \frac{\partial^2 U}{\partial \eta^2}, \tag{7.2.29}$$

where β is the pressure gradient parameter

$$\beta = \frac{2\xi}{U_e} \frac{dU_e}{d\xi}. \qquad (7.2.30)$$

Inviscid flow about a parabola whose nose is a distance k above the stagnation point if measured along the perpendicular to the axis of the parabola gives (Fig. 7.7)

$$\beta = \frac{1 - y^k}{y^2 + 1}. \qquad (7.2.31)$$

Note that whereas the zero-angle-of-attack case sees a favorable pressure gradient all along the surface of the parabola, for any angle of attack, no matter how small, there exists a region of adverse pressure gradient. The flow about a parabola provides very mild adverse pressure gradients. Nevertheless, numerical integration demonstrated that the features of the Goldstein singularity are not affected by the pressure distribution, if of course separation occurs all together. In particular, as the station of separation is approached, quantities like $\partial u / \partial x$, v, etc., start growing sharply and eventually blow up at the station where the wall shear vanishes. The computation requires progressively a larger number of iterations for convergence, until a station is reached, very close to the point of separation, where convergence is not possible. The singular behavior has been demonstrated, in fact, to follow a square root behavior in accordance with the analysis of Goldstein (1948).

Extension of this work to flows over moving walls was undertaken by Telionis and Werle (1973). The no-penetration condition was imposed on the parabola, but the skin of the body was given a velocity at fixed ratio to the local value of the outer-flow velocity.

$$V = 0, \qquad \frac{U}{U_e} = U_w = \text{const}, \qquad \text{at } \eta = 0. \qquad (7.2.32)$$

These conditions imply that the moving skin has a nonuniform velocity and therefore it must be thought of as stretchable.

For positive values of U/U_e, the integration encounters a point on the skin of the parabola where the shear vanishes. Calculations then proceed without any evidence of singular behavior into a region of partially reversed

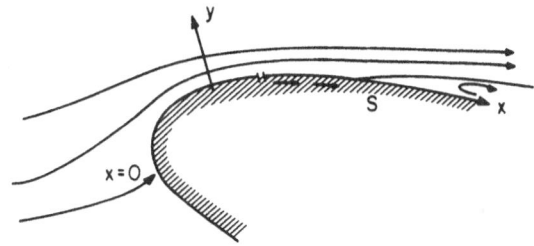

Fig. 7.7 The flow over a parabola at an angle. of attack with a sliding skin.

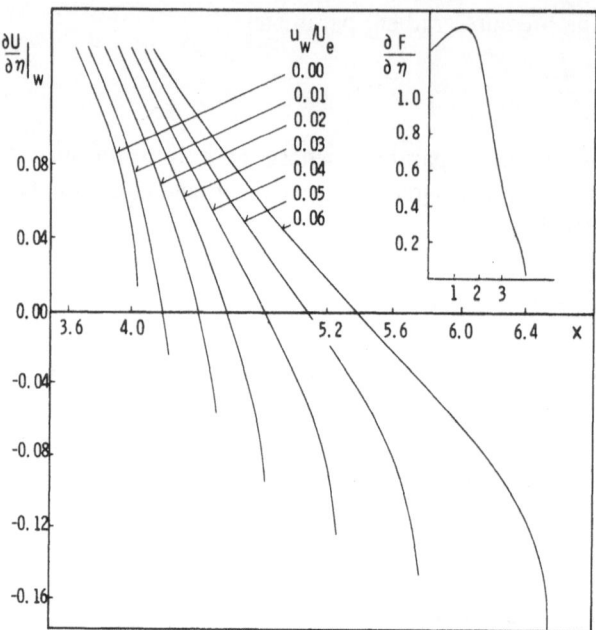

Fig. 7.8 The wall shear function versus the distance x in the neighborhood of the separation singularity.

flow. Eventually, near the station where the MRS criterion would hold, typical features of the Goldstein singularity become apparent. This is clearly shown in Fig. 7.8 where the skin friction variation with the distance along the wall is plotted for different values of the parameter U_w/U_e. Note that for a fixed skin, separation occurs at approximately $x \cong 4$. The singular behavior becomes evident at a considerable distance upstream of separation. It shows at first as a tendency of slope increase that eventually forces the curve to dive almost vertically to the point where the calculation is terminated. To verify that this singularity is a Goldstein-type of a singularity, the square of the quantity $\partial U/\partial \eta$ was plotted versus x and found to be very nearly a straight line. It should be emphasized that the singular behavior described is not an uncontrolled instability that could be easily confused with numerical instabilities. In fact, the first portion of the turning of the curves in Fig. 7.8 has been calculated with no evidence of any difficulties for convergence.

Details of the velocity profiles have been plotted in Fig. 7.9 to indicate how closely the MRS profile can be approached. This set of profiles was obtained with five iterations at each station. Regardless of the value of the increment Δx, the calculation fails to converge beyond the point of $U = \partial U/\partial \eta = 0$. In alternative efforts, conditions for convergence are built in the code to force the necessary number of iterations. The number of iterations is found to increase as separation is approached, but beyond the MRS station 100 iterations were found insufficient for convergence.

Fig. 7.9 Detail of the U-velocity profile in the neighborhood of the MRS station for U_w/U_e = 0.06 and Δt = 0.002. The integration was forced to 5 iterations at each station.

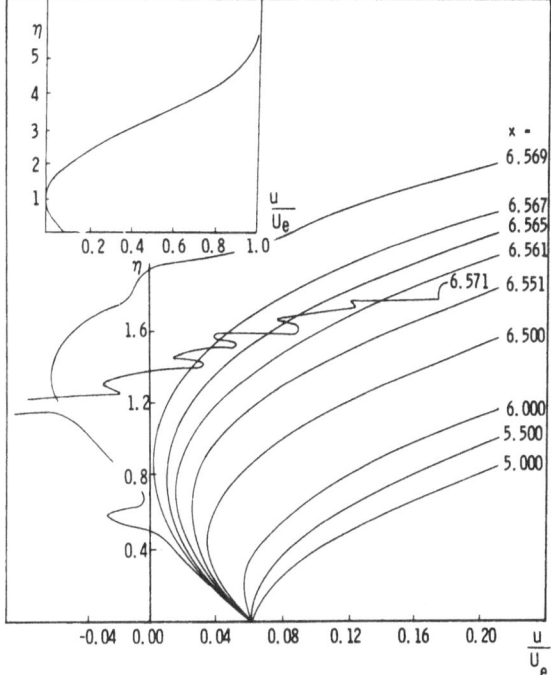

The case of upstream moving walls poses much greater difficulties to the investigator. Separation over upstream moving walls corresponds to a downstream moving separation over a fixed wall. The MRS criterion is expected to hold in this case as well (see Fig. 7.2). However, to begin with, boundary-layer calculations would never produce results similar to the ones described above for the downstream case simply because the boundary-layer equations do not accept an MRS profile for an upstream moving wall. With $u < 0$ at the wall, the MRS profile as postulated by Moore, Rott, and Sears has an inflection at the point of stagnation. That is, at the point $u = \partial u/\partial y = 0$, the quantity $\partial^2 u/\partial y^2$ also vanishes. The equation of momentum reveals that this is possible only if the pressure gradient is zero there, a very severe restriction that is unrealistic for separating flows.

A second serious difficulty arises from the fact that with negative velocities at the wall, regions of reversing flow are encountered and the direction of marching for the integration of the parabolic equation is reversed. In fact, Williams and Johnson (1974a) report that their calculations encountered instabilities when their code was used for integration over upstream moving walls. To overcome this difficulty Tsahalis (1977) converted the problem to a transient. The steady separating flow over a Howarth body was first calculated to serve as an initial condition. Upstream motion of the wall was then gradually introduced and the problem was integrated as an unsteady flow problem by the code described briefly in Chapter 2 (Telionis et al., 1973).

The calculations were marched in time until all transient characteristics were eliminated and a stable steady flow field with upstream moving skin was at hand. The results at separation were encouraging. It appears that the flow characteristics approach those of an MRS profile; the Goldstein singularity is present for sure. However, Tsahalis emphasizes that boundary-layer calculations will never produce the desired results and the full Navier–Stokes equations will have to be brought into play, at least in the neighborhood of separation.

The behavior of integral quantities like the momentum thickness and the form factor in the neighborhood of separation is of great interest for various reasons. First, such quantities are free from slight changes in the velocity profiles. Second, quite often in practice, the problem is solved in terms of integral equations in the form suggested by Kármán and Polhausen. Finally, such quantities have proved very helpful in investigations of turbulent boundary layers due to the complexity of analyzing the detailed mean velocity profiles in the presence of random oscillations. The significance of the outer layer of the boundary layer and especially its integral quantities to the phenomenon of separation has been demonstrated by Shen (1978) and briefly argued in the next section.

The displacement thickness δ_1, momentum thickness θ, energy thickness θ_1, form factor H, and energy factor K are defined by the equations

$$\delta_1 = \int_0^\infty \left(1 - \frac{u}{U_e}\right) dy, \tag{7.2.33}$$

$$\theta = \int_0^\infty \frac{u}{U_e}\left(1 - \frac{u}{U_e}\right) dy, \tag{7.2.34}$$

$$\theta_1 = \int_0^\infty \left(\frac{u}{U_e}\right)^2\left(1 - \frac{u}{U_e}\right) dy, \tag{7.2.35}$$

$$H = \delta_1/\theta, \qquad K = \theta_1/\theta. \tag{7.2.36}$$

The integral momentum and energy equations then read (Fansler and Danberg, 1977)

$$U_e \frac{d\theta^2}{dx} = 2\left[2T - (2 + H)\theta^2 \frac{dU_e}{dx}\right], \tag{7.2.37}$$

$$U_e \frac{d(K^2\theta^2)}{dx} = 2\left[4K(L + u_w T/U_e) - 3K^2\theta^2 \frac{dU_e}{dx}\right], \tag{7.2.38}$$

where L is the dissipation integral multiplied by the dimensionless part of $\theta/\mu U_e^2$, and T is a skin friction factor

$$T = \frac{\theta}{U_e} \frac{\partial u}{\partial y}\bigg|_{y=0}. \tag{7.2.39}$$

Fansler and Danberg use the self-similar profiles to calculate shape parameters and integrate the equations. Their results are shown in Fig. 7.10, where H is plotted versus K for the flow around a rotating cylinder. All curves

Fig. 7.10 The form factor
H (ratio of displacement
thickness to momentum
thickness) plotted against
K (energy thickness di-
vided by momentum
thickness) for flow over
moving walls.

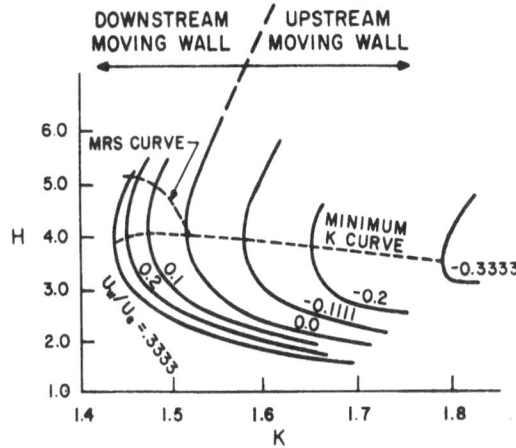

display double values but the lower branch corresponds to a developing
boundary layer. The curves are traced from right to left and K and H
decrease and increase, respectively, as separation is approached. It is most
interesting that in the neighborhood of separation a singularity develops in
the $H-K$ curves. For fixed walls this corresponds exactly to a zero skin
friction profile. However, for downstream moving walls the locus of minimum K and the locus of points that correspond to an MRS profile deviate
considerably. This is a disappointing discrepancy in tune with earlier features
of self-similar separating flows. How far the two points are has not yet been
estimated, although this would rather be immaterial at this point. Fansler
and Danberg claim agreement of the location of their singular point with the
experimental data of Vidal (1959), and Ludwig (1964) for separation over
moving walls.

If the momentum equation is used to eliminate $d\theta^2/dx$, one obtains from
(7.2.37) and (7.2.38)

$$\frac{dH}{dx} = E\left[2\theta^2 U_e \frac{\partial K}{\partial H}\right]^{-1}, \tag{7.2.40}$$

where E is a finite nonzero quantity. It is then deduced that since the $K-H$
critical point is approached in a singular manner, H varies with a square root
singularity, reminiscent of the Goldstein singularity. Fansler and Danberg
propose to use this singular behavior as a criterion of separation for steady
flows over moving walls.

7.3 Asymptotic Methods

A variety of asymptotic solutions have been reported in literature. However,
looking back and studying the work in perspective, one finds that there is no
continuation in the contributions. The theory does not build up to a

comprehensive structure. In fact some of the theories presented in this section may be considered peripheral, since they do not address the problem of unsteady separation per se. Other developments have been considered incomplete.

Asymptotic techniques have been used to study the behavior of the boundary-layer equations in the immediate neighborhood of separation. Pioneering in this area is the work of Goldstein (1948), who showed that for steady flow the boundary-layer equations accept a singular solution at the point of zero skin friction. This work has been extended by Stewartson and others as described in Brown and Stewartson (1969). More recently it was demonstrated that the singular behavior is actually due to the approximate form of the boundary-layer equation and can be removed by a variety of methods; for example, if the displacement thickness of the wall shear is prescribed instead of the pressure gradient. Successful integrations through the point of zero skin friction in steady flow were reported by Catherall and Mangler (1966), Klemp and Acrivos (1971), Inger (1974), Klineberg and Steger (1974), Carter (1974), and others. The mathematical implications of formally removing the separation singularity were addressed by Stewartson (1970).

There is no doubt that the appearance of the singularity is due to the inadequacy of the model. The actual flow, of course, and the correct mathematical model, the Navier–Stokes equations, are free of singularities. However, although no formal proof has yet been presented, it is widely accepted that the steady boundary-layer equations behave in a singular fashion at the point of zero skin friction. The question is now posed as follows. Is this singularity appearing at the point of zero skin friction or is it accompanying always the point of separation? In other words, is the singular behavior the response of the boundary-layer equations at the point of zero skin friction, or is it the exaggerated behavior of the solution in the neighborhood where the actual flow separates from the wall? This chapter addresses the questions by a number of different methods. In the present section we briefly describe attempts to generalize asymptotic theories.

Perturbation techniques have been also used to solve the unsteady Navier–Stokes equations for separating flows. Interest in boundary-layer techniques is however stronger since numerical integration of boundary-layer equations can be used to attack realistic problems of arbitrary two- or three-dimensional shapes. Moreover, such methods can be readily extended to include models of transition to turbulence and turbulent boundary layers.

Slowly varying flows

Asymptotic methods are based on the assumption that a physical property of the problem is relatively small. Unsteady separation has been approached by assuming that the time variation is weak, or that the extent of the solution beyond the point of separation is small, etc. Consider, for example, the case of a point of separation moving upstream with a constant speed εU_s, where U_s is a characteristic velocity of the problem and ε is a small parameter, both constant. We assume that this is the result of an outer-flow distribution with a weak unsteadiness confined to order ε

$$U_e(x,t) = U_0(x) + \varepsilon U_1(x,t). \tag{7.3.1}$$

The flow as seen by an observer moving with the point of separation is given in terms of moving coordinates

$$\bar{x} = x + \varepsilon U_s t, \qquad \bar{y} = y, \qquad \bar{t} = t, \tag{7.3.2}$$

and the dependent variables

$$\bar{u} = u + \varepsilon U_s, \qquad \bar{v} = v, \qquad \bar{U}_e = U_e + \varepsilon U_s. \tag{7.3.3}$$

Consider further a solution in powers of the small parameter ε

$$\bar{u} = u_0 + \varepsilon u_1 + \ldots, \tag{7.3.4}$$

$$\bar{v} = v_0 + \varepsilon v_1 + \ldots. \tag{7.3.5}$$

The quantities u_0, v_0 here should not be confused with the initial conditions of the previous section. Substitution of these expressions in Eq. (7.2.1) and (7.2.2) and collecting coefficients of powers of ε yields

Order ε^0

$$\frac{\partial u_0}{\partial \bar{x}} + \frac{\partial v_0}{\partial \bar{y}} = 0, \tag{7.3.6}$$

$$\frac{\partial u_0}{\partial \bar{t}} + u_0 \frac{\partial u_0}{\partial \bar{x}} + v_0 \frac{\partial u_0}{\partial \bar{y}} = U_0 \frac{\partial U_0}{\partial \bar{x}} + \frac{1}{Re} \frac{\partial^2 u_0}{\partial \bar{y}^2}. \tag{7.3.7}$$

Order ε^1

$$\frac{\partial u_1}{\partial \bar{x}} + \frac{\partial v_1}{\partial \bar{y}} = 0, \tag{7.3.8}$$

$$\frac{\partial u_1}{\partial \bar{t}} + U_s \frac{\partial u_0}{\partial \bar{x}} + u_0 \frac{\partial u_1}{\partial \bar{x}} + (u_1 - U_s) \frac{\partial u_0}{\partial \bar{x}} + v_0 \frac{\partial u_1}{\partial \bar{y}} + v_1 \frac{\partial u_0}{\partial \bar{y}}$$
$$= U_0 \frac{\partial U_1}{\partial x} + (U_1 - U_s) \frac{\partial U_0}{\partial \bar{x}} + \frac{1}{Re} \frac{\partial^2 u_1}{\partial \bar{y}^2}. \tag{7.3.9}$$

To meet the boundary conditions, we now chose to satisfy the nonhomogeneous wall condition at the ε^0 level. Thus Eqs. (7.2.3) and (7.2.4) lead to two sets of boundary conditions for the functions u_0, v_0, and u_1, v_1:

$$u_0 = \varepsilon U_s, \ v = 0, \quad \text{at } y = 0, \tag{7.3.10}$$

$$u_0 = U_0, \qquad \text{as } y \to \infty, \tag{7.3.11}$$

$$u_1 = v_1 = 0, \qquad \text{at } y = 0, \tag{7.3.12}$$

$$u_1 = U_1, \qquad \text{as } y \to \infty. \tag{7.3.13}$$

The problem of order ε^0 has boundary conditions independent of time. We may therefore assume that for a constant speed of the point of separation, the first problem admits a solution independent of time. This is the solution representing steady separating flow over a wall that is moving downstream with a constant speed εU_s. The second problem represents the unsteady perturbation. The equations and arguments presented above are essentially a paraphrase of those of Moore (1958), who frames his problem in terms of a stream function. Moore actually assumes $U_1(x,t) = 0$, which renders the boundary conditions of the second problem homogeneous. The weak time variations should then be contained in U_0 and enter the boundary conditions of the first problem thus rendering it quasi-steady. In both problems the point of separation is fixed. Near the position of separation, therefore, there are two essential problems to be considered. The first is a quasi-steady boundary layer over a surface that is moving with a constant speed, and the second is an unsteady boundary layer involving a fixed wall and a steady point of separation. Moore (1958) proceeds further to investigate model problems of these two kinds but does not propose a composite asymptotic solution for a specific slowly varying separating flow. The first problem has been discussed to some extent in the previous section. The second, Moore attacks by an approximate method, showing that a singular solution, similar to the one studied by Goldstein (1948), is possible.

Slowly varying separating flows were studied also by Buckmaster (1973) via a parameter expansion tailored after Kaplun's (1967) method. With dimensionless variables

$$x = \frac{x^*}{x_s^*}, \qquad y = \frac{Re^{1/2}y^*}{x_s^*}, \qquad t = \frac{\varepsilon t^* U_0}{x_s^*}, \tag{7.3.14}$$

where x_s^* is the distance to separation, and ε an artificially introduced small parameter representing the small-time domain under consideration, the solution in the neighborhood of the point of zero skin friction is sought in the form of an expansion

$$\psi = \frac{1}{6} y^3 + \varepsilon \psi_1(x, y, t) + \varepsilon^2 \psi_2(x, y, t) + \ldots. \tag{7.3.15}$$

Substituting in the stream-function equation and collecting powers of ε, we derive first

$$\frac{1}{2} y^2 \frac{\partial^2 \psi_1}{\partial x \partial y} - y \frac{\partial \psi_1}{\partial x} = \frac{\partial^3 \psi_1}{\partial y^3}, \tag{7.3.16}$$

which is independent of the small time scale, t. This equation does not have an acceptable solution for arbitrary initial data but, for special initial conditions, it admits a solution in the form

$$\psi_1 = y^2 F(x; t), \qquad F(0, t) = 1. \tag{7.3.17}$$

The function ψ_2 is found to have a double zero at the wall and, at worst, algebraic growth as $y \to \infty$, only if a certain integral constraint is satisfied. Buckmaster finally derives a complicated integro-differential equation for $F(x, t)$ from which he is able to show that if $F(x, 0) = (1 - x^2)^{1/2}$, then $F(x, \infty) = (1 - bx^2)^{1/2}$, where b is a constant. Thus the square root singularity eventually moves to a new position. More interesting is the behavior of the solution for small times. It is found that if the point of zero skin friction is displaced, the singularity is initially left behind. This is in agreement with results of Moore (1958) and compatible with the results of Sears and Telionis (1971). In fact, most recent experimental data (Koromilas and Telionis, 1980) indicate that this is exactly the behavior of separation if a transient or even impulsive change of the pressure gradient is imposed. The point of separation displays considerable inertia in its response to external disturbances. The point of zero skin friction instantly moves away from the neighborhood of separation, but separation itself remains unaffected for a considerable amount of time following the initiation of the impulse.

Weak temporal variations and their effect on separation were also investigated by Shen and Nenni (1975). Their analysis proceeds in a phase plane in terms of new independent variables

$$\xi = \int_0^x U_e \, dx, \qquad \eta = U_e y, \tag{7.3.18}$$

and a new dependent variable

$$Z = U_e(u - U_e). \tag{7.3.19}$$

Inspired by Tollmien's (1946) work, they investigate the behavior of the outer layer of the boundary layer. Shen and Nenni propose to associate separation directly with the condition that the boundary layer becomes incompatible, in their terminology, "unmatchable" with the outer flow. This, for example, is the situation if the induced normal velocity at the outer edge of the boundary layer attains such a magnitude as to invalidate the basic boundary-layer assumption that the ratio v/u is of order $(Re)^{-1/2}$. It is then readily deduced that an equivalent condition is

$$\frac{\partial}{\partial x}(U_e \delta_1) > O(1). \tag{7.3.20}$$

It can be proved that for steady flow, the above condition leads directly to a Goldstein-type of singularity, which for this special case coincides with the station of vanishing wall shear. Extending the theory to weakly unsteady flows leads to a differential equation of the Burger's type for the wall shear. Singularities then form in the x–t plane exactly in the way that shock waves form in gas dynamics. The method is therefore capable of explaining the emergence of separation at finite time under suitable conditions. The analysis of Shen and Nenni is very involved and cumbersome. More in tune with the style of this monograph is a simplified but crude development due to Shen (1978) which, however, contains most of the basic elements of the theory.

Very useful in qualitative investigations of boundary layers is a set of compatibility conditions derived by successive differentiations of the momentum equation and evaluations at the wall. If the terms of the momentum equation are evaluated at the wall, with $u = v = 0$ at $y = 0$, we obtain, for example,

$$0 = -\frac{\partial p}{\partial x} + \frac{1}{Re} \frac{\partial^2 u}{\partial y^2}\bigg|_w, \tag{7.3.21}$$

where the subscript w denotes evaluation at $y = 0$. Taking one derivative of the momentum equation and noting that by virtue of the continuity equation

$$\frac{\partial v}{\partial y}\bigg|_w = -\frac{\partial u}{\partial x}\bigg|_w = 0, \tag{7.3.22}$$

we get

$$\frac{\partial^2 u}{\partial y \, \partial t}\bigg|_w = \frac{1}{Re} \frac{\partial^3 u}{\partial y^3}\bigg|_w. \tag{7.3.23}$$

Similarly, making use of the relation

$$\frac{\partial}{\partial y}\left(\frac{\partial v}{\partial y}\right)\bigg|_w = -\frac{\partial^2 u}{\partial x \, \partial y}\bigg|_w, \tag{7.3.24}$$

which is again an immediate conclusion from continuity, we reduce the second derivative of the momentum equation at the wall to

$$\frac{\partial^3 u}{\partial y^2 \partial t}\bigg|_w + \frac{\partial u}{\partial y}\bigg|_w \frac{\partial}{\partial x}\left(\frac{\partial u}{\partial y}\right)\bigg|_w = \frac{1}{Re} \frac{\partial^4 u}{\partial y^4}\bigg|_w. \tag{7.3.25}$$

Consider now the special case of steady flow and assume a crude profile for u (Shen, 1978) in terms of a stretched variable $Y = y\sqrt{Re}$

$$u(x, y) = a_1(x)Y + a_2(x)Y^2 + a_3(x)Y^3 + a_4(x)Y^4. \tag{7.3.26}$$

Conditions (7.3.22)–(7.3.25) for steady flow then dictate that

$$a_2 = \frac{1}{2}\frac{\partial p}{\partial x} = -\frac{1}{2}U_e\frac{dU_e}{dx}, \qquad a_3 = 0, \qquad a_4 = \frac{a_1 a_1'}{24}. \tag{7.3.27}$$

A very crude patching of this profile with the outer flow would require that

$$u(x, Y) = U_e(x), \quad \frac{\partial u}{\partial Y} = 0, \quad \text{at } Y = \delta. \tag{7.3.28}$$

The above conditions can be used to determine the quantities δ and a_4 near the station of zero skin friction, in terms of U_e and a_1:

$$\delta = \sqrt{-4/U_e'}, \tag{7.3.29}$$

$$a_1 \frac{da_1}{dx} = -6U_e U_e'^2. \tag{7.3.30}$$

For the special case of Howarth's flow, that is, for a linearly decelerating outer-flow velocity $U_e = 1 - x/8$, separation was found to occur at $x_s = 0.959$. Numerical solutions indicate that the wall shear, that is, the function a_1 behaves in this neighborhood in a singular manner:

$$a_1 = 1.33(x_s - x)^{1/2}. \tag{7.3.31}$$

It can be proved that the crude approximation represented by (7.3.30) results in

$$a_1 = 1.62(x_s - x)^{1/2}. \tag{7.3.32}$$

It is most interesting that the proper singular behavior is captured by Shen's crude example and the numerical coefficient is not far from its proper value.

Extending this approach to unsteady flow and assuming that the functions a_i depend on x and t, we deduce from the compatibility conditions

$$\dot{a}_0 + a_0 a_0' = -\frac{\partial p}{\partial x} + 2a_2, \tag{7.3.33}$$

$$\dot{a}_1 + a_0 a_1' = 6a_3, \tag{7.3.34}$$

$$2\dot{a}_2 + 2a_0 a_2 + a_1 a_1' = 12a_4, \tag{7.3.35}$$

where a dot and a prime denote differentiation with respect to time and space, respectively. The patching conditions (7.3.28) now take the form

$$a_3 \delta^3 + a_4 \delta^4 = U_e - a_1 \delta - a_2 \delta^2, \tag{7.3.36}$$

$$3a_3 \delta^2 + 4a_4 \delta^3 = -a_1 - 2a_2 \delta. \tag{7.3.37}$$

The analysis soon becomes very complex but the main point can be seen by inspecting the above equations. The functions a_2, a_3, and a_4 can be expressed in terms of a_1 and δ by virtue of Eqs. (7.3.33)–(7.3.35) as in the case of steady flow. The system of (7.3.36) and (7.3.37) can then be used to eliminate δ and finally yield

$$a_4 = G(\dot{a}_1, a_1, \ldots), \tag{7.3.38}$$

where the dots in the parenthesis refer to quantities obtainable from the prescribed free stream. For a weak dependence on time, Shen proposes to expand as follows:

$$a_4 = G_0(a_1, \ldots) + G_1(a_1, \ldots)\dot{a}_1 + G_2(a_1, \ldots)\dot{a}_1^2 + \cdots, \tag{7.3.39}$$

and, keeping only the first term, he obtains

$$h_1 \dot{a}_1 + a_1 a_1' = k_1, \tag{7.3.40}$$

where h_1 and k_1 depend on a_1 and other quantities defined by the outer flow. This equation resembles Burger's equation in gas dynamics, and therefore this mathematical model opens the possibility for the wall shear a_1 to develop a singularity in the x–t plane, much like in one-dimensional gas dynamics a shock wave is developed. The shock path then would correspond to the temporal path of the point of singularity. Moreover, since it is well known that shocks may develop in time within a domain that was completely free of any discontinuities, this model demonstrates that the same is possible for unsteady boundary-layer flows: The separation singularity may develop in a finite time in a field that started with no singularities. Equations similar to (7.3.40) and in fact a nonlinear version that takes into account the term $\dot{a}_1^2$ was derived formally by Nenni (1976).

Coordinate expansions

The traditional way of approaching the problem of separation is by expanding in powers of the distances parallel and perpendicular to the wall from the point of separation (Goldstein, 1948; Brown and Stewartson, 1969). Such methods of course do not prove that the boundary-layer equations are singular at the point of separation. They can only reveal a possible behavior of the solution in the neighborhood of separation. The main ideas can be lucidly described in terms of simple mathematical observations due to Landau and Lifshitz (1959). Extensions of the arguments to unsteady flow were attempted by Sears and Telionis (1972).

Based on our definition of separation, we seek to investigate the behavior of the boundary-layer equation in the neighborhood of a point where the v component of the velocity very sharply grows. It has been a common numerical result that indeed in the neighborhood of steady separation quantities like v, $\partial u/\partial x$, etc., grow out of bounds. Indeed with v and $\partial v/\partial y$ blowing up, so does $\partial u/\partial x$ by virtue of the continuity equation. This behavior can in fact be defined per se as separation in agreement with Prandtl's original concept of flow break-away from the wall. Then $\partial x/\partial u = 0$ at separation, and a possible expression of $x(u, y, t)$ close to separation is

$$x_0 - x = (u_0 - u)^2 f(y, t) + \dots, \tag{7.3.41}$$

where f is some function of y and t,

$$u_0 = u_0(y, t) = u(x_0, y, t) \tag{7.3.42}$$

and $x_0 = x_0(t)$ is the value of x at separation.

Let us introduce a new independent variable $\chi = x_0(t) - x$. Also, for reasons that will become clear, we prefer to replace y in terms of another new independent variable $\psi = y - y_0(t)$. The significance of the quantity $y_0(t)$ will be explained later. In terms of these new variables, following Goldstein (1948), and in agreement with Eq. (7.3.41), we assume now that the u component of velocity can be expanded as follows:

$$u(x, y, t) = u_0(\psi, t) + f_0'\chi^{1/2} + f_1'\chi^{3/4} + f_2'\chi$$
$$+ \tilde{f}_2'\chi \ln \chi + f_3'\chi^{5/4} + \dots, \tag{7.3.43}$$

where primes denote differentiation with respect to y and the functions $u_0, f_0, \ldots,$ are functions, undetermined as yet, of y and t. By continuity then,

$$v = \frac{1}{2} f_0 x^{-1/2} + \frac{3}{4} f_1 x^{-1/4} + (f_2 + \tilde{f}_2) + \tilde{f}_2 \ln x$$
$$+ \frac{5}{4} f_3 x^{1/4} + \frac{3}{2} f_4 x^{1/2} + \ldots . \tag{7.3.44}$$

This is only to say that if $\partial u/\partial x$ blows up at a certain station $x = x_0(t)$, a plausible singularity for the blowup is the square root, which Goldstein (1948) also assumed a priori. We have not yet made use of the boundary-layer equations; we now do so, substituting Eqs. (7.3.43) and (7.3.44) into the momentum equation Eq. (7.2.2), assuming that the Reynolds number has been absorbed by appropriate stretching. The result pertains only to small values of x; it is a rather lengthy equation in which the coefficients of various powers of x can be collected and individually equated.

The dominant terms are those in $x^{-1/2}$, and they give

$$f_0' \frac{dx_0}{dt} - u_0 f_0' + f_0 u_0' = 0. \tag{7.3.45}$$

The derivative dx_0/dt is the u-component of the velocity of the point of separation, which we shall denote by

$$U_s = U_s(t) = \frac{dx_0}{dt} . \tag{7.3.46}$$

Then Eq. (7.3.45) can be written in the form

$$-f_0'(u_0 - U_s) + f_0 \frac{\partial}{\partial y} (u_0 - U_s) = 0 \tag{7.3.47}$$

whence

$$f_0(\psi, t) = A(t)\left[u_0(\psi, t) - U_s(t) \right], \tag{7.3.48}$$

where $A(t)$ is undetermined. Thus Eqs. (7.3.43) and (7.3.44), describing the flow near separation, become

$$u(x, y, t) = u_0(\psi, t) + A(t) \frac{\partial u_0(\psi, t)}{\partial y} x^{1/2} + O(x^{3/4}), \tag{7.3.49}$$

$$v(x, y, t) = \frac{1}{2} A(t)\left[u_0(\psi, t) - U_s(t) \right] x^{-1/2} + O(x^{-1/4}). \tag{7.3.50}$$

We note that, far from the wall, the boundary-layer velocity component $u(x, y, t)$ should merge smoothly into the outer-flow velocity, say $U_e(x, t)$, which is nonsingular. Thus,

$$\frac{\partial u_0}{\partial y} \to 0, \quad f_1' \to 0 \quad \text{and} \quad f_2' \to -\frac{\partial U_e}{\partial x}\bigg|_{x = x_0}, \quad \text{as } y \to \infty. \tag{7.3.51}$$

We then collect coefficients of successively higher orders of magnitude in terms of χ, that is, $\chi^{-1/4}$, $\ln\chi$, χ^0, and $\chi^{1/4}$, and obtain the following equations:

$$f_1'U_s - f_1'u_0 + f_1u_0' = 0,\tag{7.3.52}$$

$$\tilde{f}_2'U_s - \tilde{f}_2'u_0 + \tilde{f}_2u_0' = 0,\tag{7.3.53}$$

$$(f_2' + \tilde{f}_2')U_s + \frac{\partial u_0}{\partial t} - \frac{\partial u_0}{\partial y}V_s - (f_2' + \tilde{f}_2')u_0$$

$$-\frac{1}{2}f_0'^2 + \frac{1}{2}f_0f_0'' + (f_2 + \tilde{f}_2)u_0' = -\frac{\partial p}{\partial x} + u_0'',\tag{7.3.54}$$

$$5f_3'U_s - 5f_3'u_0 - 3f_1'f_0' - 2f_0'f_1' + 2f_1''f_0 + 3f_1f_0'' + 5f_3u_0' = 0.\tag{7.3.55}$$

These equations can be solved successively for the unknown functions f_1, f_2, $\tilde{f}_2$, and f_3, while the calculation of higher-order terms would require the inclusion of terms of the form $(\ln\chi)^2$ or $\ln(\ln\chi)$ in order that the subsequent equations could be solved successively for the functions f_4, f_5, etc.

Up to now we have studied only the possible form of the singularity that may appear at separation. Our model in this case involves a singularity of Goldstein's type, moving in response to time-varying boundary conditions according to Eqs. (7.3.49) and (7.3.50). This model clearly reduces to Goldstein's when the time derivatives, including U_s and V_s, are put equal to zero; y_0 is then constant and may conveniently be taken equal to zero. But the model is incomplete. In the classical case, one uses the wall condition $u = 0$ at $y = 0$, for all x, to obtain Prandtl's famous criterion Eq. (7.1.1) from relationship (7.3.49).

In the more general cases treated here, we do not have an analogous condition. In fact, we cannot accept the condition $u(x, 0, t) = 0$ in Eq. (7.3.49), for it would lead to Eq. (7.1.1), which is incorrect for this type of flow, as we have pointed out (Sears and Telionis, 1972). We must conclude, therefore, that Eqs. (7.3.49) and (7.3.50) are not valid for all values of y, but only for y greater than the value $y_0(t)$, heretofore undetermined; that is, for $\psi \geqslant 0$. We call the point x_0, y_0 the "center of separation," and, as already mentioned, we specifically assume that v and $\partial u/\partial x$ are nonsingular near this point. Hence, from Eq. (7.3.50),

$$u_0(0, t) = U_s(t).\tag{7.3.56}$$

Thus, we visualize the "center of separation" as a kind of moving stagnation or bifurcation point, and we accept Eqs. (7.3.49) and (7.3.50) as a boundary-layer simulation of this phenomenon.

A second immediate consequence of this assumption is

$$\frac{\partial u_0(0, t)}{\partial y} = \frac{\partial u}{\partial y}\bigg|_{y = y_0(t)} = 0,\tag{7.3.57}$$

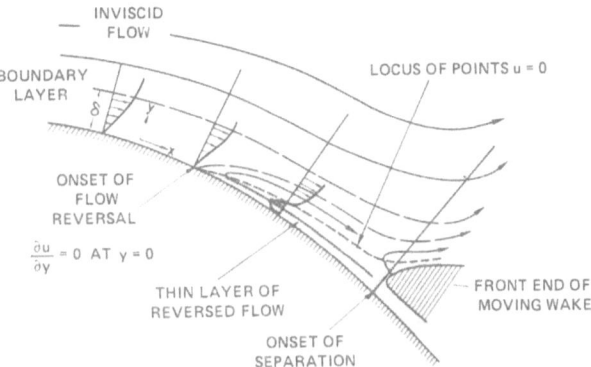

Fig. 7.11 Schematic representation of the flow in the neighborhood of upstream moving separation.

that is, there is a point of vanishing shear, not at the wall in general. It is clear that in the special case of steady flow past a moving wall, this becomes the MRS criterion of Eq. (7.1.2). The situation is shown schematically in Fig. 7.11, which is essentially the flow field of Fig. 7.2 viewed by an observer moving with the wall.

The analysis presented in this section can only serve as a qualitative indicator of the possible forms of the solution in the neighborhood of unsteady separation. What is missing here is the analogy to Goldstein's inner singular expansion for the stream function in powers of $x^{1/4}$. It is widely accepted that it is this expansion in Goldstein's work that establishes the viability of the whole scheme. This task however has been accomplished for semisimilar flows by Shen (1978).

Asymptotic methods cannot in general provide proof that a singularity will or will not appear. They can only indicate what form of a singularity may be acceptable. To this end it is quite often helpful to consider constraint equations derived from the governing equations. The compatibility equation given by Eq. (7.3.25), for example, seems to play a vital role in the control of the behavior near separation. Shen (1978) points out in fact that failure of integral Polhausen profiles to satisfy this compatibility condition is the reason for the poor performance of such methods in the neighborhood of separation.

By virtue of the momentum equation, Eq. (7.3.25) can be written as

$$\frac{\partial}{\partial x}\left[\frac{1}{2}\left(\frac{\partial u}{\partial y}\right)^2\Big|_w\right] = \frac{1}{Re}\left[\frac{\partial^4 u}{\partial y^4}\Big|_w - \frac{\partial^2 p}{\partial x\,\partial t}\right]. \tag{7.3.58}$$

Bradshaw (1979) notes that integration of this equation yields

$$\frac{\partial u}{\partial y} = C(x_z - x)^{1/2}, \tag{7.3.59}$$

where x_z is the distance to the point of zero skin friction and C is proportional to the right-hand side of Eq. (7.3.58) and therefore is independent of y. Clearly there is a singularity, unless C vanishes at this point. Integrating with respect to y and differentiating with respect to t, we obtain the dominant terms of the acceleration in the immediate neighborhood of the wall.

$$\frac{\partial u}{\partial t} = (x_z - x)^{1/2} \frac{\partial C}{\partial t} y + \frac{\frac{1}{2}Cy}{(x_z - x)^{1/2}} \frac{dx_z}{dt} + \dots \qquad (7.3.60)$$

Infinite accelerations would result unless the position of zero skin friction is stationary, which is a very restrictive requirement, or, alternatively, if C vanishes. However, this line of thought cannot offer insight into the behavior in the neighborhood of a generalized MRS point, since the expansion in (7.3.60) is about the point of zero skin friction.

7.4 Semisimilar Solutions

In special cases, it is possible to combine independent variables and reduce their number, thus simplifying the differential equation. The solutions to such equations are known as self-similar solutions and some classical examples are familiar to fluid dynamicists. The existence of such solutions signifies the fact that appropriate stretching or displacement may collapse the properties of the flow that are evaluated at different points in space. Such information is also very valuable to experimentalists, since it may guide them to obtain data at the appropriate positions and help them reduce the number of necessary measurements. The term "self-similarity" has been actually accepted in literature to denote problems that reduce to ordinary differential equations. If an appropriate transformation results in reduction of the number of coordinates from 3 to 2, then it is usually termed a semisimilar transformation.

Semisimilar transformations are particularly helpful in the case of unsteady separation because they contribute substantially to the understanding of the physics involved. Mapping an unsteady two-dimensional flow into a steady flow will permit us to transfer results and information on steady separation to the unsteady regime.

A semisimilar transformation for a flow that involves adverse pressure gradients and separation was proposed first by Tani (1958). Introducing new similarity variables

$$\xi = \frac{x}{1 - \lambda t}, \qquad \eta = \frac{y}{2} \sqrt{\frac{Re}{x}}, \qquad (7.4.1)$$

we can replace derivatives with respect to x, y, and t by

$$\frac{\partial}{\partial t} = \frac{\lambda \xi}{(1 - \lambda t)} \frac{\partial}{\partial \xi},$$
(7.4.2)

$$\frac{\partial}{\partial x} = \frac{1}{1 - \lambda t} \frac{\partial}{\partial \xi} - \frac{\eta}{2x} \frac{\partial}{\partial \eta},$$
(7.4.3)

$$\frac{\partial}{\partial y} = \frac{1}{2} \sqrt{\frac{Re}{x}} \frac{\partial}{\partial \eta},$$
(7.4.4)

$$\frac{\partial^2}{\partial y^2} = \frac{1}{4} \frac{Re}{x} \frac{\partial^2}{\partial \eta^2}.$$
(7.4.5)

Expressing Eqs. (7.2.1) and (7.2.2) in terms of the variables ξ and η immediately indicates that the variable x can be absorbed in v and thus all the coordinates x, y, and t are eliminated and the problem is expressed in terms of ξ and η alone. With

$$V = v\sqrt{x \, Re},$$
(7.4.6)

Eqs. (7.4.1) and (7.4.2) become

$$2\xi \frac{\partial u}{\partial \xi} - \eta \frac{\partial u}{\partial \eta} + \frac{\partial V}{\partial \eta} = 0,$$
(7.4.7)

$$4\xi(\lambda \xi + u) \frac{\partial u}{\partial \xi} + 2(V - u\eta) \frac{\partial u}{\partial \eta} = 4\xi(\lambda \xi + U_e) \frac{\partial U_e}{\partial \xi} + \frac{\partial^2 u}{\partial \eta^2}.$$
(7.4.8)

The problem is of course semisimilar only if the boundary conditions as well are expressed in terms of ξ alone:

$$u = v = 0, \quad \text{at } \eta = 0,$$
(7.4.9)

$$u = U_e(\xi), \quad \text{as } \eta \to \infty.$$
(7.4.10)

Tani actually assumed a Howarth-type flow

$$U_e(\xi) = 1 - \xi$$
(7.4.11)

and solved the problem in terms of a coordinate expansion in power of ξ:

$$u = \frac{1}{2} \sum_{n=1}^{5} \xi^n f_n'(\eta) - \frac{1}{2} F(\xi) G(\eta),$$
(7.4.12)

where $G(\eta)$ is determined according to Howarth (1938) and given in tabulated form.

A more general approach to the problem of obtaining semisimilar solutions is based on new variables given by the equations

$$\xi = \xi(x, t),$$
(7.4.13)

$$\eta = \frac{y}{g(x, t)} \sqrt{Re},$$
(7.4.14)

where $\xi(x, t)$ and $g(x, t)$ are functions to be determined later. A straightforward substitution in the equations of continuity and momentum, (7.2.1) and (7.2.2) yields

$$\frac{\partial u}{\partial \xi} \frac{\partial \xi}{\partial x} - \frac{\eta}{g} \frac{\partial g}{\partial x} \frac{\partial u}{\partial \eta} + \frac{1}{g} \sqrt{Re} \frac{\partial v}{\partial \eta} = 0, \tag{7.4.15}$$

$$\left(\frac{\partial \xi}{\partial t} + u \frac{\partial \xi}{\partial x} \right) \frac{\partial u}{\partial \xi} + \left[\frac{v}{g} \sqrt{Re} - \eta \left(\frac{1}{g} \frac{\partial g}{\partial t} + \frac{u}{g} \frac{\partial g}{\partial x} \right) \right] \frac{\partial u}{\partial \eta}$$

$$= \left(\frac{\partial \xi}{\partial t} + U_e \frac{\partial \xi}{\partial x} \right) \frac{\partial U_e}{\partial \xi} + \frac{1}{g^2} \frac{\partial^2 u}{\partial \eta^2}. \tag{7.4.16}$$

The requirement that the coefficients of these equations be functions of ξ alone results in sets of equations that can be used to determine the unknown functions ξ and g. Dividing through both equations by g^{-2} and absorbing the function g in a new dependent variable

$$V = vg\sqrt{Re}, \tag{7.4.17}$$

we may rewrite (7.4.15) and (7.4.16) in the form

$$G_2 \frac{\partial u}{\partial \xi} - G_4 \eta \frac{\partial u}{\partial \eta} + \frac{\partial V}{\partial \eta} = 0, \tag{7.4.18}$$

$$(G_1 + uG_2) \frac{\partial u}{\partial \xi} + [V - \eta(G_3 + uG_4)] \frac{\partial u}{\partial \eta}$$

$$= (G_1 + U_e G_2) \frac{\partial U_e}{\partial \xi} + \frac{\partial^2 u}{\partial \eta^2}. \tag{7.4.19}$$

In the above equations, the quantities G_1 through G_4 must be functions of ξ alone and they are related to ξ and g via

$$G_1 = g^2 \frac{\partial \xi}{\partial t}, \qquad G_2 = g^2 \frac{\partial \xi}{\partial x}, \qquad G_3 = g \frac{\partial g}{\partial t}, \qquad G_4 = g \frac{\partial g}{\partial x}. \tag{7.4.20}$$

The first systematic approach toward the development of a general semi-similar solution in the spirit of the above development was reported by Hayasi (1962). In this report, however, ξ was confined to dependence on ξ or x alone. Later Williams (1974) and Williams and Johnson (1974a, 1975) reconsidered the problem and suggested more general solutions. Formulating the problem in terms of a streamfunction, Williams and Johnson (1974a) assumed arbitrarily some of the functions they introduced to arrive at

$$\xi = \frac{x + Kt}{1 - Bt}, \qquad g = \sqrt{\frac{x + Kt}{U_e(\xi)}}. \tag{7.4.21}$$

In their work, Williams and Johnson insisted in a final form of the differential equation as close as possible to its corresponding steady-state form. In terms of the variables (7.4.21), the functions G_i are indeed functions of ξ alone

$$G_1 = \frac{\xi}{U_e}(K + B\xi), \qquad G_2 = \frac{\xi}{U_e}, \qquad G_3 = \frac{K}{2U_e}, \qquad G_4 = \frac{1}{2U_e},$$

$$(7.4.22)$$

provided of course U_e is a function of ξ alone as well.

Formulating the problem in terms of the equations of continuity and momentum, Shen (1978) was able to prove that the above choice of variables is in fact the most general one for semisimilar boundary layers with $U_e = U_e(\xi)$. To this end, Shen considered the ratios G_1/G_2 and G_3/G_4 and demanded, instead of the condition (7.4.22), that the functions $F_1 = G_1/G_2$, $F_2 = G_3/G_4$, and G_2, G_4 be functions of ξ alone:

$$F_1(\xi) = \frac{G_1}{G_2} = \frac{\partial \xi/\partial t}{\partial \xi/\partial x}, \qquad F_2(\xi) = \frac{G_3}{G_4} = \frac{\partial g/\partial t}{\partial g/\partial x}, \qquad (7.4.23)$$

$$G_2(\xi) = g^2 \frac{\partial \xi}{\partial x}, \qquad G_4(\xi) = g \frac{\partial g}{\partial x}. \qquad (7.4.24)$$

Interchanging the role of dependent and independent variables in (7.4.22), Shen wrote

$$\frac{\partial x}{\partial \xi} + F_2 \frac{\partial t}{\partial \xi} = 0, \qquad (7.4.25)$$

$$\frac{\partial x}{\partial g} + F_1 \frac{\partial t}{\partial g} = 0, \qquad (7.4.26)$$

and eliminating x by cross differentiation,

$$(F_2 - F_1) \frac{\partial^2 t}{\partial g} - \frac{\partial F_1}{\partial \xi} \frac{\partial t}{\partial g} = 0, \qquad (7.4.27)$$

he was able to integrate and derive a general solution

$$t = A_1(g)\exp\left(\int \frac{F_1' d\xi}{F_2 - F_1}\right) + B_1(\xi), \qquad (7.4.28)$$

$$x = -F_1 t + B_2(\xi), \qquad (7.4.29)$$

where A_1, B_1, and B_2 are arbitrary functions of ξ. It can be shown then by virtue of Eqs. (7.4.23–26) that

$$F_1 = \frac{x - \alpha_2}{\alpha_1 - t}, \qquad \delta^2 = (\alpha_1 - t)G_2. \qquad (7.4.30)$$

With $\alpha_1 = 1/\lambda$, $\alpha_2 = 0$, $F_1 = \lambda\xi$, and $G_2 = 4\xi$, we recover the Tani variables and the governing equations reduce to (7.4.7) and (7.4.8), subject to fixed wall boundary conditions

$$u(\xi,\eta) = v(\xi,\eta) = 0, \quad \text{at } \eta = 0, \tag{7.4.31}$$

$$u(\xi,\eta) \to U_e(\xi), \quad \text{as } \eta \to \infty, \tag{7.4.32}$$

$$u(\xi,\eta) = u_0(\eta), \quad \text{at } \xi = 0 \tag{7.4.33}$$

where $\bar{u}_0$ is a known function.

It is observed that (7.4.8) is a parabolic equation with only one parabolic variable, the coordinate ξ. Its numerical integration will therefore run into instabilities at the point in the ξ, η plane where the coefficient of $\partial u/\partial \xi$ vanishes. This has been demonstrated analytically and numerically in earlier chapters, especially in Chapter 2. It turns out that the coefficient of $\partial u/\partial \xi$, the quantity

$$K = 4\xi(\lambda\xi + u) \tag{7.4.34}$$

plays a much more significant role (Shen, 1978). Assuming that the curve $K = 0$ in the ξ, η plane is well behaved and that there exists a region where $K > 0$, then the minimum ξ at which $K = 0$ is a critical point at which the derivative $\partial K/\partial \eta$ vanishes. Let ξ_c, η_c be the coordinates of such a critical point,

$$K(\xi_c, \eta_c) = 0, \qquad \left.\frac{\partial K}{\partial \eta}\right|_c = 0. \tag{7.4.35}$$

If more than one such point exists, then the calculation will have to be terminated at the point with the minimum ξ_c. The situation is shown schematically in Fig. 7.12. Inspecting Eq. (7.4.16) reveals that the coefficient of $\partial u/\partial \xi$ is actually the total time rate of particle displacement. K therefore is nothing but the true forward velocity of fluid particles. If K decreases, then the fluid particles decelerate and eventually they lose completely their momentum at the point where condition (7.4.35) is met. The last statement is almost identical to Prandtl's description of separation. In other words, it is proposed that this critical point coincides with the point of separation. Condition (7.4.35) is indeed compatible with the MRS criterion for unsteady separation. The shear $\partial u/\partial \eta$ vanishes at a certain height η_c above the wall. In fact, this condition now yields the speed of the critical point that is not specified in the MRS definition.

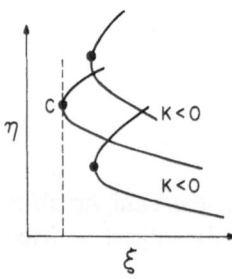

Fig. 7.12 Schematic representation of $K = 0$ curves. K is defined by Eq. (7.4.37).

To test the validity of this argument, Shen compares with the numerical results of Williams and Johnson (1974a) for the separation point ξ_s of Howarth's flow $U_e = 1 - \xi$:

λ	0	0.5	1.0
ξ_s	0.117	0.161	0.221
$\bar{u}_s$	0	-0.0805	-0.221

The above data are in full agreement with (7.4.35), which provides evidence that indeed $\xi_c = \xi_s$. However, the vanishing of K does not necessarily signal separation. Williams (1981) points out that this condition is satisfied at the position of Stewartson's singularity for a semiinfinite plate started from rest (see Section 3.5).

7.5 Direct Numerical Integration—Impulsive and Transient Changes

Direct numerical integrations in the three-coordinate space x, y, and t have been described on many occasions in earlier chapters. Crucial to the investigation of the present problem is a finite-difference scheme capable of integrating through regions of reverse flow. Zig-zag schemes or upwind differencing schemes or alternative formulations are needed as described in Chapter 2.

Transient changes of the outer-flow conditions provide the most unambiguous cases of unsteady separation. This is because an established separating steady flow is used as an initial condition. Separation, therefore, and all the well-known properties that accompany it are part of the flow field at $t = 0$. The task is then to find how, with a transient or impulsive change of the outer flow, the boundary layer readjusts itself and passes over to a new flow field. Unlike in cases where initially the flow is attached, here we have a separation point at $t = 0$ and we can focus our attention on its behavior in subsequent times.

The problem is better described in terms of simple examples worked out by Telionis et al. (1973), Telionis and Tsahalis (1974), and Wang (1979). Consider, for example, a Howarth flow

$$U_e(x,t) = 1 - T(t)x, \tag{7.5.1}$$

where the function $T(t)$ is given by

$$
\begin{aligned}
T(t) &= C_1 + C_2 t^2, && \text{for } 0 < t < t_0, \\
T(t) &= C_3 + C_4 t + C_5 t^2, && \text{for } t > t_0.
\end{aligned}
\tag{7.5.2}
$$

The idea behind this choice is that the time rate of change of U_e at $t = 0$ is zero, so that unsteadiness is introduced in a smooth way. The constants are chosen in such a way that a smooth transition occurs at $t = t_0$ and the outer-flow distribution is reached asymptotically.

The most interesting finding is that the separation singularity responds to the changes of the outer flow and displaces upstream at a finite rate. It is emphasized again that this is not simply a region of instabilities we are identifying as a singularity, but a well-behaved increase of certain properties and slopes accompanied by a well-ordered but increasing number of iterations required for convergence. The v component of the velocity at $\eta = 0.5$, which clearly exhibits the singular behavior, is plotted in Fig. 7.13 and the accompanying skin friction distributions are shown in Fig. 7.14. The time domain represented is the latter part of the motion that shows how the flow tends to a new steady-state position. The circles in Fig. 7.14 represent actual computational points and the numbers in parentheses are the number of iterations required for convergence at each point.

An equally interesting unsteady flow problem is the impulsive change of a Howarth flow. Let the outer flow be given by

$$U_{e_1} = 1 - A_1 x, \quad \text{for } t \leqslant 0, \tag{7.5.3}$$

but assume that it is impulsively changed to

$$U_{e_2} = 1 - A_2 x, \quad \text{for } t > 0, \tag{7.5.4}$$

with $A_2 > A_1$. As described in Chapter 2, impulsive changes of the outer flow generate potential corrections throughout the entire flow field and give rise to vortex sheets that cover solid boundaries.

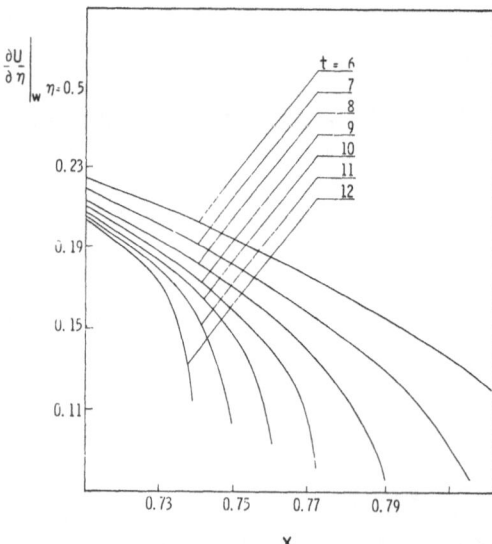

Fig. 7.14 Shear distribution at $\eta = 0.5$ plotted against distance x for unsteady flow over fixed walls for $t = 6$–12, corresponding to the free-stream distribution of Fig. 7.14.

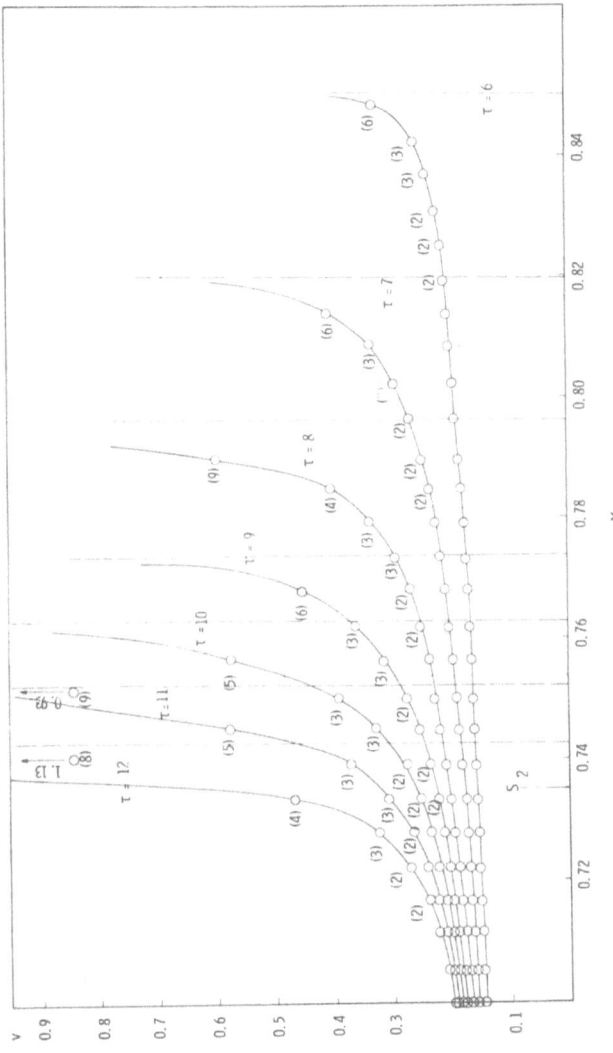

Fig. 7.13 The v component of velocity at $\eta = 0.5$ plotted against x for various instances t according to outer-flow distribution described in text. Numbers in parentheses denote the number of iterations required for convergence. S_2 is the location of steady-state separation.

According to the theorem proved in Chapter 2, the change $\Delta U_e = (A_2 - A_1)x$ is inviscid and should be superimposed on the established viscous solution. If $u_H(x, y)$ and $v_H(x, y)$ correspond to the steady-state solution with an outer flow given by Eq. (7.5.3), then at $t = 0^+$ the flow field should be given by

$$u(x, y, 0^+) = u_H(x, y) - \Delta U_e, \tag{7.5.5}$$

$$v(x, y, 0^+) = v_H(x, y). \tag{7.5.6}$$

Figure 7.15 depicts schematically this situation. The velocity field given by the Eqs. (7.5.5) and (7.5.6) will be considered as the initial flow field for the unsteady problem. At each position x, these equations represent a uniform displacement ΔU_e of the established profile to the left and therefore a slip velocity at the wall equal to ΔU_e.

One of the advantages of this particular outer-flow distribution is that for $t < 0$ and $t \to \infty$, the flow field corresponds to two solutions of the family considered by Howarth. These are steady-state solutions for outer-flow distributions given by Eqs. (7.5.3) and (7.5.4), respectively, and can therefore be checked against Howarth (1938) or calculated with a steady-state scheme of integration. For these two extreme cases, the location of separation is unambiguously defined as the point of zero skin friction. Howarth's calculations have shown that the distance to separation is given by

$$S_{zf} \simeq \frac{0.120}{A} . \tag{7.5.7}$$

The numerical analysis encounters a difficulty for small times and very close to the wall. This is due to the inability of the numerical scheme to represent a vortex sheet. An approximate analytical solution for small times is therefore derived (Telionis and Tsahalis, 1974b) and incorporated in the computer program to help start the calculations. It is known that for impulsive changes and for very small times, the convection terms of the momentum equation are much smaller than the unsteady and viscous terms. This is the Blasius idea described in Section 3.6. We can therefore approxi-

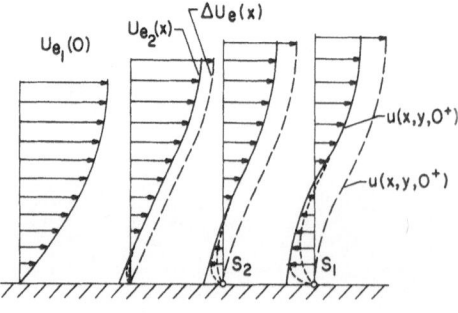

Fig. 7.15 Schematic representation of velocity profiles before an impulsive change ($t \leqslant 0$) and immediately after an impulsive change ($t = 0^+$).

mate the problem in its nondimensional form with

$$\partial u/\partial t = \partial^2 u/\partial y^2, \tag{7.5.8}$$

$$u(x, y, 0) = u_H(x, y) - \Delta U_e, \tag{7.5.9}$$

$$u(x, 0, t) = 0, \tag{7.5.10}$$

$$u(x, y, t) \rightarrow U_{e2}(x), \quad \text{as } y \rightarrow \infty, \tag{7.5.11}$$

where u_H is the solution of the steady-state problem corresponding to a Howarth outer-flow distribution given by Eq. (7.5.3). The solution to the problem posed by Eqs. (7.5.8) to (7.5.11) is

$$u(x, y, t) = u_H(x, y) - (A_2 - A_1)x \operatorname{erf}\left[\frac{y}{2\sqrt{t}}\right]. \tag{7.5.12}$$

The steady-state equations with U_{e1} given by Eq. (7.5.3) were first integrated numerically in order to store the quantities $u_H(s, N)$ and $v_H(x, y)$. Then $u_H(x, y)$ was replaced by Eq. (7.5.12) evaluated at a very small t, U_{e1} was replaced by U_{e2}, and the unsteady program was left to continue the numerical integration.

One of the most interesting features of the time-dependent flow field under consideration is that at $t = 0^+$ the point of zero skin friction "jumps" from S_1 to $x = 0$. As time tends to 0^+ from above, we recover the vortex sheet that is equivalent to a wall shear tending to $-\infty$. We shall now prove that the first point where the shear will increase to zero is indeed the point $x = 0$. The wall shear by virtue of Eq. (7.5.12) is given by

$$\left.\frac{\partial u}{\partial y}\right|_{y=0} = \left.\frac{\partial u_H}{\partial y}\right|_{y=0} - (A_2 - A_1)\frac{x}{2\sqrt{t}}. \tag{7.5.13}$$

We therefore observe that for a fixed station x, the wall shear can be made as small as we please for small enough time t. That is,

$$\left.\frac{\partial u}{\partial y}\right|_{y=0} \rightarrow -\infty, \quad \text{as } t \rightarrow 0^+. \tag{7.5.14}$$

It remains to compare two values of the wall shear for fixed time t_0 and two stations x_I and x_II, where $x_\mathrm{I} < x_\mathrm{II}$.

The difference of the quantities

$$\frac{\partial u(x_\mathrm{I}, 0, t_0)}{\partial y} - \frac{\partial u(x_\mathrm{II}, 0, t_0)}{\partial y} \tag{7.5.15}$$

can be written according to Eq. (7.5.13) as

$$\left.\frac{\partial u_H}{\partial y}\right|_{x_\mathrm{I}} - \left.\frac{\partial u_H}{\partial y}\right|_{x_\mathrm{II}} - \frac{A_2 - A_1}{2(t)^{1/2}}(x_\mathrm{I} - x_\mathrm{II}). \tag{7.5.16}$$

But the solution of the problem for u_H has indicated that the shear

$\partial u_H / \partial y|_{y=0}$ varies with a power r of x less than unity. This implies that the quantity (7.5.16) becomes positive as $x_I - x_{II}$ tends to zero. We therefore conclude that for $y = 0$ and $t \to 0^+$, the wall shear is everywhere negative and absolutely smaller at x_I than at x_{II}, if $x_I < x_{II}$. The zero of the wall shear therefore tends to $x = 0$ as $t \to 0$ from above. Any claim that this should be the behavior of separation should definitely disturb the conscience of any fluid dynamicist, because it would imply that the wake suddenly expands and embraces the whole body. We should mention here that this problem and the description of this section were originally conceived by W. R. Sears who often used this example in order to point out how misleading the criterion of zero skin friction can be in unsteady flows.

In contrast to the point of zero skin friction, separation stays behind and at first remains unaffected by the impulsive change in agreement with the analytical prediction of Buckmaster (1973) and recent experimental evidence (Koromilas and Telionis, 1980).

The temporal path of the point of zero skin friction and the point of separation were calculated by Telionis and Tsahalis (1974a) and Wang (1979). Results for the unsteady Howarth problem are shown in Fig. 7.16. There is an unacceptable discrepancy in the time scale between the results of Telionis *et al.* and Wang. This is indeed puzzling, since calculations by the same authors of other problems, for example, the impulsive start of a circular cylinder, are in excellent agreement, at least with respect to the location of the point of zero skin friction and its upstream displacement in time. Moreover, discrepancies are also obvious in the location of the initial and final positions of separation. The first group of discrepancies may be due to errors in the definition of the dimensionless time scale. It is obvious that a factor of 2 in time would collapse the data on one curve. The second group of discrepancies is not as disturbing and may be due to inaccuracies in the actual numerical calculations. However, at least qualitatively, the trend seems to be the same. Unsteady separation follows with some delay the point

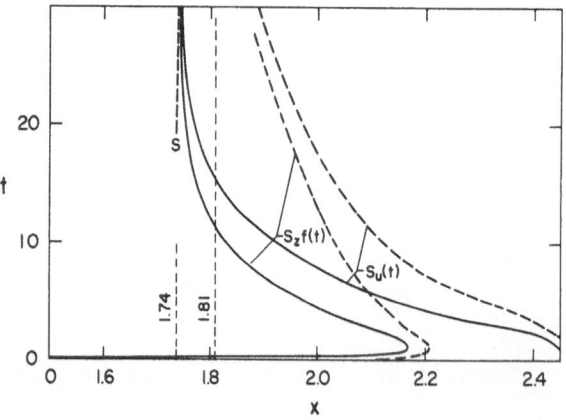

Fig. 7.16 The temporal excursions of the points of unsteady separation (S_u) and zero skin friction (S_{zf}) for an impulsive change of a Howarth flow. ———, Telionis and Tsahalis, 1974a; – – –, Wang, 1979.

of zero skin friction until both asymptotically arrive at the position of separation that corresponds to the outer flow given by (7.5.3). It should be emphasized here that Telionis *et al.* use the appearance of the Goldstein singularity to signal separation, whereas Wang traces the skin friction lines in an x–t plane and defines separation as an envelope of such lines in analogy to steady three-dimensional flow, that is, definition (7.1.4).

Other methods of direct integration have been employed to solve unsteady separating flows, but all are concerned with flows started from rest and therefore search for the emergence of separation rather than its displacement. Description of the computational techniques for most of these methods have been included in earlier chapters. In Section 7.7 we shall concentrate on the results and their physical significance.

7.6 Lagrangian Formulation

There is no doubt that a Lagrangian description of motion is intuitively much easier to comprehend. It is because of our inability to follow the fluid particles experimentally and at the same time because of much more tractable mathematical models that fluid mechanics is formulated in terms of Eulerian frames. It is in this frame that we define, for example, as steady fields, velocity fields that actually involve fluid particle accelerations. It is therefore natural in cases of physical ambiguity to resort to the Lagrangian description. The method offers in fact a completely different point of view for the problem of unsteady separation, especially since in Lagrangian formulations there is no distinction between steady and unsteady flows. This approach has been followed most recently and very successfully by Van Dommelen and Shen (1977) and was clearly described in a detailed review article by Shen (1979). The method deserves special attention and this section is entirely devoted to it. Inevitably, this will follow closely the development and description of the single source.

Let a particle be denoted by its coordinates ξ, η at $t = 0$. Particle paths then will be functions of the initial position of the particle and time t:

$$x = x(\xi, \eta, t), \qquad y = y(\xi, \eta, t), \tag{7.6.1}$$

and therefore by definition,

$$\xi = x(\xi, \eta, 0), \qquad \eta = y(\xi, \eta, 0). \tag{7.6.2}$$

Velocity and acceleration components then take very simple forms,

$$u = \frac{\partial x}{\partial t}, \qquad v = \frac{\partial y}{\partial t}, \tag{7.6.3}$$

$$a_x = \frac{\partial^2 x}{\partial t^2} = \frac{\partial u}{\partial t}, \qquad a_y = \frac{\partial^2 y}{\partial t^2} = \frac{\partial v}{\partial t}, \tag{7.6.4}$$

where of course partial time derivatives imply that the other variables ξ and η, that is, the particles under consideration remain fixed.

A formal transformation from the system x, y, and t to the system ξ, η, and τ with $\tau = t$ requires the inversion of a matrix

$$\begin{bmatrix} \xi_x & \xi_y & \xi_t \\ \eta_x & \eta_y & \eta_t \\ \tau_x & \tau_y & \tau_t \end{bmatrix} = \begin{bmatrix} x_\xi & x_\eta & x_\tau \\ y_\xi & y_\eta & y_\tau \\ t_\xi & t_\eta & t_\tau \end{bmatrix}^{-1}, \tag{7.6.5}$$

where, for brevity, differentiation is denoted by a subscript. Equation (7.6.5) with a little algebra yields

$$\begin{bmatrix} \xi_x & \xi_y & \xi_t \\ \eta_x & \eta_y & \eta_t \\ \tau_x & \tau_y & \tau_t \end{bmatrix} = \frac{1}{J} \begin{bmatrix} y_\eta & -x_\eta & vx_\eta - uy_\eta \\ -y_\xi & x_\xi & uy_\xi - vx_\xi \\ 0 & 0 & J \end{bmatrix}, \tag{7.6.6}$$

where the Jacobian J is given by

$$J = x_\xi y_\eta - y_\xi x_\eta. \tag{7.6.7}$$

To translate derivatives from one system to the other, we employ the chain rule and (7.6.6)

$$\frac{\partial}{\partial x} = \frac{1}{J} \left(y_\eta \frac{\partial}{\partial \xi} - y_\xi \frac{\partial}{\partial \eta} \right), \tag{7.6.8}$$

$$\frac{\partial}{\partial y} = \frac{1}{J} \left(-x_\eta \frac{\partial}{\partial \xi} + x_\xi \frac{\partial}{\partial \eta} \right). \tag{7.6.9}$$

Employing these expressions, we can write the continuity equation in the form

$$u_\xi y_\eta + u_\eta(-y_\xi) - v_\xi x_\eta + x_\xi v_\eta = 0, \tag{7.6.10}$$

which can be readily identified as the time derivative of the Jacobian J. The continuity equation is therefore equivalent to

$$J = x_\xi y_\eta - y_\xi x_\eta = 1. \tag{7.6.11}$$

To write now the x component of the momentum equation in its boundary-layer form, we need to apply operator (7.6.9) on itself. The convective part of the equation of course is nothing but the acceleration in the x direction. In terms of stretched variables, whereby the Reynolds number is absorbed in the vertical coordinate, the momentum equation becomes

$$\begin{aligned} u_\tau = {} & -p_x + x_\xi^2 u_{\eta\eta} - 2x_\xi x_\eta u_{\xi\eta} + x_\eta^2 u_{\xi\xi} - x_\xi u_\xi x_{\eta\eta} \\ & + (x_\xi u_\eta + x_\eta u_\xi)x_{\xi\eta} - x_\eta u_\eta x_{\xi\xi} \end{aligned} \tag{7.6.12}$$

The system of Eqs. (7.6.3) and (7.6.12) can be solved numerically subject to the appropriate initial conditions

$$x(\xi, \eta, t) = \xi, y(\xi, \eta, t) = \eta, \quad \text{at } t = 0, \tag{7.6.13}$$

$$u(\xi, \eta, t) = u_0(\xi, \eta), v(\xi, \eta, t) = v_0(\xi, \eta), \quad \text{at } t = 0, \tag{7.6.14}$$

where u_0, v_0 is the initial velocity distribution.

Among other examples, Shen describes a model problem

$$u_0 = f'(\eta)\sin\xi, \quad v_0 = -f(\eta)\cos\xi, \tag{7.6.15}$$

where $f(\eta)$ is the Hiemenz profile. With $f'(\eta)$ tending to 1 at infinity, we recognize the outer-flow distribution around a circular cylinder. Equations (7.6.15) therefore represent a model boundary layer, uniform in thickness and profile shape. Van Dommelen and Shen use a 33×73 rectangular grid for their numerical calculation. They then follow the subsequent displacement of the particles, which are situated at the nodes of the initial lattice. Results for a few selected points that make up a 5×7 grid are shown in Fig. 7.17. It is most intriguing that the distorting $\eta = $ constant curves develop in time in exactly the same way that a series of hydrogen bubbles would move if released by a vertical wire at $t = 0$.

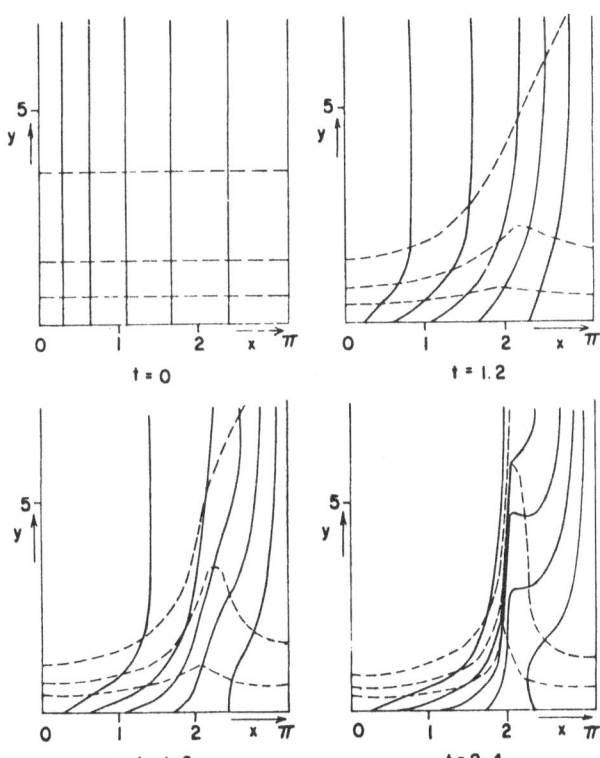

Fig. 7.17 The deformation in time of an initially rectangular mesh attached to fluid particles (van Dommelen and Shen, 1977).

The numerical results indicate that up to $t = 2.4$, both x and u remain bounded, but y_ξ, y_η, and u_x tend to blow up at approximately $x = 2$. It is significant that u_ξ and u_η remain finite and, therefore, it is due to the large growth of the terms y_ξ and y_η that the phenomenon appears singular in the Eulerian frame. The shape of the distorted lattice in Fig. 7.17 clearly indicates the singular character of the flow.

Characteristic of the phenomenon is the Eulerian coordinate y, which is given by

$$y = \int_0^s \frac{ds}{\left(x_\xi^2 + x_\eta^2\right)^{1/2}}, \tag{7.6.16}$$

where s is the arc length along $x = $ constant and $s = 0$ at $y = 0$. As the singular point is approached, both x_ξ and x_η tend to zero and the coordinate y tends to infinity. The vanishing of these two terms Shen identifies as the conditions for the formation of an envelope of the paths $x(\xi, \eta, \tau)$ for different particles ξ, η. This in turn implies that the position x is reached by different particles ξ simultaneously since $\partial x / \partial \xi = 0$. Therefore the distance between two consecutive particles tends to zero as the critical point is approached. Considering the fact that the flow is assumed incompressible, this is possible if the vertical size of a rectangular fluid element blows up, which is precisely what Eq. (7.6.16) yields. The particles therefore literally run up against a virtual barrier and further penetration becomes impossible, as Shen very descriptively notes. This singular behavior is identified as separation.

The results of this investigation are discussed in more detail in the next section together with the results of other methods.

7.7 The Emergence of Separation

The case whereby some external disturbance forces separation to displace itself upstream or downstream or even disappear from the flow field, as the outer flow reattaches is common in real life. Conceptually and analytically, the problem is straightforward since there is a good starting point. However, perhaps even more common in practice are problems that involve the appearance of separation in a field that is fully attached. This is possible, for example, if the flow about a blunt body is started impulsively from rest or, alternatively, if with an established flow the configuration of the body changes to generate regions of adverse pressure gradients. A classic case that belongs to the first category is the impulsive start of a circular cylinder that has been studied by almost all investigators working on unsteady boundary layers.

Blasius (1908) and Goldstein and Rosenhead (1936) recognized the fact that immediately after the impulsive start, the flow around the cylinder is potential and calculated the early stages of the boundary-layer development and the upstream displacement of the point of zero skin friction, which at the time was believed to accompany separation. Proudman and Johnson (1962) clearly demonstrated with their asymptotic analysis that reversing flow may have absolutely no effect on the outer flow and that a reversing layer is compatible with attached external flow at the rear stagnation point of an impulsively started cylinder.

A number of contributions based on exact solutions of the Navier–Stokes equations appeared in the late 1960s and their results are reviewed adequately by Collins and Dennis (1973). In this section we shall concentrate more on recent contributions, addressing specifically the problem of unsteady separation for large values of the Reynolds number. Most of the contributions mentioned here have been discussed adequately in Chapters 2 and 3, at least with respect to the methodology. Only a brief outline of the techniques employed is provided here for completeness.

Belcher *et al.* (1971) describe two methods of numerical calculations of the boundary-layer equation: the first is a straightforward implicit method; the second employs data from two previous stations of t and neighboring points on the x grid both upstream and downstream, much like, perhaps, in Eqs. (2.8.2) and (2.8.3). Collins and Dennis (1973) extend their method for solving the full Navier–Stokes equations to compute the impulsively started flow around a circular cylinder for different values of the Reynolds number. Their method is based on expansions of the Fourier type, which reduces to ordinary differential equations in terms of the radial distance and time. Telionis and Tsahalis (1974b) integrate the boundary-layer equations by an implicit method using a zig-zag scheme for partially reversed flows to study the impulsive start of a circular cylinder and an ellipse at an angle of attack. Bar-Lev and Yang (1975) solve the full vorticity equation by the method of matched asymptotic expansions while Wang (1979) and Cebeci (1979) employ again their numerical schemes to solve the boundary-layer equations in a three-coordinate space. Van Dommelen and Shen (1977) introduce a novel method for the calculation of unsteady viscous flows based on a Lagrangian formulation.

All results are in good agreement with respect to the excursions of the point of zero skin friction and the properties of the reversing layer as shown in Fig. 3.13. Some discrepancies exist in the estimation of the time when the point of flow reversal departs from the rear stagnation point, but soon after, the results appear to merge together. This is most clearly demonstrated by comparing the dimensionless time required for zero skin friction to emerge at the point of rear stagnation. A table comparing results of different methods has been constructed by Bar-Lev and Yang (Table 3.1).

The temporal excursions of the point of zero skin friction have been calculated for a large range of Reynolds numbers by Collins and Dennis

(1973) and Bar-Lev and Yang (1975). These data consistently indicate that for smaller Reynolds numbers, that is, less viscous fluids, it takes longer for the sequence of different phases of the flow to occur (see Fig. 3.14). This is in contrast to the well-known response of unsteady viscous flows. It is known, for example, in the Rayleigh problem, that is, the impulsively started flow over an infinite flat plate, that events occur faster for less viscous fluids. Time appears in the solution multiplying viscosity and therefore for a fluid A, with viscosity twice the viscosity of a fluid B, it will take half the time to achieve a certain value of deceleration at a point in space than for the fluid B. In the flow over the impulsively started cylinder, instead, it takes longer for more viscous flows to develop a zero skin friction at the rear stagnation point, and the subsequent development of the flow follows the same trend.

Having established that the point of zero skin friction is not connected with unsteady separation, some questions of great significance are now posed to the investigator. Where and when will separation emerge in boundary-layer calculations, if it appears at all in finite times?

The exact solutions of the Navier–Stokes equations indicate that the recirculating bubbles grow smoothly without any extraordinary behavior. The first indication of major changes in the character of the flow is the appearance of secondary vortices at a dimensionless time of $t > 1.0$ depending on the method and $\theta \cong 135°$, where θ is measured from the forward stagnation point. The vorticity plots of Collins and Dennis (1973) display a kink in the neighborhood of $\theta \sim 130$–$150°$, as shown in Fig. 7.18, a clear indication of the anticipation of secondary vortices. Such vortices were observed for Reynolds numbers as low as 550. Collins and Dennis reported that their calculations break down at a finite time t_m, which is progressively smaller for larger Reynolds numbers. The kink in the vorticity plot also appears at smaller times for larger Reynolds numbers. For their largest Reynolds number, calculations are terminated at $t_m = 1.25$. Boundary-layer

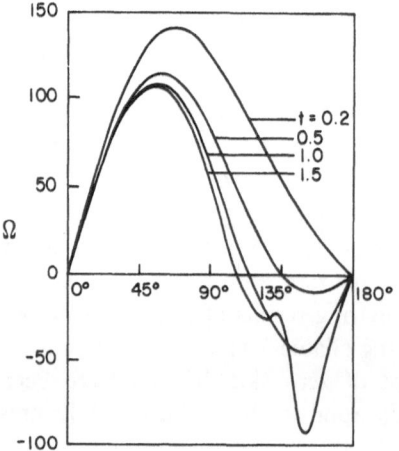

Fig. 7.18 Vorticity distribution over an impulsively started cylinder for $Re = 500$. (Collins and Dennis, 1973).

calculations reported by Belcher *et al.* (1971) were unsuccessful beyond $t = 0.45$; however, more refined differencing schemes, similar to zig-zag schemes described in Chapter 2, permitted the computation up to $t = 2$, although inaccuracies appeared near the point of zero skin friction already at $t \cong 1$.

The mesh configuration used by Cooke and Robins (Belcher *et al.*, 1972) was rather crude, containing 18 stations in the downstream direction. Boundary-layer calculations were repeated by Telionis and Tsahalis (1974a) who used a much more refined mesh configuration with 400 stations and 100–150 points at each station. Employing a zig-zag scheme, Telionis and Tsahalis found that the unexpected violent behavior of the flow is initiated approximately at $t = 0.65$ and for $\theta = 135°$. However, this behavior does not seem to become organized before approximately $t \simeq 0.95$. At this moment the abrupt growth of the v component of the velocity seems to follow the pattern of a separation singularity. On a specific time plane, the behavior of all singular quantities is monotonic, and the number of iterations required for convergence also grows monotonically until a station is reached at which no convergence is possible. A clarification is necessary here with respect to Fig. 4 of Telionis and Tsahalis (1974a). In this figure the authors have actually plotted their modified v-component of the velocity, which they define by their Eq. (12). This is a poor choice because this quantity by definition blows up at the rear stagnation point and behaves erratically in its neighborhood. However, away from the rear stagnation point, its properties are qualitatively similar to those of the true v-component. Using inner and outer expansions, Bar-Lev and Yang (1975) solved the full Navier–Stokes equations and obtained results in agreement with those of Collins and Dennis. However, their calculations for $Re = \infty$ indicate no peculiar behavior in the range $0.6 < t < 0.8$.

Cebeci (1979) undertook to repeat careful boundary-layer calculations. Using the standard box method with 41 stations and 100 points at each station, Cebeci discovered far from the wall unacceptable behavior of the velocity gradient reminiscent of the unmatchability condition of Shen (1979). This behavior was encountered first at $t = 0.65$, $\theta = 153°$, and propagated upstream with a rate comparable with the rate of propagation of the singularity found by Telionis and Tsahalis. However, Cebeci found that introduction of a zig-zag scheme completely eliminates any singularities at least up to a dimensionless time of $t = 2.8$, thus contradicting the findings of Telionis and Tsahalis.

Experience indicates that with a coarse mesh, numerical integration may go through a station of singularity without any evidence of instability. In fact, calculations can proceed sometimes beyond the point of separation into a region of well-behaved and nonreversing flow, which bears no connection to the correct solution. In the calculations reported by Belcher *et al.* (1971), the upper surface of the cylinder, from $\theta = 0°$ to $180°$, was divided into only 18 increments. Each increment is therefore $\Delta\theta = 10°$. The calculations of

Cebeci (1979) using 41 θ stations spaced by $\Delta\theta = 4.5°$ around the circular cylinder indicate no difficulties at all for much larger times. It is felt that such mesh configurations are coarse. The work of Telionis and Tsahalis (1974a) with 400 stations within the same domain of integration, that is with $\Delta\theta$ approximately equal to 0.45°, indicates that the singular behavior usually develops within a region of 2°–2.5°. Calculations with $\Delta\theta > 3°$ could therefore easily "jump over" the singularity. It appears that until today the finest mesh used for the numerical solution of this problem is the one employed by Telionis and Tsahalis. Moreover, it is emphasized that the results of a numerical calculation could never be considered as a proof that a specific solution is free of singularities since with a coarse mesh the solution may appear perfectly normal in the entire domain of integration.

The problem of whether a singularity will appear in a finite time in a domain with no initial singular behavior Cebeci (1978) has addressed earlier, investigating a Howarth-type of flow that is initiated as a Blasius flat plate flow. In this case as well, the numerical analysis was marched with a zig-zag scheme through a region of continuously growing reversed flow with no evidence of singularity. Most recently, Dwyer and Sherman (1979) imposed an unsteady Howarth-like edge condition on the boundary-layer equation

$$U_e = 1 - Ax + x^2.$$

This flow generates a small recirculating bubble that grows with time but is fully contained within the domain of integration in qualitative agreement with the flow studied by Cebeci (1978). Dwyer and Sherman found that the behavior of all properties is very smooth for a while, and the recirculating bubble grows slowly until, at a certain instant, a singularity appears in the skin friction, accompanied by an erratic oscillatory behavior of the velocity profile at this station. Dwyer and Sherman found that the "unphysical growth" is independent of the size of the reversed flow region. Moreover, they argue that Cebeci (1978) may have missed the singular behavior because he did not consider large enough values of dimensionless time.

Intrigued by the controversy over the specific case of the impulsively started cylinder, Van Dommelen and Shen (1977) and Wang (1979) decided to reconsider the problem, the first through a novel method and the second using a new definition of separation. Van Dommelen and Shen recast the boundary-layer equations in Lagrangian coordinates as described in the previous section. In this system of coordinates there is no distinction between steady and unsteady flow and therefore no difficulty should arise in interpreting the results. With ξ, η fixed coordinates attached on the body and $x(\xi,\eta,t)$, $y(\xi,\eta,t)$ the coordinates of the particle that was found at ξ, η at time $t = 0$, Van Dommelen and Shen studied the distortion in time of the x, y grid and searched for a singularity that appears as a blowup of quantities like $\partial y/\partial\xi$, $\partial y/\partial\eta$, and $\partial u/\partial x$, while quantities like x and u remain finite. They noted further that the singular behavior is accompanied by vanishing of $(\partial x/\partial \xi)^2 + (\partial x/\partial \eta)^2$, which implies the vanishing of both terms in this expression. This, in turn, implies the formation of an envelope of the $x = $ constant curves in the x–y plane, which appears as a physical barrier or as a vertical wall in Fig. 7.17. No evidence of singular behavior was found until $t_m = 2.4$. The formation of the singularity became obvious in the next time step and at the point $\theta = 110°$. The larger values of t_m may be attributed to the artificial initial boundary-layer distribution.

Wang employed his classical numerical scheme for the calculation of unsteady boundary layers, but introduced a new definition of separation. In analogy to the case of steady three-dimensional flow, whereby separation is defined as an envelope of skin friction lines, Wang searched for similar geometrical configurations in an x–t plane. The analogy to "skin friction" lines or, equivalently, "limiting streamlines" are the curves defined by Eq. (7.1.4). Wang (1979) considered again the problem of the impulsively started cylinder and calculated the limiting streamlines shown in Fig. 7.19. The formation of an envelope is clearly shown in this figure although it may not be possible to define the time of its initiation. On the same figure we plot the trace of the traveling singularity reported by Telionis and Tsahalis (1974a). The first few points, about which the authors expressed some hesitation, cut

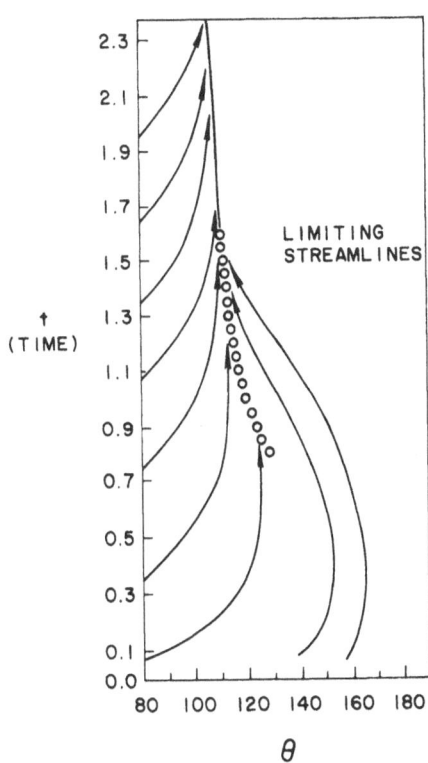

Fig. 7.19 Skin friction lines for two-dimensional flow over an impulsively started circular cylinder. ——, Wang (1979); ○, calculated singularity, Telionis and Tsahalis (1974a).

across some of Wang's limiting streamlines. However, after $t = 1.0$, it appears that the temporal path of the separation singularity is aligned with the neighboring curves and seems to be directing itself toward the initiation of the separation envelope.

The idea of envelope formation in an $x-t$ plane is not actually new. Concepts of unsteady separation and the ideas of a moving separation singularity have been considered by investigators of unsteady turbulent flows as well. Cousteix et al. (1977) presented a very similar pattern of limiting streamlines, indicating the formation of an envelope, almost two years before publication of Wang's work. In Cousteix et al. (1977), the results are compared with calculated trajectories of the separation singularity. Convincing qualitative agreement is evident but it is not possible to claim that the two definitions give identical results.

Recapitulating, we may reduce the main questions to (1) is it possible at all for a singularity to develop in the solution of the boundary-layer equation, if the domain of integration is free of such singularities at earlier times? Furthermore, (2) is such a singularity related to separation and, if so, could it be tracked and recorded to provide information on unsteady separation? Finally, (3) if the answer to the last question is negative, is there any other criterion that would signal unsteady separation? In other words, does the boundary-layer equation contain information about the extent of its validity? Most of the authors believe that the answer to the last question is YES. If the boundary-layer equation can be used to predict separation in steady flow, it should contain enough information to predict separation in unsteady flow as well.

The most controversial of the issues is centered around the first question and at this time the evidence points also to the answer YES; it is indeed possible that a singularity may develop in the solution of the unsteady boundary layer at a point in space that was free of singularities at earlier times. Such a singularity can be convected to the point of interest from downstream, as pointed out by Telionis et al. (1973) and Wang (1979). This has not been disputed up to now. But then going one step further, is it possible for a singularity to emerge at a certain instant at a point in space if the domain of integration was free of singularities at earlier times and at all neighboring points? Evidence that this is again true is provided by a variety of methods: the numerical calculations of Telionis et al. (1973) and Dwyer and Sherman (1979), as well as a large number of turbulent-boundary-layer calculations (McCroskey, 1977; Telionis, 1975); the analytical investigation of Shen and Nenni (1975); the numerical investigations in a Lagrangian system of coordinates of Van Dommelen and Shen (1977). Wang (1979) reports no evidence of singular behavior, however, the formation of an envelope in the $x-t$ plane certainly implies infinite variation of a property with distance at a fixed time.

In connection with the second question, the singularities discovered by the above investigators cannot be proved to be separation singularities for sure. However, they emerge at the point where separation is expected to appear and after some time has elapsed, they displace themselves to the point of steady-state separation. In another case, not directly related to external flow separation, Bodonyi and Stewartson (1977) present a solution to the boundary-layer equation, which is initially well behaved but breaks down after a finite interval of time.

7.8 Wakes

The determination of the point of separation in steady or unsteady flows is an important phase in the solution of problems in aerodynamics, because it controls the size and perhaps the shape of the wake and therefore the magnitude and direction of forces exerted on the solid bodies. However, very little has been done to develop methods of calculation of the wake properties. The problem of course is inherently unsteady and complexity increases if the almost random behavior is coupled with a discrete periodic or transient external disturbance.

Very little has been attempted in analyzing wakes, let alone unsteady wakes. Early efforts were based on the "free-streamline idea" whereby a potential flow solution is matched to a wake region via the condition that the pressure in the wake is constant and equal to the pressure at separation. Purely inviscid but more advanced methods like the above have been later developed that actually capture the natural periodicity of the wake. One of the most promising methods based on the release of discrete vortices which are then allowed to be convected downstream is reviewed adequately by Sarpkaya (1975) and McCroskey (1976). Most of the contributions in this area assume an arbitrary rule for the release of discrete vortices. In more refined models, some type of a heuristic Kutta-Joukowski-type of condition is imposed. For example, Clements and Maull (1975) insert vortices at a small distance downstream of a separating corner, controlling their strength by the condition that the flow must leave the body in a direction parallel to the upstream body surface. Sarpkaya (1975) controlled both the strength of the vortices as well as their position of release with a similar type of a condition.

The next phase of refinements involves the boundary layer equation. The problem attacked by Deffenbaugh and Marshall (1976) and Sarpkaya and Shoaff (1979) is more in line with the discussion of the present chapter. It is the problem of an impulsively started circular cylinder with emphasis on large dimensionless times and therefore phenomena like the development and shedding of large scale vortices. The unsteady boundary layers are calculated with the sole purpose of estimating the point of separation. The value of the outer flow at the station of separation, U_s, is then used to estimate the time rate of change of circulation (Fage and Johansen, 1928)

$$\frac{\partial \Gamma}{\partial t} = \frac{1}{2} U_{es}^2 \tag{7.8.1}$$

where U_{es} is the outer flow velocity at separation. The strength of nascent vortices assumed to be released at intervals Δt is given by

$$\Gamma = \frac{1}{2} U_{es}^2 \Delta t \tag{7.8.2}$$

In both contributions integral forms of the boundary-layer equations are used. Deffenbaugh and Marshall calculate the transient initial phase of the boundary-layer development but soon after assume that the flow is quasi-steady and separation is stationary. However, they find that their outer flow has a strong influence on the determination of the separation angle and that the accuracy of their results can be improved by incorporating a more accurate method for the calculation of separation. Sarpkaya and Shoaff retain the unsteady character of the boundary-layer calculations and discover that the position of zero-skin friction oscillates in space and reaches its minimum angle from the leading edge at the instant that a vortex sheet is cut off from the body and about to be shed in the wake.

Truly viscous effects in both methods are entered only via the unsteady boundary layer, since the convection and interaction of the vortices is modeled in an inviscid manner. The first group of investigators in fact assumes that the convective derivative of vorticity is zero, whereas the second introduces some artificial damping by assuming that every vortex in the wake loses its strength in an amount proportional to its current strength. Various methods have been employed for the analysis of discrete vortices, as, for example, vortex clouds, or identifiable sets of discrete vortices affected only by coalescence, wall proximity and cancellation of oppositely signed vortices. Most recently Sarpkaya and Shoaff (1979) employ a method of rediscretization whereby vorticity is redistributed along the vortex sheets at each time step. The shapes of such vortex sheets at different instances after the impulsive start of the motion are shown in Fig. 7.20.

An important element of the wake structure is the natural periodicity. Once such a periodic pattern is set up it can be apparently sustained by the inviscid models described above. Sarpkaya and Shoaff marched to a dimensionless time $U_\infty t / D$ equal to 200 which includes approximately 20 cycles of vortex shedding, predicting with very reasonable accuracy the Strouhal phenomenon. However, the initial triggering of the phenomenon requires some external interference. Apparently, in real life, some instability mechanism, perhaps in the developing free shear layers that emanate from separation is responsible for triggering the alternate shedding and such a phenomenon cannot be captured by an inviscid model.

Discrete-vortex shedding models have been used with moderate success to

Fig. 7.20 The onset of vortex shedding (a through e) triggered by artificial disturbances and one period of the established shedding pattern (f through j). The figures in the cylinders denote dimensionless time (Sarpkaya and Schoaff, 1979).

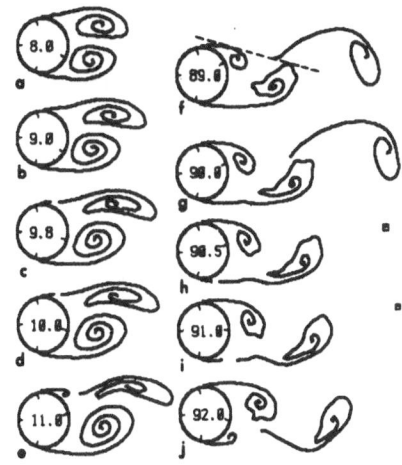

solve problems of unsteady wakes over airfoils and provide data for unsteady stall. With milder adverse pressure gradients, quasi-steady definitions of unsteady separation and quasi-steady conditions for the shedding of vorticity become insufficient. Sears (1956, 1975) points out that Eq. (7.8.1) is insufficient because the vorticity flux at the point of separation in unsteady flow is given by the integral

$$\int_0^\delta \Omega(u - U_s)\, dy \cong \int_0^\delta \frac{\partial u}{\partial y} (u - U_s)\, dy \tag{7.8.3}$$

where δ denotes the local boundary-layer thickness and U_s is the speed of separation. Upon integration we arrive at the generalized Howarth criterion

$$\left[\tfrac{1}{2} U_{es}^2 - U_s U_{es} \right]_l^u = - d\Gamma/dt \tag{7.8.4}$$

where $[\ ]_l^u$ denotes the difference between the values at the upper and lower separation point.

McCroskey (1977) describes in his review methods of solving unsteady stall problems based on purely inviscid models, or models involving weak or strong boundary layer and potential flow interaction. In some cases the full Navier–Stokes equations are employed in the immediate neighborhood of the leading-edge bubble, a very critical region for the development of the phenomenon. The full Navier–Stokes equations have also been employed for the entire flow field as described and referenced at the end of Chapters 2, 3, and 5. Great efforts have been made to increase the Reynolds numbers in such methods but even today this appears to be a serious limitation.

Here, to preserve continuity, we describe again a continuation of the work of Mehta and Lavan (1975). This is due to Mehta (1977) who undertook to study the flow about an oscillating airfoil in an effort to shed more light on the difficult problem of unsteady separation and stall.

The problem is again formulated in terms of a Joukowski transformation as described in Chapter 3. However, now the Navier–Stokes equations are expressed in terms of a frame which is attached to the airfoil. Let ω be the angular velocity of the rotating frame. Then the velocity V with respect to the rotating coordinates and the velocity V_I with respect to the inertial frame are related by

$$V_I = V + \omega \times r \tag{7.8.5}$$

where r is the position vector of the point under consideration. The Navier–Stokes equations in the rotating frame then take the form

$$\frac{\partial V}{\partial t} + \frac{1}{2} \nabla(V^2) - V \times (\nabla \times V) + \frac{d\omega}{dt} \times r + 2\omega \times V + \omega \times (\omega \times r)$$

$$= - \nabla p + \frac{L}{Re} \nabla^2 V \tag{7.8.6}$$

where L is the chord length and Re is the Reynolds number.

In the original publication of Mehta (1977) the symbols Ω and ω are used to denote the angular velocity and vorticity respectively. This is most unfortunate because in this monograph exactly the opposite notation is used. For two-dimensional flow and in the ζ-plane, Eq. (7.8.6) expressed in the terms of the vorticity, becomes

$$H^2 r^2 \frac{Re}{l} \frac{\partial \Omega}{\partial t} = r^2 \frac{\partial^2 \Omega}{\partial r^2} + r \frac{\partial \Omega}{\partial r} + \frac{\partial^2 \Omega}{\partial \theta^2} - r \frac{Re}{l} J\left(\frac{\Omega, \Psi}{r, \theta} \right) - 2 \frac{d\omega}{dt} H^2 r^2 \frac{Re}{l} \tag{7.8.7}$$

with the streamfunction defined by

$$r^2 \frac{\partial^2 \Psi}{\partial r^2} + r \frac{\partial \Psi}{\partial r} + \frac{\partial^2 \Psi}{\partial \theta^2} = - H^2 r^2 \Omega \tag{7.8.8}$$

where

$$H^2 = \left(\frac{\partial x}{\partial r} \frac{\partial y}{\partial \theta} - \frac{\partial x}{\partial \theta} \frac{\partial y}{\partial r} \right) / r \tag{7.8.9}$$

A stretched variable ρ is introduced and the equations are recast in terms of a disturbance streamfunction

$$\Psi = \psi + y + \frac{\omega}{2} (x^2 + y^2) \tag{7.8.10}$$

in a manner similar to the discussion of Chapter 3. Mehta and Lavan (1975) solved the elliptic set of equations using a successive overrelaxation procedure with an implicit finite differencing of the vorticity equation. Mehta (1977) solves the vorticity equation with an implicit factored algorithm and the stream function equation with a direct-solution approach.

To this end the vorticity equation is written in finite difference form in the stretched space ρ, θ (see, for example, Eq. 2.9.77)

$$\left(1 - \frac{2\Delta t}{AT_1} \partial^n_{\theta\theta}\right)\left(1 - \frac{2\Delta t}{AT_1} \partial^n_{\rho\rho}\right)\Omega_1 = -\frac{1}{T_1}\left(T_2\Omega_1^{n-1} + T_3\Omega_1^{n-2}\right)$$

$$+ \frac{4(\Delta t)^2}{T_1^2}\left(\frac{\partial^n_{\theta\theta}}{A}\right)\left(\frac{\partial^n_{\rho\rho}\Omega_1^n}{A}\right) \tag{7.8.11}$$

where the symbols $\partial_{\theta\theta}$ and $\partial_{\rho\rho}$ represent complex nonlinear operators involving second derivatives in terms of θ and ρ respectively, the constants T_1, T_2 and T_3 take integer values (see Eqs. 3.7.83 and 84) for two- and three-point backward time differences, the superscript n is the time index and A, Ω_I are

$$A = H^2 r^2 Re/l, \quad \Omega_1 = \Omega + 2\omega \tag{7.8.12}$$

Instead of linearizing the right-hand side by expressing the nonlinear operator in terms of quantities at the t_{n-1} plane, Mehta retains the derivatives at the n-time level. The operator $\partial_{\theta\theta}$ is then replaced by fourth order-accurate rational fractions or Padé forms. Very briefly, the computation scheme proceeds as follows: vorticity Ω_1^n at the time plane n is calculated by solving Eq. (7.8.11). Then $\partial\psi/\partial\rho$ is calculated by solving an equation resulting from the boundary condition at the downstream boundary, where the rotational field exits from the domain of integration. This is generalization of Eq. (2.9.34) which for a rotating frame of reference becomes

$$\frac{\partial\psi}{\partial\rho} = \frac{dr}{d\rho}\left[-(\sin\alpha - x\omega)\frac{\partial x}{\partial r} + (\cos\alpha + y\omega)\frac{\partial y}{\partial r} - \frac{\partial y}{\partial r}\right.$$

$$\left. - \frac{\omega}{2}\frac{\partial(x^2 + y^2)}{\partial r}\right] \tag{7.8.13}$$

with α the angle of attack. Next the disturbance streamfunction is determined by solving the Fourier transform form of the equation derived by substituting in Eq. (7.8.8) the expression (7.8.10)

$$\left(r\frac{d\rho}{dr}\right)^2\frac{\partial^2\hat{\psi}}{\partial\rho^2} + \left(\frac{d\rho}{dr}r + \frac{d^2\rho}{dr^2}r^2\right)\frac{\partial\hat{\psi}}{\partial\rho} - \hat{k}^2\hat{\psi} = \hat{\Omega} \tag{7.8.14}$$

where $\hat{\psi}$ and $\hat{\Omega}$ are Fourier transforms of ψ and $-H^2 r^2\Omega_1$ respectively and $\hat{k}^2$ is related to the wave number corresponding to θ variations.

$$\hat{k}^2 = \frac{12}{(\Delta\theta)^2}\left[\frac{1 - \cos(k'\Delta\theta)}{5 + \cos(k'\Delta\theta)}\right] \tag{7.8.15}$$

Finally the surface vorticity is calculated from the condition that incorporates the no-penetration and no-slip conditions which combined with Eq.

(7.8.8) gives

$$\Omega = -\frac{1}{H^2}\left(\frac{d\rho}{dr}\right)^2 \frac{\partial^2 \Psi}{\partial \rho^2} \tag{7.8.16}$$

This sequence is repeated as shown in the flow-chart of Fig. 7.21 until the surface vorticity changes less than some prescribed value. To speed up the convergence the surface vorticity iterates are estimated by

$$\Omega_{m,JL}^{n,s} = \Omega_{m,JL}^{n,s-1} + \beta\left(\Omega_{m,JL}^n - \Omega_{m,JL}^{n,s-1}\right) \tag{7.8.17}$$

where $\Omega_{m,JL}^n$ is determined from Eq. (7.8.16) and s is the iteration counter. Mehta proposes for β values varying with θ as follows

$\beta = \beta_1$ for $k = 2 - 4$
$\beta = \beta_1 k/4$ for $k = 5 - 34$
$\beta = \beta_1 \beta_2(k-1)/4$ for $k = 35 - 63$
$\beta = \beta_1 \beta_2$ for $k = 64 - 66$

Results of Navier–Stokes solutions involving unsteady separation pertain to rather low Reynolds numbers. For $Re \leqslant 10^4$ no evidence is available of unsteady separation with the characteristics described earlier in this chapter. For an airfoil started impulsively from rest, for example, a sequence of streamlines and vorticity lines (Fig. 7.22) indicate that a very thin bubble appears and grows near the leading edge of the airfoil. The bubble originates at $x_c = 0.205$ and approximately at $t = 1.5$ where the wall shear vanishes for the first time. It grows quickly at first in the chordwise direction but its thickness grows smoothly without any evidence of sharp changes. The criteria for unsteady separation for large Reynolds numbers discussed in the earlier sections are clearly not applicable here.

A bubble is defined as a body of fluid contained in a closed streamline. A bubble is said to open or burst (Mehta and Lavan, 1975) if its bounding contour opens retaining inside closed streamline loops. The bursting of the first bubble in Fig. 7.22 occurs at the instant its downstream boundary reaches the trailing edge. Once again no peculiar behavior is observed. However, it has been known that at higher Reynolds numbers, bubbles may burst unexpectedly and before their contour encounters any obstacle or other geometrical characteristic of the solid body.

Similar results obtained for a higher Reynolds number, $Re = 10^4$ and an airfoil performing oscillations in pitch at a reduced frequency of 0.5 are shown in Fig. 7.23. In this figure the first two rows display the streamlines with respect to a coordinate system attached to the airfoil and an inertial system respectively. The third row shows the velocity vectors and the fourth row displays the equivorticity lines. The results compare beautifully with flow visualization, not shown in this figure. Note that with respect to the inertial frame, the airfoil is not a streamline and hence there is no bifurcation or unification of streamlines on the surface. Streamlines may naturally emanate from the surface. To detect flow detachment and reattachment one

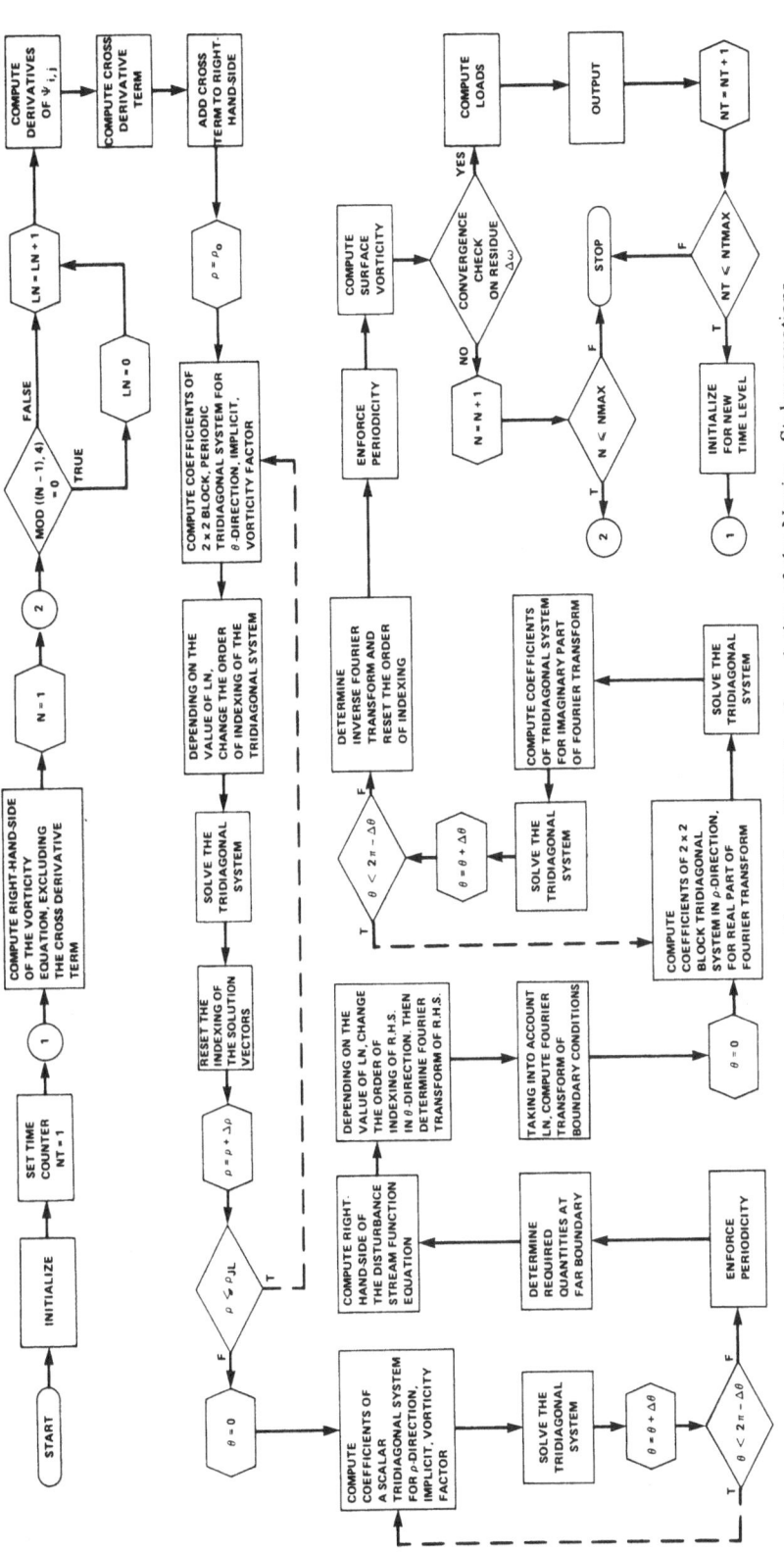

Fig. 7.21 Flowchart of the program prepared by Mehta (1977) for the solution of the Navier–Stokes equations.

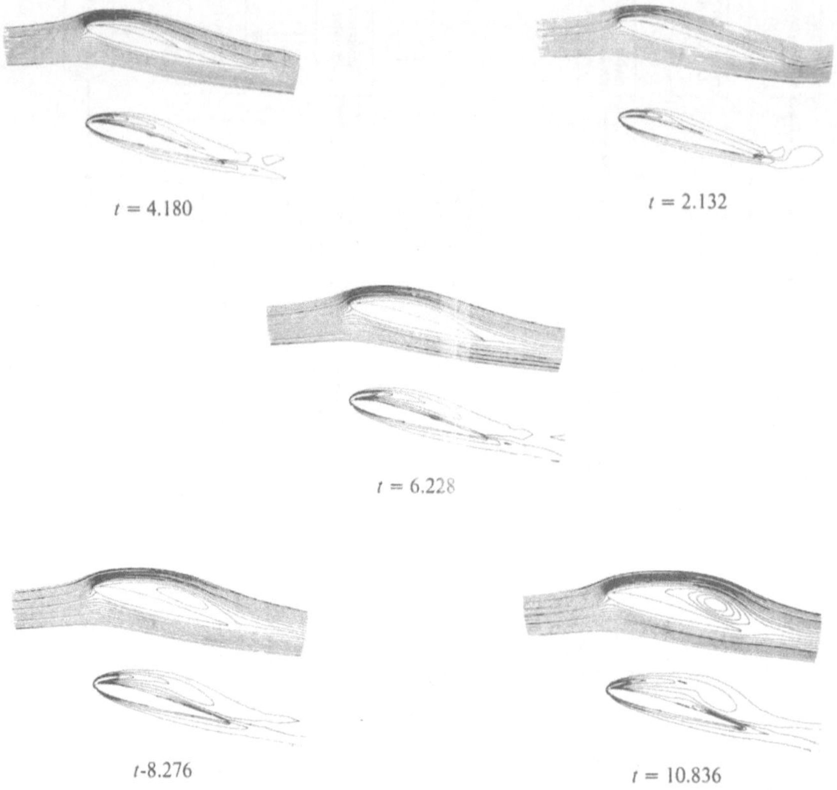

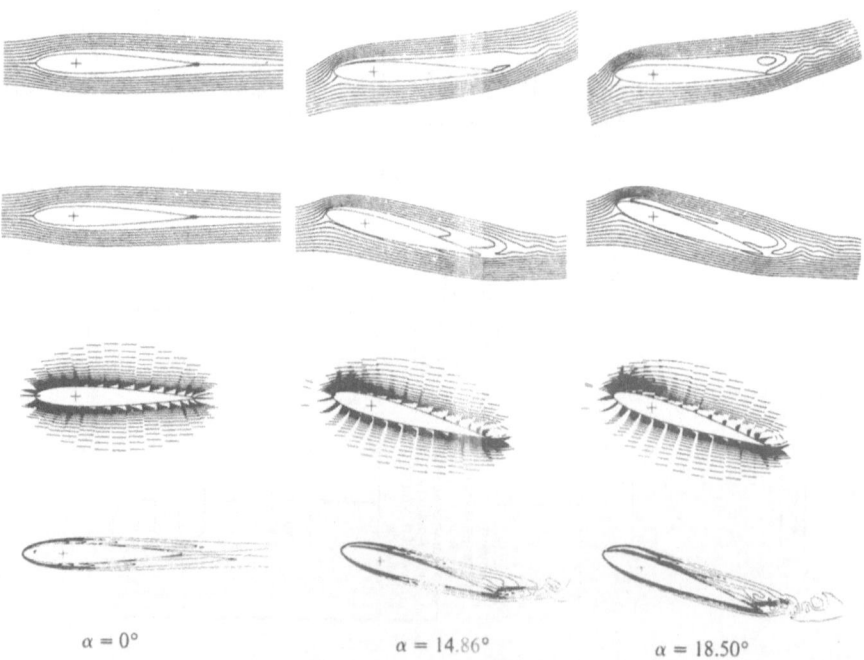

$t = 4.180$

$t = 2.132$

$t = 6.228$

t-8.276

$t = 10.836$

Fig. 7.22 Streamlines and equivorticity lines about an airfoil started impulsively from rest (Mehta and Lavan, 1975).

$\alpha = 0°$

$\alpha = 14.86°$

$\alpha = 18.50°$

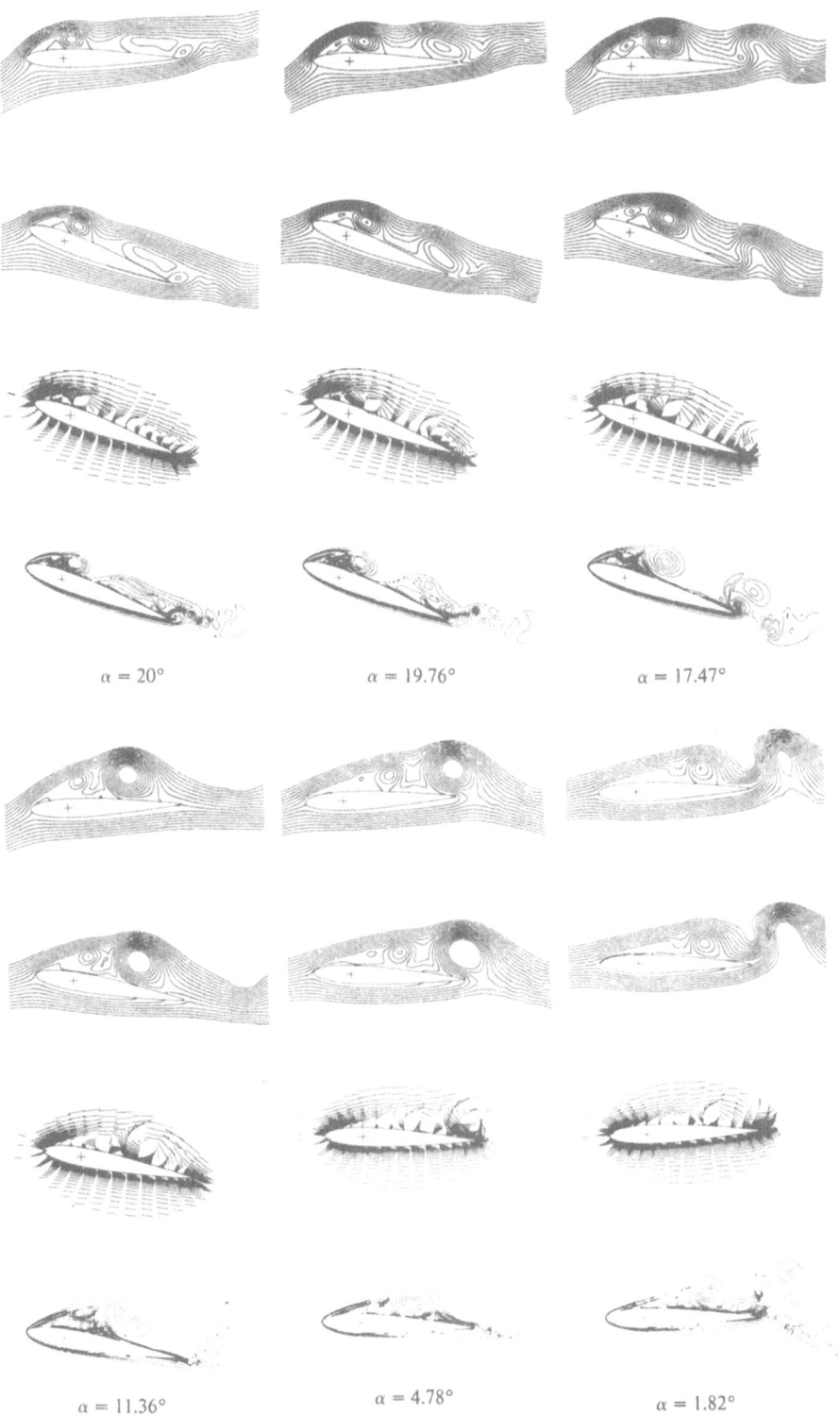

$\alpha = 20°$ $\alpha = 19.76°$ $\alpha = 17.47°$

$\alpha = 11.36°$ $\alpha = 4.78°$ $\alpha = 1.82°$

Fig. 7.23 Streamlines with respect to a frame attached on the airfoil; streamlines with respect to an inertial frame; velocity vectors and vorticity lines for an airfoil at $Re = 5000$ oscillating in pitch at a reduced frequency equal to 0.5 (Mehta, 1977).

has to inspect the streamline patterns with respect to a frame of reference attached to the airfoil. With established periodic flow the bubbles grow and they are eventually shed in the wake. Forces and moments vary periodically and the information is very useful for the understanding of the phenomenon of unsteady stall.

7.9 A Computer Program

Computer codes based on boundary-layer models have been employed for the calculation of unsteady separation. However, two such programs have been described and listed in earlier chapters and boundary-layer solutions with specific attention to separation have been plotted and discussed in the previous section. It was decided here instead to include a program due to Mehta (1977) which solves the full Navier–Stokes equations in the cases of impulsively started or oscillating airfoils that pass through unsteady stall.

Perhaps this program does not reflect the emphasis given to this chapter, namely, interpretation of boundary-layer solutions. This emphasis however has not been intentional. The description of unsteady separation as revealed by the boundary-layer equations has been actively pursued in the past decade and a lot has been learned in this direction. However, it is perhaps time to proceed further. It is not suggested here that the only method to follow is the full Navier–Stokes equations. Free shear layers emanating from the point of separation and discrete but unsteady large scale vortices, together with appropriate turbulence models can perhaps be the basis for the development of efficient engineering methods of calculation. The choice of the program described below is therefore one more attempt on the part of this author to encourage broadening of our point of view.

This computer program (see flow chart in Fig. 7.21) solves the full Navier–Stokes equations about a Joukowski airfoil or a cylinder. Impulsive starts followed by oscillations can be calculated. The operator may vary the geometry of the airfoil, as for example, thickness ratio, camber, and trailing edge curvature. Parameters like the point about which oscillations are performed, the amplitude and the frequency of oscillation can also be prescribed. Sufficient information about the program is included in the listing itself. Specific information about the input data and output information is also provided here in a slightly expanded form, in order to conform with the style adopted in this monograph.

Input

The data are read in the main program and the subroutines SETUP and START. In the main program one finds 3 READ statements and the corresponding data needed are as follows:

1st READ Statement
The size of the mesh in the $\rho - \theta$ space is defined. The number of grid points is read in a 2I5 format.

IL   Number of grid-points in the θ-direction
JL   Number of grid-points in the r-direction

2nd READ Statement
Read in I5 format:

NP   Number of problems, namely the number of sets of data cards that will define the number of different cases to be computed.

3rd READ Statement
A number of flags are read controlling for example the direction of calculations depending on the particular geometry, the stretching, or the way the output will be printed. If a flag is given a value other than the value specified below, the command is not executed. The first 14 variables are integers to be read in 14I5 format.

NMAX   Number of iterations per time step. The author recommends NMAX < 40.
NTMAX   Total number of time steps.
KL2   In case of reading data calculated earlier and stored on disk or tape, set KL2 = 1.
KL3   Flag for writing PY, VPT, V, and UP. If KL3 $\neq$ 1, the quantities are not written on a disk or tape.
KL4   Flag for Joukowski's transformation
        = 1, for circle or ellipse depending on the thickness ratio
        = 2, for airfoil and AS = 0
        = 3, for airfoil and AS > 0.
KL5   Flag for relaxation parameter β_1(BVRT) of surface vorticity (see previous section)
        = 1, BVRT varies with KCI
KL6   Transformation for stretching r
        = 1, no transformation for r
        = 2, hyperbolic stretching
        = 3, exponential stretching
KL9   Frequency of printing. Information is printed from subroutine OUTPUT every KL9 time steps.
KL11   Reading data from disk or tape, in case one needs to halve the time increment as compared to the previous calculation, set KL11 = 1. Then MT and NTAP must also be read.
KL13   The initial values of flow field and geometrical variables are printed if KL13 = 1.
KL14   Velocity information is not plotted if KL14 = 1.
KL15   Call subroutine CDCL to calculate and print forces exerted on the body if KL15 = 1.

KL16   Do not call subroutine OUTPUT which calculates velocity distributions and prints flow field and geometrical variables if KL16 = 1.

JS   Controls printing in r-direction in subroutine OUTPUT; JS = 2, prints at even values of J

The following 4 real variables are read on the next card in 4F10.5 format.

RI   Distance to the outer boundary. Author recommends RI = 0.02 – 0.04.

RNN   Reynolds number

BPSI   Relaxation parameter, β_2, mentioned in previous section

BVRT   Relaxation parameter, β_1, for surface vorticity. For a modified NACA 0012 the author recommends BVRT = 0.01–0.03 BPSI = 1.1–1.5

Four more variables follow on the same card in F10.2 and 3F10.2.

DVRTM   Maximum residue in surface vorticity. Author recommends for the circular cylinder and airfoil DVRTM $\sim 10^{-4}$ and $\sim 10^{-2}$ respectively.

TIME   Dimensionless time

DLTI   Time increment, Δt

RF   Reduced frequency

Finally on the last card for this READ statement, read in 8 more variables in F10.5, 2I5, 2F10.5 and 3I5.

AMP   Amplitude of oscillation in degrees

KL10   Check on the vorticity residue.
                = 1, skips one extra check on the size of the residue

KL21   Initializes the vorticity on a flat plate. VRT (I, JL) = 0.0 for I = 2, IM, IL if KL21 = 1 and sets RF = 0.0

DF   Dissipation factor, β_3, defined in previous section, DF should be < CL.

TIMOSC   Time at which a body started at a fixed angle of attack starts oscillating.

JBLE   Set equal to 2

KL22   Extra check of Courant number and mesh Reynolds number for accuracy; if KL22 $\neq$ 1, it calls RNCN every KL9 time steps.

JRNCN   Skip on J print-out in RNCN subroutine.

The following READ statements are found in the subroutines SETUP and START. In the subroutine SETUP:

4th READ Statement

Read the following 7 variables in 7F10.5 format.

TR   Thickness ratio for elliptic cylinder

OR   Determines thickness of Joukowski airfoil

OI	Determines camber of Joukowski airfoil
AS	Radius of curvature at the trailing edge of a Joukowski airfoil. Suggested value AS ≥ 0.04.
AT	Angle of attack in degrees.
PC	Percentage of chord from leading edge at which the origin of the physical coordinates, i.e., the center of oscillation is located.
RMAX	Distance at which the outer boundary is kept for KL6 = 3. Author suggests 8 to 10 chord lengths or cylinder diameters.
IRPI	Controls the relaxation procedure = 4 if KL5 = 1.

In the subroutine START the program is instructed either to continue a calculation or to start from the beginning a new calculation. In the first case, depending on whether the previously calculated data are stored on disk or tape, the program is instructed to search for the location and number of the source in the 5th READ statement. The integers MT, MT1 and NTAP are adjusted by the operator to allow the program to locate the previously calculated data. MT and MT1 are the file numbers on the disk pack where the solutions for two previous times are located. NTAP is the unit number for reading from disk-pack.

Depending on the value of the flag KL11 one or both of the following READ statements will be executed. This corresponds to data at one or two earlier times.

PY(I, J)	Ψ, the total streamfunction
VRTN2(I, J)	Ω_1, vorticity at the $N - 2$ time station
TIMED2	nondimensional time with respect to the radius of circular cylinder in transformed plane at the $N - 2$ time station
TIMEL	nondimensional time with respect to chord at the $N - 1$ time station
ANG	angle of attack
ANGVEL	angular velocity
VN2(I)	tangential velocity at the outer downstream boundary
UP2(I, J)	radial velocity necessary to generate the pressure upstream of the airfoil
UN2(I)	radial velocity at the outer boundary

To save computer time the author has prepared in COMPASS language subroutines that perform some elementary operations. One of these subroutines, the subroutine TRIB which solves a system of simultaneous equations in tridiagonal form, is left in the program as an example. The others have been removed to save space in this monograph. To run this program one should either reconstruct these subroutines or replace them by a conventional operation. More specifically instead of calling any of the following subroutines, the following operation must be written in conventional FOR-

TRAN notation:

ABSO	$W(I) = ABS(U(I))$
AZS	$Z(I) = A(I) + S$
AZZ	$Z(I) = A(I) + B(I)$
COPY	$Z(I) = A(I)$
MZS	$Z(I) = A(I)*S$
MZZ	$Z(I) = A(I)*B(I)$
QABS	$Z(I) = ABS(A(I))$
SETZ	$Z(I) = S$
SZS	$Z(I) = A(I) - S$
SZZ	$Z(I) = A(I) - B(I)$
FFT	Computes Fourier transforms. The arguments of this subrou-

tine are A, B, N, ISN where

A,   contains the real component of the data and result

B,   contains the imaginary component

N,   the number of complex data values

ISN,   the sign of ISN is the sign of the exponential in the transform and its magnitude is one.

Glossary and output

A IL $\times$ JL rectangular mesh is defined in the domain of integration, ρ, θ. Notice that $\rho = 1$ corresponds to the wall of the body and $\rho = 0$ to the outer physical boundary. Some of the basic symbols used in Chapter 3 and here can be identified in the program as follows:

I, J, N i, j, n, indexes running along the ρ, θ and t direction. In the notation of the present monograph these correspond to k, m and n respectively.

PY Ψ, stream function

PSI ψ, disturbance stream function

UP v_ρ, radial velocity

U   radial velocity at outer boundary

V   tangential velocity at outer boundary

VRT Ω, vorticity

X, Y x, y, physical coordinates

A typical output follows after the computer listing. This corresponds to an airfoil at zero angle of attack started impulsively from rest. At first a number of controlling parameters and flag values are printed. They are essentially the input values and their meaning has been discussed in the "Input" section.

The first few columns of output, labeled, I, K, and KM are the running index, a wave number and a modified wave number respectively, for the solution of the Fourier transform of the stream function equation, Eq. (7.8.14).

The following information printed is the displacement thickness at the first time step. The NORMALIZED SURFACE Y and the NORMALIZED CHORD LENGTH are the physical y and x coordinates that define the contour of the airfoil at zero angle of attack. The sequence of numbers represents coordinate values from the leading to the trailing edge to be read from left to right on each row and then from top to bottom for consecutive rows.

Values of the following controlling variables are then printed.

CC	c^2, a constant in the Joukowski transformation
CL	l, chord length of the airfoil
TH	thickness of the airfoil
RATIO	thickness to chord length ratio
RNNCL	Re/l, the Reynolds number divided by the chord length
XMIN	minimum value of coordinate x on the surface of the airfoil
XMAX	maximum value of coordinate x on the surface of the airfoil
COSC	the x location of the center of oscillation

The first set of calculated values are then printed in a set of columns labeled as follows:

J	j, index of radial distance (m, in the notation of this monograph)
RHO	ρ, stretched radial coordinate
RHO1	$d\rho/dr$, see Eq. (3.7.8)
RHO1S	$(d\rho/dr)^2$, see Eq. (3.7.8)
RHO2	$d^2\rho/dr^2$

Four blocks of data are then printed to provide information about the variable grid sizes. They are grid values of physical coordinates along lines of fixed I or J. The figures are printed from left to right along consecutive rows:

Y-TOP	Values of y at I = IMP, that is the "center" of the airfoil from the outer edge of the domain down to the airfoil
X/L IN THE WAKE	Reduced x coordinate from the downstream boundary back to the trailing edge
ETA	The boundary layer coordinate η corresponding to Y-top.
X-WAKE	The physical variable x corresponding to X/L

The values of the square of the Jacobian of the transformation are then printed

| H(2, JL) | H^2 (see Eq. 7.8.9) at the trailing edge |
| H(2, 1) | H^2 at the farthest downstream point |

Numerical results of quantities evaluated on the skin of the airfoil follow in column form:

| I | i, index running from the trailing edge towards the leading edge on the lower side of the airfoil and then back towards the trailing edge along the upper side of the airfoil |

XC(I)   normalized x-coordinate corresponding to I (k, in the notation of this monograph)

UPS   potential flow velocity on the airfoil calculated numerically

UPSE   potential flow velocity on the airfoil from exact solution

UPS2   the square of UPS

PPR   pressure calculated from UPS

PPRE   pressure calculated from UPSE

PRESSURE FORCE AT OUTER BOUNDARY   the pressure gradient at outer boundary

The index N in the next column represents the number of iterations and the corresponding quantity DVRTLD is the maximum residue of the vorticity on the surface of the airfoil. The numbers in the parentheses represent the value of I at which the maximum residue is found.

IS   is either the location either of the point of stagnation or of the point upstream of the point of stagnation.

PRESSURE DERIVATIVE   is the quantity $\partial p/\partial r$ from Bernoulli's equation at I = IS on the upstream boundary.

K.E. DEFECT   is the kinetic energy defect at the same point

UT(1)   is the velocity derivative $\partial U/\partial T$ at the same point

The block of numbers that follow are the values of the pressure coefficient at all grid points from upstream to the airfoil along I = IS. The last value is also printed separately. The same type of results for a point immediately downstream of the stagnation point (I = IS − 1) follow.

The following results are calculated by integrating either from the trailing to the leading edge or vice versa.

PM   the pressure coefficient at the leading edge generated by integration along the lower side of the airfoil.

PM1   the pressure coefficient at the leading edge generated by integration along the upper side of the airfoil.

CP C_p, the drag coefficient due to pressure

CF C_f, the drag coefficient due to skin friction

DRAG $C_p + C_f$, total drag coefficient

CLP C_{Lp}, the lift coefficient due to pressure distribution

CLF C_{Lf}, the lift coefficient due to friction

LIFT $C_{Lp} + C_{Lf}$, total lift coefficient

CMF C_{Mf}, moment coefficient due to friction

CMP C_{Mp}, moment coefficient due to pressure

MOMENT $C_{Mf} + C_{Mp}$, total moment coefficient

The final set of calculated results are printed in a set of columns with the following heading:

I   index, as defined above

XC(I)   normalized x-coordinate

PY(JLMN)	Ψ, stream function at the point $J = JL - 1$
VRT(JL)	Ω, vorticity at $J = JL$
VRT(JLMN)	vorticity at $J = JL - 1$
VRT(JLM2)	vorticity at $J = JL - 2$
VRTB(JL)	vorticity in body coordinates at $J = JL$ (in this case VRTB = VRT)
VRT1	$\partial\Omega/\partial\rho$, vorticity gradient
PR(L – T)	pressure coefficient, relative to zero pressure at the trailing edge, obtained by integrating from the leading edge to the trailing edge
PR1(T – L)	pressure coefficient, relative to zero pressure at the trailing edge, obtained by integrating from the trailing edge to the leading edge
V	the θ-component of the velocity at the outer boundary
CPT(L – T)	pressure coefficient obtained by integration from the leading edge to the trailing edge
CPT1(T – L)	pressure coefficient obtained by integration from the trailing edge to the leading edge.
THETA	the coordinate θ in radians

At this point the locations of stagnation points are calculated, time is incremented and new data are obtained and printed in the same format.

```
      PROGRAM MAIN(INPUT,OUTPUT,TAPE1=INPUT,TAPE3=OUTPUT,TAPE10,TAPE11, MAIN    2
     1TAPE14)                                                           MAIN    3
C                                                                       MAIN    4
C                       A RESEARCH PROGRAM FOR                          MAIN    5
C             UNSTEADY LAMINAR INCOMPRESSIBLE NAVIER-STOKES             MAIN    6
C             EQUATIONS SOLVED AROUND A TWO-DIMENSIONAL BODY            MAIN    7
C                                                                       MAIN    8
C     SOURCE - NASA AMES RESEARCH CENTER, MOFFETT FIELD, CA 94035, USA  MAIN    9
C     CONTACT - DR. UNMEEL B. MEHTA                                     MAIN   10
C     REFERENCE - MEHTA, U. B., DYNAMIC STALL OF AN OSCILLATING AIRFOIL,MAIN   11
C                 AGARD CP-227, PAPER NO. 23, SEPTEMBER 1977            MAIN   12
C                                                                       MAIN   13
C     PROGRAM COMPILED ON CDC 7600 WITH SCOPE 2.1.3                     MAIN   14
C                    USING OPT=2, AND ROUNDED ARITHMETIC                MAIN   15
C                                                                       MAIN   16
C     AUGUST 5, 1977 (3/6/81)                                          MAIN   17
C                                                                       MAIN   18
      COMMON /FIX/ KL20,DLTRI2,DLTRI3,DLTTI3,JS,JLM2,JLM3,              FIX     2
     *DVRTM,IMM,IMMN,IM,IMPL,IMP,ILMN,ILPL,ILPLP,JLMN,NP,NT,NTMAX,KL0,KLF IX    3
     11,KL2,KL3,KL4,KL5,KL6,KL9,KL10,KL14,KL15,KL16,ITS,JTS,ITV,JTV,NMAXF IX    4
     2,LN,LNT,PI,XN,XPN,YN,YPN,MARK,DLTI2,DLTI3,RH1DR,RH1DR2,RI,RNN,RNNCF IX    5
     3L,SRNN,DLHRNC,BPSI,BVRT,DVRTLD,R1TR2,ADVRT,TIME,DLTI,MIJF,        FIX     6
     4DLTT2I,DLTR2I,DLTR,DLTR2,DTR2,DLTT,DLTT2,DTT2,C,C2,OI,AS,AT,RATIO,FIX     7
     5CL,DLH,DEN,SIAT,COAT,AP,OR,ORI,CC,P,RHO2,DLTT4,IL,JL,ILM2,JLPL,IUPFIX    8
     6,DLTR24,DLHR,PS,IS,DLTI2I,COSC,ANGVEL,ANGACC,RF,AMP,AVEL1,AVA,RH1SFIX    9
     76,CLI,R1R6,DLRN,DLCL,NTI,T,T1,T2,KL11,IUPI,IUPFL,ANG,NSW,MN,DLT4RIFIX   10
     8,R11,KL21,R1R3,AV2,ILM3,DIT2D,DF,DR,DR2,DR2MDR,DEN2,FAC           FIX    11
     9,ILM4,DI2TR2,D24TR2,JLM4,JM,JM1,JBLE,JDIF1,MIJ,JTI,JVI,JRNCN,KL22 FIX    12
      COMMON /THER/ THETA(130),R(84),SINE(130),COSINE(130),RP(130)      THER    2
      COMMON        PY(130,84),VRT(130,84),RR(84),RHO1S(84),            MAIN   21
     1A5(84),HRR(130),ARDT(130),HD(130)                                MAIN   22
      COMMON /BCF/XC(130),V(130),VN2(130),PSIR(130,2)                   BCF     2
      COMMON        /SMALLD/ F(130,84)                                  MAIN   24
      COMMON /DUM1/AD(132),BD(132),CD(132),DD(132)                      DUM1    2
      COMMON /WN/ WAV(130,4),WAV2(130,4),WAVF(130,4),WAV2F(130,4)       WN      2
      COMMON /BPF/ ABF(84),BBF(84,2,2),CBF(84)                          BPF     2
      COMMON /LOADS/YTL(132),RHO1(84),XTF(132),PR(132),VRT1(132),       MAIN   28
```

```
      1XYT(132),YXT(132),DISS(130)                               MAIN  29
      COMMON /LDMSTOR/  TM(40),FDM(40),FLM(40),FMM(40),PDM(40,2),PLM(40,LDMSTOR2
     12),PMM(40,2),CDM(40,2),CLM(40,2),CMM(40,2)                 LDMSTOR3
      COMMON /LMAIN/XY2(130,84),VRTN1(130,84),VN1(130),XY2RO(130,84),  MAIN  31
     1XY2T(130,84),AXYYRO(130,84),AXYYT(130,84),RHO1I(84)        MAIN  32
      COMMON /LSTR/ PSI(130,84),Y(130,84),VRTN2(130,84),H(130,84)  LSTR   2
     1,YY(130,84),PSIN1(130,84)                                  LSTR   3
      LEVEL 2,PSI,Y,VRTN2,H,YY,PSIN1                             LSTR   4
      COMMON /LBOUND/ YR(130,84),YT(130,84),SH(130,84)           MAIN  34
      COMMON /LSTAGP/ UP(33,84),UP1(33,84),UP2(33,84),RNRI(84)   MAIN  35
      COMMON /LOUT/ X(130,84),XR(130,84),XT(130,84)              LOUT   2
      LEVEL 2,X,XR,XT                                            LOUT   3
      COMMON /HS/HA5T(130,84),HRRDLT(130,84),HTRR1(130,84)       MAIN  37
      COMMON /SIDER/ SIR(130,84),SIT(130,84),SIRN1(130,84),SITN1(130,84)MAIN  38
      COMMON /SU/ U(130),UN1(130),UN2(130),SQH1(130)            MAIN  39
      COMMON /LCROSS/ CROSST(130,84),VJLN2(130)                  LCROSS 2
      LEVEL 2,CROSST,VJLN2                                       LCROSS 3
      LEVEL 2,YTL,RHO1,XTF,PR,VRT1,XYT,YXT,DISS                  MAIN  41
      LEVEL 2,XY2,VRTN1,VN1,XY2RO,XY2T,AXYYRO,AXYYT,RHO1I        MAIN  42
      LEVEL 2, YR,YT,SH                                          MAIN  43
      LEVEL 2, UP,UP1,UP2,RNRI                                   MAIN  44
      LEVEL 2,HA5T,HRRDLT,HTRR1                                  MAIN  45
      LEVEL 2, SIR,SIT,SIRN1,SITN1                               MAIN  46
      DOUBLE PRECISION FRETIM,ARG                                MAIN  47
C                                                                MAIN  48
C     ANG - ANGLE OF ATTACK IN RADIANS                          MAIN  49
C     ANGVEL -  ANGULAR VELOCITY                                MAIN  50
C     H - JACOBIAN OF TRANSFORMATION DIVIDED BY R               MAIN  51
C     IL - NUMBER OF GRID-POINTS IN THETA-DIRECTION             MAIN  52
C     IL MUST BE 2**P+2.  THIS RESTRICTION IS DICTATED BY SUBROUTINE FFTMAIN  53
C     JL - NUMBER OF GRID-POINTS IN R-DIRECTION                 MAIN  54
C     JL MUST BE EVEN                                           MAIN  55
C     NP - NUMBER OF PROBLEMS                                   MAIN  56
C     NT - TIME INDEX                                           MAIN  57
C     PSI - DISTURBANCE STREAM FUNCTION                         MAIN  58
C     PY - STREAM FUNCTION                                      MAIN  59
C     PSIN1 - PSI AT NT-1                                       MAIN  60
C     RHO - STRECTHED RADIAL COORDINATE                         MAIN  61
C     RHO1 - FIRST DERIVATIVE OF TRANSFORMATION IN R-DIRECTION  MAIN  62
C     RHO2 - SECOND DERIVATIVE OF TRANSFORMATION IN R-DIRECTION MAIN  63
C     RP - RELAXATION PARAMETER ARRAY                           MAIN  64
C     TIME - NONDIMENSIONAL TIME, NORMALIZED BY THE RADIUS OF CIRCULAR  MAIN  65
C            CYLINDER  IN THE TRANSFORMED PLANE                 MAIN  66
C     TIMEL - NONDIMENSIONAL TIME, NORMALIZED BY THE CHORD LENGTH  MAIN  67
C     U - RADIAL (R-DIRECTION) VELOCITY AT OUTER BOUNDARY       MAIN  68
C     UN1 - U AT NT-1                                           MAIN  69
C     UN2 - U AT NT-2                                           MAIN  70
C     UP - RADIAL VELOCITY                                      MAIN  71
C     UP1 - UP AT NT-1                                          MAIN  72
C     UP2 - UP AT NT-2                                          MAIN  73
C     V - TANGENTIAL (THETA-DIRECTION) VELOCITY AT OUTER BOUNDARY  MAIN  74
C     VN1 - V AT NT-1                                           MAIN  75
C     VRT - VORTICITY                                           MAIN  76
C     VRTN1 - VRT AT NT-1                                       MAIN  77
C     WAV - MODIFIED WAVE NUMBER USED TO DETERMINE FOURTH-ORDER ACCURATEMAIN  78
C           FIRST DERIVATIVE                                    MAIN  79
C     WAV2 - MODIFIED WAVE NUMBER SQUARED USED TO DETERMINE FOURTH-ORDERMAIN  80
C            ACCURATE SECOND DERIVATIVE                         MAIN  81
C     X,Y - PHYSICAL COORDINATES                                MAIN  82
C     XC - NORMALIZED DISTANCE ALONG CHORD                      MAIN  83
C     XR,YR - DERIVATIVES OF PHYSICAL COORDINATES WITH RESPECT TO R  MAIN  84
C     XT,YT - DERIVATIVES OF PHYSICAL COORDINATES WITH RESPECT TO THETA MAIN  85
C                                                                MAIN  86
C     HIERARCHY-CHART OF MAJOR SUBROUTINES                      MAIN  87
C              MAIN                                             MAIN  88
C                    SETUP                                      MAIN  89
C                          SUFLE1                               MAIN  90
C                    TRANS                                      MAIN  91
C                    COEF                                       MAIN  92
C                    START                                      MAIN  93
C                          STREAM                               MAIN  94
C                                SUFLE1                         MAIN  95
C                                FFT                            MAIN  96
C                                B2TRI                          MAIN  97
C                                B2F                            MAIN  98
C                          DERSI                                MAIN  99
C                                TRIB                           MAIN 100
C                          LOCATE                               MAIN 101
```

```
C                           FFT                                    MAIN 102
C                     FFT                                          MAIN 103
C                     BOUND                                        MAIN 104
C                           SUFLE4                                 MAIN 105
C                           TRIPER                                 MAIN 106
C                           SUFLE1                                 MAIN 107
C                     ADI                                          MAIN 108
C                     FFT                                          MAIN 109
C                     DERSI                                        MAIN 110
C                           TRIB                                   MAIN 111
C                     CROSS                                        MAIN 112
C                           SUFLE1                                 MAIN 113
C                           FFT                                    MAIN 114
C                     BTRIP                                        MAIN 115
C                     SUFLE1                                       MAIN 116
C                     SUFLE4                                       MAIN 117
C                     TRIB                                         MAIN 118
C                     BOUND                                        MAIN 119
C                           SUFLE4                                 MAIN 120
C                           TRIPER                                 MAIN 121
C                           SUFLE1                                 MAIN 122
C                     STREAM                                       MAIN 123
C                           SUFLE1                                 MAIN 124
C                           FFT                                    MAIN 125
C                           B2TRI                                  MAIN 126
C                           B2F                                    MAIN 127
C               STAGP                                              MAIN 128
C               FORCES                                             MAIN 129
C               OUTPUT                                             MAIN 130
C               RNCN                                               MAIN 131
C                     FFT                                          MAIN 132
C                                                                  MAIN 133
C                                                                  MAIN 134
C     INPUT PARAMETERS READ IN MAIN PROGRAM AND IN SUBROUTINES SETUP ANDMAIN 135
C                     START                                        MAIN 136
C                                                                  MAIN 137
      WRITE (3,245)                                                MAIN 138
      READ(1,260) IL,JL                                           MAIN 139
      WRITE (3,250) IL,JL                                         MAIN 140
      READ (1,260) NP                                            MAIN 141
      NB=0                                                        MAIN 142
  240 CONTINUE                                                    MAIN 143
      NB=NB+1                                                     MAIN 144
      READ(1,260) NMAX,NTMAX,KL2,KL3,KL4,KL5,KL6,KL9,KL11,KL13,KL14,  MAIN 145
     1KL15,KL16,JS,RI,RNN,BPSI,BVRT,DVRTM,TIME,DLTI              MAIN 146
     2,RF,AMP,KL10,KL21,DF,TIMOSC,JBLE,KL22,JRNCN                MAIN 147
C                                                                  MAIN 148
C     IF KL2=1,  READ  PY AND VRTN1,VRTN2,VN1,VN2,UP1,UP2         MAIN 149
C                FOR CONTINUATION OF SOLUTION                      MAIN 150
C     IF KL2=1,  INPUT MT, MT1 AND NTAP                           MAIN 151
C     IF KL3=1,WRITE  PY AND VRT, V,UP                            MAIN 152
C     IF KL4=1,  JOUKOWSKI TRANSFORMATION FOR ELLIPSE, INPUT TR.  MAIN 153
C                OR, OI, AS MUST BE ZERO.                         MAIN 154
C     IF KL4=2,  JOUKOWSKI TRANSFORMATION FOR AIRFOIL             MAIN 155
C     IF KL4=3,  JOUKOWSKI TRANSFORMATION FOR AIRFOIL AND AS.GT.0.0  MAIN 156
C     IF KL5=1,  BVRT VARIES  WITH I                              MAIN 157
C     IF KL6=1,  NO TRANSFORMATION IN R                           MAIN 158
C     IF KL6=2,  HYPERBOLIC TANGENT TRANSFORMATION IN R           MAIN 159
C     IF KL6=3,  EXPONENTIAL TRANSFORMATION IN R                  MAIN 160
C     KL9 - FREQUENCY OF PRINT-OUT FROM SUBROUTINE OUTPUT         MAIN 161
C     IF KL10=1, SKIP ADVRT CHECK IN MAIN                         MAIN 162
C     IF KL11=1, HALF THE PREVIOUS TIME STEP. INPUT MT AND NTAP   MAIN 163
C     IF KL13=1, INITIAL VALUES OF FLOW FIELD VARIABLES AND GEOMETRICAL MAIN 164
C                VARIABLES ARE PRINTED                            MAIN 165
C     IF KL14=1, DO NOT PRINT VELOCITIES                          MAIN 166
C     IF KL15=1, CALL CDCL                                        MAIN 167
C     IF KL16=1, DO NOT CALL OUTPUT                               MAIN 168
C     IF KL21=1, VRT(I,JL)=0.0, I=2,IM,IL                         MAIN 169
C     IF KL21=1, RF=0.0                                           MAIN 170
C     FOR FLAT PLATE, KL21=1                                      MAIN 171
C      ZERO  THICKNESS FLAT PLATE MUST BE AT ZERO ANGLE OF ATTACK MAIN 172
C     NMAX - NUMBER OF ITERATIONS PER TIME STEP                   MAIN 173
C     NTMAX - NUMBER OF TIME STEPS                                MAIN 174
C     JS - DETERMINES PRINT-OUT IN R-DIRECTION FROM SUBROUTINE OUTPUT  MAIN 175
C          E.G. IF JS = 2. PRINT-OUT IS AT EVEN VALUES OF J       MAIN 176
C     RI - FIXES THE DISTANCE AT WHICH OUTER BOUNDARY IS LOCATED  MAIN 177
C          RI = 0.02-0.04 (A RECOMMENDATION)                      MAIN 178
C     RNN - REYNOLDS NUMBER                                       MAIN 179
```

```
C     BPSI - FOR VARIABLE RP. BVRT IS MULTIPLED BY BPSI AWAY FROM BOTH   MAIN 180
C            LEADING AND TRAILING EDGES                                  MAIN 181
C     BVRT - RELAXATION PARAMETER FOR SURFACE VORTICITY                  MAIN 182
C     VALUE OF BVRT DEPENDS MAINLY UPON TRAILING EDGE RADIUS OF          MAIN 183
C                 CURVATURE.  SMALLER THIS RADIUS SMALLER BVRT.          MAIN 184
C                 FOR MODIFIED NACA 0012, BVRT=0.01-0.03 AND             MAIN 185
C                 BPSI=1.1-1.5                                           MAIN 186
C                 FOR IMPULSIVELY STARTED AIRFOIL AT AT.NE.0.0, BVRT     MAIN 187
C                                    WOULD BE SMALLER THAN ABOVE         MAIN 188
C                                    VALUE.                             MAIN 189
C                 A THEORETICAL STUDY IS NEEDED TO DETERMINE OPTIMUM     MAIN 190
C                 VALUE OF BVRT.                                         MAIN 191
C     DVRTM - MAXIMUM RESIDUE IN SURFACE VORTICITY                       MAIN 192
C     TIME - TIME                                                        MAIN 193
C     DLTI - TIME INCREMENT                                              MAIN 194
C     RF - REDUCED FREQUENCY                                             MAIN 195
C     AMP - AMPLITUDE OF OSCILLATION IN DEGREES                          MAIN 196
C     DF - DISSIPATION FACTOR LESS THAN OR EQUAL TO CL                   MAIN 197
C     TIMOSC - TIME AT WHICH BODY IS OSCILLATED                          MAIN 198
C     JBLE MUST BE EVEN.  USE JBLE=2, UNLESS YOU CAN FIGURE OUT HOW TO   MAIN 199
C                 USE IT.  IF JBLE.NE.2, MOD(NMAX-3,5).EQ.0 FOR          MAIN 200
C                 NMAX.GT.3                                              MAIN 201
C     IF KL22.NE.1, CALL RNCN EVERY KL9 TIMES                            MAIN 202
C     JRNCN - SKIP ON J PRINT-OUT IN RNCN                                MAIN 203
C                                                                        MAIN 204
      WRITE(3,265) NMAX,NTMAX,KL2,KL3,KL4,KL5,KL6,KL9,KL11,KL13,KL14,    MAIN 205
     1KL15,KL16,JS,RI,RNN,BPSI,BVRT,DVRTM,TIME,DLTI,                     MAIN 206
     2 ,RF,AMP ,KL10,KL21,DF,TIMOSC                                      MAIN 207
      WRITE(3,266) JBLE,KL22,JRNCN                                       MAIN 208
      IF (IL.NE.34.AND.IL.NE.66.AND.IL.NE.130) STOP                      MAIN 209
      IF (MOD(JL,2).NE.0) STOP                                           MAIN 210
      IF (MOD(JBLE,2).NE.0) STOP                                         MAIN 211
      PI=3.1415926535898                                                 MAIN 212
      CALL SETUP                                                         MAIN 213
      AMP=AMP*PI/180.0                                                   MAIN 214
      FREQ=2.0*RF*CLI                                                    MAIN 215
      CALL TRANS                                                         MAIN 216
      CALL COEF                                                          MAIN 217
      WRITE(3,290) H(2,JL),H(2,1)                                        MAIN 218
  290 FORMAT(/3X,8HH(2,JL)=,1PE12.5,3X,7HH(2,1)=,E12.5,/)                MAIN 219
      DEN=2.0+3.0*(1.0-R(JLMN))                                          MAIN 220
      RH1S6=RHO1S(JL)*6.0/DTR2                                           MAIN 221
      R1R6=RHO1(JL)/(DLTR*RNNCL)                                         MAIN 222
      R1R3=RHO1(JL)*DLTR2I                                               MAIN 223
      DLT4RI=1.0/(DLTT4*R(1))                                            MAIN 224
      R1I=DF/R(1)                                                        MAIN 225
      DO 350 J=1,JL                                                      MAIN 226
  350 RNRI(J)=1.0/(RNNCL*DLTT2*R(J))                                     MAIN 227
      DO 10 I=2,IL                                                       MAIN 228
      SQH1(I)=SQRT(H(I,1))                                               MAIN 229
          HRR(I)=H(I,JLMN)*RR(JLMN)                                      MAIN 230
   10 CONTINUE                                                           MAIN 231
      SQH1(1)=SQH1(ILMN)                                                 MAIN 232
      IF (KL21.EQ.1) GO TO 7                                             MAIN 233
      DO 6 I=2,IL                                                        MAIN 234
    6     HD(I)=1.0/(H(I,JL)*DEN)                                        MAIN 235
      GO TO 9                                                            MAIN 236
    7 ID=IM                                                              MAIN 237
      DO 8 I=3,IMMN                                                      MAIN 238
      ID=ID+1                                                            MAIN 239
      HD(I)=1.0/(H(I,JL)*DEN)                                            MAIN 240
    8 HD(ID)=1.0/( H(ID,JL)*DEN)                                         MAIN 241
      HD(2)=0.0                                                          MAIN 242
      HD(IM)=0.0                                                         MAIN 243
      HD(IL)=0.0                                                         MAIN 244
    9 CONTINUE                                                           MAIN 245
      DEN=6.0/DR                                                         MAIN 246
      IF (RF.NE.0.0) GO TO 14                                            MAIN 247
      DO 11 I=2,IL                                                       MAIN 248
      XTF(I)=XT(I,JL)                                                    MAIN 249
      YTL(I)=YT(I,JL)                                                    MAIN 250
      XYT(I)=X(I,JL)*XT(I,JL)+Y(I,JL)*YT(I,JL)                           MAIN 251
   11 YXT(I)=-Y(I,JL)*XT(I,JL)+X(I,JL)*YT(I,JL)                          MAIN 252
      XTF(ILPL)=XTF(3)                                                   MAIN 253
      XTF(ILPLP)=XTF(4)                                                  MAIN 254
      XTF(1)=XTF(ILMN)                                                   MAIN 255
      YTL(ILPL)=YTL(3)                                                   MAIN 256
      YTL(ILPLP)=YTL(4)                                                  MAIN 257
```

```
              YTL(1)=YTL(ILMN)                                    MAIN 258
              XYT(ILPL)=XYT(3)                                    MAIN 259
              XYT(ILPLP)=XYT(4)                                   MAIN 260
              XYT(1)=XYT(ILMN)                                    MAIN 261
              YXT(ILPL)=YXT(3)                                    MAIN 262
              YXT(ILPLP)=YXT(4)                                   MAIN 263
              YXT(1)=YXT(ILMN)                                    MAIN 264
      14      DO 15 J=1,JL                                        MAIN 265
      15        Y(1,J)=Y(ILMN,J)                                  MAIN 266
              RH1DR=RHO1(1) /DLTR                                 MAIN 267
              RH1DR2=RH1DR*0.5                                    MAIN 268
              R1TR2=RH1DR2*DLTT21                                 MAIN 269
            CALL START                                           MAIN 270
            IF (KL13.EQ.1) CALL OUTPUT                           MAIN 271
            WRITE(3,270)                                         MAIN 272
C                                                                MAIN 273
      60    NT=NT+1                                              MAIN 274
            TIME=TIME+DLTI                                       MAIN 275
            TIMEL=(TIME-TIMOSC)*CLI                              MAIN 276
            TM(NT)=TIMEL                                         MAIN 277
            DO 630 I=2,ILM2,2                                    MAIN 278
            BD(I)=0.0                                            MAIN 279
            BD(I+1)=0.0                                          MAIN 280
            AD(I)=VN1(I)*VN1(I)+UN1(I)*UN1(I)                    MAIN 281
      630   AD(I+1)=VN1(I+1)*VN1(I+1)+UN1(I+1)*UN1(I+1)          MAIN 282
            CALL FFT(AD(2),BD(2),ILM2,-1)                        MAIN 283
            CALL MZZ(AD(2),WAVF(2,1),AD(2),ILM2)                 MAIN 284
            CALL MZZ(BD(2),WAVF(2,1),BD(2),ILM2)                 MAIN 285
            CALL MZS(BD(2),-1.0,BD(2),ILM2)                      MAIN 286
            CALL FFT(BD(2),AD(2),ILM2,1)                         MAIN 287
            DO 650 I=2,ILMN                                      MAIN 288
      650   VN2(I)=T1*VN1(I)-T2*VN2(I)+(DLTI*SH(I,1)/R(1))*BD(I)+DLTI2*   MAIN 289
          1VRTN1(I,1)*UN1(I)                                     MAIN 290
            T1T=(T1*DLTI*DF)/(64.0*CL)                           MAIN 291
            DO 660 I=2,ILMN                                      MAIN 292
      660   AD(I+9)=ABS(VN1(I))*VN1(I)                           MAIN 293
            CALL COPY(AD(11),AD(IL+9),10)                        MAIN 294
            CALL COPY(AD(IL   ),AD(2),9)                         MAIN 295
             DO 670 I=2,ILMN                                     MAIN 296
            VN2(I)=VN2(I)+(T1T/3.0)*(AD( I+6)-6.0*(AD(I+7)+AD(I+11))+15.0*   MAIN 297
          1(AD(I+8)+AD(I+10))-20.0*AD(I+9 )+AD(I+12)+AD(I+3)-6.0*(AD(I+5)+   MAIN 298
          2AD(I+13))+15.0*(AD(I+7)+AD(I+11))-20.0*AD(I+9)+AD(I+15)+AD(I)-6.0*MAIN 299
          3(AD(I+3)+AD(I+15))+15.0*(AD(I+6)+AD(I+12))-20.0*AD(I+9)+AD(I+18)) MAIN 300
      670 CONTINUE                                               MAIN 301
            VN2(IL)=VN2(2)                                       MAIN 302
C                                                                MAIN 303
            IF(RF.EQ.0.0) GO TO 375                              MAIN 304
            FRETIM=FREQ*(TIME-TIMOSC)                            MAIN 305
            COSFT=DCOS(FRETIM)                                   MAIN 306
            AMPF=AMP*FREQ                                        MAIN 307
C                                                                MAIN 308
C     COUNTER-CLOCKWISE ROTATION POSITIVE                        MAIN 309
C                                                                MAIN 310
            ANG=AT-AMP*COSFT+AMP                                 MAIN 311
            ARG=ANG-AT                                           MAIN 312
            COAT=DCOS(ARG)                                       MAIN 313
            SIAT=DSIN(ARG)                                       MAIN 314
            ANGVEL=-AMPF*DSIN(FRETIM)                            MAIN 315
            ANGACC=-AMPF*FREQ*COSFT                              MAIN 316
            AVA=ANGVEL                                           MAIN 317
            AV2=ANGVEL*2.0                                       MAIN 318
            ANGD=ANG*180.0/PI                                    MAIN 319
            WRITE(3,380) ANGD,ANGVEL,ANGACC,FRETIM               MAIN 320
      380 FORMAT(/3X7HANGD   =1PE21.14,3X7HANGVEL=E21.14,3X7HANGACC=E21.14,3XMAIN 321
          17HFRETIM=,E21.14/)                                    MAIN 322
C                                                                MAIN 323
C     PY IN BODY CO-ORDINATES                                    MAIN 324
C                                                                MAIN 325
            DO 385 J=1,JLMN                                      MAIN 326
            DO 385 I=2,IL                                        MAIN 327
            YY(I,J)=Y(I,J)+AVA*XY2(I,J)                          MAIN 328
      385 PY(I,J)=PSI(I,J)+YY(I,J)                               MAIN 329
            DO 386 J=1,JLMN                                      MAIN 330
      386 PY(1,J)=PY(IL,J)                                       MAIN 331
            DO 610 J=1,JL                                        MAIN 332
            DO 610 I=2,IL                                        MAIN 333
            AXYYRO(I,J)=ANGVEL*XY2RO(I,J)+YR(I,J)*RHO1I(J)       MAIN 334
      610 AXYYT(I,J)=ANGVEL*XY2T(I,J)+YT(I,J)                    MAIN 335
```

```
      DO 615 I=2,IL                                             MAIN 336
      AXR=ANGACC*XY2RO(I,JL)*RHO1(JL)                           MAIN 337
      AAX=(ANGVEL*ANGVEL)*XY2T(I,JL)                            MAIN 338
      VRT1(I)=-AXR+AAX                                          MAIN 339
      AXR=ANGACC*XY2RO(I,1)*RHO1(1)                             MAIN 340
      AAX=(ANGVEL*ANGVEL)*XY2T(I,1)                             MAIN 341
      VN2(I)=VN2(I)+DLTI2*(-AXR+AAX/R(1))*SH(I,1)               MAIN 342
      PSI(I,JL)=-Y(I,JL)-ANGVEL*XY2(I,JL)                       MAIN 343
  615 CONTINUE                                                  MAIN 344
      PSI(1,JL)=PSI(ILMN,JL)                                    MAIN 345
      DO 620 I=2,IL                                             MAIN 346
      YTL(I)=YT(I,JL)*COAT-XT(I,JL)*SIAT                        MAIN 347
      XTF(I)=XT(I,JL)*COAT+YT(I,JL)*SIAT                        MAIN 348
      XI=X(I,JL)*COAT+Y(I,JL)*SIAT                              MAIN 349
      YI=Y(I,JL)*COAT-X(I,JL)*SIAT                              MAIN 350
      XYT(I)=XI*XTF(I)+YI*YTL(I)                                MAIN 351
  620 VXT(I)=-YI*XTF(I)+XI*YTL(I)                               MAIN 352
      XTF(ILPL)=XTF(3)                                          MAIN 353
      XTF(ILPLP)=XTF(4)                                         MAIN 354
      XTF(1)=XTF(ILMN)                                          MAIN 355
      YTL(ILPL)=YTL(3)                                          MAIN 356
      YTL(ILPLP)=YTL(4)                                         MAIN 357
      YTL(1)=YTL(ILMN)                                          MAIN 358
      XYT(ILPL)=XYT(3)                                          MAIN 359
      XYT(ILPLP)=XYT(4)                                         MAIN 360
      XYT(1)=XYT(ILMN)                                          MAIN 361
      VXT(ILPL)=VXT(3)                                          MAIN 362
      VXT(ILPLP)=VXT(4)                                         MAIN 363
      VXT(1)=VXT(ILMN)                                          MAIN 364
  375 CONTINUE                                                  MAIN 365
      VN2(1)=VN2(ILMN)                                          MAIN 366
      CALL BOUND                                                MAIN 367
C                                                               MAIN 368
C     VRT IN INERTIAL FRAME OF REFERENCE                        MAIN 369
C                                                               MAIN 370
      TI=1.0/T                                                  MAIN 371
      DO 210 J=1,JL                                             MAIN 372
      DO 210 I=2,IL                                             MAIN 373
  210 VRTN2(I,J)=(T1*VRTN1(I,J)-T2*VRTN2(I,J))*TI               MAIN 374
      DO 215 J=1,JL                                             MAIN 375
  215 VRTN2(1,J)=VRTN2(ILMN,J)                                  MAIN 376
      CALL ADI                                                  MAIN 377
      DO 405 J=1,JL                                             MAIN 378
      ID=0                                                      MAIN 379
      DO 405 I=IUPI,IUPFL                                       MAIN 380
      ID=ID+1                                                   MAIN 381
  405 UP(ID,J)=SIT(I,J)*SH(I,J)/R(J)                            MAIN 382
      CALL QABS(VRT(1,1),F(1,1),MIJF)                           MAIN 383
      DO 390 J=1,JL                                             MAIN 384
      DO 390 I=1,IL                                             MAIN 385
      IF(F(I,J).LT.1.0E-06) VRT(I,J)=0.0                        MAIN 386
  390 CONTINUE                                                  MAIN 387
      IF (KL15.NE.1) GO TO 150                                  MAIN 388
      CALL STAGP                                                MAIN 389
      CALL FORCES                                               MAIN 390
  150 CONTINUE                                                  MAIN 391
      IF (KL3.NE.1) GO TO 160                                   MAIN 392
      WRITE (11) ((PY(I,J),I=2,IL),J=1,JL),((VRT(I,J),I=2,IL),J=1,JL),MAIN 393
     1TIME,TIMEL,ANG,ANGVEL,(V(I),I=2,IL), ((UP(I,J),I=1,IUP),J=1,JL) MAIN 394
     2,(U(I),I=2,IL),(PR(I),I=2,IL),(DISS(I),I=2,IL)            MAIN 395
      END FILE 11                                               MAIN 396
      MARK=MARK+1                                               MAIN 397
  160 CONTINUE                                                  MAIN 398
      IF (MOD(NT,KL9).NE.0) GO TO 175                           MAIN 399
      IF (KL16.EQ.1) GO TO 165                                  MAIN 400
      CALL OUTPUT                                               MAIN 401
  165 IF (KL22.EQ.1) GO TO 175                                  MAIN 402
      CALL RNCN                                                 MAIN 403
      NT=NTMAX                                                  MAIN 404
  175 WRITE(3,280) TIME,TIMEL                                   MAIN 405
      IF (KL10.EQ.1) GO TO 176                                  MAIN 406
      IF (DVRTM.LT.ADVRT) GO TO 230                             MAIN 407
  176 IF (NT.GE.NTMAX) GO TO 230                                MAIN 408
      AVEL1=ANGVEL                                              MAIN 409
      DO 185 J=1,JL                                             MAIN 410
      DO 185 I=1,ILM4,5                                         MAIN 411
      VRTN2(I,J)=VRTN1(I,J)                                     MAIN 412
      VRTN2(I+1,J)=VRTN1(I+1,J)                                 MAIN 413
```

```
      VRTN2(I+2,J)=VRTN1(I+2,J)                                    MAIN 414
      VRTN2(I+3,J)=VRTN1(I+3,J)                                    MAIN 415
185   VRTN2(I+4,J)=VRTN1(I+4,J)                                    MAIN 416
      DO 186 I=2,ILM4,4                                            MAIN 417
      VJLN2(I)=VRTN1(I,JL)                                         MAIN 418
      VJLN2(I+1)=VRTN1(I+1,JL)                                     MAIN 419
      VJLN2(I+2)=VRTN1(I+2,JL)                                     MAIN 420
186   VJLN2(I+3)=VRTN1(I+3,JL)                                     MAIN 421
      CALL MOVLEV(VRT(1,1),VRTN1(1,1),MIJF)                        MAIN 422
      DO 187 I=1,IL                                                MAIN 423
      VN2(I)=VN1(I)                                                MAIN 424
      VN1(I)=V(I)                                                  MAIN 425
      UN2(I)=UN1(I)                                                MAIN 426
187   UN1(I)=U(I)                                                  MAIN 427
      IF (MOD(NT+1,KL9).NE.0.AND.KL22.EQ.1) GO TO 190             MAIN 428
      WRITE(14)((VRTN2(I,J),I=2,ILMN),J=1,JL)                      MAIN 429
      END FILE 14                                                 MAIN 430
      REWIND 14                                                   MAIN 431
190   CONTINUE                                                    MAIN 432
      IF (NT-NTI) 205,203,204                                     MAIN 433
203      T=3.0                                                    MAIN 434
         T1=4.0                                                   MAIN 435
         T2=1.0                                                   MAIN 436
      DO 550 J=1,JL                                               MAIN 437
      D2=(T*DTR2)*A5(J)                                           MAIN 438
      DO 550 I=1,ILM4,5                                           MAIN 439
      HA5T(I,J)=D2*H(I,J)                                         MAIN 440
      HA5T(I+1,J)=D2*H(I+1,J)                                     MAIN 441
      HA5T(I+2,J)=D2*H(I+2,J)                                     MAIN 442
      HA5T(I+3,J)=D2*H(I+3,J)                                     MAIN 443
550   HA5T(I+4,J)=D2*H(I+4,J)                                     MAIN 444
      DO 530 J=1,JL                                               MAIN 445
      D4=4.0*DLTI*RR(J)*RHO1S(J)                                  MAIN 446
      DO 530 I=1,ILM4,5                                           MAIN 447
      HTRR1(I,J)=HA5T(I,J)+D4                                     MAIN 448
      HTRR1(I+1,J)=HA5T(I+1,J)+D4                                 MAIN 449
      HTRR1(I+2,J)=HA5T(I+2,J)+D4                                 MAIN 450
      HTRR1(I+3,J)=HA5T(I+3,J)+D4                                 MAIN 451
530   HTRR1(I+4,J)=HA5T(I+4,J)+D4                                 MAIN 452
204      CONTINUE                                                 MAIN 453
      DO 227 J=1,JL                                               MAIN 454
      DO 227 I=1,ILMN,2                                           MAIN 455
      VRT(I,J)=VRTN1(I,J)+VRTN1(I,J)-VRTN2(I,J)                   MAIN 456
227   VRT(I+1,J)=VRTN1(I+1,J)+VRTN1(I+1,J)-VRTN2(I+1,J)           MAIN 457
      DO 228 J=1,JLMN                                             MAIN 458
      DO 228 I=1,IL                                               MAIN 459
      F(I,J)=PSI(I,J)+PSI(I,J)-PSIN1(I,J)                         MAIN 460
      PSIN1(I,J)=PSI(I,J)                                         MAIN 461
228   CONTINUE                                                    MAIN 462
C                                                                 MAIN 463
C     CHANGE BELOW 130 IF THE FIRST DIMENSION OF TWO-DIMENSIONAL ARRAY  MAIN 464
C     IS CHANGED                                                  MAIN 465
C                                                                 MAIN 466
      CALL MOVLEV(F(1,1),PSI(1,1),MIJF-130)                       MAIN 467
      DO 229 I=2,IL                                               MAIN 468
      V(I)=VN1(I)+VN1(I)-VN2(I)                                   MAIN 469
229   U(I)=UN1(I)+UN1(I)-UN2(I)                                   MAIN 470
      V(1)=V(ILMN)                                                MAIN 471
      U(1)=U(ILMN)                                                MAIN 472
205   CONTINUE                                                    MAIN 473
      DO 420 J=1,JL                                               MAIN 474
      DO 420 I=1,IUP                                              MAIN 475
      UP2(I,J)=UP1(I,J)                                           MAIN 476
420   UP1(I,J)= UP(I,J)                                           MAIN 477
      GO TO 60                                                    MAIN 478
230   CONTINUE                                                    MAIN 479
      WRITE(3,700)                                                MAIN 480
700   FORMAT(*                        PRESSURE INTEGRATED FROMAIN 481
     1M T.E. TO L.E.    PRESSURE INTEGRATED FORM L.E. TO T. E*//  MAIN 482
     2     *     TIME      FD       FL       FM      PD      PL       PM    DRAGMAIN 483
     3  LIFT MOMENT     PD     PL     PM      DRAG   LIFT MOMENT   A.D.   AMAIN 484
     4.L.   A.M.*/)                                               MAIN 485
C                                                                 MAIN 486
C     FD - FRICTION DRAG                                          MAIN 487
C     FL - FRICTION LIFT                                          MAIN 488
C     FM - FRICTION MOMENT                                        MAIN 489
C     PD - PRESSURE DRAG                                          MAIN 490
C     PL - PRESSURE LIFT                                          MAIN 491
```

```
C     PM - PRESSURE MOMENT                                          MAIN 492
C     A.D. - AVERAGE DRAG                                           MAIN 493
C     A.L. - AVERAGE LIFT                                           MAIN 494
C     A.M. - AVERAGE MOMENT                                         MAIN 495
C                                                                   MAIN 496
      DO 710 N=1,NTMAX                                              MAIN 497
      ADRAG=(CDM(N,1)+CDM(N,2))*0.5                                 MAIN 498
      ALIFT=(CLM(N,1)+CLM(N,2))*0.5                                 MAIN 499
      AMOM =(CMM(N,1)+CMM(N,2))*0.5                                 MAIN 500
      WRITE(3,720)  TM(N),FDM(N),FLM(N),FMM(N),PDM(N,1),PLM(N,1),PMM(N,1MAIN 501
     1),CDM(N,1),CLM(N,1),CMM(N,1),PDM(N,2),PLM(N,2),PMM(N,2),CDM(N,2),MAIN 502
     2CLM(N,2),CMM(N,2),ADRAG,ALIFT,AMOM                            MAIN 503
  720 FORMAT(F8.4,4F7.3,F6.3,5F7.3,F6.3,7F7.3)                      MAIN 504
  710 CONTINUE                                                      MAIN 505
      IF (NB.LT.NP) GO TO 240                                       MAIN 506
C                                                                   MAIN 507
  245 FORMAT (1H1,///,60X,13HA I R F O I L,////)                    MAIN 508
  250 FORMAT (//3X,7HIL    =,I3,3X,7HJL    =,I3)                    MAIN 509
  260 FORMAT(1415/4F10.5,E10.2,3F10.5/F10.5,2I5,2F10.5,3I5)         MAIN 510
  265 FORMAT((//3X,*NMAX  =*I3,3X*NTMAX =*I3,3X,           7HKL2MAIN 511
     1  =,I3,3X,7HKL3   =,I3,3X,7HKL4   =,I3,3X,7HKL5   =,I3,//,3X,7HKLMAIN 512
     26  =,I3,3X,7HKL9   =,I3,3X,7HKL11  =,I3,3X,7HKL13  =,I3,3X,7HKL14MAIN 513
     3  =,I3,3X*KL15 =*I3,3X*KL16 =*I3,3X*JS  =*I3,//3X*RI =*1PE12.5,MAIN 514
     43X*RNN =*E12.5,3X*BPSI =*E12.5,3X*BVRT =*E12.5//3X*DVRTM =*  MAIN 515
     5E12.5,3X,                                         7HTMAIN 516
     6IME  =,E12.5,3X7HDLTI =,E12.5//3X7HRF    =E12.5,3X7HAMP  =E12.5,MAIN 517
     73X7HKL10 =,I2,3X,7HKL21 =,I2,3X7HDF    =E12.5,3X7HTIMOSC=,E12.5)MAIN 518
  266 FORMAT(/3X,*JBLE=*,I3,5X,*KL22=*,I3,5X,*JRNCN=*,I3,/)          MAIN 519
  270 FORMAT (/)                                                    MAIN 520
  280 FORMAT (/3X,7HTIME   =,1PE12.5,3X,20HTIME BASED ON CHORD=,E12.5//)MAIN 521
      END                                                           MAIN 522
                                                                   SETUP   2
                                                                   SETUP   3
      SUBROUTINE SETUP                                             SETUP   4
C                                                                   SETUP   5
C     THIS SUBROUTINE SETS UP VALUES OF MOST OF THE VARIABLES WHICH ARE SETUP   6
C     EXPECTED TO RETAIN THESE VALUES DURING EXECUTION             SETUP   7
C                                                                   SETUP   8
      COMMON /FIX/ KL20,DLTRI2,DLTRI3,DLTTI3,JS,JLM2,JLM3,          FIX     2
     *DVRTM,IMM,IMMN,IM,IMPL,IMP,ILMN,ILPL,ILPLP,JLMN,NP,NT,NTMAX,KL0,KLFIX 3
     11,KL2,KL3,KL4,KL5,KL6,KL9,KL10,KL14,KL15,KL16,ITS,JTS,ITV,JTV,NMAXFIX 4
     2,LN,LNT,PI,XN,XPN,YN,YPN,MARK,DLTI2,DLTI3,RH1DR,RH1DR2,RI,RNN,RNNCFIX 5
     3L,SRNN,DLHRNC,BPSI,BVRT,DVRTLD,R1TR2,ADVRT,TIME,DLTI,MIJF,     FIX     6
     4DLTT2I,DLTR2I,DLTR,DLTR2,DTR2,DLTT,DLTT2,DTT2,C,C2,OI,AS,AT,RATIO,FIX 7
     5CL,DLH,DEN,SIAT,COAT,AP,OR,ORI,CC,P,RHO2,DLTT4,IL,JL,ILM2,JLPL,IUPFIX 8
     6,DLTR24,DLHR,PS,IS,DLTI2I,COSC,ANGVEL,ANGACC,RF,AMP,AVEL1,AVA,RH1SFIX 9
     76,CLI,R1R6,DLRN,DLCL,NTI,T,T1,T2,KL11,IUP1,IUPFL,ANG,NSW,MN,DLT4RIFIX 10
     8,R1I,KL21,R1R3,AV2,ILM3,DIT2D,DF,DR,DR2,DR2MDR,DEN2,FAC       FIX     11
     9,ILM4,D12TR2,D24TR2,JLM4,JM,JM1,JBLE,JDIF1,MIJ,JTI,JVI,JRNCN,KL22 FIX 12
      COMMON /THER/ THETA(130),R(84),SINE(130),COSINE(130),RP(130)  THER    2
      COMMON /BCF/XC(130),V(130),VN2(130),PSIR(130,2)               BCF     2
      COMMON /DUM/ RHO(132),XD(132),ED(132),FD(132)                 SETUP  12
      COMMON /WN/ WAV(130,4),WAV2(130,4),WAVF(130,4),WAV2F(130,4)   WN      2
      DOUBLE PRECISION WAVE,DTT2D,COSKD                             SETUP  14
      DOUBLE PRECISION PID,PD,DLTTD                                 SETUP  15
C                                                                   SETUP  16
      PID=3.14159265358979323846264343D-00                         SETUP  17
C                                                                   SETUP  18
      SRNN=SQRT(RNN)                                                SETUP  19
      NTI=0                                                         SETUP  20
      T=3.0                                                         SETUP  21
      T1=4.0                                                        SETUP  22
      T2=1.0                                                        SETUP  23
      IF (KL2.EQ.1.AND.KL11.NE.1) GO TO 6                           SETUP  24
      NTI=2                                                         SETUP  25
      T=2.0                                                         SETUP  26
      T1=2.0                                                        SETUP  27
      T2=0.0                                                        SETUP  28
    6 JLMN=JL-1                                                     SETUP  29
      JLPL=JL+1                                                     SETUP  30
      JLM2=JL-2                                                     SETUP  31
      JLM3=JL-3                                                     SETUP  32
      JLM4=JL-4                                                     SETUP  33
      ILMN=IL-1                                                     SETUP  34
      ILM2=IL-2                                                     SETUP  35
      ILM3=IL-3                                                     SETUP  36
      ILM4=IL-4                                                     SETUP  37
      ILPL=IL+1                                                     SETUP  38
```

```
        ILPLP=IL+2                                                    SETUP 39
        IM=IL/2+1                                                     SETUP 40
        IMMN=IM-1                                                     SETUP 41
        IMPL=IM+1                                                     SETUP 42
        IMM=IM/2+1                                                    SETUP 43
        IMP=IM+IMM-2                                                  SETUP 44
        IUPI=IMM+7                                                    SETUP 45
        IUPFL=IM+7                                                    SETUP 46
        IUP=IUPFL-IUPI+1                                              SETUP 47
        FAC=1.0/ILM2                                                  SETUP 48
        LN=1                                                          SETUP 49
        LNT=0                                                         SETUP 50
        NT=0                                                          SETUP 51
        MARK=0                                                        SETUP 52
        NSW=1                                                         SETUP 53
        MN=1                                                          SETUP 54
        DLTI2=DLTI*2.0                                                SETUP 55
        DLTI3=DLTI*6.0                                                SETUP 56
        DLTI2I=1.0/DLTI2                                              SETUP 57
        JM=2                                                          SETUP 58
        JM1=1                                                         SETUP 59
        JDIF1=JLM2                                                    SETUP 60
C                                                                     SETUP 61
C       CHANGE BELOW 130 IF THE FIRST DIMENSION OF TWO-DIMENSIONAL ARRAY SETUP 62
C       IS CHANGED                                                    SETUP 63
C                                                                     SETUP 64
        MIJ=130*JL                                                    SETUP 65
        MIJF=MIJ                                                      SETUP 66
C                                                                     SETUP 67
        READ(1,80) TR,OR,OI,AS,AT,PC,RMAX,IRPI                        SETUP 68
C                                                                     SETUP 69
C       TR - THICKNESS RATION FOR ELLIPTIC CYLINDER                   SETUP 70
C       OR AND OI RESPECTIVELY DETERMINE THICKNESS AND CAMBER OF A    SETUP 71
C       OR AND OI RESPECTIVELY DETERMINE THICKNESS AND CAMBER OF A JOUKOWSSETUP 72
C             JOUKOWSKI AIRFOIL,   OR.LE.0.0,  OI.LT.1.0,             SETUP 73
C                           IF OI.NE.0.0, THEN OR.NE.0.0              SETUP 74
C       AS - DETERMINES THE RADIUS OF CURVATURE OF AN JOUKOWSKI AIRFOIL SETUP 75
C             SUGGESTED VALUE OF AS.GE.0.04                           SETUP 76
C       AT - ANGLE OF ATTACK IN DEGREES                              SETUP 77
C       FLAT PLATE (ZERO THICKNESS) MUST BE AT ZERO ANGLE OF ATTACK   SETUP 78
C       PC.LT.1.0 - PERCENTAGE OF CHORD FROM LEADING EDGE AT WHICH THE SETUP 79
C                   ORIGIN OF PHYSICAL COORDINATES IS LOCATED         SETUP 80
C       RMAX - DISTANCE AT WHICH THE OUTER BOUNDARY IS KEPT FOR KL6=3  SETUP 81
C       SUGGESTED VALUE OF IRPI IS 4, IF KL5=1                        SETUP 82
C                                                                     SETUP 83
        WRITE(3,85)TR,OR,OI,AS,AT,PC,RMAX,IRPI                        SETUP 84
        ORI=OR*OR+OI*OI                                               SETUP 85
        IF (RF.NE.0.0) AT=AT-AMP                                      SETUP 86
        MAT=AT                                                        SETUP 87
        AT=AT*PI/180.0                                                SETUP 88
        ANG=AT                                                        SETUP 89
        ANGVEL=0.0                                                    SETUP 90
        AVEL1=0.0                                                     SETUP 91
        AV2=0.0                                                       SETUP 92
        AVA=0.0                                                       SETIIP 93
        ANGACC=0.0                                                    SETUP 94
        IF (ABS(AT).GT.1.0E-06) GO TO 7                               SETUP 95
        SIAT=0.0                                                      SETUP 96
        COAT=1.0                                                      SETUP 97
        GO TO 8                                                       SETUP 98
      7 SIAT=SIN(AT)                                                  SETUP 99
        COAT=COS(AT)                                                  SETUP100
      8 IF (KL6.NE.2) GO TO 11                                        SETUP101
C                                                                     SETUP102
        READ (1,80) XP,XN                                            SETUP103
C                                                                     SETUP104
C       XP - DETERMINES GRID-SPACING IN R-DIRECTION NEAR THE SURFACE  SETUP105
C       XN - DETERMINES GRID-SPACING IN R-DIRECTION NEAR THE OUTER    SETUP106
C             BOUNDARY                                                SETUP107
C                                                                     SETUP108
        YP=TANH(XP)                                                   SETUP109
        YN=TANH(XN)                                                   SETUP110
        XPN=XP+XN                                                     SETUP111
        YPN=YP+YN                                                     SETUP112
        AA=-(RI*YPN)/(RI-1.0)                                         SETUP113
        RI=0.0                                                        SETUP114
        YN=YN+AA                                                      SETUP115
        YPN=YPN+AA                                                    SETUP116
```

```
      WRITE (3,90) XP,XN,RI                              SETUP117
C                                                        SETUP118
C     DLTR - GRID INCREMENT IN R-DIRECTION               SETUP119
C     DLTT - GRID INCREMENT IN THETA-DIRECTION           SETUP120
C                                                        SETUP121
   11 DLTR=(1.0-RI)/JLMN                                 SETUP122
      DLTT=(2.0*PI)/ILM2                                 SETUP123
      C=DLTR/DLTT                                        SETUP124
      C2=C*C                                             SETUP125
      DTR2=DLTR*DLTR                                     SETUP126
      D12TR2=12.0/DTR2                                   SETUP127
      D24TR2=24.0/DTR2                                   SETUP128
      DTT2=DLTT*DLTT                                     SETUP129
      DLTT2=DLTT+DLTT                                    SETUP130
      DLTT2I=1.0/DLTT2                                   SETUP131
      DLTT4=DLTT2+DLTT2                                  SETUP132
      DLTR2=DLTR+DLTR                                    SETUP133
      DLTR2I=1.0/DLTR2                                   SETUP134
      DLH=DLTT/720.0                                     SETUP135
      DLTR24=DLTR2/24.0                                  SETUP136
      DLHR=DLTR2/720.0                                   SETUP137
      DIT2D=DLTI*DTT2*DF                                 SETUP138
      MTAIL=0                                            SETUP139
      AAT=ABS(AT)                                        SETUP140
      IF (AAT.LT.1.0E-05) GO TO 16                       SETUP141
      MTAIL=AAT/DLTT+4                                   SETUP142
      MTAIL=ISIGN(MTAIL,MAT)                             SETUP143
      IUPI=IUPI-MTAIL                                    SETUP144
      IUPFL=IUPFL-MTAIL                                  SETUP145
   16 CONTINUE                                           SETUP146
      DLTTD=(2.0*PID)/ILM2                               SETUP147
      DO 20 I=2,IL                                       SETUP148
      PD=DLTTD*(I-2)                                     SETUP149
      THETA(I)=PD                                        SETUP150
      SINE(I)=DSIN(PD)                                   SETUP151
   20 COSINE(I)=DCOS(PD)                                 SETUP152
      DTT2D=DLTTD*DLTTD                                  SETUP153
      DO 32 I=2,ILMN                                     SETUP154
      II=(I-1)/IMMN                                      SETUP155
      WAVE   =(I-1-(II*ILM2+1))*DLTTD                    SETUP156
      ED(I)=(I-1-(II*ILM2+1))                            SETUP157
      COSKD=DCOS(WAVE)                                   SETUP158
      WAV(I,1)=(3.0D-00*DSIN(WAVE))/(DLTTD*(2.0D-00+COSKD))   SETUP159
   32 WAV2(I,1)=12.0D-00*(1.0D-00-COSKD)/(5.0D-00+COSKD)/DTT2D  SETUP160
      WAV(IM,1)=0.0                                      SETUP161
      CALL COPY(WAV(2,1),WAV(2,2),ILM2)                  SETUP162
      CALL COPY(WAV(2,1),WAV(2,3),ILM2)                  SETUP163
      CALL COPY(WAV2(2,1),WAV2(2,2),ILM2)                SETUP164
      CALL COPY(WAV2(2,1),WAV2(2,3),ILM2)                SETUP165
      CALL SUFLE1(WAV(2,2),FD(2),2,ILM2)                 SETUP166
      CALL SUFLE1(WAV2(2,2),FD(2),2,ILM2)                SETUP167
      CALL COPY(WAV(2,2),WAV(2,4),ILM2)                  SETUP168
      CALL COPY(WAV2(2,2),WAV2(2,4),ILM2)                SETUP169
      DO 670 J=1,4                                       SETUP170
      CALL MZS(WAV(2,J),FAC,WAVF(2,J),ILM2)              SETUP171
  670 CALL MZS(WAV2(2,J),FAC,WAV2F(2,J),ILM2)            SETUP172
      WRITE(3,680)                                       SETUP173
  680 FORMAT(//,*     I*,9X,*EXACT WAVE NUMBER, K*,4X,*MODIFIED WAVE NUMBSETUP174
     1ER,  KM*,8X,3HK*K,17X,*MODIFIED SQUARE OF KM*,//)  SETUP175
      WRITE(3,690)(I,ED(I),WAV(I,1),ED(I)*ED(I),WAV2(I,1),I=2,IM)  SETUP176
  690 FORMAT(I5,1P4E25.5)                                SETUP177
C                                                        SETUP178
      DO 30 J=1,JL                                       SETUP179
   30 RHO(J)=DLTR*(J-1)+RI                               SETUP180
      IF (KL6.NE.3) GO TO 150                            SETUP181
      AA=ALOG(1.0/RMAX)/(PI*(RHO(1)-1.0))                SETUP182
      AP=AA*PI                                           SETUP183
  150 IF (KL4.NE.1) GO TO 50                             SETUP184
      CC=(TR-1.0)/(TR+1.0)                               SETUP185
      GO TO 55                                           SETUP186
   50 CC=((OR+SQRT(1.0-OI*OI))*(1.0-AS))**2              SETUP187
   55 YMIN=0.0                                           SETUP188
      YMAX=0.0                                           SETUP189
      DO 60 I=2,IL                                       SETUP190
      X1=COSINE(I)+OR                                    SETUP191
      X2=SINE(I)-OI                                      SETUP192
      DENOM= CC/(X1*X1+X2*X2)                            SETUP193
      XD(I)=X1*(DENOM+1.0)                               SETUP194
```

```
         XC(I)=X2*(DENOM-1.Ø)                                          SETUP195
           YMIN=AMIN1(XC(I),YMIN)                                      SETUP196
           YMAX=AMAX1(XC(I),YMAX)                                      SETUP197
    6Ø   CONTINUE                                                      SETUP198
         CL=XD(2)-XD(IM)                                               SETUP199
         CLI=1.Ø/CL                                                    SETUP2ØØ
         COSC=XD(IM)+CL*PC                                             SETUP2Ø1
         TH=-YMIN+YMAX                                                 SETUP2Ø2
         RATIO=TH*CLI                                                  SETUP2Ø3
         RNNCL=RNN*CLI                                                 SETUP2Ø4
         DLHRNC=2.Ø*DLH                                                SETUP2Ø5
         DLRN=2.Ø*DLH/RNN                                              SETUP2Ø6
         DLCL=DLH*CLI                                                  SETUP2Ø7
         XMIN=XD(IM)-COSC                                              SETUP2Ø8
         XMAX=XD(2)-COSC                                               SETUP2Ø9
         DELTA=2.Ø*SQRT(DLTI/(PI*RNNCL))                               SETUP21Ø
         WRITE (3,1Ø5) DELTA                                           SETUP211
         DLTRI2=2.Ø/DLTR                                               SETUP212
         DLTRI3=3.Ø/DLTR                                               SETUP213
         DLTTI3=3.Ø/DLTT                                               SETUP214
         DO 62 I=IM,IL                                                 SETUP215
    62   XC(I)=XC(I)*CLI                                               SETUP216
C                                                                      SETUP217
C        FOLLOWING PRINT-OUTS ARE FOR THE AIRFOIL OR ELLIPSE AT ZERO ANGLE SETUP218
C                 OF ATTACK                                            SETUP219
C                                                                      SETUP22Ø
         WRITE(3.125)                                                  SETUP221
   125 FORMAT(/,1ØX,2ØHNORMALIZED SURFACE Y,/)                         SETUP222
         WRITE(3,115) (XC(I),I=IM,IL)                                  SETUP223
         RADIUS=XC(IMPL)*XC(IMPL)*5Ø.Ø                                 SETUP224
         XNO=XD(IM)                                                    SETUP225
         ID=IMPL                                                       SETUP226
         DO 65 I=IM,IL                                                 SETUP227
           XC(I)=(XD(I)-XNO)*CLI                                       SETUP228
           ID=ID-1                                                     SETUP229
    65   XC(ID)=XC(I)                                                  SETUP23Ø
         WRITE (3,11Ø)                                                 SETUP231
         WRITE (3,115) (XC(I),I=IM,IL)                                 SETUP232
         IF (ABS(XC(IMPL)).GT.1.ØE-Ø8) RADIUS=RADIUS/XC(IMPL)          SETUP233
         IF (ABS(XC(IMPL)).LT.1.ØE-Ø8) RADIUS=Ø.Ø                      SETUP234
         WRITE(3,1ØØ) CC,CL,TH,RATIO,RNNCL,XMIN,XMAX,COSC,RADIUS       SETUP235
C                                                                      SETUP236
C        CC - SQUARE OF THE CONSTANT IN THE JOUKOWSKI TRANSFORMATION   SETUP237
C        CL - CHORD LENGTH                                             SETUP238
C        TH - THICKNESS OF THE BODY                                    SETUP239
C        RATIO - PER CENT THICKNESS OF THE BODY                        SETUP24Ø
C        RNNCL - RNN/CL                                                SETUP241
C        XMIN - LOCATION OF LEADING EDGE MEASURED ALONG CHORD          SETUP242
C        XMAX - LOCATION OF TRAILING EDGE MEASURED ALONG CHORD         SETUP243
C        COSC - X-DISTANCE FROM THE LEADING EDGE AT WHICH THE ORIGIN OF SETUP244
C               COORDINATE SYSTEM IS LOCATED. IT IS ALSO THE CENTER OF SETUP245
C               OSCILLATION  WHEN RF.NE.Ø.                             SETUP246
C        RADIUS - NORMALIZED RADIUS OF CURVATURE OF THE LEADING EDGE   SETUP247
C                                                                      SETUP248
         IF (KL5.EQ.1) GO TO 22Ø                                       SETUP249
         DO 21Ø I=2,ILMN                                               SETUP25Ø
   21Ø RP(I)=BVRT                                                      SETUP251
         GO TO 26Ø                                                     SETUP252
   22Ø DO 225 I=2,IRPI                                                 SETUP253
   225 RP(I)=BVRT                                                      SETUP254
         BVP=RP(IRPI)/IRPI                                             SETUP255
         II=IRPI+1                                                     SETUP256
         DO 23Ø I=II,34                                                SETUP257
   23Ø RP(I)=BVP*I                                                     SETUP258
         BVP=BVP*BPSI                                                  SETUP259
         ID=35                                                         SETUP26Ø
         IFL=IM-IRPI+1                                                 SETUP261
         DO 24Ø I=35,IFL                                               SETUP262
           ID=ID-1                                                     SETUP263
   24Ø RP(I)=BVP*ID                                                    SETUP264
         II=IFL+1                                                      SETUP265
         DO 245 I=II,IM                                                SETUP266
   245 RP(I)=BVRT*BPSI                                                 SETUP267
         II=IM                                                         SETUP268
         DO 25Ø I=IMPL,ILMN                                            SETUP269
           II=II-1                                                     SETUP27Ø
   25Ø RP(I)=RP(II)                                                    SETUP271
   26Ø CONTINUE                                                        SETUP272
```

```
      IF (KL4.GT.1.AND.ABS(OR).LT.1.E-5) GO TO 70        SETUP273
      RETURN                                             SETUP274
 70   DO 75 I=IMPL,IL                                    SETUP275
 75      XD(I)=0.664/SQRT(XC(I))                         SETUP276
      XD(IM)=0.0                                         SETUP277
      WRITE (3,120)                                      SETUP278
      WRITE (3,115) (XD(I),I=IM,IL)                      SETUP279
      RETURN                                             SETUP280
C                                                        SETUP281
 80   FORMAT(7F10.5,I5)                                  SETUP282
 85   FORMAT (//3X,7HTR    =,1PE12.5,3X,7HOR    =,E12.5,3X,7HOI   =,E12SETUP283
     1.5,3X7HAS   =,E12.5,3X7HAT   =E12.5//3X7HPC   =E12.5,3X,7HRMAX   SETUP284
     2=,E12.5,3X7HIRPI  =.I3)                            SETUP285
 90   FORMAT (//3x,7HXP    =,1PE12.5,3X,7HXN    =,E12.5,3X,7HRI   =,E12SETUP286
     1.5)                                                SETUP287
 95   FORMAT (//3X,6HMTAIL=,I2,3X6HIIC  =,I2,3X6HIFC  =,I3)  SETUP288
100   FORMAT (//,3X,7HCC    =,1PE12.5,3X,7HCL   =,E12.5,3X,7HTH   =,E1SETUP289
     12.5,3X,7HRATIO =,E12.5,3X,7HRNNCL =,E12.5//3X,7HXMIN =,E12.5,3X,7SETUP290
     2HXMAX  =,E12.5,3X7HCOSC  =,E12.5,3X,*LEADING EDGE RADIUS OF CURVATSETUP291
     3URE=*,E12.5,*  (APPROX)*,//)                       SETUP292
105   FORMAT(/3X,*DISPLACEMENT THICKNESS AT FIRST TIME STEP = *,1PE12.5 SETUP293
     1/)                                                 SETUP294
110   FORMAT (/,10X,23HNORMALIZED CHORD LENGTH,/)        SETUP295
115   FORMAT (1P10E13.5)                                 SETUP296
120   FORMAT (/,10X,33HBLASIUS SKIN FRICTION * SQRT(RNN),/)  SETUP297
      END                                                SETUP298
                                                         TRANS   2
                                                         TRANS   3
      SUBROUTINE TRANS                                   TRANS   4
C                                                        TRANS   5
C     THIS SUBROUTINE DEFINES THE TRANSFORMATION IN R-DIRECTION AND    TRANS   6
C     CALCULATES PHYSICAL COORDINATES AND THEIR DERIVATIVES WITH RESPECTTRANS   7
C     TO TRANSFORMED COORDINATES                         TRANS   8
C                                                        TRANS   9
      COMMON /FIX/ KL20,DLTRI2,DLTRI3,DLTTI3,JS,JLM2,JLM3,  FIX   2
     *DVRTM,IMM,IMMN,IM,IMPL,IMP,ILMN,ILPL,ILPLP,JLMN,NP,NT,NTMAX,KL0,KLFIX  3
     11,KL2,KL3,KL4,KL5,KL6,KL9,KL10,KL14,KL15,KL16,ITS,JTS,ITV,JTV,NMAXFIX  4
     2,LN,LNT,PI,XN,XPN,YN,YPN,MARK,DLTI2,DLTI3,RH1DR,RH1DR2,RI,RNN,RNNCFIX  5
     3L,SRNN,DLHRNC,BPSI,BVRT,DVRTLD,R1TR2,ADVRT,TIME,DLTI,MIJF,      FIX   6
     4DLTT2I,DLTR2I,DLTR,DLTR2,DTR2,DLTT,DLTT2,DTT2,C,C2,OI,AS,AT,RATIO,FIX  7
     5CL,DLH,DEN,SIAT,COAT,AP,OR,ORI,CC,P,RHO2,DLTT4,IL,JL,ILM2,JLPL,IUPFIX  8
     6,DLTR24,DLHR,PS,IS,DLTI2I,COSC,ANGVEL,ANGACC,RF,AMP,AVEL1,AVA,RH1SFIX  9
     76,CLI,R1R6,DLRN,DLCL,NTI,T,T1,T2,KL11,IUPI,IUPFL,ANG,NSW,MN,DLT4RIFIX 10
     8,R1I,KL21,R1R3,AV2,ILM3,DIT2D,DF,DR,DR2,DR2MDR,DEN2,FAC         FIX  11
     9,ILM4,D12TR2,D24TR2,JLM4,JM,JM1,JBLE,JDIF1,MIJ,JTI,JVI,JRNCN,KL22 FIX 12
      COMMON /THER/ THETA(130),R(84),SINE(130),COSINE(130),RP(130)    THER   2
      COMMON           PY(130,84),VRT(130,84),RR(84),RHO1S(84),       TRANS 12
     1A5(84),HRR(130),ARDT(130),HD(130)                  TRANS 13
      COMMON /CADI/ ARDTT(84),ARDTR(84),AR(84),CR(84),SC1(84),SC2(84), CADI   2
     1SC3(84)                                            CADI   3
      COMMON /LOADS/YTL(132),RHO1(84),XTF(132),PR(132),VRT1(132),     TRANS 15
     1XYT(132),YXT(132),DISS(130)                        TRANS 16
      COMMON /DUM/ RHO(132),XD(132),QD(132),RD(132)      TRANS 17
      COMMON /LMAIN/XY2(130,84),VRTN1(130,84),VN1(130),XY2RO(130,84),  TRANS 18
     1XY2T(130,84),AXYYRO(130,84),AXYYT(130,84),RHO1I(84)  TRANS 19
      COMMON /LSTR/ PSI(130,84),Y(130,84),VRTN2(130,84),H(130,84)     LSTR   2
     1,YY(130,84),PSIN1(130,84)                          LSTR   3
      LEVEL 2,PSI,Y,VRTN2,H,YY,PSIN1                     LSTR   4
      COMMON /LBOUND/ YR(130,84),YT(130,84),SH(130,84)   TRANS 21
      COMMON /LOUT/ X(130,84),XR(130,84),XT(130,84)      LOUT   2
      LEVEL 2,X,XR,XT                                    LOUT   3
      COMMON /HS/HA5T(130,84),HRRDLT(130,84),HTRR1(130,84)  TRANS 23
      LEVEL 2,YTL,RHO1,XTF,PR,VRT1,XYT,YXT,DISS          TRANS 24
      LEVEL 2,XY2,VRTN1,VN1,XY2RO,XY2T,AXYYRO,AXYYT,RHO1I  TRANS 25
      LEVEL 2,YR,YT,SH                                   TRANS 26
      LEVEL 2,HA5T,HRRDLT,HTRR1                          TRANS 27
      COMMON /BTRI2/AB(84,2),BB(84,2,2),CB(84,2),DB(84,2),D6LTR2,DDD   TRANS 28
     1,JIB,JFLB                                          TRANS 29
      COMMON /BPF/ ABF(84),BBF(84,2,2),CBF(84)           BPF   2
      WRITE (3,60)                                       TRANS 31
      DO 30 J=1,JL                                       TRANS 32
      IF (KL6.EQ.2) GO TO 20                             TRANS 33
      IF (KL6.EQ.3) GO TO 15                             TRANS 34
         R(J)=RHO(J)                                     TRANS 35
         RHO1(J)=1.0                                     TRANS 36
         RHO2=0.0                                        TRANS 37
         RHO1S(J)=1.0                                    TRANS 38
```

```
      GO TO 25                                              TRANS 39
 20   R(J)=(TANH(XPN*RHO(J)-XN)+YN)/YPN                     TRANS 40
      SQ=(R(J)*YPN-YN)                                      TRANS 41
      RHO1(J)=YPN/(XPN*(1.0-SQ*SQ))                         TRANS 42
      RHO1S(J)=RHO1(J)*RHO1(J)                              TRANS 43
      RHO2=2.0*RHO1S(J)*XPN*SQ                              TRANS 44
      GO TO 25                                              TRANS 45
 15   R(J)=EXP(-AP*(1.0-RHO(J)))                            TRANS 46
      RHO1(J)=1.0/(AP*R(J))                                 TRANS 47
      RHO1S(J)=RHO1(J)*RHO1(J)                              TRANS 48
      RHO2=-RHO1(J)/R(J)                                    TRANS 49
 25   RR(J)=R(J)*R(J)                                       TRANS 50
      RHO1I(J)=1.0/RHO1(J)                                  TRANS 51
      SC1(J)=RR(J)*RHO1S(J)                                 TRANS 52
      SC2(J)=RHO1(J)*R(J)+RR(J)*RHO2                        TRANS 53
      BBF(J,1,1)=-1.0/SC1(J)                                TRANS 54
      BBF(J,1,2)=SC2(J)/SC1(J)                              TRANS 55
      BBF(J,2,1)=0.0                                        TRANS 56
      BBF(J,2,2)=4.0                                        TRANS 57
      AR(J)=DLTI2*(-SC1(J)+(0.5*DLTR)*SC2(J))               TRANS 58
      CR(J)=DLTI2*(-SC1(J)-(0.5*DLTR)*SC2(J))               TRANS 59
      A5(J)=RR(J)*RNNCL                                     TRANS 60
 30      WRITE (3,65) J,R(J),RHO(J),RHO1(J),RHO1S(J),RHO2   TRANS 61
      DO 27 J=1,JL                                          TRANS 62
      SC3(J)=R(J)*RHO1(J)*RNNCL                             TRANS 63
      ARDTT(J)=(R(J)*RHO1(J))*(RNNCL*DLTI2*DTR2*3.0/DLTT)   TRANS 64
 27   ARDTR(J)=(R(J)*RHO1(J))*(RNNCL*DLTI2*DLTR*0.5)        TRANS 65
      D6LTR2=6.0*DLTR2I                                     TRANS 66
      DO 33 J=2,JLMN                                        TRANS 67
       ABF(J)=BBF(J-1,1,1)                                  TRANS 68
      AB(J,2)=BBF(J-1,1,2)                                  TRANS 69
      CBF(J)=BBF(J+1,1,1)                                   TRANS 70
      CB(J,2)=BBF(J+1,1,2)                                  TRANS 71
 33   CONTINUE                                              TRANS 72
C                                                           TRANS 73
C     ABF,AB,CBF,CB,BBF,HRRDLT ARE FOR STREAM               TRANS 74
C                                                           TRANS 75
      DO 50 J=1,JL                                          TRANS 76
      D2=(T*DTR2)*A5(J)                                     TRANS 77
      D3=RR(J)                                              TRANS 78
      D4=4.0*DLTI*RR(J)*RHO1S(J)                            TRANS 79
      DO 35 I=2,IL                                          TRANS 80
       SI=SINE(I)                                           TRANS 81
       CO=COSINE(I)                                         TRANS 82
       X1=CO+R(J)*OR                                        TRANS 83
       X2=SI-R(J)*OI                                        TRANS 84
       DENOM=1.0/(X1*X1+X2*X2)                              TRANS 85
       DENOM2=DENOM*DENOM                                   TRANS 86
       X3=2.0*(R(J)*ORI+OR*CO-OI*SI)                        TRANS 87
       X4=2.0*R(J)*(OR*SI+OI*CO)                            TRANS 88
       X10=CC*R(J)*DENOM                                    TRANS 89
       X11=1.0/R(J)                                         TRANS 90
       X5=X10+X11                                           TRANS 91
       X6=X10-X11                                           TRANS 92
      X15=X1*X5                                             TRANS 93
      X26=X2*X6                                             TRANS 94
      X15=X15-COSC                                          TRANS 95
       X(I,J)=X15*COAT+X26*SIAT                             TRANS 96
       Y(I,J)=X26*COAT-X15*SIAT                             TRANS 97
      YY(I,J)=Y(I,J)                                        TRANS 98
      XY2(I,J)=(X(I,J)*X(I,J)+Y(I,J)*Y(I,J))*0.5            TRANS 99
       X12=CC*(DENOM-R(J)*X3*DENOM2)                        TRANS100
       X13=1.0/RR(J)                                        TRANS101
       DXR=OR*X5+X1*(X12-X13)                               TRANS102
       DYR=-OI*X6+X2*(X12+X13)                              TRANS103
      CRX4D2=CC*X4*DENOM2*R(J)                              TRANS104
       DXT=(-SI*X5+X1*CRX4D2)                               TRANS105
       DYT=CO*X6+X2*CRX4D2                                  TRANS106
       XR(I,J)=DXR*COAT+DYR*SIAT                            TRANS107
       XT(I,J)=DXT*COAT+DYT*SIAT                            TRANS108
       YR(I,J)=DYR*COAT-DXR*SIAT                            TRANS109
       YT(I,J)=DYT*COAT-DXT*SIAT                            TRANS110
      H(I,J)=(XR(I,J)*YT(I,J)-XT(I,J)*YR(I,J))/R(J)         TRANS111
      XY2RO(I,J)=(X(I,J)*XR(I,J)+Y(I,J)*YR(I,J))*RHO1I(J)   TRANS112
       XY2T(I,J)=(X(I,J)*XT(I,J)+Y(I,J)*YT(I,J))            TRANS113
      HA5T(I,J)=D2*H(I,J)                                   TRANS114
      HTRR1(I,J)=HA5T(I,J)+D4                               TRANS115
```

```
            HRRDLT(I,J)=D3*H(I,J)                                        TRANS116
   35 CONTINUE                                                           TRANS117
      BBF(J,1,1)=10.0*BBF(J,1,1)                                         TRANS118
      BBF(J,1,2)=10.0*BBF(J,1,2)                                         TRANS119
   50 CONTINUE                                                           TRANS120
      IF (RF.NE.0.0) GO TO 53                                            TRANS121
      DO 51 J=1,JL                                                       TRANS122
      DO 51 I=2,IL                                                       TRANS123
      AXYYRO(I,J)=YR(I,J)*RHO1I(J)                                       TRANS124
   51 AXYYT(I,J)=YT(I,J)                                                 TRANS125
   53 CONTINUE                                                           TRANS126
      DO 36 J=1,JLMN                                                     TRANS127
      DO 36 I=2,IL                                                       TRANS128
      SH(I,J)=1.0/SQRT(H(I,J))                                           TRANS129
   36 CONTINUE                                                           TRANS130
      IF (ABS(H(2,JL)).GT.1.0E-10) GO TO 37                             TRANS131
      ID=IM                                                              TRANS132
      DO 34 I=3,IMMN                                                     TRANS133
      ID=ID+1                                                            TRANS134
      SH(I,JL)=1.0/SQRT(H(I,JL))                                         TRANS135
   34 SH(ID,JL)=1.0/SQRT(H(I,JL))                                        TRANS136
      SH(2,JL)=0.0                                                       TRANS137
      SH(IL,JL)=0.0                                                      TRANS138
      SH(IM,JL)=0.0                                                      TRANS139
      GO TO 39                                                           TRANS140
   37 DO 38 I=2,IL                                                       TRANS141
   38 SH(I,JL)=1.0/SQRT(H(I,JL))                                         TRANS142
   39 CONTINUE                                                           TRANS143
      DR=(1.0-R(JLMN))*(1.0-R(JLMN))                                     TRANS144
C                                                                        TRANS145
C                                                                        TRANS146
C     FOLLOWING PRINT-OUTS ARE FOR THE AIRFOIL OR ELLIPSE AT PRESCRIBED  TRANS146
C               ANGLE OF ATTACK                                          TRANS147
C                                                                        TRANS148
      WRITE (3,70)                                                       TRANS149
      WRITE (3,75) (Y(IMP,J),J=1,JL)                                     TRANS150
      DO 40 J=1,JL                                                       TRANS151
      YT(1,J)=YT(ILMN,J)                                                 TRANS152
      H(1,J)=H(ILMN,J)                                                   TRANS153
      HA5T(1,J)=HA5T(ILMN,J)                                             TRANS154
   40    XD(J)=(X(2,J)-X(2,JL))*CLI                                      TRANS155
      WRITE (3,80)                                                       TRANS156
      WRITE (3,85) (XD(J),J=1,JL)                                        TRANS157
      SR2CLI=SRNN*1.41421356*CLI                                        TRANS158
      DO 45 J=1,JL                                                       TRANS159
   45    XD(J)=(Y(IMP,J)-Y(IMP,JL))*SR2CLI                              TRANS160
      WRITE (3,90)                                                       TRANS161
      WRITE (3,85) (XD(J),J=1,JL)                                        TRANS162
      WRITE (3,100)                                                      TRANS163
      WRITE (3,75) (X(2 ,J),J=1,JL)                                      TRANS164
      RETURN                                                             TRANS165
C                                                                        TRANS166
   60 FORMAT (/4X,1HJ,8X,1HR,14X,3HRHO,12X,4HRHO1,11X,5HRHO1S,10X,4HRHO2 TRANS167
     1,//)                                                               TRANS168
   65 FORMAT (I5,1P5E15.4)                                               TRANS169
   70 FORMAT (/10X,7HY - TOP,/)                                          TRANS170
   75 FORMAT (1P10E13.5)                                                 TRANS171
   80 FORMAT (/,10X,15HX/L IN THE WAKE,/)                                TRANS172
   85 FORMAT (1P10E13.5)                                                 TRANS173
   90 FORMAT (/,10X,3HETA,/)                                             TRANS174
  100 FORMAT (/10X,8HX - WAKE,/)                                         TRANS175
      END                                                                TRANS176
                                                                         START    2
                                                                         START    3
      SUBROUTINE START                                                   START    4
C                                                                        START    5
C     THIS SUBROUTINE EITHER INITIALIZES THE FLOW VARIABLES FOR          START    6
C     IMPULSIVE START OR READS THEM FROM EXTERNAL STORAGE TO CONTINUE     START    7
C     SOLUTION FURTHER IN TIME                                           START    8
C                                                                        START    9
      COMMON /FIX/ KL20,DLTRI2,DLTRI3,DLTTI3,JS,JLM2,JLM3,               FIX      2
     *DVRTM,IMM,IMMN,IM,IMPL,IMP,ILMN,ILPL,ILPLP,JLMN,NP,NT,NTMAX,KL0,KL FIX      3
     11,KL2,KL3,KL4,KL5,KL6,KL9,KL10,KL14,KL15,KL16,ITS,JTS,ITV,JTV,NMAX FIX      4
     2,LN,LNT,PI,XN,XPN,YN,YPN,MARK,DLTI2,DLTI3,RH1DR,RH1DR2,RI,RNN,RNNC FIX      5
     3L,SRNN,DLHRNC,BPSI,BVRT,DVRTLD,R1TR2,ADVRT,TIME,DLTI,MIJF,         FIX      6
     4DLTT2I,DLTR2I,DLTR,DLTR2,DTR2,DLTT,DLTT2,DTT2,C,C2,OI,AS,AT,RATIO, FIX      7
     5CL,DLH,DEN,SIAT,COAT,AP,OR,ORI,CC,P,RHO2,DLTT4,IL,JL,ILM2,JLPL,IUPF FIX      8
     6,DLTR24,DLHR,PS,IS,DLTI2I,COSC,ANGVEL,ANGACC,RF,AMP,AVEL1,AVA,RH1SF FIX      9
     76,CLI,R1R6,DLRN,DLCL,NTI,T,T1,T2,KL11,IUPI,IUPFL,ANG,NSW,MN,DLT4RI FIX     10
```

```
      8,R1I,KL21,R1R3,AV2,ILM3,DIT2D,DF,DR,DR2,DR2MDR,DEN2,FAC            FIX   11
      9,ILM4,D12TR2,D24TR2,JLM4,JM,JM1,JBLE,JDIF1,MIJ,JTI,JVI,JRNCN,KL22 FIX   12
       COMMON /THER/ THETA(130),R(84),SINE(130),COSINE(130),RP(130)      THER   2
       COMMON /DUM1/AD(132),BD(132),CD(132),DD(132)                      DUM1   2
       COMMON /DUM/ ED(132),FD(132),QD(132),RD(132)                      DUM    2
       COMMON         PY(130,84),VRT(130,84),RR(84),RHO1S(84),           START 14
      1A5(84),HRR(130),ARDT(130),HD(130)                                 START 15
       COMMON /BCF/XC(130),V(130),VN2(130),PSIR(130,2)                   BCF    2
       COMMON /SMALLD/F(130,84)                                          SMALLD 2
       COMMON /LOADS/YTL(132),RHO1(84),XTF(132),PR(132),VRT1(132),       START 18
      1XYT(132),YXT(132),DISS(130)                                       START 19
       COMMON /LMAIN/XY2(130,84),VRTN1(130,84),VN1(130),XY2RO(130,84),   START 20
      1XY2T(130,84),AXYVRO(130,84),AXYYT(130,84),RHO1I(84)               START 21
       COMMON /LSTR/ PSI(130,84),Y(130,84),VRTN2(130,84),H(130,84)       LSTR   2
      1,YY(130,84),PSIN1(130,84)                                         LSTR   3
       LEVEL 2,PSI,Y,VRTN2,H,YY,PSIN1                                    LSTR   4
       COMMON /LBOUND/ YR(130,84),YT(130,84),SH(130,84)                  START 23
       COMMON /LSTAGP/ UP(33,84),UP1(33,84),UP2(33,84),RNRI(84)          START 24
       COMMON /SIDER/ SIR(130,84),SIT(130,84),SIRN1(130,84),SITN1(130,84)START 25
       COMMON /SU/ U(130),UN1(130),UN2(130),SQH1(130)                    START 26
       COMMON /WN/ WAV(130,4),WAV2(130,4),WAVF(130,4),WAV2F(130,4)       WN     2
       COMMON /LCROSS/ CROSST(130,84),VJLN2(130)                         LCROSS 2
       LEVEL 2,CROSST,VJLN2                                              LCROSS 3
       LEVEL 2,YTL,RHO1,XTF,PR,VRT1,XYT,YXT,DISS                         START 29
       LEVEL 2,XY2,VRTN1,VN1,XY2RO,XY2T,AXYVRO,AXYYT,RHO1I               START 30
       LEVEL 2, YR,YT,SH                                                 START 31
       LEVEL 2, UP,UP1,UP2,RNRI                                          START 32
       LEVEL 2, SIR,SIT,SIRN1,SITN1                                      START 33
       IF (KL2.EQ.1) GO TO 25                                            START 34
       DO 5 J=1,JL                                                       START 35
         DO 5 I=1,IL                                                     START 36
       VRTN1(I,J)=0.0                                                    START 37
         VRTN2(I,J)=0.0                                                  START 38
       CROSST(I,J)=0.0                                                   START 39
     5   VRT(I,J)=0.0                                                    START 40
       DO 15 I=2,IL                                                      START 41
       VJLN2(I)=0.0                                                      START 42
    15 VN2(I)=0.0                                                        START 43
       DO 350 J=1,JL                                                     START 44
       DO 350 I=1,IUP                                                    START 45
   350 UP2(I,J)=0.0                                                      START 46
       DO 70 I=2,IL                                                      START 47
    70 PSI(I,JL)=-Y(I,JL)                                                START 48
       DO 60 I=2,ILMN                                                    START 49
       AD(I)=PSI(I,JL)                                                   START 50
       CD(I)=RHO1I(JL)*(2.0*SINE(I)-YR(I,JL))                           START 51
       QD(I)=RHO1I(1)*(SINE(I)/RR(1)+SINE(I)-YR(I,1))                   START 52
       PSIR(I,1)=QD(I)                                                   START 53
    60 VN1(I)=-(YR(I,1)+RHO1(1)*QD(I))*SH(I,1)                          START 54
       CALL SETZ(F(1,1),0.0,MIJF)                                       START 55
       CALL STREAM                                                       START 56
       CALL MOVLEV(F(1,1),PSIN1(1,1),MIJF)                              START 57
       DO 65 J=2,JLMN                                                    START 58
       DO 65 I=2,ILMN                                                    START 59
       PSI(I,J)=F(I,J)                                                   START 60
    65 PY(I,J)=F(I,J)+Y(I,J)                                             START 61
       DO 66 I=2,IL                                                      START 62
       PY(I,1)=PSI(I,1)+Y(I,1)                                           START 63
    66 PY(I,JL)=0.0                                                      START 64
C      J1=JL/3                                                           START 65
C      J2=2*JL/3                                                         START 66
C      DO 1166 I=2,ILMN                                                  START 67
C      PE1=-SINE(I)*(1.0-RR(J1))/R(J1)-Y(I,J1)                          START 68
C      PE2=-SINE(I)*(1.0-RR(J2))/R(J2)-Y(I,J2)                          START 69
C      ER1=PE1-PSI(I,J1)                                                 START 70
C      ER2=PE2-PSI(I,J2)                                                 START 71
C1166 WRITE(3,9515) I,PE1,PSI(I,J1),ER1,PE2,PSI(I,J2),ER2               START 72
C9515 FORMAT(I5,1P6E20.12)                                              START 73
       DO 80 J=1,JL                                                      START 74
       PSI(IL,J)=PSI(2,J)                                                START 75
    80 PY(IL,J)=PY(2,J)                                                  START 76
       DO 75 I=2,ILMN                                                    START 77
       UN1(I)=SIT(I,1)*SH(I,1)/R(1)                                      START 78
       U(I)=UN1(I)                                                       START 79
    75 V(I)=VN1(I)                                                       START 80
       U(1)=U(ILMN)                                                      START 81
       U(IL)=U(2)                                                        START 82
       UN1(IL)=UN1(2)                                                    START 83
```

```
      VN1(IL)=VN1(2)                                           START 84
      UN1(1)=UN1(ILMN)                                         START 85
      VN1(1)=VN1(ILMN)                                         START 86
      V(1)=V(ILMN)                                             START 87
      V(IL)=V(2)                                               START 88
      DO 405 J=1,JLMN                                          START 89
      ID=0                                                     START 90
      DO 405 I=IUPI,IUPFL                                      START 91
      ID=ID+1                                                  START 92
  405 UP1(ID,J)=SIT(I,J)*SH(I,J)/R(J)                          START 93
      DO 406 I=1,IUP                                           START 94
  406 UP1(I,JL)=0.0                                            START 95
C                                                              START 96
C     UPS - POTENTIAL FLOW VELOCITY ON THE AIRFOIL CALCULATED  START 97
C           NUMERICALLY                                        START 98
C     UPSE - POTENTIAL FLOW VELOCITY ON THE AIRFOIL FROM EXACT SOLUTION START 99
C     PPR - PRESSURE CALCULATED FROM UPS                       START100
C     PPRE - PRESSURE CALCULATED FROM UPSE                     START101
C                                                              START102
      WRITE(3,89)                                              START103
   89 FORMAT(/,3X,1HI,2X,5HXC(I),5X,8HTHETA(I),4X,3HUPS,8X,4HUPSE,7X, START104
     1*UPS2*,7X,*PPR*,8X,*PPRE*,5X,*PRESSURE FORCE AT OUTER BOUNDARY*,/)START105
      CALL MZZ(U(2),U(2),AD(2),ILM2)                           START106
      CALL MZZ(V(2),V(2),BD(2),ILM2)                           START107
      CALL AZZ(AD(2),BD(2),AD(2),ILM2)                         START108
      CALL MZS(AD(2),0.5,AD(2),ILM2)                           START109
      CALL SETZ(BD(2),0.0,ILM2)                                START110
      CALL FFT(AD(2),BD(2),ILM2,-1)                            START111
      CALL MZZ(AD(2),WAVF(2,1),AD(2),ILM2)                     START112
      CALL MZZ(BD(2),WAVF(2,1),BD(2),ILM2)                     START113
      CALL MZS(BD(2),-1.0,BD(2),ILM2)                          START114
      CALL FFT(BD(2),AD(2),ILM2,1)                             START115
      DO 55 I=2,ILMN                                           START116
   55 ARDT(I)=BD(I)*SH(I,1)/R(1)                               START117
      DO 90 I=2,IL                                             START118
      UPS=-(YR(I,JL)+R1R3*(3.0*PSI(I,JL)-4.0*PSI(I,JLMN)+PSI(I,JLM2)))* START119
     1SH(I,JL)                                                 START120
      UPS2=UPS*UPS                                             START121
      PPR=1.0-UPS2                                             START122
      UPSE=-2.0*(SINE(I)*COAT+COSINE(I)*SIAT)*SH(I,JL)         START123
      PPRE=1.0-UPSE*UPSE                                       START124
   90 WRITE(3,95) I,XC(I),THETA(I),UPS,UPSE,UPS2,PPR,PPRE,ARDT(I) START125
   95 FORMAT(I4,1P2E10.3,5E11.3,13X,E12.5)                     START126
      CALL DERSI(0)                                            START127
      GO TO 85                                                 START128
C                                                              START129
   25 READ(1,120) MT,MT1,NTAP                                  START130
C                                                              START131
C     MT - NUMBER OF THE FILE TO BE READ FROM DISK-PACK        START132
C     MT1 - NUMBER OF FILES TO BE SKIPPED AFTER READING FILE MT START133
C     FILE MT IS READ FROM UNIT 10                             START134
C     NTAP - UNIT NUMBER FROM WHICH DATA IS READ.              START135
C            IF KL11.NE.1, THEN NTAP=10.                       START136
C            IF KL11.EQ.1, THEN NTAP CAN BE ANY INTEGER CONSISTENT WITH THESTART137
C                 COMPUTING SYSTEM AND MT1=MT                  START138
C                                                              START139
      CALL LOCATE(MT,10)                                       START140
      IF (KL11.EQ.1) GO TO 31                                  START141
      READ(10)((PY(I,J),I=2,IL),J=1,JL),((VRTN2(I,J),I=2,IL),J=1,JL),TIMSTART142
     1ED2,TIMEL,ANG,ANGVEL,(VN2(I),I=2,IL),((UP2(I,J),I=1,IUP),J=1,JL) START143
     2,(UN2(I),I=2,IL)                                         START144
      DO 135 J=1,JL                                            START145
      DO 135 I=2,IL                                            START146
  135 PSIN1(I,J)=PY(I,J)-Y(I,J)                                START147
      IF (RF.EQ.0.0) GO TO 140                                 START148
      DO 150 J=1,JL                                            START149
      DO 150 I=2,IL                                            START150
  150 PSIN1(I,J)=PSIN1(I,J)-ANGVEL*XY2(I,J)                    START151
  140 CONTINUE                                                 START152
      CALL SKIP(MT1,10)                                        START153
      WRITE (3,130) TIMED2                                     START154
   31 READ(NTAP)((PY(I,J),I=2,IL),J=1,JL),((  VRT(I,J),I=2,IL),J=1,JL), START155
     1TIMED1,TIMEL,ANG,AVEL1,(VN1(I),I=2,IL),((UP1(I,J),I=1,IUP),J=1,JL)START156
     2,(UN1(I),I=2,IL)                                         START157
      WRITE (3,130) TIMED1                                     START158
      IF (KL11.EQ.1) GO TO 32                                  START159
      DLTIC=TIMED1-TIMED2                                      START160
      IF (ABS(DLTI-DLTIC).GT.1.0E-05) GO TO 33                 START161
```

```
   32 IF (ABS(TIME-TIMED1).GT.1.ØE-Ø5) GO TO 34         START162
      GO TO 3Ø                                          START163
   33 WRITE(3,3ØØ)                                       START164
  3ØØ FORMAT(/3X*PROGRAM STOPPED BECAUSE INPUT DLTI DOES NOT CORRESPOND START165
     1WITH DLTI ON DISK-PACK*)                           START166
      STOP                                               START167
   34 WRITE(3,31Ø)                                       START168
  31Ø FORMAT(/3X*PROGRAM STOPPED BECAUSE INPUT TIME DOES NOT CORRESPOND START169
     1WITH TIME ON DISK-PACK*)                           START17Ø
      STOP                                               START171
   3Ø CONTINUE                                           START172
      DO 35 J=1,JL                                       START173
         DO 35 I=2,IL                                    START174
      AXYYRO(I,J)=AVEL1*XY2RO(I,J)+YR(I,J)*RHO1I(J)      START175
   35 PSI(I,J)=PY(I,J)-Y(I,J)                            START176
      IF (RF.EQ.Ø.Ø) GO TO 2ØØ                           START177
      DO 21Ø J=1,JL                                      START178
      VRT(1,J)=VRT(ILMN,J)                               START179
      DO 21Ø I=2,IL                                      START18Ø
  21Ø PSI(I,J)=PSI(I,J)-AVEL1*XY2(I,J)                   START181
  2ØØ CONTINUE                                           START182
      DO 16Ø I=2,IL                                      START183
      PSIR(I,1)=-(VN1(I)/(SH(I,1)*RHO1(1))+AVEL1*XY2RO(I,1)+YR(I,1)* START184
     1RHO1I(1))                                          START185
  16Ø CONTINUE                                           START186
      CALL DERSI(Ø)                                      START187
      DO 49 J=1,JL                                       START188
   49 VRT(1,J)=VRT(ILMN,J)                               START189
      CALL MOVLEV(VRT(1,1),VRTN1(1,1),MIJF)              START19Ø
      IF (KL11.EQ.1) GO TO 41                            START191
      DO 37 J=1,JL                                       START192
         DO 36 I=2,IL                                    START193
   36 VRT(I,J)=VRT(I,J)+VRT(I,J)-VRTN2(I,J)              START194
   37 VRT(1,J)=VRT(ILMN,J)                               START195
      DO 195 J=1,JLMN                                    START196
      DO 175 I=2,IL                                      START197
      AD(I)=PSI(I,J)                                     START198
      PSI(I,J)=PSI(I,J)+PSI(I,J)-PSIN1(I,J)              START199
      PSIN1(I,J)=AD(I)                                   START2ØØ
      PY(I,J)=PSI(I,J)+Y(I,J)                            START2Ø1
  175 CONTINUE                                           START2Ø2
  195 PSIN1(1,J)=PSIN1(ILMN,J)                           START2Ø3
      DO 4Ø I=2,IL                                       START2Ø4
      V(I)=VN1(I)+VN1(I)-VN2(I)                          START2Ø5
      VJLN2(I)=VRTN2(I,JL)                               START2Ø6
   4Ø U(I)=UN1(I)+UN1(I)-UN2(I)                          START2Ø7
      GO TO 51                                           START2Ø8
   41 DO 39 I=2,IL                                       START2Ø9
      VJLN2(I)=Ø.Ø                                       START21Ø
      V(I)=VN1(I)                                        START211
   39 VN2(I)=Ø.Ø                                         START212
      DO 5Ø I=2,IL                                       START213
   5Ø U(I)=UN1(I)                                        START214
      DO 38 J=1,JL                                       START215
      DO 38 I=1,IUP                                      START216
   38 UP2(I,J)=Ø.Ø                                       START217
      DO 27Ø J=1,JL                                      START218
      DO 26Ø I=2,IL                                      START219
      VRTN2(I,J)=Ø.Ø                                     START22Ø
  26Ø PSIN1(I,J)=PSI(I,J)                                START221
      VRTN2(1,J)=VRTN2(ILMN,J)                           START222
  27Ø PSIN1(1,J)=PSIN1(ILMN,J)                           START223
   51 V(1)=V(ILMN)                                       START224
      U(1)=U(ILMN)                                       START225
      VN1(1)=VN1(ILMN)                                   START226
      UN1(1)=UN1(ILMN)                                   START227
C                                                        START228
      DO 25Ø I=2,IL                                      START229
      PY(I,JL)=Ø.Ø                                       START23Ø
  25Ø PSI(I,JL)=-Y(I,JL)                                 START231
C                                                        START232
   85 DO 1Ø5 J=1,JL                                      START233
      PSI(1,J)=PSI(ILMN,J)                               START234
  1Ø5 PY(1,J)=PY(ILMN,J)                                 START235
      DO 378 I=1,IL                                      START236
      VRT1(I)=Ø.Ø                                        START237
      SIR(I,JL)=Ø.Ø                                      START238
  378 SIT(I,JL)=Ø.Ø                                      START239
```

```
        RETURN                                              START240
C                                                           START241
  120 FORMAT (3I5)                                          START242
  130 FORMAT (/,10X,24HREAD FROM TAPE FOR TIME=,1PE12.5/)   START243
        END                                                 START244
                                                            LOCATE  2
                                                            LOCATE  3
        SUBROUTINE LOCATE(M,N)                              LOCATE  4
C                                                           LOCATE  5
C       THIS SUBROUTINE LOCATES THE FILE TO BE READ FROM A DISK-PACK  LOCATE  6
C                                                           LOCATE  7
C       M=FILE # YOU WISH TO READ                           LOCATE  8
C       N=TAPE UNIT #                                       LOCATE  9
        MSKIP=M                                             LOCATE10
        NTAPE=N                                             LOCATE11
        ISKP=0                                              LOCATE12
        REWIND NTAPE                                        LOCATE13
        MCNT=1                                              LOCATE14
  1     IF (NTAPE.GT.0) GO TO 2                             LOCATE15
        WRITE(3,90) NTAPE                                   LOCATE16
  90    FORMAT(1H0,25HERROR:YOU REQUESTED TAPE ,I5)         LOCATE17
        RETURN                                              LOCATE18
  2     IF (MSKIP.GT.0) GO TO 3                             LOCATE19
        WRITE(3,91) MSKIP,NTAPE                             LOCATE20
  91    FORMAT(1H0,28HERROR:YOU REQUESTED SKIP OF ,I5,9H ON TAPE ,I5)  LOCATE21
  3     CONTINUE                                            LOCATE22
        IF (MCNT.LT.MSKIP) GO TO 4                          LOCATE23
        IF (ISKP.EQ.1) GO TO 6                              LOCATE24
        WRITE(3,92) MSKIP,NTAPE                             LOCATE25
  92    FORMAT(1H0,21HYOU ARE READING FILE ,I5,9H ON TAPE ,I5)  LOCATE26
  6     RETURN                                              LOCATE27
  4     READ (NTAPE) DUM                                    LOCATE28
        IF (EOF(NTAPE)) 5,4                                 LOCATE29
  5     MCNT=MCNT+1                                         LOCATE30
        GO TO 3                                             LOCATE31
        ENTRY SKIP                                          LOCATE32
C       M=# OF FILES YOU WISH TO SKIP                       LOCATE33
C       N=TAPE UNIT #                                       LOCATE34
        MSKIP=M                                             LOCATE35
        NTAPE=N                                             LOCATE36
        MCNT=0                                              LOCATE37
        ISKP=1                                              LOCATE38
        WRITE(3,93) MSKIP,NTAPE                             LOCATE39
  93    FORMAT(1H0,17HYOU HAVE SKIPPED ,I5,15H FILES ON TAPE ,I5)  LOCATE40
        GO TO 1                                             LOCATE41
        END                                                 LOCATE42
                                                            SUFLE4 2
                                                            SUFLE4 3
        SUBROUTINE SUFLE4(A,B,C,F,LN,M)                     SUFLE4 4
C                                                           SUFLE4 5
C       THIS SUBROUTINE CHANGES THE ORDER OF INDEXING OF FOUR ARRAYS  SUFLE4 6
C                                                           SUFLE4 7
        DIMENSION A(128),B(128),C(128),F(128)               SUFLE4 8
        COMMON /DUM/ AD(132),BD(132),CD(132),FD(132)        SUFLE4 9
C                                                           SUFLE410
        IF (LN.NE.1) GO TO 100                              SUFLE411
        RETURN                                              SUFLE412
  100   CALL COPY(A(1),AD(1),M)                             SUFLE413
        CALL COPY(B(1),BD(1),M)                             SUFLE414
        CALL COPY(C(1),CD(1),M)                             SUFLE415
        CALL COPY(F(1),FD(1),M)                             SUFLE416
        MH=M/2                                              SUFLE417
        MHP=MH+1                                            SUFLE418
        GO TO (311,300,320,350),LN                          SUFLE419
  300   CONTINUE                                            SUFLE420
        A(1)=CD(1)                                          SUFLE421
        C(1)=AD(1)                                          SUFLE422
        ID=M+1                                              SUFLE423
        DO 110 I=2,M                                        SUFLE424
        ID=ID-1                                             SUFLE425
        A(ID)=CD(I)                                         SUFLE426
        B(ID)=BD(I)                                         SUFLE427
        C(ID)=AD(I)                                         SUFLE428
  110   F(ID)=FD(I)                                         SUFLE429
  311   RETURN                                              SUFLE430
  320   CALL COPY(AD(1),A(MHP),MH)                          SUFLE431
        CALL COPY(BD(1),B(MHP),MH)                          SUFLE432
        CALL COPY(CD(1),C(MHP),MH)                          SUFLE433
```

```
        CALL COPY(FD(1),F(MHP),MH)                          SUFLE434
        CALL COPY(AD(MHP),A(1),MH)                          SUFLE435
        CALL COPY(BD(MHP),B(1),MH)                          SUFLE436
        CALL COPY(CD(MHP),C(1),MH)                          SUFLE437
        CALL COPY(FD(MHP),F(1),MH)                          SUFLE438
        RETURN                                              SUFLE439
 350    II=1                                                SUFLE440
        IFL=MHP                                             SUFLE441
        ID=MHP+1                                            SUFLE442
 160    DO 170 I=II,IFL                                     SUFLE443
        ID=ID-1                                             SUFLE444
        A(ID)=CD(I)                                         SUFLE445
        B(ID)=BD(I)                                         SUFLE446
        C(ID)=AD(I)                                         SUFLE447
 170    F(ID)=FD(I)                                         SUFLE448
        IF (IFL.EQ.M) RETURN                                SUFLE449
        II=MHP+1                                            SUFLE450
        IFL=M                                               SUFLE451
        ID=M+1                                              SUFLE452
        GO TO 160                                           SUFLE453
        END                                                 SUFLE454
                                                            SUFLE1 2
                                                            SUFLE1 3
        SUBROUTINE SUFLE1(F,FD,LN,M)                        SUFLE1 4
C                                                           SUFLE1 5
C       THIS SUBROUTINE CHANGES THE ORDER OF INDEXING OF ONE ARRAY   SUFLE1 6
C                                                           SUFLE1 7
        DIMENSION F(128),FD(128)                            SUFLE1 8
        MH=M/2                                              SUFLE1 9
        MHP=MH+1                                            SUFLE110
        GO TO (311,300,320,350),LN                          SUFLE111
 300    CALL COPY(F(1),FD(1),M)                             SUFLE112
        ID=M+1                                              SUFLE113
        DO 310 I=2,M                                        SUFLE114
        ID=ID-1                                             SUFLE115
 310    F(I)=FD(ID)                                         SUFLE116
 311    RETURN                                              SUFLE117
 320    CALL COPY(F(1),FD(1),M)                             SUFLE118
        CALL COPY(FD(1),F(MHP),MH)                          SUFLE119
        CALL COPY(FD(MHP),F(1),MH)                          SUFLE120
        RETURN                                              SUFLE121
 350    CALL COPY(F(1),FD(1),M)                             SUFLE122
        II=1                                                SUFLE123
        IFL=MHP                                             SUFLE124
        ID=MHP+1                                            SUFLE125
 360    DO 370 I=II,IFL                                     SUFLE126
        ID=ID-1                                             SUFLE127
 370    F(I)=FD(ID)                                         SUFLE128
        IF (IFL.EQ.M) RETURN                                SUFLE129
        II=MHP+1                                            SUFLE130
        IFL=M                                               SUFLE131
        ID=M+1                                              SUFLE132
        GO TO 360                                           SUFLE133
        END                                                 SUFLE134
                                                            DERSI  2
                                                            DERSI  3
        SUBROUTINE DERSI(IT)                                DERSI  4
C                                                           DERSI  5
C       THIS SUBROUTINE CALCULATES DERIVATIVES OF STREAM FUNCTION    DERSI  6
C                                                           DERSI  7
        COMMON /FIX/ KL20,DLTRI2,DLTRI3,DLTTI3,JS,JLM2,JLM3,         FIX   2
       *DVRTM,IMM,IMMN,IM,IMPL,IMP,ILMN,ILPL,ILPLP,JLMN,NP,NT,NTMAX,KL0,KLFIX   3
       11,KL2,KL3,KL4,KL5,KL6,KL9,KL10,KL14,KL15,KL16,ITS,JTS,ITV,JTV,NMAXFIX   4
       2,LN,LNT,PI,XN,XPN,YN,YPN,MARK,DLTI2,DLTI3,RH1DR,RH1DR2,RI,RNN,RNNCFIX   5
       3L,SRNN,DLHRNC,BPSI,BVRT,DVRTLD,R1TR2,ADVRT,TIME,DLTI,MIJF,          FIX   6
       4DLTT2I,DLTR2I,DLTR,DLTR2,DTR2,DLTT,DLTT2,DTT2,C,C2,OI,AS,AT,RATIO,FIX   7
       5CL,DLH,DEN,SIAT,COAT,AP,OR,ORI,CC,P,RHO2,DLTT4,IL,JL,ILM2,JLPL,IUPFIX   8
       6,DLTR24,DLHR,PS,IS,DLTI2I,COSC,ANGVEL,ANGACC,RF,AMP,AVEL1,AVA,RH1SFIX   9
       76,CLI,R1R6,DLRN,DLCL,NTI,T,T1,T2,KL11,IUPI,IUPFL,ANG,NSW,MN,DLT4RIFIX  10
       8,R1I,KL21,R1R3,AV2,ILM3,DIT2D,DF,DR,DR2,DR2MDR,DEN2,FAC           FIX  11
       9,ILM4,D12TR2,D24TR2,JLM4,JM,JM1,JBLE,JDIF1,MIJ,JTI,JVI,JRNCN,KL22 FIX  12
        COMMON /DUM1/AD(132),BD(132),CD(132),DD(132)        DUM1  2
        COMMON /DUM/ ED(132),FD(132),QD(132),RD(132)        DUM   2
        COMMON PY(130,84),VRT(130,84),RR(84),RHO1S(84),     DERSI 11
       1A5(84),HRR(130),ARDT(130),HD(130)                   DERSI 12
        COMMON /LOADS/YTL(132),RHO1(84),XTF(132),PR(132),VRT1(132),   DERSI 13
       1XYT(132),YXT(132),DISS(130)                         DERSI 14
        COMMON /LMAIN/XY2(130,84),VRTN1(130,84),VN1(130,84),XY2RO(130,84),  DERSI 15
```

```
      1XY2T(130,84),AXYYRO(130,84),AXYYT(130,84),RHO1I(84)              DERSI 16
       COMMON /LSTR/ PSI(130,84),Y(130,84),VRTN2(130,84),H(130,84)      LSTR   2
      1,YY(130,84),PSIN1(130,84)                                        LSTR   3
       LEVEL 2,PSI,Y,VRTN2,H,YY,PSIN1                                   LSTR   4
       COMMON /SIDER/ SIR(130,84),SIT(130,84),SIRN1(130,84),SITN1(130,84)SIDER 2
       LEVEL 2,SIR,SIT,SIRN1,SITN1                                      SIDER  3
       COMMON /LBOUND/ YR(130,84),YT(130,84),SH(130,84)                 DERSI 19
       COMMON /BCF/XC(130),V(130),VN2(130),PSIR(130,2)                  BCF    2
       LEVEL 2,YTL,RHO1,XTF,PR,VRT1,XYT,YXT,DISS                        DERSI 21
       LEVEL 2,XY2,VRTN1,VN1,XY2RO,XY2T,AXYYRO,AXYYT,RHO1I              DERSI 22
       LEVEL 2, YR,YT,SH                                                DERSI 23
C                                                                       DERSI 24
       K=1                                                              DERSI 25
       IF(JM.NE.2) K=2                                                  DERSI 26
       CALL SETZ(AD(JM),1.0,JDIF1)                                      DERSI 27
       CALL SETZ(BD(JM),4.0,JDIF1)                                      DERSI 28
       CALL SETZ(CD(JM),1.0,JDIF1)                                      DERSI 29
       AD(JM)=0.0                                                       DERSI 30
       CD(JLMN)=0.0                                                     DERSI 31
       DO 110 I=2,ILMN                                                  DERSI 32
       DD(JM1)=PSIR(I,K)                                                DERSI 33
       DO 260 J=JM,JLM2,2                                               DERSI 34
       DD(J)=(PSI(I,J+1)-PSI(I,J-1))*DLTRI3                             DERSI 35
  260  DD(J+1)=(PSI(I,J+2)-PSI(I,J))*DLTRI3                             DERSI 36
       DD(JM)=DD(JM)-DD(JM1)                                            DERSI 37
       DD(JLMN)=DD(JLMN)+AXYYRO(I,JL)                                   DERSI 38
       CALL TRIB(AD(JM),BD(JM),CD(JM),ED(JM),DD(JM),JDIF1)              DERSI 39
       PSIR(I,2)=DD(JBLE-1)                                             DERSI 40
       DO 80 J=JM1,JLMN                                                 DERSI 41
   80  SIR(I,J)=DD(J)+AXYYRO(I,J)                                       DERSI 42
       IF (IT.NE.2) GO TO 110                                           DERSI 43
       DO 230 J=1,JLMN                                                  DERSI 44
  230  SIRN1(I,J)=DD(J)*RHO1(J)+YR(I,J)                                 DERSI 45
       SIRN1(I,JL)=-ANGVEL*RHO1(JL)*XY2RO(I,JL)                         DERSI 46
  110  CONTINUE                                                         DERSI 47
       DO 120 J=JM1,JLMN                                                DERSI 48
       SIR(IL,J)=SIR(2,J)                                               DERSI 49
  120  SIR(1,J)=SIR(ILMN,J)                                             DERSI 50
       RETURN                                                           DERSI 51
       END                                                             DERSI 52
                                                                        ADI    2
                                                                        ADI    3
       SUBROUTINE ADI                                                   ADI    4
C                                                                       ADI    5
C      THIS SUBROUTINE SOLVES THE VORTICITY EQUATION                    ADI    6
C                                                                       ADI    7
       COMMON /FIX/ KL20,DLTRI2,DLTRI3,DLTTI3,JS,JLM2,JLM3,             FIX    2
      *DVRTM,IMM,IMMN,IM,IMPL,IMP,ILMN,ILPL,ILPLP,JLMN,NP,NT,NTMAX,KL0,KLFIX 3
      11,KL2,KL3,KL4,KL5,KL6,KL9,KL10,KL14,KL15,KL16,ITS,JTS,ITV,JTV,NMAXFIX 4
      2,LN,LNT,PI,XN,XPN,YN,YPN,MARK,DLTI2,DLTI3,RH1DR,RH1DR2,RI,RNN,RNNCFIX 5
      3L,SRNN,DLHRNC,BPSI,BVRT,DVRTLD,R1TR2,ADVRT,TIME,DLTI,MIJF,        FIX   6
      4DLTT2I,DLTR2I,DLTR,DLTR2,DTR2,DLTT,DLTT2,DTT2,C,C2,OI,AS,AT,RATIO,FIX  7
      5CL,DLH,DEN,SIAT,COAT,AP,OR,ORI,CC,P,RHO2,DLTT4,IL,JL,ILM2,JLPL,IUPFIX 8
      6,DLTR24,DLHR,PS,IS,DLTI2I,COSC,ANGVEL,ANGACC,RF,AMP,AVEL1,AVA,RH1SFIX 9
      76,CLI,R1R6,DLRN,DLCL,NTI,T,T1,T2,KL11,IUPI,IUPFL,ANG,NSW,MN,DLT4RIFIX 10
      8,R1I,KL21,R1R3,AV2,ILM3,DIT2D,DF,DR,DR2,DR2MDR,DEN2,FAC           FIX  11
      9,ILM4,DI2TR2,D24TR2,JLM4,JM,JM1,JBLE,JDIF1,MIJ,JTI,JVI,JRNCN,KL22 FIX  12
       COMMON        PY(130,84),VRT(130,84),RR(84),RHO1S(84),           ADI    9
      1A5(84),HRR(130),ARDT(130),HD(130)                                ADI   10
       COMMON /BPF/ ABF(84),BBF(84,2,2),CBF(84)                         BPF    2
       COMMON /CADI/ ARDTT(84),ARDTR(84),AR(84),CR(84),SC1(84),SC2(84), CADI   2
      1SC3(84)                                                          CADI   3
       COMMON /BCF/XC(130),V(130),VN2(130),PSIR(130,2)                  BCF    2
       COMMON /THER/ THETA(130),R(84),SINE(130),COSINE(130),RP(130)     THER   2
       COMMON /SU/ U(130),UN1(130),UN2(130),SQH1(130)                   ADI   15
       COMMON /DUM/ ED(132),FD(132),QD(132),RD(132)                     DUM    2
       COMMON /DUM1/AD(132),BD(132),CD(132),DD(132)                     DUM1   2
       COMMON /CTRIP/ AI(130),BI(130),CI(130),DB(130,2),AF(2,2),BF(2,2),ADI   18
      1CF(2,2)                                                          ADI   19
       COMMON /SMALLD/ F(130,84)                                        ADI   20
       COMMON /LOADS/YTL(132),RHO1(84),XTF(132),PR(132),VRT1(132),      ADI   21
      1YXT(132),DISS(130)                                               ADI   22
       COMMON /LMAIN/XY2(130,84),VRTN1(130,84),VN1(130),XY2RO(130,84),  ADI   23
      1XY2T(130,84),AXYYRO(130,84),AXYYT(130,84),RHO1I(84)              ADI   24
       COMMON /LSTR/ PSI(130,84),Y(130,84),VRTN2(130,84),H(130,84)      LSTR   2
      1,YY(130,84),PSIN1(130,84)                                        LSTR   3
       LEVEL 2,PSI,Y,VRTN2,H,YY,PSIN1                                   LSTR   4
       COMMON /HS/HA5T(130,84),HRRDLT(130,84),HTRR1(130,84)             ADI   26
```

```
        COMMON /SIDER/ SIR(130,84),SIT(130,84),VR(130,84),SITN1(130,84)   ADI   27
        COMMON /LBOUND/ YR(130,84),YT(130,84),SH(130,84)                  ADI   28
        COMMON /WN/ WAV(130,4),WAV2(130,4),WAVF(130,4),WAV2F(130,4)       WN     2
        COMMON /LCROSS/ CROSST(130,84),VJLN2(130)                         LCROSS 2
        LEVEL 2,CROSST,VJLN2                                              LCROSS 3
        LEVEL 2,YTL,RHO1,XTF,PR,VRT1,XYT,YXT,DISS                         ADI   31
        LEVEL 2,XY2,VRTN1,VN1,XY2RO,XY2T,AXYYRO,AXYYT,RHO1I               ADI   32
        LEVEL 2,HA5T,HRRDLT,HTRR1                                         ADI   33
        LEVEL 2,SIR,SIT,VR,SITN1                                          ADI   34
        LEVEL 2,YR,YT,SH                                                  ADI   35
        T1T=(T1*DLTI*DF)/(64.0*T*CL)                                      ADI   36
        DO 750 J=1,JLMN                                                   ADI   37
        DO 710 I=2,ILMN                                                   ADI   38
  710   AD(I+2)=ABS(SIR(I,J))*VRTN1(I,J)*SH(I,J)*RHO1(J)                  ADI   39
        AD(1)=AD(ILMN)                                                    ADI   40
        AD(2)=AD(IL)                                                      ADI   41
        AD(3)=AD(ILPL)                                                    ADI   42
        AD(ILPLP)=AD(4)                                                   ADI   43
        AD(IL+3)=AD(5)                                                    ADI   44
        AD(IL+4)=AD(6)                                                    ADI   45
        DO 720 I=2,ILMN                                                   ADI   46
  720   CD(I)=(AD(I-1)-6.0*(AD(I)+AD(I+4))+15.0*(AD(I+1)+AD(I+3))-20.0*   ADI   47
       2AD(I+2)+AD(I+5))*T1T                                             ADI   48
        DO 730 I=2,ILMN                                                   ADI   49
  730   VRTN2(I,J)=CD(I)+VRTN2(I,J)                                       ADI   50
        VRTN2(IL,J)=VRTN2(2,J)                                            ADI   51
        VRTN2(1,J)=VRTN2(ILMN,J)                                          ADI   52
  750   CONTINUE                                                          ADI   53
        DO 200 J=1,JL                                                     ADI   54
        DO 200 I=2,ILMN                                                   ADI   55
  200   F(I,J)=HA5T(I-1,J)*VRTN2(I-1,J)+4.0*HA5T(I,J)*VRTN2(I,J)+        ADI   56
       1HA5T(I+1,J)*VRTN2(I+1,J)                                         ADI   57
        CALL MOVLEV(F(1,1),VRTN2(1,1),MIJF)                              ADI   58
C                                                                        ADI   59
C                                                                        ADI   60
C       CHANGE BELOW 130 IF THE FIRST DIMENSION OF TWO-DIMENSIONAL ARRAY ADI   61
C       IS CHANGED                                                       ADI   62
C                                                                        ADI   63
        CALL MOVLEV(PSI(1,1),F(1,1),MIJF-130)                            ADI   64
        DO 260 J=1,JLMN                                                   ADI   65
        CALL SETZ(FD(2),0.0,ILM2)                                        ADI   66
        CALL FFT(F(2,J),FD(2),ILM2,-1)                                   ADI   67
        CALL MZZ(F(2,J),WAVF(2,1),F(2,J),ILM2)                           ADI   68
        CALL MZZ(FD(2),WAVF(2,1),FD(2),ILM2)                             ADI   69
        CALL MZS(FD(2),-1.0,FD(2),ILM2)                                  ADI   70
        CALL FFT(FD(2),F(2,J),ILM2,1)                                    ADI   71
        DO 250 I=2,ILMN                                                   ADI   72
  250   SIT(I,J)=FD(I)+AXYYT(I,J)                                         ADI   73
        SIT(1,J)=SIT(ILMN,J)                                              ADI   74
  260   SIT(IL,J)=SIT(2,J)                                                ADI   75
C                                                                        ADI   76
        DO 300 N=1,NMAX                                                   ADI   77
C                                                                        ADI   78
        MN=MOD(N,2)                                                       ADI   79
        IF (MOD((N-1),4).NE.0) GO TO 210                                 ADI   80
        LN=0                                                              ADI   81
  210   LN=LN+1                                                           ADI   82
        GO TO (220,220,230,230),LN                                       ADI   83
  220   NSW=1                                                             ADI   84
        GO TO 240                                                         ADI   85
  230   NSW=2                                                             ADI   86
  240   CONTINUE                                                          ADI   87
C                                                                        ADI   88
        IF (N.LT.3) GO TO 209                                            ADI   89
        IF (MOD((N+2),5).NE.0) GO TO 245                                 ADI   90
  209   CONTINUE                                                          ADI   91
        JM=2                                                              ADI   92
        JI=1                                                              ADI   93
        GO TO 246                                                         ADI   94
  245   JM=JBLE                                                           ADI   95
        JI=JM                                                             ADI   96
        IF (JM.EQ.2) JI=1                                                ADI   97
  246   JM1=JM-1                                                          ADI   98
        JDIF=JLMN-JI +1                                                  ADI   99
        JDIF1=JLMN-JM+1                                                  ADI  100
        MIJ=IL*(JL-JM1+1)                                                ADI  101
        JTI=JM                                                           ADI  102
        JVI=JI                                                           ADI  103
```

```
C                                                                      ADI  104
      CALL DERSI(0)                                                    ADI  105
C                                                                      ADI  106
      CALL CROSS                                                       ADI  107
      DO 270 J=JM1,JL                                                  ADI  108
      DO 270 I=2,ILMN                                                  ADI  109
 270 CROSST(I,J)=HA5T(I-1,J)*F(I-1,J)+4.0*HA5T(I,J)*F(I,J)+HA5T(I+1,J)*ADI 110
     1F(I+1,J)                                                         ADI  111
      DO 150 J=JM1,JL                                                  ADI  112
      DO 150 I=2,ILM2,2                                                ADI  113
      F(I,J)=VRTN2(I,J)+CROSST(I,J)                                    ADI  114
 150 F(I+1,J)=VRTN2(I+1,J)+CROSST(I+1,J)                               ADI  115
C                                                                      ADI  116
      JFL=JLMN                                                         ADI  117
      DO 350 J=JI,JFL                                                  ADI  118
      DO 330 I=2,ILMN                                                  ADI  119
      AI(I)=HA5T(I-1,J)+ARDTT(J)*SIR(I-1,J)                            ADI  120
      BI(I)=4.0*HA5T(I,J)                                              ADI  121
      CI(I)=HA5T(I+1,J)-ARDTT(J)*SIR(I+1,J)                            ADI  122
       DB(I,2)=0.0                                                     ADI  123
 330 DB(I,1)=F(I,J)                                                    ADI  124
      CALL SUFLE4(AI(2),BI(2),CI(2),DB(2,1),LN,ILM2)                   ADI  125
      CALL BTRIP(ILMN)                                                 ADI  126
      CALL SUFLE1(DB(2,1),RD(2),LN,ILM2)                               ADI  127
      CALL COPY(DB(2,1),VRT(2,J),ILM2)                                 ADI  128
 350 CONTINUE                                                          ADI  129
C                                                                      ADI  130
      DO 380 I=2,ILMN                                                  ADI  131
      JII=JI                                                           ADI  132
      JDIFL=JDIF                                                       ADI  133
      DO 410 J=JM,JLMN                                                 ADI  134
      AD(J)=AR(J)  -ARDTR(J) *SIT(I,J-1)                               ADI  135
      BD(J)=HTRRI1(I,J)                                                ADI  136
      CD(J)=CR(J)  +ARDTR(J) *SIT(I,J+1)                               ADI  137
 410 DD(J)=HA5T(I,J)*VRT(I,J)                                          ADI  138
      DD(JLMN)=DD(JLMN)-CD(JLMN)*VRT(I,JL)                             ADI  139
      CD(JLMN)=0.0                                                     ADI  140
      IF (JM.NE.2) GO TO 415                                           ADI  141
      IF (I.GE.IMM.AND.I.LE.IMP) GO TO 417                             ADI  142
      AD(JM1)=0.0                                                      ADI  143
      BD(JM1)=HA5T(I,JM1)-2.0*ARDTR(JM1)*SIT(I,JM1)                    ADI  144
      CD(JM1)=2.0*ARDTR(JM1)*SIT(I,JM)                                 ADI  145
      DD(JM1)=HA5T(I,JM1)*VRT(I,JM1)                                   ADI  146
      GO TO 416                                                        ADI  147
 417 JII=JI+1                                                          ADI  148
      JDIFL=JDIF-1                                                     ADI  149
      AD(JII)=0.0                                                      ADI  150
      VRT(I,1)=0.0                                                     ADI  151
      GO TO 416                                                        ADI  152
 415 CONTINUE                                                          ADI  153
      DD(JM)=DD(JM)-AD(JM)*VRT(I,JM1)                                  ADI  154
      AD(JM)=0.0                                                       ADI  155
 416 CALL TRIB(AD(JII),BD(JII),CD(JII),ED(JII),DD(JII),JDIFL)          ADI  156
      DO 420 J=JII,JLMN                                                ADI  157
 420 VRT(I,J)=DD(J)                                                    ADI  158
 380 CONTINUE                                                          ADI  159
C                                                                      ADI  160
      DO 450 J=JI,JLMN                                                 ADI  161
      VRT(IL,J)=VRT(2,J)                                               ADI  162
 450 VRT(1,J)=VRT(ILMN,J)                                              ADI  163
      IF (JM.NE.2) GO TO 510                                           ADI  164
      CALL BOUND                                                       ADI  165
      GO TO 520                                                        ADI  166
 510 CALL MOVLEV(PSI(2,JM1),ED(2),ILM2)                               ADI  167
      CALL COPY(PSIR(2,2),QD(2),ILM2)                                  ADI  168
 520 CALL MOVLEV(PSI(2,JL),AD(2),ILM2)                                 ADI  169
      DO 500 I=2,ILM4,4                                                ADI  170
      CD(I)=-AXYYRO(I,JL)                                              ADI  171
      CD(I+1)=-AXYYRO(I+1,JL)                                          ADI  172
      CD(I+2)=-AXYYRO(I+2,JL)                                          ADI  173
 500 CD(I+3)=-AXYYRO(I+3,JL)                                           ADI  174
      DO 700 J=JM1,JL                                                  ADI  175
      DO 700 I=2,ILM2,2                                                ADI  176
      F(I,J)=-HRRDLT(I,J)*VRT(I,J)                                     ADI  177
      F(I+1,J)=-HRRDLT(I+1,J)*VRT(I+1,J)                               ADI  178
 700 CONTINUE                                                          ADI  179
      CALL STREAM                                                      ADI  180
      DO 795 J=JM,JLMN                                                 ADI  181
```

```
      795 CALL MOVLEV(F(2,J),PSI(2,J),ILM2)                          ADI 182
          CALL SETZ(PY(1,JL),Ø.Ø,IL)                                 ADI 183
          DO 8ØØ J=JM,JLMN                                           ADI 184
          PSI(1,J)=PSI(ILMN,J)                                       ADI 185
      8ØØ PSI(IL,J)=PSI(2,J)                                         ADI 186
          DO 1Ø4Ø J=1,JLMN                                           ADI 187
          DO 1Ø4Ø I=2,ILM4,4                                        ADI 188
          PY(I,J)=PSI(I,J)+YY(I,J)                                   ADI 189
          PY(I+1,J)=PSI(I+1,J)+YY(I+1,J)                             ADI 19Ø
          PY(I+2,J)=PSI(I+2,J)+YY(I+2,J)                             ADI 191
     1Ø4Ø PY(I+3,J)=PSI(I+3,J)+YY(I+3,J)                             ADI 192
          DO 1Ø6Ø J=1,JLMN                                           ADI 193
          PY(IL,J)=PY(2,J)                                           ADI 194
     1Ø6Ø PY(1,J)=PY(ILMN,J)                                         ADI 195
          IF (JM.NE.2) GO TO 47Ø                                     ADI 196
          DO 46Ø I=2,ILM2,2                                         ADI 197
          U(I)=SIT(I,1)*SH(I,1)/R(1)                                 ADI 198
      46Ø U(I+1)=SIT(I+1,1)*SH(I+1,1)/R(1)                           ADI 199
          U(IL)=U(2)                                                 ADI 2ØØ
          U(1)=U(ILMN)                                               ADI 2Ø1
      47Ø CONTINUE                                                   ADI 2Ø2
    C                                                                ADI 2Ø3
    C     PSI AND VRT OF BODY CO-ORDINATE                            ADI 2Ø4
    C                                                                ADI 2Ø5
          ADVRT=Ø.Ø                                                  ADI 2Ø6
          CALL SZS(VRT(2,JLMN),AV2,AD(2),ILM2)                       ADI 2Ø7
          CALL MZZ(HRR(2),AD(2),CD(1),ILM2)                          ADI 2Ø8
     1Ø8Ø DO 1Ø65 I=2,ILMN                                           ADI 2Ø9
     1Ø65 AD(I-1)=( -PY(I,JLMN)*DEN-CD(I-1))                         ADI 21Ø
          CALL MZZ(AD(1),HD(2),BD(1),ILM2)                           ADI 211
          II=2                                                       ADI 212
          III=II-1                                                   ADI 213
          IFL=IL-II                                                  ADI 214
          IF (RF.NE.Ø.Ø) CALL AZS(BD(III),AV2,BD(III),IFL)          ADI 215
          CALL SZZ(BD(III),VRT(II,JL),CD(III),IFL)                   ADI 216
          CALL MZZ(CD(III),RP(II),AD(III),IFL)                       ADI 217
          CALL AZZ(VRT(II,JL),AD(III),VRT(II,JL),IFL)                ADI 218
          CALL QABS(CD(III),DD(III),IFL)                             ADI 219
          DO 45 I=II,ILMN                                            ADI 22Ø
          IF (ADVRT.GT.DD(I-1)) GO TO 45                             ADI 221
          DVRTLD=CD(I-1)                                             ADI 222
          ADVRT=DD(I-1)                                              ADI 223
          ITV=I                                                      ADI 224
       45 CONTINUE                                                   ADI 225
          VRT(IL,JL)=VRT(2,JL)                                       ADI 226
          VRT(1,JL)=VRT(ILMN,JL)                                     ADI 227
    C                                                                ADI 228
          IF (JM.NE.2) GO TO 3ØØ                                     ADI 229
          WRITE(3,275) N,ITV,DVRTLD                                  ADI 23Ø
      275 FORMAT(4X,2HN=,I4,5X,7HDVRTLD(,I3,2H)=,1PE12.5)            ADI 231
    C                                                                ADI 232
    C     N - ITERATION COUNTER                                      ADI 233
    C     ITV - VALUE OF I AT WHICH THE MAXIMUM RESIDUE IS FOUND     ADI 234
    C     DVRTLD - MAXIMUM RESIDUE OF THE VORTICITY ON THE SURFACE OF THE ADI 235
    C              AIRFOIL                                           ADI 236
    C                                                                ADI 237
          IF (DVRTM-ADVRT) 295,295,31Ø                              ADI 238
      295 IF (N.EQ.NMAX) GO TO 31Ø                                   ADI 239
      3ØØ CONTINUE                                                   ADI 24Ø
      31Ø CONTINUE                                                   ADI 241
          DO 61Ø I=2,IL                                             ADI 242
      61Ø DISS(I)=R1R6*(VRT(I,JL)-VRT(I,JLMN))                       ADI 243
          JM=2                                                       ADI 244
          JM1=1                                                      ADI 245
          CALL DERSI(2)                                              ADI 246
          IF (RF.NE.Ø.Ø) GO TO 65Ø                                   ADI 247
          DO 62Ø J=1,JL                                             ADI 248
          DO 62Ø I=2,ILM4,4                                         ADI 249
          SITN1(I,J)=SIT(I,J)                                        ADI 25Ø
          SITN1(I+1,J)=SIT(I+1,J)                                    ADI 251
          SITN1(I+2,J)=SIT(I+2,J)                                    ADI 252
      62Ø SITN1(I+3,J)=SIT(I+3,J)                                    ADI 253
          RETURN                                                     ADI 254
      65Ø DO 67Ø J=1,JLMN                                            ADI 255
          DO 67Ø I=2,ILM2,2                                         ADI 256
          SITN1(I,J)=SIT(I,J)-ANGVEL*XY2T(I,J)                       ADI 257
      67Ø SITN1(I+1,J)=SIT(I+1,J)-ANGVEL*XY2T(I+1,J)                 ADI 258
          DO 68Ø I=2,ILM2,2                                         ADI 259
```

```
          SITN1(I,JL)=-ANGVEL*XY2T(I,JL)                                  ADI   26Ø
    68Ø SITN1(I+1,JL)=-ANGVEL*XY2T(I+1,JL)                                 ADI   261
          RETURN                                                          ADI   262
          END                                                             ADI   263
                                                                          BOUND   2
                                                                          BOUND   3
          SUBROUTINE BOUND                                                BOUND   4
    C                                                                     BOUND   5
    C     THIS SUBROUTINE COMPUTES THETA-COMPONENT OF VELOCITY AND        BOUND   6
    C     DISTURBANCE STREAM FUNCTION AT OUTER BOUNDARY                   BOUND   7
    C                                                                     BOUND   8
          COMMON /FIX/ KL2Ø,DLTRI2,DLTRI3,DLTTI3,JS,JLM2,JLM3,            FIX    2
         *DVRTM,IMM,IMMN,IM,IMPL,IMP,ILMN,ILPL,ILPLP,JLMN,NP,NT,NTMAX,KLØ,KLFIX  3
         11,KL2,KL3,KL4,KL5,KL6,KL9,KL1Ø,KL14,KL15,KL16,ITS,JTS,ITV,JTV,NMAXFIX  4
         2,LN,LNT,PI,XN,XPN,YN,YPN,MARK,DLTI2,DLTI3,RH1DR,RH1DR2,RI,RNN,RNNCFIX  5
         3L,SRNN,DLHRNC,BPSI,BVRT,DVRTLD,R1TR2,ADVRT,TIME,DLTI,MIJF,      FIX    6
         4DLTT2I,DLTR2I,DLTR,DLTR2,DTR2,DLTT,DLTT2,DTT2,C,C2,OI,AS,AT,RATIO,FIX   7
         5CL,DLH,DEN,SIAT,COAT,AP,OR,ORI,CC,P,RHO2,DLTT4,IL,JL,ILM2,JLPL,IUPFIX   8
         6,DLTR24,DLHR,PS,IS,DLTI2I,COSC,ANGVEL,ANGACC,RF,AMP,AVEL,AVA,RH1SFIX   9
         76,CLI,R1R6,DLRN,DLCL,NTI,T,T1,T2,KL11,IUPI,IUPFL,ANG,NSW,MN,DLT4RIFIX 1Ø
         8,R1I,KL21,R1R3,AV2,ILM3,DIT2D,DF,DR,DR2,DR2MDR,DEN2,FAC         FIX   11
         9,ILM4,D12TR2,D24TR2,JLM4,JM,JM1,JBLE,JDIF1,MIJ,JTI,JVI,JRNCN,KL22 FIX  12
          COMMON            PY(13Ø,84),VRT(13Ø,84),RR(84),RHO1S(84),      BOUND  1Ø
         1A5(84),HRR(13Ø),ARDT(13Ø),HD(13Ø)                              BOUND  11
          COMMON /THER/ THETA(13Ø),R(84),SINE(13Ø),COSINE(13Ø),RP(13Ø)   THER   2
          COMMON /DUM1/AD(132),BD(132),CD(132),DD(132)                   DUM1   2
          COMMON /BCF/XC(13Ø),V(13Ø),VN2(13Ø),PSIR(13Ø,2)                BCF    2
          COMMON /DUM/ ED(132),FD(132),QD(132),RD(132)                   DUM    2
          COMMON /LOADS/YTL(132),RHO1(84),XTF(132),PR(132),VRT1(132),    BOUND  16
         1XYT(132),YXT(132),DISS(13Ø)                                    BOUND  17
          COMMON /LMAIN/XY2(13Ø,84),VRTN1(13Ø,84),VN1(13Ø),XY2RO(13Ø,84),BOUND  18
         1XY2T(13Ø,84),AXYYRO(13Ø,84),AXYYT(13Ø,84),RHO1I(84)            BOUND  19
          COMMON /LSTR/ PSI(13Ø,84),Y(13Ø,84),VRTN2(13Ø,84),H(13Ø,84)    LSTR   2
         1,YY(13Ø,84),PSIN1(13Ø,84)                                      LSTR   3
          LEVEL 2,PSI,Y,VRTN2,H,YY,PSIN1                                 LSTR   4
          COMMON /LBOUND/ YR(13Ø,84),YT(13Ø,84),SH(13Ø,84)               BOUND  21
          COMMON /SU/ U(13Ø),UN1(13Ø),UN2(13Ø),SQH1(13Ø)                 BOUND  22
          COMMON /WN/ WAV(13Ø,4),WAV2(13Ø,4),WAVF(13Ø,4),WAV2F(13Ø,4)    WN     2
          COMMON /LOUT/ X(13Ø,84),XR(13Ø,84),XT(13Ø,84)                  BOUND  24
          LEVEL 2,X,XR,XT                                                BOUND  25
          LEVEL 2,XY2,VRTN1,VN1,XY2RO,XY2T,AXYYRO,AXYYT,RHO1I            BOUND  26
          LEVEL 2,YTL,RHO1,XTF,PR,VRT1,XYT,YXT,DISS                      BOUND  27
          LEVEL 2, YR,YT,SH                                              BOUND  28
    C                                                                     BOUND  29
    C     U AND V IN BODY CO-ORDINATES                                    BOUND  3Ø
    C                                                                     BOUND  31
          DO 24Ø I=2,ILMN                                                 BOUND  32
          AD(I)=R(1)*SQH1(I-1)*T-(6.Ø*DLTI/DLTT)*VN1(I-1)                 BOUND  33
          BD(I)=4.Ø*R(1)*SQH1(I)*T                                        BOUND  34
          CD(I)=R(1)*SQH1(I+1)*T+(6.Ø*DLTI/DLTT)*VN1(I+1)                 BOUND  35
    24Ø V(I)=R(1)*SQH1(I-1)*VN2(I-1)+4.Ø*R(1)*SQH1(I)*VN2(I)+R(1)*SQH1(I+1BOUND  36
         1)*VN2(I+1)-(DLTI2*R(1))*(SQH1(I-1)*(VRTN1(I-1,1)*U(I-1)+VRT(I-1,1)BOUND 37
         2*UN1(I-1))+4.*SQH1(I)*(VRTN1(I,1)*U(I)+VRT(I,1)*UN1(I))+SQH1(I+1)*BOUND 38
         3(VRTN1(I+1,1)*U(I+1)+VRT(I+1,1)*UN1(I+1))) -(6.Ø*DLTI/DLTT)*    BOUND  39
         4(UN1(I+1)*U(I+1)-UN1(I-1)*U(I-1))                              BOUND  4Ø
          CALL SUFLE4(AD(2),BD(2),CD(2),V(2),LN,ILM2)                     BOUND  41
          CALL TRIPER(AD(2),BD(2),CD(2),V(2),ILM2)                        BOUND  42
          CALL SUFLE1(V(2),ED(2),LN,ILM2)                                 BOUND  43
          DO 155 I=2,ILMN                                                 BOUND  44
          QD(I)=-(V(I)/(SH(I,1)*RHO1(1))+AXYYRO(I,1))                     BOUND  45
          PSIR(I,1)=QD(I)                                                 BOUND  46
    155 CONTINUE                                                          BOUND  47
          V(IL)=V(2)                                                      BOUND  48
          V(1)=V(ILMN)                                                    BOUND  49
          DO 2ØØ I=IMM,IMP                                                BOUND  5Ø
          UXB =COAT     +V(I,1)*ANGVEL                                    BOUND  51
          VYB =SIAT      -X(I,1)*ANGVEL                                   BOUND  52
          PSIR(I,1)=(-VYB *XR(I,1)+UXB *YR(I,1))*RHO1I(1)-AXYYRO(I,1)     BOUND  53
          QD(I)=PSIR(I,1)                                                 BOUND  54
          V(I)=-(PSIR(I,1)+AXYYRO(I,1))*RHO1(1)*SH(I,1)                   BOUND  55
    2ØØ CONTINUE                                                          BOUND  56
          RETURN                                                          BOUND  57
          END                                                             BOUND  58
                                                                          STAGP   2
                                                                          STAGP   3
          SUBROUTINE STAGP                                                STAGP   4
    C                                                                     STAGP   5
    C     THIS SUBROUTINE DETERMINES PRESSURE COEFFICIENTS IN THE         STAGP   6
```

```
C      NEIGHBORHOOD OF UPSTREAM STAGNATION POINT                          STAGP  7
C                                                                         STAGP  8
       COMMON /FIX/ KL2Ø,DLTRI2,DLTRI3,DLTTI3,JS,JLM2,JLM3,               FIX    2
      *DVRTM,IMM,IMMN,IM,IMPL,IMP,ILMN,ILPL,ILPLP,JLMN,NP,NT,NTMAX,KLØ,KLFFIX    3
      11,KL2,KL3,KL4,KL5,KL6,KL9,KL1Ø,KL14,KL15,KL16,ITS,JTS,ITV,JTV,NMAXFIX    4
      2,LN,LNT,PI,XN,XPN,VN,VPN,MARK,DLTI2,DLTI3,RH1DR,RH1DR2,RI,RNN,RNNCFIX    5
      3L,SRNN,DLHRNC,BPSI,BVRT,DVRTLD,R1TR2,ADVRT,TIME,DLTI,MIJF,          FIX    6
      4DLTT2I,DLTR2I,DLTR,DLTR2,DTR2,DLTT,DLTT2,DTT2,C,C2,OI,AS,AT,RATIO,  FIX    7
      5CL,DLH,DEN,SIAT,COAT,AP,OR,ORI,CC,P,RHO2,DLTT4,IL,JL,ILM2,JLPL,IUPFFIX    8
      6,DLTR24,DLHR,PS,IS,DLTI2I,COSC,ANGVEL,ANGACC,RF,AMP,AVEL1,AVA,RH1SFIX    9
      76,CLI,R1R6,DLRN,DLCL,NTI,T,T1,T2,KL11,IUPI,IUPFL,ANG,NSW,MN,DLT4RIFIX   1Ø
      8,R1I,KL21,R1R3,AV2,ILM3,DIT2D,DF,DR,DR2,DR2MDR,DEN2,FAC             FIX   11
      9,ILM4,D12TR2,D24TR2,JLM4,JM,JM1,JBLE,JDIF1,MIJ,JTI,JVI,JRNCN,KL22  FIX   12
       COMMON             PY(13Ø,84),VRT(13Ø,84),RR(84),RHO1S(84),        STAGP 1Ø
      1A5(84),HRR(13Ø),ARDT(13Ø),HD(13Ø)                                  STAGP 11
       COMMON  /DUM1/ UT(132),DPR(132),VV(132),EK(132)                    STAGP 12
       COMMON  /BCF/XC(13Ø),V(13Ø),VN2(13Ø),PSIR(13Ø,2)                   BCF    2
       COMMON /LOADS/YTL(132),RHO1(84),XTF(132),PR(132),VRT1(132),        STAGP 14
      1XYT(132),YXT(132),DISS(13Ø)                                        STAGP 15
       COMMON /THER/ THETA(13Ø),R(84),SINE(13Ø),COSINE(13Ø),RP(13Ø)       THER   2
       COMMON /LMAIN/XY2(13Ø,84),VRTN1(13Ø,84),VN1(13Ø),XY2RO(13Ø,84),    STAGP 17
      1XY2T(13Ø,84),AXYYRO(13Ø,84),AXYYT(13Ø,84),RHO1I(84)                STAGP 18
       COMMON /LSTR/ PSI(13Ø,84),Y(13Ø,84),VRTN2(13Ø,84),H(13Ø,84),       LSTR   2
      1,YY(13Ø,84),PSIN1(13Ø,84)                                          LSTR   3
       LEVEL 2,PSI,Y,VRTN2,H,YY,PSIN1                                     LSTR   4
       COMMON /LBOUND/ YR(13Ø,84),YT(13Ø,84),SH(13Ø,84)                   STAGP 2Ø
       COMMON /LSTAGP/ UP(33,84),UP1(33,84),UP2(33,84),RNRI(84)           STAGP 21
       COMMON /SIDER/ SIR(13Ø,84),SIT(13Ø,84),SIRN1(13Ø,84),SITN1(13Ø,84)STAGP 22
       COMMON /LOUT/ X(13Ø,84),XR(13Ø,84),XT(13Ø,84)                      LOUT   2
       LEVEL 2,X,XR,XT                                                    LOUT   3
       LEVEL 2,YTL,RHO1,XTF,PR,VRT1,XYT,YXT,DISS                          STAGP 24
       LEVEL 2,XY2,VRTN1,VN1,XY2RO,XY2T,AXYYRO,AXYYT,RHO1I                STAGP 25
       LEVEL 2, YR,YT,SH                                                  STAGP 26
       LEVEL 2, UP,UP1,UP2,RNRI                                           STAGP 27
       LEVEL 2, SIR,SIT,SIRN1,SITN1                                       STAGP 28
C                                                                         STAGP 29
C                                                                         STAGP 3Ø
C      R MOMENTUM EQUATION IN BODY CO-ORDINATES                           STAGP 31
C      PR(1) IN INERTIAL FRAME OF REFERENCE                               STAGP 32
C                                                                         STAGP 33
       DO 1Ø I=IUPI,IUPFL                                                 STAGP 34
       IF (VRT(I,JL).LT.1.ØE-Ø7) GO TO 2Ø                                 STAGP 35
   1Ø CONTINUE                                                            STAGP 36
       WRITE(3,21Ø)                                                       STAGP 37
  21Ø FORMAT(/3X,48HPROGRAM STOPPED BECAUSE STAG. POINT NOT LOCATED,/)    STAGP 38
       STOP                                                               STAGP 39
   2Ø IS=I                                                                STAGP 4Ø
       WRITE(3,2ØØ) IS                                                    STAGP 41
  2ØØ FORMAT(/3X,7HIS    =I2,/)                                           STAGP 42
       ISTAG=IS                                                           STAGP 43
   4Ø ISD=IS-IUPI+1                                                       STAGP 44
       IF (ISD.GT.Ø) GO TO 45                                             STAGP 45
       WRITE(3,25Ø)                                                       STAGP 46
  25Ø FORMAT(/3X,32HPROGRAM STOPPED BECAUSE ISD.LT.1,/)                   STAGP 47
       STOP                                                               STAGP 48
   45 JFL=JLMN                                                            STAGP 49
       JFLMN=JFL-1                                                        STAGP 5Ø
       IF (RF.EQ.Ø.Ø) GO TO 46                                            STAGP 51
       DO 44 J=1,JL                                                       STAGP 52
   44 DPR(J)=(ANGACC*XY2T(IS,J)/R(J)+(ANGVEL*ANGVEL)*XY2RO(IS,J)*         STAGP 53
      1RHO1(J))                                                           STAGP 54
       XRI=XR(IS,1)*COAT+YR(IS,1)*SIAT                                    STAGP 55
       XTI=XT(IS,1)*COAT+YT(IS,1)*SIAT                                    STAGP 56
       YRI=YR(IS,1)*COAT-XR(IS,1)*SIAT                                    STAGP 57
       YTI=YT(IS,1)*COAT-XT(IS,1)*SIAT                                    STAGP 58
       HI=(XRI*YTI-YRI*XTI)/R(1)                                          STAGP 59
       PR(1)=1.Ø-((SITN1(IS,1)/R(1))**2+SIRN1(IS,1)*SIRN1(IS,1))/HI       STAGP 6Ø
       GO TO 47                                                           STAGP 61
   46 CALL SETZ(DPR(1),Ø.Ø,ILPLP)                                         STAGP 62
       PR(1)=1.Ø-((SITN1(IS,1)/R(1))**2+SIRN1(IS,1)*SIRN1(IS,1))/         STAGP 63
      1H(IS,1)                                                            STAGP 64
   47 DO 5Ø J=1,JFL                                                       STAGP 65
       VV(J)=-RHO1(J)*SIR(IS,J)*SH(IS,J)                                  STAGP 66
       EK(J)=(UP(ISD,J)*UP(ISD,J)+VV(J)*VV(J))*Ø.5                        STAGP 67
   5Ø UT(J)=(T*UP(ISD,J)-T1*UP1(ISD,J)+T2*UP2(ISD,J))*DLTI2I              STAGP 68
       DPR(1)=-(UT(1)/SH(IS,1)-DPR(1))*RHO1I(1)-(-3.Ø*EK(1)+4.Ø*          STAGP 69
      1EK(2)-EK(3))*DLTR2I                                                STAGP 7Ø
       DO 7Ø J=2,JFLMN                                                    STAGP 71
   7Ø DPR(J)=-((UT(J)-VV(J)*VRT(IS,J))/SH(IS,J)-DPR(J)+RNRI(J)*(VRT(      STAGP 71
```

```
      1IS+1,J)-VRT(IS-1,J)))*RHO1I(J)-(EK(J+1)-EK(J-1))*DLTR2I          STAGP 72
      J=JFL                                                            STAGP 73
      DPR(J)=-((UT(J)-VV(J)*VRT(IS,J))/SH(IS,J)-DPR(J)+RNRI(J)*(VRT(   STAGP 74
      1IS+1,J)-VRT(IS-1,J)))*RHO1I(J)-(3.0*EK(J)-4.0*EK(J-1)+EK(J-2))* STAGP 75
      2DLTR2I                                                          STAGP 76
      DPR(JL   )=-(RNRI(JL)*(VRT(IS+1,JL)-VRT(IS-1,JL))-DPR(JL)        STAGP 77
      1)*RHO1I(JL)                                                     STAGP 78
C                                                                      STAGP 79
C                                                                      STAGP 80
C        RADIAL PRESSURE DERIVATIVE, KINETIC ENERGY DEFECT AND TIME    STAGP 80
C             DERIVATIVE OF RADIAL VELOCITY AT  I=IS ARE PRINTED       STAGP 81
C                                                                      STAGP 82
      WRITE(3,240) DPR(1),PR(1),UT(1)                                  STAGP 83
  240 FORMAT(/3X20HPRESSURE DERIVATIVE=1PE12.5,5X12HK.E. DEFECT=,E12.5,STAGP 84
     1 5X6HUT(1)=E12.5/)                                               STAGP 85
      PR(2)=PR(1)+DLTR24*(9.0*DPR(1)+19.0*DPR(2)-5.0*DPR(3)+DPR(4))    STAGP 86
      JFL=JL  -2                                                       STAGP 87
      DO 80 J=3,JFL                                                    STAGP 88
      P1=9.0*DPR(J-1)+19.0*DPR(J)-5.0*DPR(J+1)+DPR(J+2)                STAGP 89
      P2=-DPR(J-2)+13.0*(DPR(J-1)+DPR(J))-DPR(J+1)                     STAGP 90
   80 PR(J)=PR(J-1)           +DLHR*(19.0*P2+11.0*P1)                  STAGP 91
      DO 90 J=JLMN,JL                                                  STAGP 92
   90 PR(J)=PR(J-1)+DLTR24*(9.*DPR(J)+19.*DPR(J-1)-5.*DPR(J-2)+DPR(J-3))STAGP 93
      PS=PR(JL)                                                        STAGP 94
      WRITE(3,300) (PR(J),J=1,JL)                                      STAGP 95
  300 FORMAT(1P10E13.5)                                                STAGP 96
      IF (ISTAG.NE.IS) GO TO 100                                       STAGP 97
      WRITE(3,220) PS                                                  STAGP 98
  220 FORMAT(/3X51HPRESSURE COEFFICIENT UPSTREAM OF STAGNATION POINT =,STAGP 99
     11PE14.5/)                                                        STAGP100
      PSD=PS                                                           STAGP101
      IS=IS-1                                                          STAGP102
      GO TO 40                                                         STAGP103
  100 WRITE(3,230) PS                                                  STAGP104
  230 FORMAT(/3X53HPRESSURE COEFFICIENT DOWNSTREAM OF STAGNATION POINT =STAGP105
     1, 1PE14.5/)                                                      STAGP106
      XCD=XC(IS)-XC(ISTAG)                                             STAGP107
      VRTST=VRT(IS,JL)-VRT(ISTAG,JL)                                   STAGP108
      XSTAG=XC(ISTAG)-(XCD*VRT(ISTAG,JL))/VRTST                        STAGP109
C     PSTAG=PSD+(PS-PSD)*(XSTAG-XC(ISTAG))/XCD                         STAGP110
C           - LINEAR. PSTAG MUST BE MAX AT XSTAG"                      STAGP111
      WRITE(3,290)    XSTAG                                            STAGP112
  290 FORMAT(/3X,*STAGNATION POINT AT =*1PE13.5/)                      STAGP113
  270 IS=ISTAG                                                         STAGP114
      PS=PSD                                                           STAGP115
      RETURN                                                           STAGP116
      END                                                             STAGP117
                                                                      FORCES 2
                                                                      FORCES 3
      SUBROUTINE FORCES                                               FORCES 4
C                                                                      FORCES 5
C        THIS SUBROUTINE DETERMINES FORCES ACTING ON THE BODY         FORCES 6
C                                                                      FORCES 7
      COMMON /FIX/ KL20,DLTRI2,DLTRI3,DLTTI3,JS,JLM2,JLM3,            FIX    2
     *DVRTM,IMM,IMMN,IM,IMPL,IMP,ILMN,ILPL,ILPLP,JLMN,NP,NT,NTMAX,KL0,KLFIX  3
     11,KL2,KL3,KL4,KL5,KL6,KL9,KL10,KL14,KL15,KL16,ITS,JTS,ITV,JTV,NMAXFIX  4
     2,LN,LNT,PI,XN,XPN,YN,YPN,MARK,DLTI2,DLTI3,RH1DR,RH1DR2,RI,RNN,RNNCFIX  5
     3L,SRNN,DLHRNC,BPSI,BVRT,DVRTLD,R1TR2,ADVRT,TIME,DLTI,MIJF,       FIX    6
     4DLTT2I,DLTR,DLTR2,DTR2,DLTT,DLTT2,DTT2,C,C2,OI,AS,AT,RATIO,FIX    FIX    7
     5CL,DLH,DEN,SIAT,COAT,AP,OR,ORI,CC,P,RHO2,DLTT4,IL,JL,ILM2,JLPL,IUPFIX   8
     6,DLTR24,DLHR,PS,IS,DLTI2I,COSC,ANGVEL,ANGACC,RF,AMP,AVEL1,AVA,RH1SFIX   9
     76,CLI,R1R6,DLRN,DLCL,NTI,T,T1,T2,KL11,IUPI,IUPFL,ANG,NSW,MN,DLT4RIFIX  10
     8,R1I,KL21,R1R3,AV2,ILM3,DIT2D,DF,DR,DR2,DR2MDR,DEN2,FAC          FIX   11
     9,ILM4,D12TR2,D24TR2,JLM4,JM,JM1,J8LE,JDIF1,MIJ,JTI,JVI,JRNCN,KL22 FIX  12
      COMMON /THER/ THETA(130),R(84),SINE(130),COSINE(130),RP(130)    THER   2
      COMMON /LDMSTOR/   TM(40),FDM(40),FLM(40),FMM(40),PDM(40),PLM(40,LDMSTOR2
     12),PMM(40,2),CDM(40,2),CLM(40,2),CMM(40,2)                       LDMSTOR3
      COMMON          PY(130,84),VRT(130,84),RR(84),RHO1S(84),        FORCES11
     1A5(84),HRR(130),ARDT(130),HD(130)                               FORCES12
      COMMON /DUM1/ PD(132),VD(132),PR1(132),DD(132)                  FORCES13
      COMMON /BCF/XC(130),V(130),VN2(130),PSIR(130,2)                 BCF    2
      COMMON /LOADS/YTL(132),RHO1(84),XTF(132),PR(132),VRT1(132),     FORCES15
     1XYT(132),YXT(132),DISS(130)                                     FORCES16
      LEVEL 2,YTL,RHO1,XTF,PR,VRT1,XYT,YXT,DISS                       FORCES17
      REAL LIFT,MOMENT                                                FORCES18
      CF=0.0                                                          FORCES19
      CLF=0.0                                                         FORCES20
      CMF=0.0                                                         FORCES21
      DO 5 I=1,IL                                                     FORCES22
```

```
            VD(I)=VRT(I,JL)                                              FORCES23
      5     PR(I)=0.0                                                    FORCES24
            VD(ILPL)=VRT(3,JL)                                          FORCES25
            VD(ILPLP)=VRT(4,JL)                                         FORCES26
            DO 10 I=2,IL                                                FORCES27
     10     VRT1(I)=DISS(I)+VRT1(I)                                     FORCES28
            K=1                                                         FORCES29
     15     DO 20 I=2,IL                                                FORCES30
     20     PD(I)=VRT1(I)                                               FORCES31
            PD(1)=PD(ILMN)                                              FORCES32
            PD(ILPL)=PD(3)                                              FORCES33
            PD(ILPLP)=PD(4)                                             FORCES34
            IF (K.EQ.2) GO TO 40                                        FORCES35
            DO 30 I=3,IM                                                FORCES36
              P1=9.0*PD(I-1)+19.0*PD(I)-5.0*PD(I+1)+PD(I+2)             FORCES37
              P2=-PD(I-2)+13.0*(PD(I-1)+PD(I))-PD(I+1)                  FORCES38
     30       PR(I)=PR(I-1)+DLHRNC*(19.0*P2+11.0*P1)                    FORCES39
            PM=PR(IM)                                                   FORCES40
            ILPMM=ILPL+IMMN                                             FORCES41
            DO 35 IR=IMPL,IL                                            FORCES42
            I=ILPMM-IR                                                  FORCES43
              P1=-(9.0*PD(I+1)+19.0*PD(I)-5.0*PD(I-1)+PD(I-2))          FORCES44
              P2=-(-PD(I+2)+13.0*(PD(I+1)+PD(I))-PD(I-1))               FORCES45
     35       PR(I)=PR(I+1)+DLHRNC*(19.0*P2+11.0*P1)                    FORCES46
            PM1=PR(IM)                                                  FORCES47
            NI=1                                                        FORCES48
            K=2                                                         FORCES49
            PR(IM)=(PM+PM1)*0.5                                         FORCES50
            WRITE (3,175)                                               FORCES51
            GO TO 60                                                    FORCES52
     40     DO 45 IR=3,IMMN                                             FORCES53
            I=IMPL-IR+1                                                 FORCES54
              P1=-(9.0*PD(I+1)+19.0*PD(I)-5.0*PD(I-1)+PD(I-2))          FORCES55
              P2=-(-PD(I+2)+13.0*(PD(I+1)+PD(I))-PD(I-1))               FORCES56
     45       PR(I)=PR(I+1)+DLHRNC*(19.0*P2+11.0*P1)                    FORCES57
            P1=-(9.0*PD(3)+19.0*PD(2)-5.0*PD(1)+PD(IL-2))               FORCES58
            P2=-(-PD(4)+13.0*(PD(3)+PD(2))-PD(1))                       FORCES59
            PR(2)=PR(3)+DLHRNC*(19.0*P2+11.0*P1)                        FORCES60
            PM=PR(2)                                                    FORCES61
            DO 50 I=IMPL,IL                                             FORCES62
              P1=9.0*PD(I-1)+19.0*PD(I)-5.0*PD(I+1)+PD(I+2)             FORCES63
              P2=-PD(I-2)+13.0*(PD(I-1)+PD(I))-PD(I+1)                  FORCES64
     50       PR(I)=PR(I-1)+DLHRNC*(19.0*P2+11.0*P1)                    FORCES65
            PM1=PR(IL)                                                  FORCES66
            NI=4                                                        FORCES67
            K=3                                                         FORCES68
            PR(2)=(PM+PM1)*0.5                                          FORCES69
            PR(IL)=PR(2)                                                FORCES70
            DO 55 I=2,IL                                                FORCES71
     55     PR(I)=PR(I)-PR(IL)                                          FORCES72
            DDRAG=DRAG                                                  FORCES73
            DLIFT=LIFT                                                  FORCES74
            DMOM=MOMENT                                                 FORCES75
            WRITE (3,180)                                               FORCES76
     60     APMM1=ABS(PM-PM1)                                          FORCES77
            PERR=200.0*APMM1/(PM+PM1)                                   FORCES78
            PMIN=0.0                                                    FORCES79
            PMAX=0.0                                                    FORCES80
            DO 61 I=2,IL                                                FORCES81
            PMIN=AMIN1(PMIN,PR(I))                                      FORCES82
     61     PMAX=AMAX1(PMAX,PR(I))                                     FORCES83
            PERRMM=100.0*APMM1/(PMAX+ABS(PMIN))                         FORCES84
C                                                                       FORCES85
C           PM - THE PRESSURE COEFFICIENT AT THE LEADING OR TRAILING EDGE FORCES86
C               DETERMINED BY INTEGRATING ALONG THE LOWER SURFACE       FORCES87
C           PM1 - THE PRESSURE COEFFICIENT AT THE LEADING OR TRAILING EDGE FORCES88
C               DETERMINED BY INTEGRATING ALONG THE UPPER SURFACE       FORCES89
            WRITE(3,185) PM,PM1,PERR,PERRMM                             FORCES90
            IF (PERRMM.LE.10.0) GO TO 62                                FORCES91
C           IF (PERRMM.LE. 5.0) GO TO 62                                FORCES92
            WRITE(3,210)                                                FORCES93
    210     FORMAT(/3X*PROGRAM STOPPED BECAUSE PRESSURE INTEGRATION ERROR GREAFORCES94
           1TER THAN 10 PER CENT*/)                                     FORCES95
            NTMAX=NT                                                    FORCES96
            KL16=1                                                      FORCES97
            KL22=1                                                      FORCES98
     62     CONTINUE                                                   FORCES99
            PR(1)=PR(ILMN)                                             FORCE100
```

```
      PR(ILPL)=PR(3)                                               FORCE101
      PR(ILPLP)=PR(4)                                              FORCE102
      CP=0.0                                                       FORCE103
      CLP=0.0                                                      FORCE104
      CMP=0.0                                                      FORCE105
      DO 135 N=NI,6                                                FORCE106
        DO 95 I=1,ILPLP                                            FORCE107
          GO TO (65,70,75,80,85,90), N                            FORCE108
   65     PD(I)=VD(I)*XTF(I)                                       FORCE109
          GO TO 95                                                 FORCE110
   70     PD(I)=VD(I)*YTL(I)                                       FORCE111
          GO TO 95                                                 FORCE112
   75     PD(I)=VD(I)*YXT(I)                                       FORCE113
          GO TO 95                                                 FORCE114
   80     PD(I)=PR(I)*YTL(I)                                       FORCE115
          GO TO 95                                                 FORCE116
   85     PD(I)=PR(I)*XTF(I)                                       FORCE117
          GO TO 95                                                 FORCE118
   90     PD(I)=PR(I)*XYT(I)                                       FORCE119
   95     CONTINUE                                                 FORCE120
        DO 130 I=3,IL                                              FORCE121
        P1=9.0*PD(I-1)+19.0*PD(I)-5.0*PD(I+1)+PD(I+2)              FORCE122
        P2=-PD(I-2)+13.0*(PD(I-1)+PD(I))-PD(I+1)                   FORCE123
        GO TO (100,105,110,115,120,125), N                        FORCE124
  100   CF=CF-(19.0*P2+11.0*P1)                                    FORCE125
        GO TO 130                                                  FORCE126
  105   CLF=CLF-(19.0*P2+11.0*P1)                                  FORCE127
        GO TO 130                                                  FORCE128
  110   CMF=CMF-(19.0*P2+11.0*P1)                                  FORCE129
        GO TO 130                                                  FORCE130
  115   CP=CP+(19.0*P2+11.0*P1)                                    FORCE131
        GO TO 130                                                  FORCE132
  120   CLP=CLP-(19.0*P2+11.0*P1)                                  FORCE133
        GO TO 130                                                  FORCE134
  125   CMP=CMP-(19.0*P2+11.0*P1))                                 FORCE135
  130   CONTINUE                                                   FORCE136
  135   CONTINUE                                                   FORCE137
      IF (NI.EQ.4) GO TO 140                                       FORCE138
      CF=CF*DLRN                                                   FORCE139
      CLF=CLF*DLRN                                                 FORCE140
      CMF=CMF*DLRN*CLI                                             FORCE141
      CMF=-CMF                                                     FORCE142
      FDM(NT)=CF                                                   FORCE143
      FLM(NT)=CLF                                                  FORCE144
      FMM(NT)=CMF                                                  FORCE145
  140 CP=CP*DLCL                                                   FORCE146
      CLP=CLP*DLCL                                                 FORCE147
      CMP=CMP*DLCL*CLI                                             FORCE148
      DRAG=CP+CF                                                   FORCE149
      LIFT=CLF+CLP                                                 FORCE150
      CMP=-CMP                                                     FORCE151
      MOMENT=CMF+CMP                                               FORCE152
      PDM(NT,K-1)=CP                                               FORCE153
      PLM(NT,K-1)=CLP                                              FORCE154
      PMM(NT,K-1)=CMP                                              FORCE155
      CDM(NT,K-1)=DRAG                                             FORCE156
      CLM(NT,K-1)=LIFT                                             FORCE157
      CMM(NT,K-1)=MOMENT                                           FORCE158
      WRITE (3,190) CP,CF,DRAG,CLP,CLF,LIFT,CMF,CMP,MOMENT         FORCE159
C       '                                                         FORCE160
C     CP - THE DRAG COEFFICIENT DUE TO PRESSURE                   FORCE161
C     CF - THE DRAG COEFFICIENT DUE TO SKIN FRICTION              FORCE162
C     CLP - THE LIFT COEFFICIENT DUE TO PRESSURE                  FORCE163
C     CLF - THE LIFT COEFFICIENT DUE TO SKIN FRICTION             FORCE164
C     CMF - THE MOMENT COEFFICIENT DUE TO SKIN FRICTION           FORCE165
C     CMP - THE MOMENT COEFFICIENT DUE TO PRESSURE                FORCE166
C                                                                 FORCE167
      IF (K.EQ.3) GO TO 150                                       FORCE168
      DO 145 I=1,ILPLP                                            FORCE169
        PR1(I)=PR(I)                                              FORCE170
  145   PR(I)=0.0                                                 FORCE171
      GO TO 15                                                    FORCE172
  150 DRAG=(DRAG+DDRAG)*0.5                                       FORCE173
      LIFT=(DLIFT+LIFT)*0.5                                       FORCE174
      MOMENT=(DMOM+MOMENT)*0.5                                    FORCE175
      WRITE (3,195) DRAG,LIFT,MOMENT                              FORCE176
      WRITE (3,200)                                               FORCE177
      PSD=PR(IS)-PS                                               FORCE178
```

```
        PSD1=PR1(IS)-PS                                                  FORCE179
        DO 170 I=2,IL                                                    FORCE180
        CPT=PR(I)-PSD                                                    FORCE181
        CPT1=PR1(I)-PSD1                                                 FORCE182
        VRTB=VRT(I,JL)-AV2                                               FORCE183
        WRITE(3,205) I,XC(I),PY(I,JLMN),VRT(I,JL),VRT(I,JLMN),VRT(I,JLM2),FORCE184
       1VRTB,VRT1(I),PR(I),PR1(I),V(I),CPT,CPT1,THETA(I)                 FORCE185
        PR(I)=CPT                                                        FORCE186
        DISS(I)=CPT1                                                     FORCE187
  170   VRT1(I)=0.0                                                      FORCE188
        DO 300 I=2,ILMN                                                  FORCE189
        IF (VRT(I,JL).GE.0.0.AND.VRT(I+1,JL).LT.0.0) GO TO 250           FORCE190
        IF (VRT(I,JL).LE.0.0.AND.VRT(I+1,JL).GT.0.0) GO TO 260           FORCE191
        GO TO 300                                                        FORCE192
  250   DV=VRT(I,JL)-VRT(I+1,JL)                                         FORCE193
        DX=XC(I+1)-XC(I)                                                 FORCE194
        ST=XC(I)+DX*VRT(I,JL)/DV                                         FORCE195
        GO TO 270                                                        FORCE196
  260   DV=-VRT(I,JL)+VRT(I+1,JL)                                        FORCE197
        DX=XC(I+1)-XC(I)                                                 FORCE198
        ST=XC(I+1)-DX*VRT(I+1,JL)/DV                                     FORCE199
  270   WRITE(3,280) I,I+1,ST                                           FORCE200
  280   FORMAT(/,3X,*STAGNATION POINT BETWEEN I=*,I3,*  AND I=*,I3,* AT *FORCE201
       1,F7.5)                                                          FORCE202
  300   CONTINUE                                                         FORCE203
        RETURN                                                           FORCE204
C                                                                        FORCE205
  175   FORMAT (//10X,61HPRESSURE INTEGRATED FROM GEOMETRICAL TRAILING TO FORCE206
       1LEADING EDGE,/)                                                  FORCE207
  180   FORMAT (/10X,61HPRESSURE INTEGRATED FROM GEOMETRICAL LEADING TO TRFORCE208
       1AILING EDGE,/)                                                   FORCE209
  185   FORMAT (3X,7HPM   =,1PE12.5,3X,7HPM1   =,E12.5,3X,14HPERCENT ERROFORCE210
       1R=,E12.5.3X,30HPERCENT ERROR BASED ON MINMAX=,E12.5/)            FORCE211
  190   FORMAT (3X,7HCP   =,1PE12.5,3X,7HCF    =,E12.5,3X,7HDRAG =,E12.5 FORCE212
       1,3X,7HCLP  =,E12.5,3X,7HCLF  =,E12.5,3X,7HLIFT =,E12.5,//3X,7HCFFORCE213
       2MF  =,E12.5,3X,7HCMP  =,E12.5,3X,7HMOMENT=,E12.5,/)              FORCE214
  195   FORMAT (/3X,13HAVERAGE DRAG=,1PE12.5,3X,13HAVERAGE LIFT=,E12.5,3X,FORCE215
       115HAVERAGE MOMENT=,E12.5//)                                      FORCE216
  200   FORMAT(3X,1HI,2X,*XC(I)*,5X,*PY(JLMN)*,1X,*VRT(JL)*,3X,*VRT(JLMN)*FORCE217
       1,1X,*VRT(JLM2)*,1X,*VRTB(JL)*,2X,*VRT1*,5X,*PR (L-T)*,2X,*PR1 (T-LFORCE218
       2)*,2X,*V*,9X,*CPT (L-T)*,2X,*CPT1 (T-L)*,1X,*THETA*,/)           FORCE219
  205   FORMAT(I4,1PE10.3,E9.1,E11.3 ,2E9.1,E11.3,E9.1,2E10.2,3E11.3,    FORCE220
       10PF7.4)                                                          FORCE221
        END                                                             FORCE222
                                                                        OUTPUT 2
                                                                        OUTPUT 3
                                                                        OUTPUT 4
        SUBROUTINE OUTPUT                                               OUTPUT 5
C                                                                        OUTPUT 6
C       THIS SUBROUTINE PRINTS FLOW FIELD AND GEOMETRICAL VARIABLES     OUTPUT 7
C                                                                        OUTPUT 9
        COMMON /FIX/ KL20,DLTRI2,DLTRI3,DLTTI3,JS,JLM2,JLM3,            FIX    2
       *DVRTM,IMM,IMMN,IM,IMPL,IMP,ILMN,ILPL,ILPLP,JLMN,NP,NT,NTMAX,KL0,KLFIX 3
       11,KL2,KL3,KL4,KL5,KL6,KL9,KL10,KL14,KL15,KL16,ITS,JTS,ITV,JTV,NMAXFIX 4
       2,LN,LNT,PI,XN,XPN,YN,YPN,MARK,DLTI2,DLTI3,RH1DR,RH1DR2,RI,RNN,RNNCFIX 5
       3L,SRNN,DLHRNC,BPSI,BVRT,DVRTLD,R1TR2,ADVRT,TIME,DLTI,MIJF,      FIX    6
       4DLTT2I,DLTR2I,DLTR,DLTR2,DTR2,DLTT,DLTT2,DTT2,C,C2,OI,AS,AT,RATIO,FIX 7
       5CL,DLH,DEN,SIAT,COAT,AP,OR,ORI,CC,P,RHO2,DLTT4,IL,JL,ILM2,JLPL,IUPFIX 8
       6,DLTR24,DLHR,PS,IS,DLTI2I,COSC,ANGVEL,ANGACC,RF,AMP,AVEL1,AVA,RH1SFIX 9
       76,CLI,R1R6,DLRN,DLCL,NTI,T,T1,T2,KL11,IUPI,IUPFL,ANG,NSW,MN,DLT4RIFIX 10
       8,R1I,KL21,R1R3,AV2,ILM3,DIT2D,DF,DR,DR2,DR2MDR,DEN2,FAC          FIX   11
       9,ILM4,D12TR2,D24TR2,JLM4,JM,JM1,JBLE,JDIF1,MIJ,JTI,JVI,JRNCN,KL22 FIX 12
        COMMON /THER/ THETA(130),R(84),SINE(130),COSINE(130),RP(130)    THER   2
        COMMON /DUM1/ U(132),V(132),UZ(132),VV(132)                    OUTPUT10
        COMMON /DUM/ VZ(132),FD(132),QD(132),RD(132)                   OUTPUT11
        COMMON       PY(130,84),VRT(130,84),RR(84),RHO1S(84),          OUTPUT12
       1A5(84),HRR(130),ARDT(130),HD(130)                              OUTPUT13
        COMMON /LOADS/YTL(132),RHO1(84),XTF(132),PR(132),VRT1(132),    OUTPUT14
       1XYT(132),YXT(132),DISS(130)                                    OUTPUT15
        COMMON /LSTR/ PSI(130,84),Y(130,84),VRTN2(130,84),H(130,84)    LSTR   2
       1,YY(130,84),PSIN1(130,84)                                      LSTR   3
        LEVEL 2,PSI,Y,VRTN2,H,YY,PSIN1                                 LSTR   4
        COMMON /LBOUND/ YR(130,84),YT(130,84),SH(130,84)               OUTPUT17
        COMMON /LOUT/ X(130,84),XR(130,84),XT(130,84)                  LOUT   2
        LEVEL 2,X,XR,XT                                                LOUT   3
        COMMON /SIDER/ SIR(130,84),SIT(130,84),SIRN1(130,84),SITN1(130,84)OUTPUT19
        DIMENSION XRI(132),XTI(132),YRI(132),YTI(132),HI(132)          OUTPUT20
        EQUIVALENCE (XRI,U),(XTI,V),(HI,UZ), (YRI,VV),(YTI,VZ)          OUTPUT21
        LEVEL 2,YTL,RHO1,XTF,PR,VRT1,XYT,YXT,DISS                      OUTPUT22
```

```
      LEVEL 2, YR,YT,SH                                              OUTPUT23
      LEVEL 2, SIR,SIT,SIRN1,SITN1                                  OUTPUT24
C                                                                   OUTPUT25
C     EXCEPT PY AND H, ALL VARIABLES IN INERTIAL FRAME OF REFERENCE OUTPUT26
C     UZ - VELOCITY IN X-DIRECTION                                  OUTPUT27
C     VZ - VELOCITY IN Y-DIRECTION                                  OUTPUT28
C                                                                   OUTPUT29
      WRITE (3,95)                                                  OUTPUT30
      IF (JS.LE.4) GO TO 10                                         OUTPUT31
      IF (JS.LT.40) JS=40                                           OUTPUT32
      INC=1                                                         OUTPUT33
      GO TO 15                                                      OUTPUT34
   10 INC=JS                                                        OUTPUT35
      IF (INC.EQ.2.OR.INC.EQ.4) JS=2                                OUTPUT36
      IF (INC.EQ.3) JS=1                                            OUTPUT37
   15 CONTINUE                                                      OUTPUT38
      ILI1=2                                                        OUTPUT39
      ILI2=14                                                       OUTPUT40
   20 IF (ILI2.GT.ILMN) ILI2=ILMN                                  OUTPUT41
      WRITE (3,105) (THETA(I),I=ILI1,ILI2)                          OUTPUT42
      WRITE (3,110)                                                 OUTPUT43
      IF (JS.EQ.1) GO TO 25                                         OUTPUT44
      JI=1                                                          OUTPUT45
      JFL=1                                                         OUTPUT46
      INCD=1                                                        OUTPUT47
      GO TO 30                                                      OUTPUT48
   25 JI=JS                                                         OUTPUT49
      JFL=JL                                                        OUTPUT50
      INCD=INC                                                      OUTPUT51
   30 DO 80 J=JI,JFL,INCD                                           OUTPUT52
      WRITE (3,115) J,R(J),(PY(I,J),I=ILI1,ILI2)                    OUTPUT53
      WRITE (3,120) (VRT(I,J),I=ILI1,ILI2)                          OUTPUT54
      WRITE (3,125) (PSI(I,J),I=ILI1,ILI2)                          OUTPUT55
      IF (NT.GT.KL9) GO TO 35                                       OUTPUT56
      IF (RF.EQ.0.0) GO TO 33                                       OUTPUT57
      DO 32 I=ILI1,ILI2                                             OUTPUT58
      U(I)=X(I,J)*COAT+Y(I,J)*SIAT                                  OUTPUT59
   32 V(I)=Y(I,J)*COAT-X(I,J)*SIAT                                  OUTPUT60
      WRITE(3,130) (V(I),I=ILI1,ILI2)                               OUTPUT61
      WRITE(3,135) (U(I),I=ILI1,ILI2)                               OUTPUT62
      GO TO 40                                                      OUTPUT63
   33 WRITE (3,130) (Y(I,J),I=ILI1,ILI2)                            OUTPUT64
      WRITE (3,135) (X(I,J),I=ILI1,ILI2)                            OUTPUT65
      WRITE (3,140) (H(I,J),I=ILI1,ILI2)                            OUTPUT66
   35 IF (J.EQ.JL) GO TO 65                                         OUTPUT67
   40 IF (KL14.EQ.1) GO TO 80                                       OUTPUT68
      IF (RF.EQ.0.0) GO TO 50                                       OUTPUT69
      DO 36 I=ILI1,ILI2                                             OUTPUT70
      XRI(I)=XR(I,J)*COAT+YR(I,J)*SIAT                              OUTPUT71
      XTI(I)=XT(I,J)*COAT+VT(I,J)*SIAT                              OUTPUT72
      YRI(I)=YR(I,J)*COAT-XR(I,J)*SIAT                              OUTPUT73
      YTI(I)=YT(I,J)*COAT-XT(I,J)*SIAT                              OUTPUT74
   36 HI(I)= XRI(I)*YTI(I)-YRI(I)*XTI(I)                            OUTPUT75
      DO 45 I=ILI1,ILI2                                             OUTPUT76
      W1=SIRN1(I,J)                                                 OUTPUT77
      W2=SITN1(I,J)                                                 OUTPUT78
      HRIJ=1.0/HI(I)                                                OUTPUT79
      UZ(I)=(W2*XRI(I)-W1*XTI(I))*HRIJ                              OUTPUT80
   45 VZ(I)=-(W1*YTI(I)-W2*YRI(I))*HRIJ                             OUTPUT81
      GO TO 56                                                      OUTPUT82
   50 DO 55 I=ILI1 ,ILI2                                            OUTPUT83
C     SH= SQRT(H(I,J))                                              OUTPUT84
      W1=SIRN1(I,J)                                                 OUTPUT85
      W2=SITN1(I,J)                                                 OUTPUT86
C     U(I) =W2/(R(J)*SH)                                            OUTPUT87
C     V(I) =-W1/SH                                                  OUTPUT88
      HRIJ=1.0/(H(I,J)*R(J))                                        OUTPUT89
      UZ(I)=(W2*XR(I,J)-W1*XT(I,J))*HRIJ                            OUTPUT90
   55 VZ(I)=-(W1*YT(I,J)-W2*YR(I,J))*HRIJ                          OUTPUT91
   56 WRITE (3,145) (VZ(I),I=ILI1,ILI2)                             OUTPUT92
      WRITE (3,150) (UZ(I),I=ILI1,ILI2)                             OUTPUT93
      DO 58 I=ILI1,ILI2                                             OUTPUT94
      U(I)=SQRT(UZ(I)*UZ(I)+VZ(I)*VZ(I))                            OUTPUT95
   58 V(I)=ATAN2(VZ(I),UZ(I))                                       OUTPUT96
      WRITE(3,210) (U(I),I=ILI1,ILI2)                               OUTPUT97
      WRITE(3,220) (V(I),I=ILI1,ILI2)                               OUTPUT98
      GO TO 80                                                      OUTPUT99
   65 DO 75 I=ILI1,ILI2                                             OUTPU100
```

```
    75    U(I)=-2.Ø*CL*VRT(I,JL)                                  OUTPU1Ø1
          WRITE (3,165) (U(I),I=ILI1,ILI2)                         OUTPU1Ø2
    8Ø    WRITE (3,17Ø)                                            OUTPU1Ø3
          IF (JFL.EQ.1) GO TO 25                                   OUTPU1Ø4
          IF (ILI2.GE.ILMN) RETURN                                 OUTPU1Ø5
          ILI1=ILI2+1                                              OUTPU1Ø6
          ILI2=ILI2+13                                             OUTPU1Ø7
          WRITE (3,175)                                            OUTPU1Ø8
          GO TO 2Ø                                                 OUTPU1Ø9
C                                                                  OUTPU11Ø
    95 FORMAT (1H1,5ØX,3ØHRESULTS IN NONDIMENSIONAL FORM,//)       OUTPU111
   1Ø5 FORMAT(1ØX,*THETA*,13F9.5)                                  OUTPU112
   11Ø FORMAT (/,2X,1HJ,3X,4HR(J)/)                                OUTPU113
   115 FORMAT(1X,I2,F8.5,1X,3HSTR,1P13E9.1)                        OUTPU114
   12Ø FORMAT(12X,3HVRT,1P13E9.1)                                  OUTPU115
   125 FORMAT(12X,3HPSI,13F9.5)                                    OUTPU116
   13Ø FORMAT(13X,1HY,1X,13F9.5)                                   OUTPU117
   135 FORMAT(13X,1HX,1X,13F9.5)                                   OUTPU118
   14Ø FORMAT(13X,1HH,1X,1P13E9.1)                                 OUTPU119
   145 FORMAT(13X,2HVZ,13F9.5)                                     OUTPU12Ø
   15Ø FORMAT(13X,2HUZ,13F9.5)                                     OUTPU121
   165 FORMAT(13X,2HCF,1P13E9.1)                                   OUTPU122
   17Ø FORMAT ()                                                   OUTPU123
   175 FORMAT (1H1)                                                OUTPU124
   21Ø FORMAT(12X,3H/U/,13F9.5)                                    OUTPU125
   22Ø FORMAT(12X,3HANG,13F9.5)                                    OUTPU126
       END                                                         OUTPU127
                                                                   COEF    2
                                                                   COEF    3
       SUBROUTINE COEF                                             COEF    4
C                                                                  COEF    5
C      THIS SUBROUTINE DEFINES ELEMENTS THAT DO NOT CHANGE FOR 2*2 BLOCKSCOEF 6
C      RESULTING FROM THE SOLUTION PROCEDURE FOR THE VORTICITY EQUATION COEF 7
C                                                                  COEF    8
       COMMON /FIX/ KL2Ø,DLTRI2,DLTRI3,DLTTI3,JS,JLM2,JLM3,        FIX     2
      *DVRTM,IMM,IMMN,IM,IMPL,IMP,ILMN,ILPL,ILPLP,JLMN,NP,NT,NTMAX,KLØ,KLFIX 3
      11,KL2,KL3,KL4,KL5,KL6,KL9,KL1Ø,KL14,KL15,KL16,ITS,JTS,ITV,JTV,NMAXFIX 4
      2,LN,LNT,PI,XN,XPN,YN,YPN,MARK,DLTI2,DLTI3,RH1DR,RH1DR2,RI,RNN,RNNCFIX 5
      3L,SRNN,DLHRNC,BPSI,BVRT,DVRTLD,R1TR2,ADVRT,TIME,DLTI,MIJF,  FIX     6
      4DLTT2I,DLTR2I,DLTR,DLTR2,DTR2,DLTT,DLTT2,DTT2,C,C2,OI,AS,AT,RATIO,FIX 7
      5CL,DLH,DEN,SIAT,COAT,AP,OR,ORI,CC,P,RHO2,DLTT4,IL,JL,ILM2,JLPL,IUPFIX 8
      6,DLTR24,DLHR,PS,IS,DLTI2I,COSC,ANGVEL,ANGACC,RF,AMP,AVEL1,AVA,RH1SFIX 9
      76,CLI,R1R6,DLRN,DLCL,NTI,T,T1,T2,KL11,IUPI,IUPFL,ANG,NSW,MN,DLT4RIFIX 1Ø
      8,R1I,KL21,R1R3,AV2,ILM3,DIT2D,DF,DR,DR2,DR2MDR,DEN2,FAC     FIX    11
      9,ILM4,D12TR2,D24TR2,JLM4,JM,JM1,JBLE,JDIF1,MIJ,JTI,JVI,JRNCN,KL22 FIX 12
       COMMON /CTRIP/ AI(13Ø),BI(13Ø),CI(13Ø),DB(13Ø,2),AF(2,2),BF(2,2), COEF 1Ø
      1CF(2,2)                                                     COEF   11
C                                                                  COEF   12
       AF(2,1)=-(12.Ø/DTT2)                                        COEF   13
       BF(2,1)= (24.Ø/DTT2)                                        COEF   14
       CF(2,1)=-(12.Ø/DTT2)                                        COEF   15
       AF(2,2)=1.Ø                                                 COEF   16
       BF(2,2)=1Ø.Ø                                                COEF   17
       CF(2,2)=1.Ø                                                 COEF   18
       AF(1,2)=-(DLTI2*DTR2)                                       COEF   19
       BF(1,2)=-(DLTI2*DTR2)*4.Ø                                   COEF   2Ø
       CF(1,2)=-(DLTI2*DTR2)                                       COEF   21
       RETURN                                                      COEF   22
       END                                                         COEF   23
                                                                   CROSS   2
                                                                   CROSS   3
       SUBROUTINE CROSS                                            CROSS   4
C                                                                  CROSS   5
C      THIS SUBROUTINE COMPLETES THE COMPUTATION OF CROSS DERIVATIVE CROSS 6
C      TERM ARISING DUE TO FACTORIZATION OF THE VORTICITY EQUATION THAT CROSS 7
C      WAS STARTED IN SUBROUTINE ADI                               CROSS   8
C                                                                  CROSS   9
       COMMON /FIX/ KL2Ø,DLTRI2,DLTRI3,DLTTI3,JS,JLM2,JLM3,        FIX     2
      *DVRTM,IMM,IMMN,IM,IMPL,IMP,ILMN,ILPL,ILPLP,JLMN,NP,NT,NTMAX,KLØ,KLFIX 3
      11,KL2,KL3,KL4,KL5,KL6,KL9,KL1Ø,KL14,KL15,KL16,ITS,JTS,ITV,JTV,NMAXFIX 4
      2,LN,LNT,PI,XN,XPN,YN,YPN,MARK,DLTI2,DLTI3,RH1DR,RH1DR2,RI,RNN,RNNCFIX 5
      3L,SRNN,DLHRNC,BPSI,BVRT,DVRTLD,R1TR2,ADVRT,TIME,DLTI,MIJF,  FIX     6
      4DLTT2I,DLTR2I,DLTR,DLTR2,DTR2,DLTT,DLTT2,DTT2,C,C2,OI,AS,AT,RATIO,FIX 7
      5CL,DLH,DEN,SIAT,COAT,AP,OR,ORI,CC,P,RHO2,DLTT4,IL,JL,ILM2,JLPL,IUPFIX 8
      6,DLTR24,DLHR,PS,IS,DLTI2I,COSC,ANGVEL,ANGACC,RF,AMP,AVEL1,AVA,RH1SFIX 9
      76,CLI,R1R6,DLRN,DLCL,NTI,T,T1,T2,KL11,IUPI,IUPFL,ANG,NSW,MN,DLT4RIFIX 1Ø
      8,R1I,KL21,R1R3,AV2,ILM3,DIT2D,DF,DR,DR2,DR2MDR,DEN2,FAC     FIX    11
      9,ILM4,D12TR2,D24TR2,JLM4,JM,JM1,JBLE,JDIF1,MIJ,JTI,JVI,JRNCN,KL22 FIX 12
```

```
      COMMON /DUM1/AD(132),BD(132),CD(132),DD(132)                    DUM1    2
      COMMON /DUM/ ED(132),FD(132),QD(132),RD(132)                    DUM     2
      COMMON /WN/ WAV(130,4),WAV2(130,4),WAVF(130,4),WAV2F(130,4)      WN      2
      COMMON PY(130,84),VRT(130,84),RR(84),RHO1S(84),                 CROSS  14
     1A5(84), HRR(130),ARDT(130),HD(130)                              CROSS  15
      COMMON /SMALLD/ F(130,84)                                        CROSS  16
      COMMON /CADI/ ARDTT(84),ARDTR(84),AR(84),CR(84),SC1(84),SC2(84), CADI    2
     1SC3(84)                                                          CADI    3
      COMMON /LSTR/ PSI(130,84),Y(130,84),VRTN2(130,84),H(130,84)      LSTR    2
     1,YY(130,84),PSIN1(130,84)                                        LSTR    3
      LEVEL 2,PSI,Y,VRTN2,H,YY,PSIN1                                   LSTR    4
      COMMON /SIDER/ SIR(130,84),SIT(130,84),VR(130,84),VT(130,84)     CROSS  19
      COMMON /HS/HA5T(130,84),HRRDLT(130,84),HTRR1(130,84)             CROSS  20
      LEVEL 2,SIR,SIT,VR,VT                                            CROSS  21
      LEVEL 2,HA5T,HRRDLT,HTRR1                                        CROSS  22
      DO 200 I=2,ILMN                                                  CROSS  23
 200  F(I,1)=          -((-2.0*ARDTR(1)*SIT(I,1))*VRT(I,1)+(2.0*ARDTR(1)* CROSS  24
     1SIT(I,2))*VRT(I,2))/HA5T(I,1)                                    CROSS  25
      DO 210 J=JM,JLMN                                                 CROSS  26
      D4=(4.0*DLTI)*RR(J)*RHO1S(J)                                     CROSS  27
      DO 210 I=2,ILMN                                                  CROSS  28
 210  F(I,J)=-((AR(J)-ARDTR(J)*SIT(I,J-1))*VRT(I,J-1)+D4*VRT(I,J)+(CR(J) CROSS  29
     1+ARDTR(J)*SIT(I,J+1))*VRT(I,J+1))/HA5T(I,J)                      CROSS  30
      CALL SETZ(F(2,JL),0.0,ILM2)                                      CROSS  31
C                                                                      CROSS  32
      DO 360 J=JVI,JLMN                                                CROSS  33
      DO 330 I=2,ILMN                                                  CROSS  34
      AD(I)=F(I,J)                                                     CROSS  35
      BD(I)=0.0                                                        CROSS  36
      CD(I)=AD(I)*SIR(I,J)                                             CROSS  37
 330  DD(I)=0.0                                                        CROSS  38
      CALL SUFLE1(AD(2),ED(2),LN,ILM2)                                 CROSS  39
      CALL SUFLE1(CD(2),ED(2),LN,ILM2)                                 CROSS  40
      CALL FFT(AD(2),BD(2),ILM2,-1)                                    CROSS  41
      CALL FFT(CD(2),DD(2),ILM2,-1)                                    CROSS  42
      CALL MZZ(AD(2),WAV2F(2,LN),AD(2),ILM2)                           CROSS  43
      CALL MZZ(BD(2),WAV2F(2,LN),BD(2),ILM2)                           CROSS  44
      CALL MZS(AD(2),-1.0,AD(2),ILM2)                                  CROSS  45
      CALL MZS(BD(2),-1.0,BD(2),ILM2)                                  CROSS  46
      CALL MZZ(CD(2), WAVF(2,LN),CD(2),ILM2)                           CROSS  47
      CALL MZZ(DD(2), WAVF(2,LN),DD(2),ILM2)                           CROSS  48
      CALL MZS(DD(2),-1.0,DD(2),ILM2)                                  CROSS  49
      CALL FFT(AD(2),BD(2),ILM2,1)                                     CROSS  50
      CALL FFT(DD(2),CD(2),ILM2,1)                                     CROSS  51
      CALL SUFLE1(AD(2),ED(2),LN,ILM2)                                 CROSS  52
      CALL SUFLE1(DD(2),ED(2),LN,ILM2)                                 CROSS  53
      DO 350 I=2,ILMN                                                  CROSS  54
 350  F(I,J)=((DLTI2*DTR2)*AD(I)+ARDTT(J)*DD(I)*(DLTT/3.0))/HA5T(I,J)  CROSS  55
 360  CONTINUE                                                         CROSS  56
      DO 430 J=JM1,JL                                                  CROSS  57
      F(IL,J)=F(2,J)                                                   CROSS  58
 430  F(1,J)=F(ILMN,J)                                                 CROSS  59
      RETURN                                                           CROSS  60
      END                                                              CROSS  61
                                                                      STREAM  2
                                                                      STREAM  3
      SUBROUTINE STREAM                                               STREAM  4
C                                                                     STREAM  5
C     THIS SUBROUTINE SOLVES THE DISTURBANCE STREAM FUNCTION EQUATION STREAM  6
C                                                                     STREAM  7
      COMMON /FIX/ KL20,DLTRI2,DLTRI3,DLTTI3,JS,JLM2,JLM3,            FIX     2
     *DVRTM,IMM,IMMN,IM,IMPL,IMP,ILMN,ILPL,ILPLP,JLMN,NP,NT,NTMAX,KL0,KLFIX 3
     11,KL2,KL3,KL4,KL5,KL6,KL9,KL10,KL14,KL15,JTS,JTV,ITV,JTV,NMAXCFIX    4
     2,LN,LNT,PI,XN,XPN,YN,YPN,MARK,DLTI2,DLTI3,RH1DR,RH1DR2,RI,RNN,RNNCFIX 5
     3L,SRNN,DLHRNC,BPSI,BVRT,DVRTLD,R1TR2,ADVRT,TIME,DLTI,MIJF,      FIX     6
     4DLTT2I,DLTR2I,DLTR,DLTR2,DTR2,DLTT,DLTT2,DTT2,C,C2,OI,AS,AT,RATIO,FIX  7
     5CL,DLH,DEN,SIAT,COAT,AP,OR,ORI,CC,P,RHO2,DLTT4,IL,JL,ILM2,JLPL,IUPFIX 8
     6,DLTR24,DLHR,PS,IS,DLTI2I,COSC,ANGVEL,ANGACC,RF,AMP,AVEL1,AVA,RH1SFIX 9
     76,CLI,R1R6,DLRN,DLCL,NTI,T,T1,T2,KL11,IUPI,IUPFL,ANG,NSW,MN,DLT4RIFIX 10
     8,R1I,KL21,R1R3,AV2,ILM3,DIT2D,DF,DR,DR2,DR2MDR,DEN2,FAC         FIX    11
     9,ILM4,D12TR2,D24TR2,JLM4,JM,JM1,JBLE,JDIF1,MIJ,JTI,JVI,JRNCN,KL22 FIX  12
      COMMON /DUM1/AD(132),BD(132),CD(132),DD(132)                    DUM1    2
      COMMON /SMALLD/FF(130,84)                                       STREAM 10
      DIMENSION Q(130,84)                                             STREAM 11
      COMMON PY(130,84),VRT(130,84),RR(84),RHO1S(84),A5(84),HRR(130), STREAM 12
     1ARDT(130),HD(130)                                               STREAM 13
      COMMON /BTRI2/AB(84,2),BB(84,2,2),CB(84,2),DB(84,2),D6LTR2,DDD  STREAM 14
     1,JIB,JFLB                                                       STREAM 15
```

```
      COMMON /BPF/ ABF(84),BBF(84,2,2),CBF(84)                     BPF    2
      COMMON /WN/ WAV(130,4),WAV2(130,4),WAVF(130,4),WAV2F(130,4)  WN     2
      COMMON /DUM/ ED(132),FD(132),QD(132),RD(132)                 DUM    2
      COMMON /SIDER/ SIR(130,84),SIT(130,84),VR(130,84),VT(130,84) STREAM19
      EQUIVALENCE (PY(1,1),Q(1,1))                                 STREAM20
      COMMON /LMAIN/XY2(130,84),VRTN1(130,84),VN1(130),XY2RO(130,84), STREAM21
     1XY2T(130,84),AXYYRO(130,84),AXYYT(130,84),RHO1I(84)          STREAM22
      COMMON /LSTR/ PSI(130,84),V(130,84),VRTN2(130,84),H(130,84)  LSTR   2
     1,YY(130,84),PSIN1(130,84)                                    LSTR   3
      LEVEL 2,PSI,V,VRTN2,H,YY,PSIN1                               LSTR   4
      LEVEL 2,XY2,VRTN1,VN1,XY2RO,XY2T,AXYYRO,AXYYT,RHO1I          STREAM24
      LEVEL 2,SIR,SIT,VR,VT                                        STREAM25
      CALL SETZ(Q(1,JM1),0.0,MIJ)                                  STREAM26
      DO 710 J=JM1,JL                                              STREAM27
      CALL SUFLE1(FF(2,J),RD(2),LN,ILM2)                           STREAM28
      CALL FFT(FF(2,J),Q(2,J),ILM2,-1)                             STREAM29
      CALL MZS(FF(2,J),FAC,FF(2,J),ILM2)                           STREAM30
      CALL MZS( Q(2,J),FAC, Q(2,J),ILM2)                           STREAM31
  710 CONTINUE                                                     STREAM32
      CALL SETZ(BD(2),0.0,ILM2)                                    STREAM33
      CALL SUFLE1(AD(2),RD(2),LN,ILM2)                             STREAM34
      CALL FFT(AD(2),BD(2),ILM2,-1)                                STREAM35
      CALL MZS(AD(2),FAC,AD(2),ILM2)                               STREAM36
      CALL MZS(BD(2),FAC,BD(2),ILM2)                               STREAM37
      CALL SETZ(DD(2),0.0,ILM2)                                    STREAM38
      CALL SUFLE1(CD(2),RD(2),LN,ILM2)                             STREAM39
      CALL FFT(CD(2),DD(2),ILM2,-1)                                STREAM40
      CALL MZS(CD(2),FAC,CD(2),ILM2)                               STREAM41
      CALL MZS(DD(2),FAC,DD(2),ILM2)                               STREAM42
      IF (JM.EQ.2) GO TO 550                                       STREAM43
      CALL SETZ(FD(2),0.0,ILM2)                                    STREAM44
      CALL SUFLE1(ED(2),RD(2),LN,ILM2)                             STREAM45
      CALL FFT(ED(2),FD(2),ILM2,-1)                                STREAM46
      CALL MZS(ED(2),FAC,ED(2),ILM2)                               STREAM47
      CALL MZS(FD(2),FAC,FD(2),ILM2)                               STREAM48
  550 CONTINUE                                                     STREAM49
      CALL SUFLE1(QD(2),RD(2),LN,ILM2)                             STREAM50
      CALL SETZ(RD(2),0.0,ILM2)                                    STREAM51
      CALL FFT(QD(2),RD(2),ILM2,-1)                                STREAM52
      CALL MZS(QD(2),FAC,QD(2),ILM2)                               STREAM53
      CALL MZS(RD(2),FAC,RD(2),ILM2)                               STREAM54
C                                                                  STREAM55
      DO 780 I=2,ILMN                                              STREAM56
      CALL COPY(BBF(JM,1,2),BB(JM,1,2),JDIF1)                      STREAM57
      CALL COPY(BBF(JM,2,1),BB(JM,2,1),JDIF1)                      STREAM58
      CALL COPY(BBF(JM,2,2),BB(JM,2,2),JDIF1)                      STREAM59
      DO 755 J=JM,JLMN                                             STREAM60
      AB(J,1)=WAV2(I,LN)*ABF(J)+D12TR2                             STREAM61
      BB(J,1,1)=WAV2(I,LN)*BBF(J,1,1)-D24TR2                       STREAM62
  755 CB(J,1)=WAV2(I,LN)*CBF(J)+D12TR2                             STREAM63
      CALL SETZ(DB(JM,2),0.0,JDIF1)                                STREAM64
      DO 760 J=JM,JLM2,2                                           STREAM65
      DB(J,1)=-(ABF(J)*FF(I,J-1)+BBF(J,1,1)*FF(I,J)+CBF(J)*FF(I,J+1)) STREAM66
  760 DB(J+1,1)=-(ABF(J+1)*FF(I,J)+BBF(J+1,1,1)*FF(I,J+1)+CBF(J+1)* STREAM67
     1FF(I,J+2))                                                   STREAM68
      DDD=D6LTR2                                                   STREAM69
      IF (JM.NE.2) GO TO 600                                       STREAM70
      BB(2,1,1)=BB(2,1,1)+0.8*AB(2,1)                              STREAM71
      BB(2,1,2)=BB(2,1,2)-(0.8*DLTR)*AB(2,1)                       STREAM72
      DDD=12.0/(5.0*DLTR)                                          STREAM73
      BB(2,2,1)=DDD                                                STREAM74
      BB(2,2,2)=1.6                                                STREAM75
      CB(2,1)=CB(2,1)+0.2*AB(2,1)                                  STREAM76
      DB(2,1)=DB(2,1)+(-AB(2,2)+(DLTR2/5.0)*AB(2,1))*QD(I)         STREAM77
      DB(2,2)=0.2*QD(I)                                            STREAM78
      GO TO 610                                                    STREAM79
  600 CONTINUE                                                     STREAM80
      DB(JM,1)=DB(JM,1)-AB(JM,2)*QD(I)-AB(JM,1)*ED(I)              STREAM81
      DB(JM,2)=-(D6LTR2*ED(I)+QD(I))                               STREAM82
  610 CONTINUE                                                     STREAM83
      DB(JLMN,1)=DB(JLMN,1)-CB(JLMN,2)*CD(I)-CB(JLMN,1)*AD(I)      STREAM84
      DB(JLMN,2)=-(-D6LTR2*AD(I)+CD(I))                            STREAM85
      JIB=JM                                                       STREAM86
      JFLB=JLMN                                                    STREAM87
      CALL B2TRI                                                   STREAM88
      DO 765 J=JM,JLM2,2                                           STREAM89
      FF(I,J)=DB(J,1)                                              STREAM90
  765 FF(I,J+1)=DB(J+1,1)                                          STREAM91
```

```
        IF (JM.NE.2) GO TO 620                                         STREAM92
        ED(I)=0.2*(-DLTR*(2.0*QD(I)+4.0*DB(2,2))+4.0*DB(2,1)+DB(3,1))   STREAM93
620     CONTINUE                                                       STREAM94
        CALL SETZ(DB(JM,2),0.0,JDIF1)                                   STREAM95
        DO 770 J=JM,JLM2,2                                              STREAM96
        DB(J,1)=-(ABF(J)* Q(I,J-1)+BBF(J,1,1)* Q(I,J)+CBF(J)* Q(I,J+1)) STREAM97
770     DB(J+1,1)=-(ABF(J+1)* Q(I,J)+BBF(J+1,1,1)* Q(I,J+1)+CBF(J+1)*   STREAM98
      1 Q(I,J+2))                                                      STREAM99
        IF (JM.NE.2) GO TO 650                                         STREA100
        DB(2,1)=DB(2,1)+(-AB(2,2)+(DLTR2/5.0)*AB(2,1))*RD(I)           STREA101
        DB(2,2)=0.2*RD(I)                                              STREA102
        GO TO 660                                                      STREA103
650     CONTINUE                                                       STREA104
        DB(JM,1)=DB(JM,1)-AB(JM,2)*RD(I)-AB(JM,1)*FD(I)                STREA105
        DB(JM,2)=-(D6LTR2*FD(I)+RD(I))                                 STREA106
660     CONTINUE                                                       STREA107
        DB(JLMN,1)=DB(JLMN,1)-CB(JLMN,2)*DD(I)-CB(JLMN,1)*BD(I)        STREA108
        DB(JLMN,2)=-(-D6LTR2*BD(I)+DD(I))                             STREA109
        CALL B2F(I,LN,D24TR2)                                          STREA110
        DO 775 J=JM,JLM2,2                                             STREA111
        Q(I,J)=DB(J,1)                                                 STREA112
775     Q(I,J+1)=DB(J+1,1)                                            STREA113
        IF (JM.NE.2) GO TO 780                                         STREA114
        FD(I)=0.2*(-DLTR*(2.0*RD(I)+4.0*DB(2,2))+4.0*DB(2,1)+DB(3,1))  STREA115
780     CONTINUE                                                       STREA116
C                                                                      STREA117
        IF (JM.NE.2) GO TO 670                                         STREA118
        CALL COPY(ED(2),AD(2),ILM2)                                    STREA119
        CALL COPY(FD(2),BD(2),ILM2)                                    STREA120
        CALL FFT(AD(2),BD(2),ILM2,1)                                   STREA121
        CALL SUFLE1(AD(2),BD(2),LN,ILM2)                              STREA122
        CALL MOVLEV(AD(2),PSI(2,1),ILM2)                              STREA123
        PSI(1,1)=PSI(ILMN,1)                                          STREA124
        PSI(IL,1)=PSI(2,1)                                            STREA125
670     CONTINUE                                                       STREA126
        CALL MZZ(ED(2),WAV(2,LN),ED(2),ILM2)                          STREA127
        CALL MZZ(FD(2),WAV(2,LN),FD(2),ILM2)                          STREA128
        CALL MZS(FD(2),-1.0,FD(2),ILM2)                               STREA129
        CALL FFT(FD(2),ED(2),ILM2,1)                                  STREA130
        CALL SUFLE1(FD(2),ED(2),LN,ILM2)                              STREA131
        DO 750 I=2,ILM2,2                                             STREA132
        SIT(I,JM1)=FD(I)+AXYYT(I,JM1)                                 STREA133
750     SIT(I+1,JM1)=FD(I+1)+AXYYT(I+1,JM1)                           STREA134
        DO 790 J=JM,JLMN                                              STREA135
        CALL MZZ(FF(2,J),WAV(2,LN),AD(2),ILM2)                        STREA136
        CALL MZZ( Q(2,J),WAV(2,LN),BD(2),ILM2)                        STREA137
        CALL MZS(BD(2),-1.0,BD(2),ILM2)                               STREA138
        CALL FFT(BD(2),AD(2),ILM2,1)                                  STREA139
        CALL SUFLE1(BD(2),RD(2),LN,ILM2)                              STREA140
        DO 785 I=2,ILM2,2                                             STREA141
        SIT(I,J)=BD(I)+AXYYT(I,J)                                     STREA142
785     SIT(I+1,J)=BD(I+1)+AXYYT(I+1,J)                               STREA143
790     CALL FFT(FF(2,J),Q(2,J),ILM2,1)                               STREA144
        DO 795 J=JM,JLMN                                              STREA145
        CALL SUFLE1(FF(2,J),RD(2),LN,ILM2)                            STREA146
795     CONTINUE                                                       STREA147
        DO 800 J=JM1,JLMN                                             STREA148
        SIT(IL,J)=SIT(2,J)                                            STREA149
800     SIT(1,J)=SIT(ILMN,J)                                         STREA150
        RETURN                                                        STREA151
        END                                                           STREA152
                                                                      B2TRI   2
                                                                      B2TRI   3
        SUBROUTINE B2TRI                                              B2TRI   4
        COMMON /BTRI2/A(84,2),B(84,2,2),C(84,2),F(84,2),D6LTR2,DDD    B2TRI   5
      1,IS,IFL                                                        B2TRI   6
        REAL L21                                                      B2TRI   7
C                                                                     B2TRI   8
C SOLUTION OF BLOCK TRIDIAGONAL SYSTEM.  A,B,C ARE 2*2 BLOCKS         B2TRI   9
C A AND C HAVE SPECIAL FORMS.                                         B2TRI  10
C F IS FORCING FUNCTION AND SOLUTION IS OUTPUT IN F,                  B2TRI  11
C BLOCK INVERSIONS USE NONPIVOTED LU DECOMPOSITION                    B2TRI  12
C IS AND IFL ARE STARTING AND ENDING INDICES                         B2TRI  13
C                                                                     B2TRI  14
        IS1=IS+1                                                      B2TRI  15
        V1=1.0/B(IS,1,1)                                              B2TRI  16
        U12=V1*B(IS,1,2)                                              B2TRI  17
        L21=B(IS,2,1)                                                 B2TRI  18
```

```
        V2=1.0/(B(IS,2,2)-L21*U12)                              B2TRI 19
        D1=V1*F(IS,1)                                           B2TRI 20
        F(IS,2)=V2*(F(IS,2)-L21*D1)                             B2TRI 21
        F(IS,1)=D1-U12*F(IS,2)                                  B2TRI 22
        D1=V1*C(IS,1)                                           B2TRI 23
        B(IS,2,1)=V2*(-DDD-L21*D1)                              B2TRI 24
        B(IS,1,1)=D1-U12*B(IS,2,1)                              B2TRI 25
        D1=V1*C(IS,2)                                           B2TRI 26
        B(IS,2,2)=V2*(1.0-L21*D1)                               B2TRI 27
        B(IS,1,2)=D1-U12*B(IS,2,2)                              B2TRI 28
C                                                               B2TRI 29
C     FORWARD SWEEP                                             B2TRI 30
C                                                               B2TRI 31
        DO 13 I=IS1,IFL                                         B2TRI 32
        F(I,1)=F(I,1)-A(I,1)*F(I-1,1)-A(I,2)*F(I-1,2)           B2TRI 33
        F(I,2)=F(I,2)-D6LTR2*F(I-1,1)-F(I-1,2)                  B2TRI 34
        V1=1.0/(B(I,1,1)-A(I,1)*B(I-1,1,1)-A(I,2)*B(I-1,2,1))   B2TRI 35
        U12=V1*(B(I,1,2)-A(I,1)*B(I-1,1,2)-A(I,2)*B(I-1,2,2))   B2TRI 36
        L21=B(I,2,1)-D6LTR2*B(I-1,1,1)-B(I-1,2,1)               B2TRI 37
        V2=1.0/(B(I,2,2)-D6LTR2*B(I-1,1,2)-B(I-1,2,2)-L21*U12)  B2TRI 38
        D1 = V1*F(I,1)                                          B2TRI 39
        F(I,2)=V2*(F(I,2)-L21*D1)                               B2TRI 40
        F(I,1) = D1         - U12*F(I,2)                        B2TRI 41
        D1 = V1*C(I,1)                                          B2TRI 42
        B(I,2,1)=V2*(-D6LTR2-L21*D1)                            B2TRI 43
        B(I,1,1)=D1-U12*B(I,2,1)                                B2TRI 44
        D1 = V1*C(I,2)                                          B2TRI 45
        B(I,2,2)=V2*(1.0-L21*D1)                                B2TRI 46
        B(I,1,2)=D1-U12*B(I,2,2)                                B2TRI 47
   13 CONTINUE                                                  B2TRI 48
C                                                               B2TRI 49
C     BACK SUBSTITUTION                                         B2TRI 50
C                                                               B2TRI 51
        IT=IS+IFL                                               B2TRI 52
        DO 21 II=IS1,IFL                                        B2TRI 53
        I=IT-II                                                 B2TRI 54
        F(I,1)=F(I,1)-B(I,1,1)*F(I+1,1)-B(I,1,2)*F(I+1,2)       B2TRI 55
        F(I,2)=F(I,2)-B(I,2,1)*F(I+1,1)-B(I,2,2)*F(I+1,2)       B2TRI 56
   21 CONTINUE                                                  B2TRI 57
        RETURN                                                  B2TRI 58
        END                                                     B2TRI 59
                                                                BTRIP  2
                                                                BTRIP  3
        SUBROUTINE BTRIP(IU)                                    BTRIP  4
        COMMON /CTRIP/ AI(130),BI(130),CI(130), F(130,2),AF(2,2),BF(2,2), BTRIP 5
       1CF(2,2)                                                 BTRIP  6
        COMMON /DUM1/ H(130,2,2),G(2,2),E(2,2)                  BTRIP  7
        COMMON /DUM/ U(130,2,2),Q(2,2),S(2,2)                   BTRIP  8
        DIMENSION FN(2)                                         BTRIP  9
        REAL L21                                                BTRIP 10
C                                                               BTRIP 11
C     ALGORITHM FOR BLOCK PERIODIC TRIDIAGONAL SYSTEM           BTRIP 12
C     THIS VERSION DOES NOT OVERLOAD THE BLOCK MATRICES         BTRIP 13
C     A,B,C, ARE 2*2 BLOCKS... F IS THE FORCING FUNCTION AND THE BTRIP 14
C     SOLUTION IS RETURNED IN F... IU IS THE LENGTH OF TRIDIAGONAL SYSTEM BTRIP 15
C     INDICES... BLOCK INVERSIONS USE NON-PIVOTED LU DECOMPOSITION BTRIP 16
C                                                               BTRIP 17
        IL=2                                                    BTRIP 18
        IS = IL +1                                              BTRIP 19
        IM = IU -1                                              BTRIP 20
        IN = IU -2                                              BTRIP 21
C                                                               BTRIP 22
C     FORWARD SWEEP                                             BTRIP 23
C                                                               BTRIP 24
        I = IL                                                  BTRIP 25
        E(1,1)=CI(IU)                                           BTRIP 26
        E(1,2)=CF(1,2)                                          BTRIP 27
        E(2,1)=CF(2,1)                                          BTRIP 28
        E(2,2)=CF(2,2)                                          BTRIP 29
        V1=1.0/BI(I)                                            BTRIP 30
        U12=V1*BF(1,2)                                          BTRIP 31
        L21=BF(2,1)                                             BTRIP 32
        V2=1.0/(BF(2,2)-L21*U12)                                BTRIP 33
        D1 = V1*F(I,1)                                          BTRIP 34
        F(I,2)=V2*(F(I,2)-L21*D1)                               BTRIP 35
        F(I,1) = D1 - U12*F(I,2)                                BTRIP 36
        D1=V1*CI(I)                                             BTRIP 37
        U(I,2,1)=V2*(CF(2,1)-L21*D1)                            BTRIP 38
```

```
      U(I,1,1)=D1-U12*U(I,2,1)                                  BTRIP 39
      D1=V1*AI(I)                                               BTRIP 40
      H(I,2,1)=V2*(AF(2,1)-L21*D1)                              BTRIP 41
      H(I,1,1)=D1-U12*H(I,2,1)                                  BTRIP 42
      D1=V1*CF(1,2)                                             BTRIP 43
      U(I,2,2)=V2*(CF(2,2)-L21*D1)                              BTRIP 44
      U(I,1,2)=D1-U12*U(I,2,2)                                  BTRIP 45
      D1=V1*AF(1,2)                                             BTRIP 46
      H(I,2,2)=V2*(AF(2,2)-L21*D1)                              BTRIP 47
      H(I,1,2)=D1-U12*H(I,2,2)                                  BTRIP 48
      FN(1)=F(IU,1)-E(1,1)*F(I,1)-E(1,2)*F(I,2)                 BTRIP 49
      FN(2)=F(IU,2)-E(2,1)*F(I,1)-E(2,2)*F(I,2)                 BTRIP 50
      Q(1,1)=BI(IU)-E(1,1)*H(I,1,1)-E(1,2)*H(I,2,1)            BTRIP 51
      Q(2,1)=BF(2,1)-E(2,1)*H(I,1,1)  -E(2,2)*H(I,2,1)         BTRIP 52
      Q(1,2)=BF(1,2)-E(1,1)*H(I,1,2)  -E(1,2)*H(I,2,2)         BTRIP 53
      Q(2,2)=BF(2,2)-E(2,1)*H(I,1,2)  -E(2,2)*H(I,2,2)         BTRIP 54
C                                                               BTRIP 55
      DO 14 I=IS,IN                                             BTRIP 56
      F(I,1)=F(I,1)-AI(I)*F(I-1,1)-AF(1,2)*F(I-1,2)            BTRIP 57
      F(I,2)=F(I,2)-AF(2,1)*F(I-1,1)-AF(2,2)*F(I-1,2)          BTRIP 58
      G(1,1)=BI(I)-AI(I)*U(I-1,1,1)-AF(1,2)*U(I-1,2,1)         BTRIP 59
      G(2,1)=BF(2,1)-AF(2,1)*U(I-1,1,1)-AF(2,2)*U(I-1,2,1)     BTRIP 60
      G(1,2)=BF(1,2)-AI(I)*U(I-1,1,2)-AF(1,2)*U(I-1,2,2)       BTRIP 61
      G(2,2)=BF(2,2)-AF(2,1)*U(I-1,1,2)-AF(2,2)*U(I-1,2,2)     BTRIP 62
      DO 15 M=1,2                                               BTRIP 63
      H(I,1,M)=-AI(I)*H(I-1,1,M)-AF(1,2)*H(I-1,2,M)            BTRIP 64
      H(I,2,M)=-AF(2,1)*H(I-1,1,M)-AF(2,2)*H(I-1,2,M)          BTRIP 65
      DO 15 N=1,2                                               BTRIP 66
      S(N,M)=-E(N,1)*U(I-1,1,M)-E(N,2)*U(I-1,2,M)             BTRIP 67
   15 CONTINUE                                                  BTRIP 68
      DO 7 N=1,2                                                BTRIP 69
      E(N,1)=S(N,1)                                             BTRIP 70
    7 E(N,2)=S(N,2)                                             BTRIP 71
      V1=1.0/G(1,1)                                             BTRIP 72
      U12=V1*G(1,2)                                             BTRIP 73
      L21=G(2,1)                                                BTRIP 74
      V2=1.0/(G(2,2)-L21*U12)                                  BTRIP 75
      D1 = V1*F(I,1)                                            BTRIP 76
      F(I,2)=V2*(F(I,2)-L21*D1)                                BTRIP 77
      F(I,1) = D1 - U12*F(I,2)                                  BTRIP 78
      D1=V1*CI(I)                                               BTRIP 79
      U(I,2,1)=V2*(CF(2,1)-L21*D1)                             BTRIP 80
      U(I,1,1)=D1-U12*U(I,2,1)                                 BTRIP 81
      D1=V1*H(I,1,1)                                            BTRIP 82
      H(I,2,1)=V2*(H(I,2,1)-L21*D1)                            BTRIP 83
      H(I,1,1)=D1-U12*H(I,2,1)                                 BTRIP 84
      D1=V1*CF(1,2)                                             BTRIP 85
      U(I,2,2)=V2*(CF(2,2)-L21*D1)                             BTRIP 86
      U(I,1,2)=D1-U12*U(I,2,2)                                 BTRIP 87
      D1=V1*H(I,1,2)                                            BTRIP 88
      H(I,2,2)=V2*(H(I,2,2)-L21*D1)                            BTRIP 89
      H(I,1,2)=D1-U12*H(I,2,2)                                 BTRIP 90
      FN(1)=FN(1)-E(1,1)*F(I,1)-E(1,2)*F(I,2)                 BTRIP 91
      FN(2)=FN(2)-E(2,1)*F(I,1)-E(2,2)*F(I,2)                 BTRIP 92
      DO 12 N=1,2                                               BTRIP 93
      Q(N,1)=Q(N,1)   -E(N,1)*H(I,1,1)-E(N,2)*H(I,2,1)        BTRIP 94
      Q(N,2)=Q(N,2)   -E(N,1)*H(I,1,2)-E(N,2)*H(I,2,2)        BTRIP 95
   12 CONTINUE                                                  BTRIP 96
   14 CONTINUE                                                  BTRIP 97
C                                                               BTRIP 98
      I = IM                                                    BTRIP 99
      F(I,1)=F(I,1)-AI(I)*F(I-1,1)-AF(1,2)*F(I-1,2)            BTRIP100
      F(I,2)=F(I,2)-AF(2,1)*F(I-1,1)-AF(2,2)*F(I-1,2)          BTRIP101
      G(1,1)=BI(I)-AI(I)*U(I-1,1,1)-AF(1,2)*U(I-1,2,1)         BTRIP102
      G(2,1)=BF(2,1)-AF(2,1)*U(I-1,1,1)-AF(2,2)*U(I-1,2,1)     BTRIP103
      G(1,2)=BF(1,2)-AI(I)*U(I-1,1,2)-AF(1,2)*U(I-1,2,2)       BTRIP104
      G(2,2)=BF(2,2)-AF(2,1)*U(I-1,1,2)-AF(2,2)*U(I-1,2,2)     BTRIP105
      S(1,1)=-E(1,1)*U(I-1,1,1)-E(1,2)*U(I-1,2,1)+AI(IU)      BTRIP106
      S(2,1)=-E(2,1)*U(I-1,1,1)-E(2,2)*U(I-1,2,1)+AF(2,1)     BTRIP107
      S(1,2)=-E(1,1)*U(I-1,1,2)-E(1,2)*U(I-1,2,2)+AF(1,2)     BTRIP108
      S(2,2)=-E(2,1)*U(I-1,1,2)-E(2,2)*U(I-1,2,2)+AF(2,2)     BTRIP109
      H(I,1,1)=-AI(I)*H(I-1,1,1)-AF(1,2)*H(I-1,2,1)+CI(I)     BTRIP110
      H(I,2,1)=-AF(2,1)*H(I-1,1,1)-AF(2,2)*H(I-1,2,1)+CF(2,1)  BTRIP111
      H(I,1,2)=-AI(I)*H(I-1,1,2)-AF(1,2)*H(I-1,2,2)+CF(1,2)   BTRIP112
      H(I,2,2)=-AF(2,1)*H(I-1,1,2)-AF(2,2)*H(I-1,2,2)+CF(2,2)  BTRIP113
   17 CONTINUE                                                  BTRIP114
      DO 8 N=1,2                                                BTRIP115
      E(N,1)=S(N,1)                                             BTRIP116
```

```
      8 E(N,2)=S(N,2)                                             BTRIP117
        V1=1.0/G(1,1)                                             BTRIP118
        U12=V1*G(1,2)                                             BTRIP119
        L21=G(2,1)                                                BTRIP120
        V2=1.0/(G(2,2)-L21*U12)                                  BTRIP121
        D1 = V1*F(I,1)                                            BTRIP122
        F(I,2)=V2*(F(I,2)-L21*D1)                                 BTRIP123
        F(I,1) = D1 - U12*F(I,2)                                  BTRIP124
        DO 18 M=1,2                                               BTRIP125
        D1 = V1*H(I,1,M)                                          BTRIP126
        H(I,2,M)=V2*(H(I,2,M)-L21*D1)                             BTRIP127
        H(I,1,M) = D1 - U12*H(I,2,M)                              BTRIP128
     18 CONTINUE                                                  BTRIP129
        FN(1)=FN(1)-E(1,1)*F(I,1)-E(1,2)*F(I,2)                   BTRIP130
        FN(2)=FN(2)-E(2,1)*F(I,1)-E(2,2)*F(I,2)                   BTRIP131
        DO 19 N=1,2                                               BTRIP132
        Q(N,1)=Q(N,1)   -E(N,1)*H(I,1,1)-E(N,2)*H(I,2,1)          BTRIP133
        Q(N,2)=Q(N,2)   -E(N,1)*H(I,1,2)-E(N,2)*H(I,2,2)          BTRIP134
     19 CONTINUE                                                  BTRIP135
C                                                                 BTRIP136
        V1=1.0/Q(1,1)                                             BTRIP137
        U12=V1*Q(1,2)                                             BTRIP138
        L21=Q(2,1)                                                BTRIP139
        V2=1.0/(Q(2,2)-L21*U12)                                  BTRIP140
        D1 = V1*FN( 1)                                            BTRIP141
        F(IU,2)=V2*(FN(2)-L21*D1)                                 BTRIP142
        F(IU,1)=D1-U12*F(IU,2)                                    BTRIP143
C                                                                 BTRIP144
C     BACKWARD SWEEP                                              BTRIP145
C                                                                 BTRIP146
        F(IM,1)=F(IM,1)-H(IM,1,1)*F(IU,1)-H(IM,1,2)*F(IU,2)       BTRIP147
        F(IM,2)=F(IM,2)-H(IM,2,1)*F(IU,1)-H(IM,2,2)*F(IU,2)       BTRIP148
        IT = IL + IN                                              BTRIP149
        DO 22 J=IL,IN                                             BTRIP150
        I = IT - J                                                BTRIP151
        DO 21 N=1,2                                               BTRIP152
        F(I,N)=F(I,N)-U(I,N,1)*F(I+1,1)-U(I,N,2)*F(I+1,2)-H(I,N,1)*F(IU,1)BTRIP153
       1-H(I,N,2)*F(IU,2)                                         BTRIP154
     21 CONTINUE                                                  BTRIP155
     22 CONTINUE                                                  BTRIP156
        RETURN                                                    BTRIP157
        END                                                       BTRIP158
                                                                  B2F     2
                                                                  B2F     3
        SUBROUTINE B2F(K,LN,D24TR2)                               B2F     4
C                                                                 B2F     5
C       THIS SUBROUTINE SOLVES SPECIAL FORM OF BLOCK TRIDIAGONAL SYSTEM  B2F 6
C       WITH 2*2 BLOCKS                                           B2F     7
C                                                                 B2F     8
        COMMON /BTRI2/A(84,2),B(84,2,2),C(84,2),F(84,2),D6LTR2,DDD  B2F  9
       1,IS,IFL                                                   B2F    10
        COMMON /WN/ WAV(130,4),WAV2(130,4),WAVF(130,4),WAV2F(130,4)  WN   2
        COMMON /BPF/ ABF(84),BBF(84,2,2),CBF(84)                  BPF     2
        REAL L21                                                  B2F    13
C                                                                 B2F    14
        IS1=IS+1                                                  B2F    15
        IF (IS.NE.2)GO TO 10                                      B2F    16
        V1=1.0/(BBF(IS,1,1)*WAV2(K,LN)-D24TR2+0.8*A(IS,1))        B2F    17
        U12=V1*(BBF(IS,1,2)-(12.0/(5.0*D6LTR2))*A(IS,1))          B2F    18
        L21=DDD                                                   B2F    19
        V2=1.0/(1.6-L21*U12)                                      B2F    20
        GO TO 15                                                  B2F    21
     10 CONTINUE                                                  B2F    22
        V1=1.0/(BBF(IS,1,1)*WAV2(K,LN)-D24TR2)                    B2F    23
        U12=V1*BBF(IS,1,2)                                        B2F    24
        L21=BBF(IS,2,1)                                           B2F    25
        V2=1.0/(BBF(IS,2,2)-L21*U12)                              B2F    26
     15 CONTINUE                                                  B2F    27
        D1=V1*F(IS,1)                                             B2F    28
        F(IS,2)=V2*(F(IS,2)-L21*D1)                               B2F    29
        F(IS,1)=D1-U12*F(IS,2)                                    B2F    30
C                                                                 B2F    31
        DO 13 I=IS1,IFL                                           B2F    32
        F(I,1)=F(I,1)-A(I,1)*F(I-1,1)-A(I,2)*F(I-1,2)             B2F    33
        F(I,2)=F(I,2)-D6LTR2*F(I-1,1)-F(I-1,2)                    B2F    34
        V1=1.0/(BBF(I,1,1)*WAV2(K,LN)-D24TR2-A(I,1)*B(I-1,1,1)-A(I,2)*  B2F  35
       1B(I-1,2,1))                                               B2F    36
        U12=V1*(BBF(I,1,2)-A(I,1)*B(I-1,1,2)-A(I,2)*B(I-1,2,2))   B2F    37
```

```
       L21=BBF(I,2,1)-D6LTR2*B(I-1,1,1)-B(I-1,2,1)                     B2F   38
       V2=1.0/(BBF(I,2,2)-D6LTR2*B(I-1,1,2)-B(I-1,2,2)-L21*U12)        B2F   39
       D1=V1*F(I,1)                                                    B2F   40
       F(I,2)=V2*(F(I,2)-L21*D1)                                       B2F   41
       F(I,1)=D1-U12*F(I,2)                                            B2F   42
   13  CONTINUE                                                        B2F   43
       IT=IS+IFL                                                       B2F   44
       DO 21 II=IS1,IFL                                                B2F   45
       I=IT-II                                                         B2F   46
       F(I,1)=F(I,1)-B(I,1,1)*F(I+1,1)-B(I,1,2)*F(I+1,2)               B2F   47
       F(I,2)=F(I,2)-B(I,2,1)*F(I+1,1)-B(I,2,2)*F(I+1,2)               B2F   48
   21  CONTINUE                                                        B2F   49
       RETURN                                                          B2F   50
       END                                                             B2F   51
                                                                       TRIPER 2
                                                                       TRIPER 3
       SUBROUTINE TRIPER(AD,BD,CD,F,ILM2)                              TRIPER 4
C                                                                      TRIPER 5
C      THIS SUBROUTINE SOLVES SCALAR PERIODIC TRIDIAGONAL SYSTEM       TRIPER 6
C                                                                      TRIPER 7
       COMMON /DUM/ A(132),B(132),C(132),D(132)                        TRIPER 8
       DIMENSION AD(128),BD(128),CD(128),F(128)                        TRIPER 9
C                                                                      TRIPER10
       AL=AD(ILM2)                                                     TRIPER11
       BL=BD(ILM2)                                                     TRIPER12
       CL=CD(ILM2)                                                     TRIPER13
       FL=F(ILM2)                                                      TRIPER14
C                                                                      TRIPER15
C      FORWARD ELIMINATION                                             TRIPER16
C                                                                      TRIPER17
       B(1)=1.0/BD(1)                                                  TRIPER18
       A(1)=-AD(1)*B(1)                                                TRIPER19
       C(1)=-CD(1)*B(1)                                                TRIPER20
       F(1)= F(1)*B(1)                                                 TRIPER21
       DO 10 L=2,ILM2                                                  TRIPER22
       B(L)=1.0/(BD(L)+AD(L)*C(L-1))                                   TRIPER23
       C(L)=-CD(L)*B(L)                                                TRIPER24
       F(L)=(F(L)-AD(L)*F(L-1))*B(L)                                   TRIPER25
       A(L)=-AD(L)*A(L-1)*B(L)                                         TRIPER26
   10  CONTINUE                                                        TRIPER27
C                                                                      TRIPER28
C      BACKWARD SWEEP                                                  TRIPER29
C                                                                      TRIPER30
       A(128)=1.0                                                      TRIPER31
       C(128)=0.0                                                      TRIPER32
       DO 20 LD=2,ILM2                                                 TRIPER33
       L=ILM2+1-LD                                                     TRIPER34
       A(L)=A(L)+C(L)*A(L+1)                                           TRIPER35
       C(L)=F(L)+C(L)*C(L+1)                                           TRIPER36
   20  CONTINUE                                                        TRIPER37
C                                                                      TRIPER38
C      BACKWARD ELIMINATION                                            TRIPER39
C                                                                      TRIPER40
       IFL=ILM2-1                                                      TRIPER41
       F(ILM2)=(FL-CL*C(1)-AL*C(IFL))/(CL*A(1)+AL*A(IFL)+BL)           TRIPER42
       DO 30 L=1,IFL                                                   TRIPER43
   30  F(L)=C(L)+F(ILM2)*A(L)                                          TRIPER44
       RETURN                                                          TRIPER45
       END                                                             TRIPER46
                                                                       RNCN   2
                                                                       RNCN   3
       SUBROUTINE RNCN                                                 RNCN   4
C                                                                      RNCN   5
C      THIS SUBROUTINE COMPUTES RATIO OF CONVECTIVE TERM AND VISCOUS TERMRNCN 6
C      AND THE VALUE OF COURANT NUMBER IN EACH DIRECTION               RNCN   7
C                                                                      RNCN   8
       COMMON /FIX/ KL20,DLTRI2,DLTRI3,DLTTI3,JS,JLM2,JLM3,            FIX    2
      *DVRTM,IMM,IMMN,IM,IMPL,IMP,ILMN,ILPL,ILPLP,JLMN,NP,NT,NTMAX,KL0,KLFIX 3
      11,KL2,KL3,KL4,KL5,KL6,KL9,KL10,KL14,KL15,KL16,ITS,JTS,ITV,JTV,NMAXFIX 4
      2,LN,LNT,PI,XN,XPN,YN,YPN,MARK,DLTI2,DLTI3,RH1DR,RH1DR2,RI,RNN,RNNCFIX 5
      3L,SRNN,DLHRNC,BPSI,BVRT,DVRTLD,R1TR2,ADVRT,TIME,DLTI,MIJF,      FIX    6
      4DLTT2I,DLTR2I,DLTR,DLTR2,DTR2,DLTT,DLTT2,DTT2,C,C2,OI,AS,AT,RATIO,FIX  7
      5CL,DLH,DEN,SIAT,COAT,AP,OR,ORI,CC,P,RHO2,DLTT4,IL,JL,ILM2,JLPL,IUPFIX 8
      6,DLTR24,DLHR,PS,IS,DLTI2I,COSC,ANGVEL,ANGACC,RF,AMP,AVEL1,AVA,RH1SFIX 9
      76,CLI,R1R6,DLRN,DLCL,NTI,T,T1,T2,KL11,IUPI,IUPFL,ANG,NSW,MN,DLT4RIFIX 10
      8,R1I,KL21,R1R3,AV2,ILM3,DIT2D,DF,DR,DR2,DR2MDR,DEN2,FAC         FIX    11
      9,ILM4,D12TR2,D24TR2,JLM4,JM,JM1,JBLE,JDIF1,MIJ,JTI,JVI,JRNCN,KL22 FIX  12
       COMMON /THER/ THETA(130),R(84),SINE(130),COSINE(130),RP(130)    THER   2
```

```
      COMMON           PY(13Ø,84),VRT(13Ø,84),RR(84),RHO1S(84),          RNCN  11
     1A5(84),HRR(13Ø),ARDT(13Ø),HD(13Ø)                                  RNCN  12
      COMMON /LSTR/ PSI(13Ø,84),Y(13Ø,84),VRTN2(13Ø,84),H(13Ø,84)        LSTR   2
     1,YY(13Ø,84),PSIN1(13Ø,84)                                          LSTR   3
      LEVEL 2,PSI,Y,VRTN2,H,YY,PSIN1                                     LSTR   4
      COMMON /LBOUND/ YR(13Ø,84),YT(13Ø,84),SH(13Ø,84)                   RNCN  14
      COMMON /SIDER/ SIR(13Ø,84),SIT(13Ø,84),SIRN1(13Ø,84),SITN1(13Ø,84) RNCN  15
      COMMON /DUM1/AD(132),BD(132),CD(132),DD(132)                       DUM1   2
      COMMON /DUM/ ED(132),FD(132),QD(132),RD(132)                       DUM    2
      COMMON /WN/ WAV(13Ø,4),WAV2(13Ø,4),WAVF(13Ø,4),WAV2F(13Ø,4)        WN     2
      COMMON /SMALLD/ F(13Ø,84)                                          RNCN  19
      COMMON /CADI/ ARDTT(84),ARDTR(84),AR(84),CR(84),SC1(84),SC2(84),   CADI   2
     1SC3(84)                                                            CADI   3
      COMMON /LMAIN/XY2(13Ø,84),VRTN1(13Ø,84),VN1(13Ø),XY2RO(13Ø,84),    RNCN  21
     1XY2T(13Ø,84),AXYYRO(13Ø,84),AXYYT(13Ø,84),RHO1I(84)                RNCN  22
      DIMENSION VR(13Ø,84)                                               RNCN  23
      EQUIVALENCE (SIRN1(1,1),VR(1,1))                                   RNCN  24
      LEVEL 2,VR                                                         RNCN  25
      COMMON /LOADS/YTL(132),RHO1(84),XTF(132),PR(132),VRT1(132),        RNCN  26
     1XYT(132),YXT(132),DISS(13Ø)                                        RNCN  27
      LEVEL 2,YTL,RHO1,XTF,PR,VRT1,XYT,YXT,DISS                          RNCN  28
      LEVEL 2,XY2,VRTN1,VN1,XY2RO,XY2T,AXYYRO,AXYYT,RHO1I                RNCN  29
      LEVEL 2, YR,YT,SH                                                  RNCN  30
      LEVEL 2, SIR,SIT,SIRN1,SITN1                                       RNCN  31
C                                                                        RNCN  32
C     VRTN2 - CONVECTION/VISCOUS R-DIRECTION                             RNCN  33
C     F - CONVECTION/VISCOUS THETA-DIRECTION                             RNCN  34
C     SIRN1 - COURANT NUMBER - R-DIRECTION                               RNCN  35
C     SITN1 - COURANT NUMBER -- THETA-DIRECTION                          RNCN  36
C                                                                        RNCN  37
      LN=1                                                               RNCN  38
      DO 2ØØ J=2,JLMN                                                    RNCN  39
      D4=(4.Ø*DLTI)*RR(J)*RHO1S(J)                                       RNCN  40
      DO 2ØØ I=2,ILMN                                                    RNCN  41
      VRTN2(I,J)=Ø.Ø                                                     RNCN  42
      CON=-(ARDTR(J)*(-VRT(I,J-1)*SIT(I,J-1)+VRT(I,J+1)*SIT(I,J-1)))     RNCN  43
      VIS=-(AR(J)*VRT(I,J-1)+D4*VRT(I,J)+CR(J)*VRT(I,J+1))               RNCN  44
      IF (ABS(VIS).LT.1ØE-1Ø) GO TO 2ØØ                                  RNCN  45
      VRTN2(I,J)=ABS(CON/VIS)                                            RNCN  46
  2ØØ CONTINUE                                                           RNCN  47
C                                                                        RNCN  48
      DO 36Ø J=2,JLMN                                                    RNCN  49
      DO 33Ø I=2,ILMN                                                    RNCN  50
      AD(I)=VRT(I,J)                                                     RNCN  51
      BD(I)=Ø.Ø                                                          RNCN  52
      CD(I)=AD(I)*SIR(I,J)                                               RNCN  53
  33Ø DD(I)=Ø.Ø                                                         RNCN  54
      CALL FFT(AD(2),BD(2),ILM2,-1)                                      RNCN  55
      CALL FFT(CD(2),DD(2),ILM2,-1)                                      RNCN  56
      CALL MZZ(AD(2),WAV2F(2,LN),AD(2),ILM2)                             RNCN  57
      CALL MZZ(BD(2),WAV2F(2,LN),BD(2),ILM2)                             RNCN  58
      CALL MZS(AD(2),-1.Ø,AD(2),ILM2)                                    RNCN  59
      CALL MZS(BD(2),-1.Ø,BD(2),ILM2)                                    RNCN  60
      CALL MZZ(CD(2), WAVF(2,LN),CD(2),ILM2)                             RNCN  61
      CALL MZZ(DD(2), WAVF(2,LN),DD(2),ILM2)                             RNCN  62
      CALL MZS(DD(2),-1.Ø,DD(2),ILM2)                                    RNCN  63
      CALL FFT(AD(2),BD(2),ILM2,1)                                       RNCN  64
      CALL FFT(DD(2),CD(2),ILM2,1)                                       RNCN  65
      DO 35Ø I=2,ILMN                                                    RNCN  66
      IF (ABS(AD(I)).LT.1ØE-1Ø) GO TO 35Ø                               RNCN  67
      F(I,J)=ABS(SC3(J)*DD(I)/AD(I))                                     RNCN  68
  35Ø CONTINUE                                                           RNCN  69
  36Ø CONTINUE                                                           RNCN  70
C                                                                        RNCN  71
      DO 55 J=2,JLMN                                                     RNCN  72
      DO 55 I=2,ILMN                                                     RNCN  73
      HRI=DLTI2/(H(I,J)*R(J))                                            RNCN  74
      SIRN1(I,J)=ABS(SIT(I,J)*HRI/(R(J+1)-R(J-1)))                       RNCN  75
   55 SITN1(I,J)=ABS(-SIR(I,J)*RHO1(J)*HRI*DLTT2I)                       RNCN  76
C                                                                        RNCN  77
      READ(14) ((PSI(I,J),I=2,ILMN),J=1,JL)                             RNCN  78
      REWIND 14                                                          RNCN  79
      TDERVM=Ø.Ø                                                         RNCN  80
      DO 37Ø J=1,JL                                                      RNCN  81
      DO 37Ø I=2,ILMN                                                    RNCN  82
      TDERV=(T*VRT(I,J)-T1*VRTN1(I,J)+T2*PSI(I,J))*DLTI2I                RNCN  83
      ATDERV=ABS(TDERV)                                                  RNCN  84
      IF (ATDERV    .LT.TDERVM) GO TO 37Ø                               RNCN  85
```

```
      TDERVM=ATDERV                                                RNCN  86
      ITVM=I                                                       RNCN  87
      JTVM=J                                                       RNCN  88
 370  CONTINUE                                                     RNCN  89
      WRITE(3,150) TDERVM,ITVM,JTVM                                RNCN  90
 150  FORMAT( /,3X,*MAXIMUM TIME DERIVATIVE OF VORTICITY  *,1PE14.5,* RNCN  91
     1AT I=*, I3.*  J=*,I3,/)                                      RNCN  92
C                                                                  RNCN  93
      JI=2                                                         RNCN  94
      JFL=JL/2                                                     RNCN  95
      J2=JL*3/4                                                    RNCN  96
      WRITE(3,460)                                                 RNCN  97
 460  FORMAT(///20X,*STATISTICS FROM J=2 TO 43*,/)                 RNCN  98
 405  CONTINUE                                                     RNCN  99
      RNRM=0.0                                                     RNCN 100
      RNTM=0.0                                                     RNCN 101
      RCOM=0.0                                                     RNCN 102
      TCOM=0.0                                                     RNCN 103
      DO 500 J=JI,JFL                                              RNCN 104
      DO 500 I=2,ILMN                                              RNCN 105
      IF (VRTN2(I,J).LT.RNRM) GO TO 410                            RNCN 106
      RNRM=VRTN2(I,J)                                              RNCN 107
      IRNRM=I                                                      RNCN 108
      JRNRM=J                                                      RNCN 109
 410  IF(F(I,J).LT.RNTM) GO TO 420                                 RNCN 110
      RNTM=F(I,J)                                                  RNCN 111
      IRNTM=I                                                      RNCN 112
      JRNTM=J                                                      RNCN 113
 420  IF (SIRN1(I,J).LT.RCOM) GO TO 450                            RNCN 114
      RCOM=SIRN1(I,J)                                              RNCN 115
      IRCOM=I                                                      RNCN 116
      JRCOM=J                                                      RNCN 117
 450  IF (SITN1(I,J).LT.TCOM) GO TO 500                            RNCN 118
      TCOM=SITN1(I,J)                                              RNCN 119
      ITCOM=I                                                      RNCN 120
      JTCOM=J                                                      RNCN 121
 500  CONTINUE                                                     RNCN 122
      WRITE(3,510) RNRM,IRNRM,JRNRM                                RNCN 123
      WRITE(3,520) RNTM,IRNTM,JRNTM                                RNCN 124
      WRITE(3,540)RCOM,IRCOM,JRCOM                                 RNCN 125
      WRITE(3,550)TCOM,ITCOM,JTCOM                                 RNCN 126
 510  FORMAT(/3X,*MAXIMUM CON/VIS IN R-DIERCTION *,1PE14.5,*  AT I=*,I3, RNCN 127
     1*   J=*,I3/)                                                 RNCN 128
 520  FORMAT(/3X,*MAXIMUM CON/VIS IN T-DIERCTION *,1PE14.5,*  AT I=*,I3, RNCN 129
     1*   J=*,I3/)                                                 RNCN 130
 540  FORMAT(/3X,*MAXIMUM COURANT NUMBER IN R-DIRECTION *,1PE14.5,* AT IRNCN 131
     1=*,I3,*  J=*,I3/)                                            RNCN 132
 550  FORMAT(/3X,*MAXIMUM COURANT NUMBER IN T-DIRECTION *,1PE14.5,* AT IRNCN 133
     1=*,I3,*   J=*,I3/)                                           RNCN 134
      IF(JFL-J2) 560,570,580                                       RNCN 135
 560  JI=JFL+1                                                     RNCN 136
      JFL=J2                                                       RNCN 137
      WRITE(3,470)                                                 RNCN 138
 470  FORMAT(  /20X,*STATISTICS FROM J=44 TO 63*,/)                RNCN 139
      GO TO 405                                                    RNCN 140
 570  JI=JFL+1                                                     RNCN 141
      JFL=JLMN                                                     RNCN 142
      WRITE(3,480)                                                 RNCN 143
 480  FORMAT(  /20X,*STATISTICS FROM J=64 TO JLMN*,/)              RNCN 144
      GO TO 405                                                    RNCN 145
 580  CONTINUE                                                     RNCN 146
C                                                                  RNCN 147
      WRITE (3,95)                                                 RNCN 148
      ILI1=2                                                       RNCN 149
      ILI2=14                                                      RNCN 150
  20  IF (ILI2.GT.ILMN) ILI2=ILMN                                  RNCN 151
      WRITE (3,105) (THETA(I),I=ILI1,ILI2)                         RNCN 152
      WRITE (3,110)                                                RNCN 153
      DO 80 J=2,JLMN,JRNCN                                         RNCN 154
        WRITE (3,115) J,R(J),(VRTN2(I,J),I=ILI1,ILI2)              RNCN 155
        WRITE (3,120) ( F(I,J),I=ILI1,ILI2)                        RNCN 156
      WRITE(3,210) (SIRN1(I,J),I=ILI1,ILI2)                        RNCN 157
      WRITE(3,220) (SITN1(I,J),I=ILI1,ILI2)                        RNCN 158
      WRITE(3,230) (VRT(I,J),I=ILI1,ILI2)                          RNCN 159
  80    WRITE (3,170)                                              RNCN 160
      IF (ILI2.GE.ILMN) RETURN                                     RNCN 161
      ILI1=ILI2+1                                                  RNCN 162
      ILI2=ILI2+13                                                 RNCN 163
```

```
       WRITE (3,175)                                                RNCN 164
       GO TO 20                                                     RNCN 165
c                                                                   RNCN 166
  95 FORMAT (1H1,50X,30HRESULTS IN NONDIMENSIONAL FORM,//)          RNCN 167
 105 FORMAT(10X,*THETA*,13F9.5)                                     RNCN 168
 110 FORMAT (/,2X,1HJ,3X,4HR(J)/)                                   RNCN 169
 115 FORMAT (1X,I2,F8.5  ,1X,3HRNR,1P13E9.1)                        RNCN 170
 120 FORMAT (12X,3HRNT,1P13E9.1)                                    RNCN 171
 170 FORMAT ()                                                      RNCN 172
 175 FORMAT (1H1)                                                   RNCN 173
 210 FORMAT(12X,3HRCO,1P13E9.1)                                     RNCN 174
 220 FORMAT(12X,3HTCO,1P13E9.1)                                     RNCN 175
 230 FORMAT(12X,3HVRT,1P13E9.1)                                     RNCN 176
       END                                                         RNCN 177
                                                                   TRIB   2
                                                                   TRIB   3
           IDENT   TRIB                                            TRIB   4
           ENTRY   TRIB                                            TRIB   5
*                                                                  TRIB   6
*      SUBROUTINE TRIB(A,B,C,X,F,NU)                               TRIB   7
*                                                                  TRIB   8
*      ALL ELEMENTS IN SCM                                         TRIB   9
*      NU - NUMBER OF SIMULTANEOUS EQUATIONS                       TRIB  10
*      A, B, C - COEFFICIENTS OF TRIDIAGONAL MATRIX                TRIB  11
*      X - DUMMY ARRAY                                             TRIB  12
*      F - RIGHT-HAND SIDE OF SYSTEM                               TRIB  13
*                                                                  TRIB  14
****************************                                       TRIB  15
  TRIB     CON     0                                               TRIB  16
           SB7     B0+*F         GET *F                            TRIB  17
           SB7     B7-1                                            TRIB  18
           LE      B7,B0,RUN                                       TRIB  19
*          FOR FTN COMPILERS                                       TRIB  20
           SB1     X1            SET A                             TRIB  21
           SA2     A1+1          FETCH B                           TRIB  22
           SA3     A1+2          FETCH C                           TRIB  23
           SA4     A1+3          FETCH D                           TRIB  24
           SA5     A1+4          FETCH E                           TRIB  25
           SA1     A1+5          FETCH F                           TRIB  26
           SX6     A0            MOVE A0                           TRIB  27
           SA6     SAVE          SAVE A0                           TRIB  28
           SB2     X2            SET B                             TRIB  29
           SB3     X3            SET C                             TRIB  30
           SB4     X4            SET D                             TRIB  31
           SB5     X5            SET E                             TRIB  32
           SB6     X1            SET F                             TRIB  33
  RUN      BSS     0                                               TRIB  34
*          FOR RUN COMPILERS                                       TRIB  35
           SA3     B2            FETCH B1                          TRIB  36
           SA0     17204 0B                                        TRIB  37
           SA1     B6            FETCH NU                          TRIB  38
           SX0     A0                                              TRIB  39
           LX0     42            X0=1.0                            TRIB  40
           SB6     1             B6=1                              TRIB  41
           RX3     X0/X3         Z1=1./B1                          TRIB  42
           SB7     X1            B7=NU                             TRIB  43
           SA4     B3            FETCH C1                          TRIB  44
           SA2     B1+1          FETCH A2                          TRIB  45
           SB1     1                                               TRIB  46
           SA5     A3+B1         FETCH B2                          TRIB  47
           RX7     X4*X2         C1*A2                             TRIB  48
           SA2     A2-B1         FETCH A1 (FOR INDEX ONLY)         TRIB  49
           RX1     X7*X3         A2*X1                             TRIB  50
           MX6     0             DON'T USE F0                      TRIB  51
           RX7     X4*X3         C1Z1=X1                           TRIB  52
           SA4     B5            F1                                TRIB  53
           MX2     0             DON'T USE A1                      TRIB  54
           RX1     X5-X1         B2-A2X1                           TRIB  55
           SA7     B4            STORE X1                          TRIB  56
           SB4     B1+B1                                           TRIB  57
           NX1     X1                                              TRIB  58
           SB2     B2+B4                                           TRIB  59
  LOOP1    RX1     X0/X1         DIVIDE TO GET Z2                  TRIB  60
           RX6     X2*X6         A1F0                              TRIB  61
           SA2     A2+B4         FETCH A3                          TRIB  62
           SB3     B3+B1         BUMP C COUNTER                    TRIB  63
           RX6     X4-X6         F1-A1F0                           TRIB  64
           SA4     B3            FETCH C2                          TRIB  65
```

```
        NX6   X6                                              TRIB   66
        RX6   X3*X6        F1=Z1(F1-A1FØ)                      TRIB   67
        SB6   B6+B1        COUNTER                             TRIB   68
        SA6   B5           STORE F1                            TRIB   69
        GE    B6,B7,OUT                                        TRIB   7Ø
        RX7   X1*X4        X2=Z2C2                             TRIB   71
        SA5   B2           B3                                  TRIB   72
        SB5   B5+B1        BUMP F2                             TRIB   73
        BX3   X1                                               TRIB   74
        SA4   B5           F2                                  TRIB   75
        RX1   X7*X2        A3X2                                TRIB   76
        SA2   A2-B1        A2                                  TRIB   77
        SA7   A7+B1        STORE X2                            TRIB   78
        SB2   B2+B1                                            TRIB   79
        RX1   X5-X1        B3-A3X2                             TRIB   8Ø
        NX1   X1                                               TRIB   81
        JP    BØ+LOOP1                                         TRIB   82
OUT     SA2   A2-B1        FETCH A2                            TRIB   83
        SB5   B5+B1        BUMP F COUNTER                      TRIB   84
        SA4   B5           FETCH F2                            TRIB   85
        RX6   X2*X6        A2F1                                TRIB   86
        RX6   X4-X6        F2-A2F1                             TRIB   87
        SA4   B5-B1        FETCH F1 FOR LOOP2                  TRIB   88
        SA5   A7           FETCH X1 FOR LOOP2                  TRIB   89
        NX6   X6                                               TRIB   9Ø
        SB6   B4           SET COUNTER FOR LOOP 2              TRIB   91
        RX6   X6*X1        F2=Z2(F2-A2F1)                      TRIB   92
        SA6   A4+B1        STORE F2                            TRIB   93
LOOP2   RX6   X5*X6        X1F2                                TRIB   94
        SA5   A5-B1        FETCH XØ                            TRIB   95
        RX6   X4-X6        F1=F1-X1F2                          TRIB   96
        SA4   A4-B1        FETCH FØ                            TRIB   97
        SB6   B6+B1        COUNTER                             TRIB   98
        NX6   X6                                               TRIB   99
        SA6   A6-B1        STORE F1                            TRIB  1ØØ
        GE    B7,B6,LOOP2                                      TRIB  1Ø1
RAØ     SA1   SAVE                                             TRIB  1Ø2
        SAØ   X1                                               TRIB  1Ø3
        EQ    TRIB                                             TRIB  1Ø4
SAVE    BSSZ  1                                                TRIB  1Ø5
        END                                                   TRIB  1Ø6
```

```
IL  =13Ø   JL  = 84

NMAX = 38   NTMAX = 2   KL2 = Ø   KL3 = Ø   KL4 = 3   KL5 = 1
KL6 = 2   KL9 = 5   KL11 = Ø   KL13 = Ø   KL14 = Ø   KL15 = 1   KL16 = Ø   JS = 4
RI = 4.ØØØØØE-Ø2   RNN = 5.ØØØØØE+Ø3   BPSI = 1.3ØØØØE+ØØ   BVRT = 3.ØØØØØE-Ø2
DVRTM = 1.5ØØØØE-Ø4   TIME = Ø.   DLTI = 1.ØØØØØE-Ø3
RF = Ø.   AMP = Ø.   KL1Ø = 1   KL21 = 1   DF = 3.619Ø6E+ØØ   TIMOSC= Ø.
J3LE= 2   KL22= Ø   JRNCN= 4

TR = Ø.   OR =-6.47384E-Ø2   OI = Ø.   AS = 4.ØØØØØE-Ø2   AT = Ø.
PC = 2.5ØØØØE-Ø1   RMAX = Ø.   IRPI = 4
XP = 2.ØØØØØE+ØØ   XN = 2.ØØØØØE+ØØ   RI = Ø.
```

I	EXACT WAVE NUMBER, K	MODIFIED WAVE NUMBER, KM	K*K	MODIFIED SQUARE OF KM
2	Ø.	Ø.	Ø.	Ø.
3	1.ØØØØØE+ØØ	1.ØØØØØE+ØØ	1.ØØØØØE+ØØ	1.ØØØØØE+ØØ
4	2.ØØØØØE+ØØ	2.ØØØØØE+ØØ	4.ØØØØØE+ØØ	4.ØØØØØE+ØØ
5	3.ØØØØØE+ØØ	2.99999E+ØØ	9.ØØØØØE+ØØ	8.99998E+ØØ
6	4.ØØØØØE+ØØ	3.9999ØE+ØØ	1.6ØØØØE+Ø1	1.59999E+Ø1
7	5.ØØØØØE+ØØ	4.9999ØE+ØØ	2.5ØØØØE+Ø1	2.49996E+Ø1
8	6.ØØØØØE+ØØ	5.99975E+ØØ	3.6ØØØØE+Ø1	3.59989E+Ø1
9	7.ØØØØØE+ØØ	6.99945E+ØØ	4.9ØØØØE+Ø1	4.89971E+Ø1
1Ø	8.ØØØØØE+ØØ	7.99892E+ØØ	6.4ØØØØE+Ø1	6.39936E+Ø1
11	9.ØØØØØE+ØØ	8.998Ø5E+ØØ	8.1ØØØØE+Ø1	8.Ø987ØE+Ø1
12	1.ØØØØØE+Ø1	9.99668E+ØØ	1.ØØØØØE+Ø2	9.9975ØE+Ø1
13	1.1ØØØØE+Ø1	1.Ø9946E+Ø1	1.21ØØØE+Ø2	1.2Ø957E+Ø2
14	1.2ØØØØE+Ø1	1.19916E+Ø1	1.44ØØØE+Ø2	1.43927E+Ø2
15	1.3ØØØØE+Ø1	1.2987ØE+Ø1	1.69ØØØE+Ø2	1.688Ø1E+Ø2
16	1.4ØØØØE+Ø1	1.39816E+Ø1	1.96ØØØE+Ø2	1.95815E+Ø2
17	1.5ØØØØE+Ø1	1.49739E+Ø1	2.25ØØØE+Ø2	2.24719E+Ø2
18	1.6ØØØØE+Ø1	1.59636E+Ø1	2.56ØØØE+Ø2	2.55585E+Ø2
19	1.7ØØØØE+Ø1	1.695Ø3E+Ø1	2.89ØØØE+Ø2	2.884Ø1E+Ø2
2Ø	1.8ØØØØE+Ø1	1.7933ØE+Ø1	3.24ØØØE+Ø2	3.23153E+Ø2
21	1.9ØØØØE+Ø1	1.89114E+Ø1	3.61ØØØE+Ø2	3.59826E+Ø2

22	2.00000E+01	1.98843E+01	4.00000E+02	3.98397E+02
23	2.10000E+01	2.00506E+01	4.41000E+02	4.38845E+02
24	2.20000E+01	2.18091E+01	4.84000E+02	4.81142E+02
25	2.30000E+01	2.27586E+01	5.29000E+02	5.25256E+02
26	2.40000E+01	2.36974E+01	5.76000E+02	5.71150E+02
27	2.50000E+01	2.46237E+01	6.25000E+02	6.18784E+02
28	2.60000E+01	2.55356E+01	6.76000E+02	6.68108E+02
29	2.70000E+01	2.64309E+01	7.29000E+02	7.19069E+02
30	2.80000E+01	2.73069E+01	7.84000E+02	7.71606E+02
31	2.90000E+01	2.81610E+01	8.41000E+02	8.25651E+02
32	3.00000E+01	2.89899E+01	9.00000E+02	8.81127E+02
33	3.10000E+01	2.97901E+01	9.61000E+02	9.37950E+02
34	3.20000E+01	3.05577E+01	1.02400E+03	9.96028E+02
35	3.30000E+01	3.12886E+01	1.08900E+03	1.05526E+03
36	3.40000E+01	3.19778E+01	1.15600E+03	1.11552E+03
37	3.50000E+01	3.26202E+01	1.22500E+03	1.17671E+03
38	3.60000E+01	3.32101E+01	1.29600E+03	1.23867E+03
39	3.70000E+01	3.37412E+01	1.36900E+03	1.30128E+03
40	3.80000E+01	3.42068E+01	1.44400E+03	1.36437E+03
41	3.90000E+01	3.45996E+01	1.52100E+03	1.42778E+03
42	4.00000E+01	3.49118E+01	1.60000E+03	1.49133E+03
43	4.10000E+01	3.51349E+01	1.68100E+03	1.55484E+03
44	4.20000E+01	3.52603E+01	1.76400E+03	1.61811E+03
45	4.30000E+01	3.52787E+01	1.84900E+03	1.68092E+03
46	4.40000E+01	3.51804E+01	1.93600E+03	1.74307E+03
47	4.50000E+01	3.49558E+01	2.02500E+03	1.80433E+03
48	4.60000E+01	3.45948E+01	2.11600E+03	1.86446E+03
49	4.70000E+01	3.40878E+01	2.20900E+03	1.92323E+03
50	4.80000E+01	3.34252E+01	2.30400E+03	1.98040E+03
51	4.90000E+01	3.25981E+01	2.40100E+03	2.03571E+03
52	5.00000E+01	3.15987E+01	2.50000E+03	2.09892E+03
53	5.10000E+01	3.04200E+01	2.60100E+03	2.13978E+03
54	5.20000E+01	2.90570E+01	2.70400E+03	2.20085E+03
55	5.30000E+01	2.75063E+01	2.80900E+03	2.23507E+03
56	5.40000E+01	2.57671E+01	2.91600E+03	2.27587E+03
57	5.50000E+01	2.38412E+01	3.02500E+03	2.31497E+03
58	5.60000E+01	2.17335E+01	3.13600E+03	2.35057E+03
59	5.70000E+01	1.94521E+01	3.24900E+03	2.30247E+03
60	5.80000E+01	1.70985E+01	3.36400E+03	2.41051E+03
61	5.90000E+01	1.44178E+01	3.49100E+03	2.44524E+03
62	6.00000E+01	1.16983E+01	3.63000E+03	2.45436E+03
63	6.10000E+01	8.87119E+00	3.72100E+03	2.46991E+03
64	6.20000E+01	5.96166E+00	3.84400E+03	2.48109E+03
65	6.30000E+01	2.99519E+00	3.95900E+03	2.48782E+03
66	-6.40000E+01	0.	4.09600E+03	2.49007E+03

DISPLACEMENT THICKNESS AT FIRST TIME STEP = 9.59993E-04

NORMALIZED SURFACE, Y

```
-3.05950E-26  3.91579E-03  7.81420E-03  1.16779E-02  1.54899E-02  1.92334E-02  2.28918E-02  2.64493E-02  2.98905E-02  3.32007E-02
 3.63650E-02  3.93720E-02  4.22093E-02  4.48641E-02  4.73267E-02  4.95888E-02  5.16406E-02  5.34769E-02  5.50917E-02  5.64807E-02
 5.76411E-02  5.85713E-02  5.92710E-02  5.97413E-02  5.99956E-02  5.98085E-02  5.94961E-02  5.87883E-02  5.79812E-02
 6.69895E-02  5.58212E-02  5.44970E-02  5.30203E-02  5.14090E-02  4.96769E-02  4.78362E-02  4.59247E-02  4.38963E-02  4.10274E-02
 3.97120E-02  3.75655E-02  3.54024E-02  3.32366E-02  3.00815E-02  2.89494E-02  2.68519E-02  2.47992E-02  2.28004E-02  2.33634E-02
 1.89944E-02  1.71985E-02  1.54788E-02  1.38322E-02  1.22740E-02  1.07878E-02  9.37590E-03  6.75664E-03  5.53697E-03
 4.36716E-03  3.23840E-03  2.14113E-03  1.06518E-03  1.66047E-26
```

NORMALIZED CHORD LENGTH

```
0.           5.40755E-04  2.16206E-03  4.86104E-03  8.63289E-03  1.34709E-02  1.93665E-02  2.63490E-02  3.42861E-02  4.32835E-02
 5.32848E-02  6.42720E-02  7.62251E-02  8.91222E-02  1.02940E-01  1.17652E-01  1.33231E-01  1.49648E-01  1.66873E-01  1.84871E-01
 2.03609E-01  2.23049E-01  2.43156E-01  2.63885E-01  2.85198E-01  3.07052E-01  3.29400E-01  3.52196E-01  3.75390E-01  3.93939E-01
 4.22784E-01  4.46875E-01  4.71159E-01  4.95579E-01  5.20079E-01  5.44601E-01  5.69086E-01  5.93474E-01  6.17716E-01  6.41716E-01
 5.65447E-01  6.08034E-01  7.11814E-01  7.34274E-01  7.56388E-01  7.77695E-01  7.98428E-01  8.18444E-01  8.37668E-01  8.55995E-01
 8.73613E-01  8.90187E-01  9.05764E-01  9.20294E-01  9.33776E-01  9.46027E-01  9.57145E-01  9.67046E-01  9.75699E-01
 3.99136E-01  9.93877E-01  9.97274E-01  9.99318E-01  1.00000E+00
```

CC = 8.06137E-01   CL = 3.61906E+00   TH = 3.61906E+00   RATIO = 1.20001E-01   RNNCL = 1.30157E+03

XMIN = -9.04765E-01   XMAX = 2.71429E+00   COSC = -9.17095E-01   = -4.34328E-01   = -9.17095E-01

LEADING EDGE RADIUS OF CURVATURE = 1.41778E+00 (APPROX)

J	R	RHO	RHO1	RHOIS	RHO2
1	4.0000E-02	0.	7.1067E+00	5.0506E+01	-3.8951E+02
2	4.1776E-02	1.2048E-02	6.4772E+00	4.1955E+01	-3.2237E+02
3	4.3725E-02	2.4996E-02	5.3056E+00	3.4877E+01	-2.6688E+02
4	4.5862E-02	3.6145E-02	5.3866E+00	2.9015E+01	-2.2104E+02
5	4.8205E-02	4.8193E-02	4.9153E+00	2.4168E+01	-1.8311E+02
6	5.0771E-02	6.0241E-02	4.4874E+00	2.0137E+01	-1.5181E+02
7	5.3581E-02	7.2289E-02	4.3988E+00	1.6809E+01	-1.2590E+02
8	5.6657E-02	8.4337E-02	3.7461E+00	1.4033E+01	-1.0447E+02
9	6.0022E-02	9.6386E-02	3.4258E+00	1.1736E+01	-8.6736E+01
10	6.3700E-02	1.0843E-01	3.1350E+00	9.8285E+00	-7.2057E+01
11	6.7717E-02	1.2048E-01	2.8711E+00	8.2432E+00	-5.9909E+01
12	7.2102E-02	1.3253E-01	2.6315E+00	6.9249E+00	-4.9834E+01
13	7.6884E-02	1.4458E-01	2.4141E+00	5.8277E+00	-4.1491E+01
14	8.2094E-02	1.5663E-01	2.2167E+00	4.9139E+00	-3.4573E+01
15	8.7765E-02	1.6867E-01	2.0377E+00	4.1521E+00	-2.8835E+01
16	9.3931E-02	1.8072E-01	1.8752E+00	3.5164E+00	-2.4072E+01
17	1.0063E-01	1.9277E-01	1.7279E+00	2.9855E+00	-2.0116E+01

18	1.0789E-01	2.0482E-01	1.5942E+00	2.5416E+00	-1.6829E+01
19	1.1575E-01	2.1687E-01	1.4731E+00	2.1700E+00	-1.4094E+01
20	1.2426E-01	2.2892E-01	1.3633E+00	1.8586E+00	-1.1818E+01
21	1.3344E-01	2.4096E-01	1.2639E+00	1.5974E+00	-9.9214E+00
22	1.4334E-01	2.5301E-01	1.1739E+00	1.3780E+00	-8.3394E+00
23	1.5398E-01	2.6506E-01	1.0924E+00	1.1934E+00	-7.0184E+00
24	1.6540E-01	2.7711E-01	1.0188E+00	1.0380E+00	-5.9139E+00
25	1.7764E-01	2.8916E-01	9.5235E-01	9.0698E-01	-4.9900E+00
26	1.9071E-01	3.0120E-01	8.9240E-01	7.9638E-01	-4.2134E+00
27	2.0465E-01	3.1325E-01	8.3841E-01	7.0293E-01	-3.5616E+00
28	2.1946E-01	3.2530E-01	7.8980E-01	6.2391E-01	-3.0127E+00
29	2.3516E-01	3.3735E-01	7.4635E-01	5.5705E-01	-2.5494E+00
30	2.5174E-01	3.4940E-01	7.0744E-01	5.0048E-01	-2.1571E+00
31	2.6922E-01	3.6145E-01	6.7277E-01	4.5262E-01	-1.8238E+00
32	2.8755E-01	3.7349E-01	6.4202E-01	4.1219E-01	-1.5394E+00
33	3.0674E-01	3.8554E-01	6.1490E-01	3.7811E-01	-1.2956E+00
34	3.2673E-01	3.9759E-01	5.9117E-01	3.4948E-01	-1.0853E+00
35	3.4748E-01	4.0964E-01	5.7063E-01	3.2558E-01	-9.0249E-01
36	3.6894E-01	4.2169E-01	5.5300E-01	3.0581E-01	-7.4224E-01
37	3.9103E-01	4.3373E-01	5.3821E-01	2.8967E-01	-6.0024E-01
38	4.1368E-01	4.4578E-01	5.2608E-01	2.7677E-01	-4.7278E-01
39	4.3680E-01	4.5783E-01	5.1652E-01	2.6679E-01	-3.5663E-01
40	4.6030E-01	4.6988E-01	5.0942E-01	2.5951E-01	-2.4893E-01
41	4.8407E-01	4.8193E-01	5.0473E-01	2.5475E-01	-1.4707E-01
42	5.0800E-01	4.9398E-01	5.0239E-01	2.5239E-01	-4.8645E-02
43	5.3200E-01	5.0602E-01	5.0239E-01	2.5239E-01	4.8645E-02
44	5.5593E-01	5.1807E-01	5.0473E-01	2.5475E-01	1.4707E-01
45	5.7970E-01	5.3012E-01	5.0942E-01	2.5951E-01	2.4893E-01
46	6.0320E-01	5.4217E-01	5.1652E-01	2.6679E-01	3.5663E-01
47	6.2632E-01	5.5422E-01	5.2608E-01	2.7677E-01	4.7278E-01
48	6.4897E-01	5.6627E-01	5.3821E-01	2.8967E-01	6.0024E-01
49	6.7106E-01	5.7831E-01	5.5300E-01	3.0581E-01	7.4224E-01
50	6.9252E-01	5.9036E-01	5.7063E-01	3.2558E-01	9.0249E-01
51	7.1327E-01	6.0241E-01	5.9117E-01	3.4948E-01	1.0853E+00
52	7.3326E-01	6.1446E-01	6.1490E-01	3.7811E-01	1.2956E+00
53	7.5245E-01	6.2651E-01	6.4202E-01	4.1219E-01	1.5394E+00
54	7.7078E-01	6.3855E-01	6.7277E-01	4.5262E-01	1.8238E+00
55	7.8826E-01	6.5060E-01	7.0744E-01	5.0048E-01	2.1571E+00
56	8.0484E-01	6.6265E-01	7.4635E-01	5.5705E-01	2.5494E+00
57	8.2054E-01	6.7470E-01	7.8980E-01	6.2391E-01	3.0127E+00
58	8.3535E-01	6.8675E-01	8.3841E-01	7.0293E-01	3.5616E+00
59	8.4929E-01	6.9880E-01	8.9240E-01	7.9638E-01	4.2134E+00
60	8.6236E-01	7.1084E-01	9.5235E-01	9.0698E-01	4.9900E+00
61	8.7460E-01	7.2289E-01	1.0188E+00	1.0380E+00	5.9139E+00
62	8.8602E-01	7.3494E-01	1.0924E+00	1.1934E+00	7.0184E+00
63	8.9666E-01	7.4699E-01	1.1739E+00	1.3780E+00	8.3394E+00
64	9.0656E-01	7.5904E-01	1.2639E+00	1.5974E+00	9.9214E+00
65	9.1574E-01	7.7108E-01	1.3633E+00	1.8586E+00	1.1818E+01
66	9.2425E-01	7.8313E-01	1.4731E+00	2.1700E+00	1.4094E+01

67	9.3211E-01	7.9518E-01	1.5942E+00	2.5416E+00	1.6829E+01
68	9.3937E-01	8.0723E-01	1.7279E+00	2.9055E+00	2.0116E+01
69	9.4607E-01	8.1928E-01	1.8752E+00	3.5164E+00	2.4072E+01
70	9.5223E-01	8.3133E-01	2.0377E+00	4.1521E+00	2.8835E+01
71	9.5791E-01	8.4337E-01	2.2167E+00	4.9139E+00	3.4573E+01
72	9.6312E-01	8.5542E-01	2.4141E+00	5.8277E+00	4.1491E+01
73	9.6790E-01	8.6747E-01	2.6315E+00	6.9249E+00	4.9834E+01
74	9.7228E-01	8.7952E-01	2.8711E+00	8.2432E+00	5.9902E+01
75	9.7630E-01	8.9157E-01	3.1350E+00	9.8285E+00	7.2057E+01
76	9.7993E-01	9.0361E-01	3.4258E+00	1.1736E+01	8.6736E+01
77	9.8334E-01	9.1566E-01	3.7461E+00	1.4033E+01	1.0447E+02
78	9.8642E-01	9.2771E-01	4.0988E+00	1.6804E+01	1.2590E+02
79	9.8923E-01	9.3976E-01	4.4874E+00	2.0137E+01	1.5181E+02
80	9.9180E-01	9.5181E-01	4.9153E+00	2.4160E+01	1.8314E+02
81	9.9414E-01	9.6386E-01	5.3866E+00	2.9015E+01	2.2104E+02
82	9.9627E-01	9.7590E-01	5.9056E+00	3.4877E+01	2.6689E+02
83	9.9822E-01	9.8795E-01	6.4772E+00	4.1955E+01	3.2237E+02
84	1.0000E+00	1.0000E+00	7.1067E+00	5.0506E+01	3.8951E+02

Y - TOP

2.49678E+01	2.39032E+01	2.28348E+01	2.17675E+01	2.07060E+01	1.96554E+01	1.86200E+01	1.76443E+01	1.66122E+01	1.55473E+01
1.47128E+01	1.38111E+01	1.29446E+01	1.21150E+01	1.13233E+01	1.05704E+01	9.85654E+00	9.18179E+00	8.54568E+00	7.94753E+00
7.38637E+00	6.86105E+00	6.37024E+00	5.91246E+00	5.48616E+00	5.08972E+00	4.72148E+00	4.37977E+00	4.06298E+00	3.76940E+00
3.49754E+00	3.24587E+00	3.01295E+00	2.79738E+00	2.59789E+00	2.41323E+00	2.24233E+00	2.08408E+00	1.93752E+00	1.84176E+00
1.67599E+00	1.55941E+00	1.45136E+00	1.35121E+00	1.25837E+00	1.17231E+00	1.09256E+00	1.01867E+00	9.50224E-01	8.85893E-01
8.28218E-01	7.73987E-01	7.23962E-01	6.77564E-01	6.34833E-01	5.95422E-01	5.59103E-01	5.25658E-01	4.94804E-01	4.65589E-01
4.40595E-01	4.16731E-01	3.94840E-01	3.74771E-01	3.56385E-01	3.39553E-01	3.24151E-01	3.10267E-01	2.97195E-01	2.85436E-01
2.74699E-01	2.64900E-01	2.55960E-01	2.47807E-01	2.40375E-01	2.33601E-01	2.27430E-01	2.21810E-01	2.16692E-01	2.12032E-01
2.07792E-01	2.03933E-01	2.00422E-01	1.97228E-01						

X/L IN THE WAKE

6.40233E+00	6.10898E+00	5.81463E+00	5.52066E+00	5.22842E+00	4.93925E+00	4.65442E+00	4.37514E+00	4.10252E+00	3.83755E+00
3.55111E+00	3.33394E+00	3.09665E+00	2.86973E+00	2.65352E+00	2.44824E+00	2.25401E+00	2.07082E+00	1.89853E+00	1.73712E+00
1.53619E+00	1.44548E+00	1.31465E+00	1.19330E+00	1.08102E+00	9.77372E-01	8.81907E-01	7.94177E-01	7.13731E-01	6.41124E-01
5.72922E-01	5.11700E-01	4.56048E-01	4.05573E-01	3.59896E-01	3.18659E-01	2.81516E-01	2.48143E-01	2.18233E-01	1.91493E-01
1.67650E-01	1.46446E-01	1.27640E-01	1.10005E-01	9.63190E-02	8.34235E-02	7.20986E-02	6.21697E-02	5.35425E-02	4.64158E-02
3.54809E-02	3.38207E-02	2.89294E-02	2.18226E-02	1.79658E-02	1.30864E-02	1.29366E-02	1.30816E-02	2.06889E-03	1.73687E-03
7.95994E-03	6.74787E-03	5.71613E-03	4.83839E-03	4.09197E-03	2.91783E-03	2.45998E-03	2.06889E-03	1.73687E-03	1.97677E-04
1.27151E-04	8.47939E-05	3.94131E-05	8.32275E-05	0.	4.43699E-04	3.49131E-04	4.43699E-04	2.67781E-04	

ETA

6.84447E+02	6.55032E+02	6.25511E+02	5.96018E+02	5.66089E+02	5.37657E+02	5.09049E+02	4.80984E+02	4.53571E+02	4.26910E+02
4.01686E+02	3.76173E+02	3.52236E+02	3.29305E+02	3.07429E+02	2.86625E+02	2.66925E+02	2.48257E+02	2.30684E+02	2.14152E+02
1.98647E+02	1.84131E+02	1.76570E+02	1.57921E+02	1.46141E+02	1.35187E+02	1.25012E+02	1.15570E+02	1.06816E+02	9.87043E+01

```
9.11926E+01  8.42387E+01  7.78025E+01  7.18461E+01  6.63380E+01  6.12319E+01  5.65092E+01  5.21365E+01  4.80869E+01  4.43357E+01
4.08601E+01  3.76391E+01  3.46536E+01  3.18862E+01  2.93299E+01  2.69431E+01  2.47394E+01  2.26976E+01  2.08064E+01  1.90555E+01
1.74352E+01  1.59367E+01  1.45517E+01  1.32724E+01  1.20917E+01  1.09027E+01  9.99014E+00  9.07506E+00  8.22468E+00  7.44206E+00
6.72460E+00  6.05521E+00  1.62286E+00  4.90576E+00  4.39776E+00  3.93265E+00  3.50709E+00  3.11793E+00  2.76225E+00  2.43733E+00
2.91896E+00  1.85266E-01  8.82514E-02  0.           1.19221E+00  1.03505E+00  8.34541E-01  6.79234E-01  5.37815E-01  4.69973E-01
```

X - WAKE

```
2.58847E+01  2.48230E+01  2.37578E+01  2.26939E+01  2.16362E+01  2.05897E+01  1.95589E+01  1.85482E+01  1.75615E+01  1.65026E+01
1.56745E+01  1.47800E+01  1.39213E+01  1.31000E+01  1.23175E+01  1.15746E+01  1.08717E+01  1.02007E+01  9.58537E+00  9.07103E+00
8.45480E+00  7.94558E+00  7.47209E+00  7.03293E+00  6.62657E+00  6.25146E+00  5.90597E+00  5.58847E+00  5.29730E+00  5.63794E+00
4.78773E+00  4.56617E+00  4.36476E+00  4.18209E+00  4.01678E+00  3.86754E+00  3.73312E+00  3.61234E+00  3.50403E+00  3.47032E+00
2.85718E+00  3.24429E+00  3.17623E+00  3.11603E+00  3.06292E+00  3.01621E+00  2.97522E+00  2.93936E+00  2.90807E+00  2.90883E+00
2.74319E+00  2.83669E+00  2.81099E+00  2.80373E+00  2.79931E+00  2.76964E+00  2.76136E+00  2.76136E+00  2.75429E+00  2.74825E+00
2.71956E+00  2.73872E+00  2.73498E+00  2.73180E+00  2.72910E+00  2.72681E+00  2.72319E+00  2.72319E+00  2.72174E+00  2.72501E+00
2.71479E+00  2.71468E+00  2.71444E+00  2.71731E+00  2.71129E+00  2.71676E+00  2.71590E+00  2.71556E+00  2.71526E+00  2.71501E+00
```

H(2,JL)= 6.14656E-03   H(2,1)= 3.89613E-05

I	XC(I)	THETA(I)	UPS	UPSE	UPS2	PPR	PPRE	PRESSURE FORCE AT OUTER BOUNDARY
2	1.000E+00	0.	1.209E-13	0.	1.462E-26	1.000E+00	1.000E+00	0.
3	9.993E-01	4.909E-02	-7.663E-01	-7.683E-01	5.903E-01	4.097E-01	4.097E-01	2.32914E-06
4	9.973E-01	9.817E-02	-9.074E-01	-9.073E-01	8.233E-01	1.767E-01	1.768E-01	4.63685E-06
5	9.939E-01	1.473E-01	-9.431E-01	-9.431E-01	8.895E-01	1.105E-01	1.106E-01	6.90190E-06
6	9.891E-01	1.963E-01	-9.575E-01	-9.575E-01	9.169E-01	8.312E-02	8.319E-02	9.16343E-06
7	9.831E-01	2.454E-01	-9.653E-01	-9.653E-01	9.319E-01	6.837E-02	6.813E-02	1.12212E-05
8	9.757E-01	2.945E-01	-9.707E-01	-9.707E-01	9.423E-01	5.775E-02	5.780E-02	1.32356E-05
9	9.670E-01	3.436E-01	-9.750E-01	-9.750E-01	9.506E-01	4.940E-02	4.944E-02	1.51280E-05
10	9.571E-01	3.927E-01	-9.789E-01	-9.789E-01	9.582E-01	4.182E-02	4.185E-02	1.68809E-05
11	9.460E-01	4.418E-01	-9.826E-01	-9.826E-01	9.656E-01	3.444E-02	3.445E-02	1.84781E-05
12	9.337E-01	4.905E-01	-9.864E-01	-9.864E-01	9.730E-01	2.636E-02	2.696E-02	1.99046E-05
13	9.203E-01	5.400E-01	-9.903E-01	-9.904E-01	9.808E-01	1.921E-02	1.919E-02	2.11470E-05
14	9.058E-01	5.899E-01	-9.944E-01	-9.945E-01	9.889E-01	1.109E-02	1.105E-02	2.21936E-05
15	8.902E-01	6.381E-01	-9.987E-01	-9.988E-01	9.974E-01	2.551E-03	2.486E-03	2.30343E-05
16	8.736E-01	6.872E-01	-1.003E+00	-1.003E+00	1.006E+00	-6.445E-03	-6.533E-03	2.36611E-05
17	8.561E-01	7.363E-01	-1.008E+00	-1.008E+00	1.016E+00	-1.591E-02	-1.602E-02	2.40677E-05
18	8.184E-01	7.854E-01	-1.013E+00	-1.013E+00	1.026E+00	-2.585E-02	-2.599E-02	2.42498E-05
19	8.184E-01	8.345E-01	-1.018E+00	-1.018E+00	1.036E+00	-3.626E-02	-3.643E-02	2.42051E-05
20	7.984E-01	8.836E-01	-1.023E+00	-1.023E+00	1.047E+00	-4.713E-02	-4.733E-02	2.39334E-05
21	7.777E-01	9.327E-01	-1.029E+00	-1.029E+00	1.058E+00	-5.845E-02	-5.868E-02	2.34364E-05
22	7.563E-01	9.817E-01	-1.035E+00	-1.035E+00	1.070E+00	-7.047E-02	-7.047E-02	2.27182E-05
23	7.343E-01	1.031E+00	-1.041E+00	-1.041E+00	1.082E+00	-8.238E-02	-8.267E-02	2.17845E-05
24	7.118E-01	1.080E+00	-1.047E+00	-1.047E+00	1.095E+00	-9.493E-02	-9.527E-02	2.06432E-05
25	6.888E-01	1.129E+00	-1.053E+00	-1.053E+00	1.108E+00	-1.079E-01	-1.082E-01	1.93041E-05
26	6.654E-01	1.178E+00	-1.059E+00	-1.059E+00	1.121E+00	-1.211E-01	-1.215E-01	1.77789E-05

27	6.417E-01	1.227E+00	1.065E+00	1.135E+00	-1.347E-01	-1.351E-01	1.60809E-05
28	6.177E-01	1.276E+00	1.072E+00	1.149E+00	-1.435E-01	-1.490E-01	1.42251E-05
29	5.935E-01	1.325E+00	1.078E+00	1.163E+00	-1.626E-01	-1.631E-01	1.22283E-05
30	5.691E-01	1.374E+00	1.085E+00	1.177E+00	-1.769E-01	-1.774E-01	1.01084E-05
31	5.446E-01	1.424E+00	1.092E+00	1.191E+00	-1.913E-01	-1.919E-01	7.88471E-06
32	5.201E-01	1.473E+00	1.098E+00	1.206E+00	-2.066E-01	-2.065E-01	5.57764E-06
33	4.956E-01	1.522E+00	1.105E+00	1.220E+00	-2.205E-01	-2.212E-01	5.20085E-06
34	4.712E-01	1.571E+00	1.112E+00	1.235E+00	-2.351E-01	-2.359E-01	7.99513E-07
35	4.469E-01	1.620E+00	1.118E+00	1.250E+00	-2.498E-01	-2.505E-01	-1.62677E-06
36	4.228E-01	1.669E+00	1.125E+00	1.264E+00	-2.643E-01	-2.651E-01	-4.04726E-06
37	3.989E-01	1.718E+00	1.131E+00	1.279E+00	-2.787E-01	-2.796E-01	-6.43874E-06
38	3.754E-01	1.767E+00	1.137E+00	1.293E+00	-2.929E-01	-2.938E-01	-8.77807E-06
39	3.522E-01	1.816E+00	1.144E+00	1.307E+00	-3.069E-01	-3.078E-01	-1.10424E-05
40	3.294E-01	1.865E+00	1.150E+00	1.321E+00	-3.205E-01	-3.214E-01	-1.32092E-05
41	3.071E-01	1.914E+00	1.155E+00	1.334E+00	-3.336E-01	-3.347E-01	-1.52570E-05
42	2.852E-01	1.963E+00	1.161E+00	1.346E+00	-3.463E-01	-3.474E-01	-1.71649E-05
43	2.639E-01	2.013E+00	1.166E+00	1.358E+00	-3.584E-01	-3.595E-01	-1.89135E-05
44	2.432E-01	2.062E+00	1.171E+00	1.370E+00	-3.697E-01	-3.709E-01	-2.04844E-05
45	2.230E-01	2.111E+00	1.175E+00	1.380E+00	-3.803E-01	-3.815E-01	-2.18611E-05
46	2.036E-01	2.160E+00	1.179E+00	1.390E+00	-3.898E-01	-3.911E-01	-2.30286E-05
47	1.849E-01	2.209E+00	1.182E+00	1.398E+00	-3.982E-01	-3.995E-01	-2.39742E-05
48	1.669E-01	2.258E+00	1.185E+00	1.405E+00	-4.053E-01	-4.066E-01	-2.46869E-05
49	1.496E-01	2.307E+00	1.188E+00	1.411E+00	-4.108E-01	-4.121E-01	-2.51583E-05
50	1.332E-01	2.356E+00	1.190E+00	1.416E+00	-4.144E-01	-4.157E-01	-2.53822E-05
51	1.177E-01	2.405E+00	1.190E+00	1.416E+00	-4.157E-01	-4.171E-01	-2.53549E-05
52	1.029E-01	2.454E+00	1.189E+00	1.414E+00	-4.143E-01	-4.157E-01	-2.50753E-05
53	8.912E-02	2.503E+00	1.187E+00	1.410E+00	-4.095E-01	-4.109E-01	-2.45449E-05
54	7.623E-02	2.553E+00	1.183E+00	1.406E+00	-4.034E-01	-4.019E-01	-2.37679E-05
55	6.427E-02	2.602E+00	1.178E+00	1.386E+00	-3.859E-01	-3.873E-01	-2.27589E-05
56	5.328E-02	2.651E+00	1.169E+00	1.364E+00	-3.643E-01	-3.657E-01	-2.15036E-05
57	4.328E-02	2.700E+00	1.155E+00	1.333E+00	-3.333E-01	-3.347E-01	-2.00371E-05
58	3.429E-02	2.749E+00	1.136E+00	1.289E+00	-2.895E-01	-2.908E-01	-1.83663E-05
59	2.631E-02	2.798E+00	1.108E+00	1.228E+00	-2.279E-01	-2.292E-01	-1.65775E-05
60	1.937E-02	2.847E+00	1.069E+00	1.141E+00	-1.415E-01	-1.427E-01	-1.44792E-05
61	1.347E-02	2.896E+00	1.011E+00	9.521E-01	-2.138E-02	-1.470E-01	-1.23018E-05
62	8.633E-03	2.945E+00	9.236E-01	6.274E-01	3.726E-01	3.719E-01	-9.99761E-06
63	4.861E-03	2.994E+00	7.925E-01	5.982E-01	6.425E-01	6.421E-01	-7.59914E-06
64	2.162E-03	3.043E+00	5.982E-01	7.925E-01	3.726E-01	3.719E-01	-5.10418E-06
65	5.408E-04	3.093E+00	3.281E-01	9.231E-01	1.479E-01	1.470E-01	-2.56537E-06
66	5.408E-04	3.142E+00	3.281E-13	1.075E+00	1.000E+00	8.923E-01	0.00000E+00
67	5.162E-03	3.191E+00	5.982E-01	9.925E-01	8.925E-01	8.923E-01	2.56637E-06
68	2.162E-03	3.240E+00	7.925E-01	6.425E-01	6.421E-01	6.421E-01	5.10418E-06
69	4.861E-03	3.289E+00	9.231E-01	3.726E-01	3.719E-01	3.719E-01	7.59514E-06
70	8.633E-03	3.338E+00	1.011E+00	1.479E-01	-1.427E-01	-1.470E-01	9.99761E-06
71	1.937E-02	3.387E+00	1.069E+00	1.141E+00	-1.415E-01	-1.427E-01	1.23018E-05
72	2.631E-02	3.436E+00	1.108E+00	1.228E+00	-2.279E-01	-2.292E-01	1.44792E-05
73	2.631E-02	3.485E+00	1.136E+00	1.283E+00	-2.895E-01	-2.908E-01	1.65975E-05
74	3.429E-02	3.534E+00	1.136E+00	1.333E+00	-3.333E-01	-3.347E-01	1.83663E-05
75	4.328E-02	3.583E+00	1.155E+00	1.333E+00	-3.333E-01	-3.347E-01	2.00371E-05

Line								
76	5.328E-02	3.632E+00	1.168E+00	1.169E+00	1.364E+00	-3.643E-01	-3.657E-01	2.15034E-05
77	6.427E-02	3.682E+00	1.177E+00	1.178E+00	1.386E+00	-3.859E-01	-3.873E-01	2.27509E-05
78	7.623E-02	3.731E+00	1.183E+00	1.188E+00	1.400E+00	-4.004E-01	-4.004E-01	2.37679E-05
79	8.912E-02	3.780E+00	1.187E+00	1.190E+00	1.410E+00	-4.095E-01	-4.109E-01	2.45449E-05
80	1.029E-01	3.829E+00	1.189E+00	1.190E+00	1.414E+00	-4.143E-01	-4.157E-01	2.50753E-05
81	1.177E-01	3.878E+00	1.190E+00	1.190E+00	1.416E+00	-4.157E-01	-4.171E-01	2.53549E-05
82	1.332E-01	3.927E+00	1.189E+00	1.188E+00	1.414E+00	-4.144E-01	-4.157E-01	2.53822E-05
83	1.496E-01	3.976E+00	1.188E+00	1.186E+00	1.411E+00	-4.108E-01	-4.121E-01	2.51508E-05
84	1.669E-01	4.025E+00	1.185E+00	1.183E+00	1.405E+00	-4.053E-01	-4.066E-01	2.46869E-05
85	1.849E-01	4.074E+00	1.182E+00	1.179E+00	1.398E+00	-3.982E-01	-3.995E-01	2.39742E-05
86	2.036E-01	4.123E+00	1.179E+00	1.175E+00	1.390E+00	-3.898E-01	-3.911E-01	2.30286E-05
87	2.230E-01	4.172E+00	1.175E+00	1.171E+00	1.380E+00	-3.803E-01	-3.815E-01	2.18611E-05
88	2.432E-01	4.222E+00	1.170E+00	1.166E+00	1.370E+00	-3.697E-01	-3.709E-01	2.04844E-05
89	2.639E-01	4.271E+00	1.165E+00	1.161E+00	1.350E+00	-3.584E-01	-3.595E-01	1.89135E-05
90	2.852E-01	4.320E+00	1.160E+00	1.155E+00	1.345E+00	-3.463E-01	-3.474E-01	1.71649E-05
91	3.071E-01	4.369E+00	1.155E+00	1.150E+00	1.334E+00	-3.336E-01	-3.347E-01	1.52520E-05
92	3.294E-01	4.418E+00	1.149E+00	1.144E+00	1.320E+00	-3.205E-01	-3.214E-01	1.32092E-05
93	3.522E-01	4.467E+00	1.143E+00	1.137E+00	1.307E+00	-3.069E-01	-3.078E-01	1.10424E-05
94	3.754E-01	4.516E+00	1.137E+00	1.131E+00	1.293E+00	-2.929E-01	-2.938E-01	8.77807E-06
95	3.989E-01	4.565E+00	1.131E+00	1.125E+00	1.279E+00	-2.787E-01	-2.796E-01	6.43874E-06
96	4.228E-01	4.614E+00	1.124E+00	1.118E+00	1.264E+00	-2.643E-01	-2.651E-01	4.04725E-06
97	4.469E-01	4.663E+00	1.118E+00	1.112E+00	1.253E+00	-2.533E-01	-2.505E-01	1.62677E-06
98	4.712E-01	4.712E+00	1.111E+00	1.105E+00	1.238E+00	-2.351E-01	-2.359E-01	-7.99513E-07
99	4.956E-01	4.761E+00	1.105E+00	1.098E+00	1.223E+00	-2.205E-01	-2.212E-01	-3.20554E-06
100	5.201E-01	4.811E+00	1.098E+00	1.092E+00	1.205E+00	-2.059E-01	-2.065E-01	-5.57764E-06
101	5.446E-01	4.860E+00	1.091E+00	1.085E+00	1.191E+00	-1.913E-01	-1.919E-01	-7.88171E-06
102	5.691E-01	4.909E+00	1.085E+00	1.078E+00	1.177E+00	-1.769E-01	-1.774E-01	-1.01094E-05
103	5.935E-01	4.958E+00	1.078E+00	1.072E+00	1.163E+00	-1.626E-01	-1.631E-01	-1.22283E-05
104	6.177E-01	5.007E+00	1.072E+00	1.065E+00	1.143E+00	-1.485E-01	-1.490E-01	-1.42251E-05
105	6.417E-01	5.056E+00	1.065E+00	1.059E+00	1.135E+00	-1.347E-01	-1.351E-01	-1.60309E-05
106	6.654E-01	5.105E+00	1.059E+00	1.053E+00	1.121E+00	-1.211E-01	-1.215E-01	-1.77789E-05
107	6.888E-01	5.154E+00	1.053E+00	1.046E+00	1.108E+00	-1.079E-01	-1.082E-01	-1.93041E-05
108	7.118E-01	5.203E+00	1.046E+00	1.041E+00	1.095E+00	-9.493E-02	-9.527E-02	-2.06432E-05
109	7.343E-01	5.252E+00	1.040E+00	1.035E+00	1.076E+00	-8.238E-02	-8.267E-02	-2.17045E-05
110	7.563E-01	5.301E+00	1.035E+00	1.029E+00	1.059E+00	-7.047E-02	-7.047E-02	-2.27182E-05
111	7.777E-01	5.351E+00	1.029E+00	1.023E+00	1.047E+00	-5.845E-02	-5.868E-02	-2.34364E-05
112	7.984E-01	5.400E+00	1.023E+00	1.018E+00	1.023E+00	-4.713E-02	-4.733E-02	-2.39334E-05
113	8.184E-01	5.449E+00	1.018E+00	1.013E+00	1.018E+00	-3.626E-02	-3.643E-02	-2.42051E-05
114	8.377E-01	5.498E+00	1.013E+00	1.008E+00	1.013E+00	-2.505E-02	-2.599E-02	-2.42498E-05
115	8.561E-01	5.547E+00	1.003E+00	1.003E+00	1.003E+00	-1.591E-02	-1.602E-02	-2.40677E-05
116	8.736E-01	5.596E+00	1.003E+00	9.988E-01	1.003E+00	-6.445E-03	-6.533E-03	-2.36611E-05
117	8.902E-01	5.645E+00	9.987E-01	9.945E-01	9.974E-01	2.551E-03	2.486E-02	-2.30343E-05
118	9.058E-01	5.694E+00	9.945E-01	9.904E-01	9.883E-01	1.095E-02	1.919E-02	-2.21936E-05
119	9.203E-01	5.743E+00	9.904E-01	9.864E-01	9.739E-01	1.921E-02	2.696E-02	-2.11470E-05
120	9.337E-01	5.792E+00	9.864E-01	9.826E-01	9.656E-01	2.696E-02	3.445E-02	-1.99046E-05
121	9.460E-01	5.841E+00	9.826E-01	9.789E-01	9.582E-01	3.444E-02	4.105E-02	-1.84781E-05
122	9.571E-01	5.890E+00	9.789E-01	9.756E-01	9.505E-01	4.182E-02	4.944E-02	-1.68809E-05
123	9.671E-01	5.940E+00	9.756E-01	9.756E-01	9.505E-01	4.940E-02	4.944E-02	-1.51280E-05
124	9.757E-01	5.989E+00	9.707E-01	9.707E-01	9.423E-01	5.775E-02	5.780E-02	-1.32356E-05

```
125  9.831E-01  6.038E+00  9.654E-01  9.653E-01  6.807E-02  6.813E-02  -1.12212E-05
126  9.891E-01  6.087E+00  9.575E-01  9.575E-01  8.312E-02  8.319E-02  -9.10343E-06
127  9.939E-01  6.136E+00  9.431E-01  9.431E-01  1.105E-01  1.106E-01  -6.90190E-06
128  6.185E-01  9.074E-01  8.233E-01  1.767E-01  1.768E-01  -4.63685E-06
129  9.993E-01  6.234E+00  7.683E-01  7.683E-01  5.903E-01  4.097E-01  4.097E-01  -2.32914E-06
130  1.000E+00  6.283E+00  1.289E-13  1.955E-23  1.462E-26  1.000E+00  1.000E+00  -1
```

```
N=  1    DVRTLD(   3)=  1.00448E+04
N=  2    DVRTLD(   3)=  6.87048E+03
N=  3    DVRTLD(129)= -4.71091E+03
N=  4    DVRTLD(128)= -3.63399E+03
N=  5    DVRTLD(   4)=  2.91071E+03
N=  6    DVRTLD(   4)=  2.33658E+03
N=  7    DVRTLD(128)= -1.87870E+03
N=  8    DVRTLD(   4)=  1.51232E+03
N=  9    DVRTLD(   4)=  1.21044E+03
N= 10    DVRTLD(   4)=  9.82302E+02
N= 11    DVRTLD(   4)=  7.92292E+02
N= 12    DVRTLD(128)= -6.39242E+02
N= 13    DVRTLD(   5)=  5.16021E+02
N= 14    DVRTLD(   5)=  4.26856E+02
N= 15    DVRTLD(127)= -3.53120E+02
N= 16    DVRTLD(127)= -2.92136E+02
N= 17    DVRTLD(127)= -2.41692E+02
N= 18    DVRTLD(   5)=  1.99964E+02
N= 19    DVRTLD(127)= -1.65442E+02
N= 20    DVRTLD(   5)=  1.36081E+02
N= 21    DVRTLD(   6)=  1.13722E+02
N= 22    DVRTLD(126)= -9.53160E+01
N= 23    DVRTLD(   6)=  7.98915E+01
N= 24    DVRTLD(   6)=  6.69651E+01
N= 25    DVRTLD(126)= -5.61316E+01
N= 26    DVRTLD(   6)=  4.70519E+01
N= 27    DVRTLD(   6)=  3.94419E+01
N= 28    DVRTLD(126)= -3.30634E+01
N= 29    DVRTLD(   6)=  2.77170E+01
N= 30    DVRTLD(   6)=  2.32355E+01
N= 31    DVRTLD(126)= -1.94790E+01
N= 32    DVRTLD(126)= -1.63301E+01
N= 33    DVRTLD(   6)=  1.36904E+01
N= 34    DVRTLD(   6)=  1.14776E+01
N= 35    DVRTLD(126)= -9.62260E+00
N= 36    DVRTLD(125)= -8.08743E+00
N= 37    DVRTLD(   7)=  6.82037E+00
N= 38    DVRTLD(125)= -5.75194E+00
```

IS =-66

PRESSURE DERIVATIVE= 2.35206E+01   K.E. DEFECT= 1.16977E-03   UT(1)=-2.67769E-01

1.16977E-03	5.67246E-01	1.13001E+00	1.68663E+00	2.23435E+00	2.77063E+00	3.29311E+00	3.79973E+00	4.28871E+00	4.75861E+00
5.20831E+00	5.63703E+00	6.04430E+00	6.42998E+00	6.79417E+00	7.13725E+00	7.45978E+00	7.76253E+00	8.04639E+00	8.31235E+00
8.56152E+00	8.79501E+00	9.01399E+00	9.21965E+00	9.41314E+00	9.59560E+00	9.76817E+00	9.93190E+00	1.00878E+01	1.02369E+01
1.03802E+01	1.05184E+01	1.06524E+01	1.07829E+01	1.09108E+01	1.10365E+01	1.11608E+01	1.12841E+01	1.14069E+01	1.15297E+01
1.16527E+01	1.17764E+01	1.19008E+01	1.20264E+01	1.21530E+01	1.22808E+01	1.24098E+01	1.25399E+01	1.26710E+01	1.28029E+01
1.29352E+01	1.30678E+01	1.32002E+01	1.33221E+01	1.34629E+01	1.35923E+01	1.37198E+01	1.38448E+01	1.39670E+01	1.40858E+01
1.42008E+01	1.43117E+01	1.44182E+01	1.45198E+01	1.46165E+01	1.47080E+01	1.47943E+01	1.48754E+01	1.49512E+01	1.50218E+01
1.50875E+01	1.51482E+01	1.52042E+01	1.52558E+01	1.53031E+01	1.53465E+01	1.53861E+01	1.54223E+01	1.54551E+01	1.54850E+01
1.55120E+01	1.55365E+01	1.55587E+01	1.55789E+01						

PRESSURE COEFFICIENT UPSTREAM OF STAGNATION POINT =   1.55789E+01

PRESSURE DERIVATIVE= 2.34804E+01   K.E. DEFECT= 1.16662E-03   UT(1)=-2.67309E-01

1.16662E-03	5.66269E-01	1.12806E+00	1.68369E+00	2.23044E+00	2.76575E+00	3.28729E+00	3.79298E+00	4.28106E+00	4.75009E+00
5.19894E+00	5.62686E+00	6.03337E+00	6.41832E+00	6.78183E+00	7.12426E+00	7.44620E+00	7.74839E+00	8.03173E+00	8.29721E+00
8.54559E+00	8.77900E+00	8.99760E+00	9.20290E+00	9.39605E+00	9.57820E+00	9.75046E+00	9.91391E+00	1.00696E+01	1.02184E+01
1.03614E+01	1.04993E+01	1.06330E+01	1.07633E+01	1.08908E+01	1.10162E+01	1.11401E+01	1.12631E+01	1.13854E+01	1.15077E+01
1.16302E+01	1.17532E+01	1.18779E+01	1.20016E+01	1.21273E+01	1.22544E+01	1.23818E+01	1.25105E+01	1.26399E+01	1.27698E+01
1.29000E+01	1.30301E+01	1.31598E+01	1.32885E+01	1.34159E+01	1.35414E+01	1.36646E+01	1.37850E+01	1.39021E+01	1.40154E+01
1.41246E+01	1.42293E+01	1.43292E+01	1.44241E+01	1.45137E+01	1.45980E+01	1.46770E+01	1.47506E+01	1.48189E+01	1.48821E+01
1.49401E+01	1.49930E+01	1.50428E+01	1.50875E+01	1.51282E+01	1.51652E+01	1.51989E+01	1.52294E+01	1.52572E+01	1.52827E+01
1.53062E+01	1.53263E+01	1.53459E+01	1.53668E+01						

PRESSURE COEFFICIENT DOWNSTREAM OF STAGNATION POINT =   1.53668E+01

STAGNATION POINT AT = 0.

PRESSURE INTEGRATED FROM GEOMETRICAL TRAILING TO LEADING EDGE

PM = 1.88676E+03   PM1 = 1.88676E+03   PERCENT ERROR= 1.15690E-12   PERCENT ERROR BASED ON MINMAX= 1.15690E-12

CP = 1.37079E+02   CF   = 2.42072E+00   DRAG = 1.39500E+02   CLP = 1.13833E-11   CLF =-3.30490E-15   LIFT = 1.13800E-11

CMF =-1.29830E-16   CMP =-3.53372E-12   MOMENT=-3.53385E-12

PRESSURE INTEGRATED FROM GEOMETRICAL LEADING TO TRAILING EDGE

PM =-1.88676E+03   PM1 =-1.88676E+03   PERCENT ERROR=-3.85633E-13   PERCENT ERROR BASED ON MINMAX= 3.85633E-13

CP = 1.37079E+02   CF   = 2.42072E+00   DRAG = 1.39500E+02   CLP = 1.57690E-11   CLF =-3.30490E-15   LIFT = 1.57647E-11

CMF =-1.29830E-16   CMP =-4.11903E-12   MOMENT=-4.11916E-12

AVERAGE DRAG= 1.39500E+02   AVERAGE LIFT= 1.35723E-11   AVERAGE MOMENT=-3.82651E-12

I	XC(I)	PY(JLMN)	VRT(JL)	VRT(JLMN)	VRT(JLM2)	VRT1	VRTB(JL)	VRT1	V	PR (L-T)	PRI (T-L)	CPT (L-T)	CPTI (T-L)	THETA
2	1.000E+00	-2.8E-16	0.	0.	0.	-1.5E-11	0.	-1.5E-11	1.110E-16	0.	0.	-1.871E+03	-1.871E+03	0.0000
3	9.993E-01	-2.2E-05	9.363E+02	7.1E+02	5.2E+02	9.7E+01	9.363E+02	9.7E+01	-4.921E-02	4.85E+00	4.90E+00	-1.866E+03	-1.866E+03	.0491
4	9.972E-01	-7.0E-05	1.076E+03	8.0E+02	4.1E+02	1.7E+02	1.076E+03	1.7E+02	-9.836E-02	1.60E+01	1.60E+01	-1.853E+03	-1.854E+03	.0982
5	9.939E-01	-1.4E-04	1.094E+03	5.7E+02	1.9E+02	2.6E+02	1.094E+03	2.6E+02	-1.471E-01	3.73E+01	3.73E+01	-1.810E+03	-1.834E+03	.1473
6	9.891E-01	-2.3E-04	1.086E+03	3.7E+02	1.3E+02	2.9E+02	1.086E+03	2.9E+02	-1.956E-01	6.12E+01	6.12E+01	-1.753E+03	-1.818E+03	.1963
7	9.831E-01	-3.4E-04	1.067E+03	3.9E+02	8.6E+01	3.1E+02	1.067E+03	3.1E+02	-2.436E-01	8.65E+01	8.85E+01	-1.722E+03	-1.783E+03	.2454
8	9.757E-01	-4.5E-04	1.041E+03	3.2E+02	5.9E+01	3.2E+02	1.041E+03	3.2E+02	-2.911E-01	1.18E+02	1.18E+02	-1.753E+03	-1.753E+03	.2945
9	9.670E-01	-5.8E-04	1.012E+03	2.2E+02	3.0E+01	3.2E+02	1.012E+03	3.2E+02	-3.378E-01	1.49E+02	1.49E+02	-1.722E+03	-1.722E+03	.3436
10	9.571E-01	-7.0E-04	9.816E+02	1.9E+02	2.2E+01	3.3E+02	9.816E+02	3.3E+02	-3.836E-01	1.81E+02	1.81E+02	-1.690E+03	-1.690E+03	.3927
11	9.459E-01	-8.3E-04	9.509E+02	1.9E+02	1.8E+01	3.3E+02	9.509E+02	3.3E+02	-4.285E-01	2.13E+02	2.13E+02	-1.658E+03	-1.658E+03	.4418
12	9.337E-01	-9.6E-04	9.210E+02	2.3E+01	2.3E+01	3.3E+02	9.210E+02	3.3E+02	-4.725E-01	2.45E+02	2.45E+02	-1.626E+03	-1.626E+03	.4909
13	9.203E-01	-1.1E-03	8.926E+02	1.7E+01	1.7E+01	3.2E+02	8.926E+02	3.2E+02	-5.152E-01	2.77E+02	2.77E+02	-1.594E+03	-1.594E+03	.5400
14	9.058E-01	-1.2E-03	8.660E+02	1.3E+01	1.3E+01	3.2E+02	8.660E+02	3.2E+02	-5.567E-01	3.09E+02	3.09E+02	-1.562E+03	-1.562E+03	.5890
15	8.902E-01	-1.3E-03	8.414E+02	1.0E+01	1.0E+01	3.1E+02	8.414E+02	3.1E+02	-5.969E-01	3.40E+02	3.40E+02	-1.531E+03	-1.531E+03	.6381
16	8.736E-01	-1.5E-03	8.190E+02	8.9E+00	8.4E+00	3.1E+02	8.190E+02	3.1E+02	-6.356E-01	3.71E+02	3.71E+02	-1.503E+03	-1.503E+03	.6872
17	8.561E-01	-1.6E-03	7.988E+02	7.2E+00	6.9E+00	3.1E+02	7.988E+02	3.1E+02	-6.727E-01	4.01E+02	4.01E+02	-1.470E+03	-1.470E+03	.7363
18	8.377E-01	-1.7E-03	7.808E+02	7.2E+00	7.2E+00	3.0E+02	7.808E+02	3.0E+02	-7.082E-01	4.31E+02	4.31E+02	-1.440E+03	-1.440E+03	.7854
19	8.191E-01	-1.8E-03	7.649E+02	6.5E+00	4.9E+00	3.0E+02	7.649E+02	3.0E+02	-7.420E-01	4.61E+02	4.61E+02	-1.410E+03	-1.410E+03	.8345
20	7.984E-01	-1.9E-03	7.510E+02	5.7E+00	4.2E+00	2.9E+02	7.510E+02	2.9E+02	-7.740E-01	4.90E+02	4.90E+02	-1.381E+03	-1.381E+03	.8836
21	7.777E-01	-2.0E-03	7.390E+02	5.3E+00	3.7E+00	2.9E+02	7.390E+02	2.9E+02	-8.042E-01	5.19E+02	5.19E+02	-1.352E+03	-1.352E+03	.9327
22	7.563E-01	-2.1E-03	7.289E+02	4.2E+00	3.3E+00	2.9E+02	7.289E+02	2.9E+02	-8.324E-01	5.47E+02	5.47E+02	-1.324E+03	-1.324E+03	.9817
23	7.343E-01	-2.1E-03	7.205E+02	4.2E+00	3.3E+00	2.8E+02	7.205E+02	2.8E+02	-8.586E-01	5.75E+02	5.75E+02	-1.296E+03	-1.296E+03	1.0308
24	7.118E-01	-2.2E-03	7.137E+02	4.6E+00	2.7E+00	2.8E+02	7.137E+02	2.8E+02	-8.827E-01	6.03E+02	6.03E+02	-1.268E+03	-1.268E+03	1.0799
25	6.890E-01	-2.3E-03	7.085E+02	4.6E+00	2.5E+00	2.5E+02	7.085E+02	2.5E+02	-9.047E-01	6.31E+02	6.31E+02	-1.240E+03	-1.240E+03	1.1290
26	6.654E-01	-2.3E-03	7.048E+02	4.5E+00	2.4E+00	2.4E+02	7.048E+02	2.4E+02	-9.245E-01	6.59E+02	6.59E+02	-1.212E+03	-1.212E+03	1.1791
27	6.417E-01	-2.4E-03	7.026E+02	4.3E+00	2.3E+00	2.3E+02	7.026E+02	2.3E+02	-9.421E-01	6.87E+02	6.87E+02	-1.185E+03	-1.185E+03	1.2272
28	6.175E-01	-2.4E-03	7.018E+02	4.2E+00	2.2E+00	2.2E+02	7.018E+02	2.2E+02	-9.574E-01	7.14E+02	7.14E+02	-1.157E+03	-1.157E+03	1.2763
29	5.935E-01	-2.5E-03	7.023E+02	4.2E+00	2.1E+00	2.1E+02	7.023E+02	2.1E+02	-9.705E-01	7.42E+02	7.42E+02	-1.129E+03	-1.129E+03	1.3254
30	5.691E-01	-2.5E-03	7.041E+02	4.3E+00	2.1E+00	2.1E+02	7.041E+02	2.1E+02	-9.812E-01	7.70E+02	7.70E+02	-1.102E+03	-1.102E+03	1.3744
31	5.446E-01	-2.5E-03	7.071E+02	4.4E+00	2.3E+00	2.1E+02	7.071E+02	2.1E+02	-9.895E-01	7.97E+02	7.97E+02	-1.046E+03	-1.046E+03	1.4235
32	5.201E-01	-2.5E-03	7.115E+02	4.7E+00	2.4E+00	2.1E+02	7.115E+02	2.1E+02	-9.955E-01	8.25E+02	8.25E+02	-1.018E+03	-1.018E+03	1.4726
33	4.956E-01	-2.5E-03	7.170E+02	5.7E+00	2.5E+00	2.1E+02	7.170E+02	2.1E+02	-9.991E-01	8.53E+02	8.53E+02	-1.018E+03	-1.018E+03	1.5217
34	4.712E-01	-2.5E-03	7.238E+02	6.1E+00	2.7E+00	2.2E+02	7.238E+02	2.2E+02	-1.000E+00	8.82E+02	8.82E+02	-9.893E+02	-9.893E+02	1.5738
35	4.469E-01	-2.5E-03	7.318E+02	6.5E+00	2.9E+00	2.3E+02	7.318E+02	2.3E+02	-9.980E-01	9.11E+02	9.11E+02	-9.606E+02	-9.606E+02	1.6199
36	4.228E-01	-2.5E-03	7.411E+02	5.2E+00	2.4E+00	2.4E+02	7.411E+02	2.4E+02	-9.952E-01	9.40E+02	9.40E+02	-9.317E+02	-9.317E+02	1.6690
37	3.989E-01	-2.5E-03	7.516E+02	5.4E+00	2.5E+00	2.5E+02	7.516E+02	2.5E+02	-9.892E-01	9.69E+02	9.69E+02	-8.726E+02	-8.726E+02	1.7181
38	3.754E-01	-2.5E-03	7.633E+02	5.7E+00	2.7E+00	2.7E+02	7.633E+02	2.7E+02	-9.809E-01	9.99E+02	9.99E+02	-8.425E+02	-8.425E+02	1.7671
39	3.522E-01	-2.4E-03	7.763E+02	4.9E+00	2.9E+00	2.9E+02	7.763E+02	2.9E+02	-9.702E-01	1.03E+03	1.03E+03	-8.119E+02	-8.119E+02	1.8162
40	3.294E-01	-2.4E-03	7.905E+02	5.2E+00	2.9E+00	3.1E+02	7.905E+02	3.1E+02	-9.571E-01	1.06E+03	1.06E+03	-8.119E+02	-8.119E+02	1.8653
41	3.071E-01	-2.3E-03	8.061E+02	5.7E+00	3.5E+00	3.2E+02	8.061E+02	3.2E+02	-9.418E-01	1.09E+03	1.09E+03	-7.807E+02	-7.807E+02	1.9144
42	2.852E-01	-2.3E-03	8.229E+02	6.1E+00	3.8E+00	3.5E+02	8.229E+02	3.5E+02	-9.242E-01	1.12E+03	1.12E+03	-7.491E+02	-7.491E+02	1.9635
43	2.639E-01	-2.2E-03	8.411E+02	6.5E+00	4.3E+00	3.8E+02	8.411E+02	3.8E+02	-9.044E-01	1.15E+03	1.15E+03	-7.169E+02	-7.169E+02	2.0126
44	2.432E-01	-2.1E-03	8.607E+02	7.0E+00	4.9E+00	4.3E+02	8.607E+02	4.3E+02	-8.824E-01	1.19E+03	1.19E+03	-6.840E+02	-6.840E+02	2.0617

#	(1)	(2)	(3)	(4)	(5)	(6)	(7)	(8)	(9)	(10)	(11)	(12)	(13)	(14)
45	2.230E-01	-2.1E-03	2.036E-01	7.6E+01	5.6E+00	8.815E+02	3.4E+02	1.22E+03	1.22E+03	8.583E-01	-6.506E+02	-6.506E+02	8.815E+02	2.1108
46	2.036E-01	-2.0E-03	1.849E-01	8.3E+01	6.4E+00	9.037E+02	3.5E+02	1.25E+03	1.25E+03	8.321E-01	-6.165E+02	-6.165E+02	9.037E+02	2.1598
47	1.849E-01	-1.9E-03	1.669E-01	9.0E+01	8.0E+00	9.272E+02	3.6E+02	1.29E+03	1.29E+03	8.039E-01	-5.818E+02	-5.818E+02	9.272E+02	2.2089
48	1.669E-01	-1.7E-03	1.496E-01	9.9E+01	8.0E+00	9.518E+02	3.6E+02	1.32E+03	1.32E+03	7.738E-01	-5.463E+02	-5.463E+02	9.518E+02	2.2580
49	1.496E-01	-1.6E-03	1.332E-01	1.1E+02	1.3E+01	9.776E+02	3.7E+02	1.36E+03	1.36E+03	7.418E-01	-5.103E+02	-5.103E+02	9.776E+02	2.3071
50	1.332E-01	-1.5E-03	1.177E-01	1.2E+02	1.5E+01	1.004E+03	3.8E+02	1.40E+03	1.40E+03	7.081E-01	-4.736E+02	-4.736E+02	1.004E+03	2.3562
51	1.177E-01	-1.4E-03	1.029E-01	1.4E+02	1.5E+01	1.032E+03	3.9E+02	1.43E+03	1.43E+03	6.725E-01	-4.362E+02	-4.362E+02	1.032E+03	2.4053
52	1.029E-01	-1.2E-03	8.912E-02	1.5E+02	2.3E+01	1.060E+03	3.9E+02	1.47E+03	1.47E+03	6.354E-01	-3.984E+02	-3.984E+02	1.060E+03	2.4544
53	8.912E-02	-1.1E-03	7.623E-02	1.7E+02	1.7E+01	1.087E+03	3.9E+02	1.51E+03	1.51E+03	5.967E-01	-3.601E+02	-3.601E+02	1.087E+03	2.5035
54	7.623E-02	-1.1E-03	6.428E-02	1.9E+02	2.9E+01	1.114E+03	3.9E+02	1.55E+03	1.55E+03	5.566E-01	-3.215E+02	-3.215E+02	1.114E+03	2.5525
55	6.428E-02	-8.8E-04	5.328E-02	2.1E+02	2.9E+01	1.139E+03	3.9E+02	1.59E+03	1.59E+03	5.151E-01	-2.828E+02	-2.828E+02	1.139E+03	2.6016
56	5.328E-02	-7.6E-04	4.328E-02	2.5E+02	4.6E+01	1.161E+03	3.9E+02	1.63E+03	1.63E+03	4.723E-01	-2.443E+02	-2.443E+02	1.161E+03	2.6507
57	4.328E-02	-6.5E-04	3.429E-02	2.8E+02	5.8E+01	1.179E+03	3.7E+02	1.66E+03	1.66E+03	4.285E-01	-2.062E+02	-2.062E+02	1.179E+03	2.6998
58	3.429E-02	-5.4E-04	2.631E-02	3.1E+02	7.4E+01	1.189E+03	3.6E+02	1.70E+03	1.70E+03	3.835E-01	-1.689E+02	-1.689E+02	1.189E+03	2.7489
59	2.631E-02	-4.6E-04	1.937E-02	3.5E+02	9.4E+01	1.187E+03	2.6E+02	1.74E+03	1.74E+03	3.377E-01	-1.331E+02	-1.331E+02	1.187E+03	2.7980
60	1.937E-02	-3.6E-04	1.347E-02	3.9E+02	1.4E+02	1.170E+03	1.5E+02	1.77E+03	1.77E+03	2.910E-01	-9.917E+01	-9.917E+01	1.170E+03	2.8471
61	1.347E-02	-2.4E-04	8.633E-03	4.2E+02	1.2E+02	1.128E+03	7.8E+01	1.80E+03	1.80E+03	2.436E-01	-6.797E+01	-6.797E+01	1.128E+03	2.8962
62	8.633E-03	-1.8E-04	4.861E-03	4.4E+02	1.7E+02	1.049E+03	-1.5E+01	1.85E+03	1.85E+03	1.956E-01	-4.032E+01	-4.032E+01	1.049E+03	2.9452
63	4.861E-03	-1.1E-04	2.162E-03	4.2E+02	1.8E+02	9.144E+02	-2.6E+02	1.87E+03	1.87E+03	1.471E-01	-1.714E+01	-1.714E+01	9.144E+02	2.9943
64	2.162E-03	-4.7E-05	5.408E-04	3.5E+02	1.8E+02	6.993E+02	-3.3E+02	1.88E+03	1.88E+03	9.827E-02	-5.457E+00	-5.457E+00	6.993E+02	3.0434
65	5.408E-04	-3.3E-16	5.498E-04	2.1E+02	1.0E+02	3.880E+02	-7.8E+01	1.89E+03	1.89E+03	4.919E-02	-5.457E-01	-5.457E-01	3.880E+02	3.0925
66	3.3E-16	6.3E-16	4.7E-05	-7.8E-12	0	-3.3E-16	7.8E-12	1.88E+03	1.88E+03	3.842E-25	0	0	3.3E-16	3.1416
67	5.408E-04	4.7E-05	5.498E-04	-2.1E+02	-1.0E+02	-3.880E+02	-7.8E+01	1.88E+03	1.88E+03	4.919E-02	5.457E-01	5.457E-01	3.880E+02	3.1907
68	2.162E-03	1.6E-03	2.162E-03	-3.5E+02	-1.6E+02	-6.993E+02	-1.5E+02	1.87E+03	1.87E+03	9.827E-02	5.457E+00	5.457E+00	6.993E+02	3.2398
69	4.861E-03	1.6E-03	4.861E-03	-4.2E+02	-1.8E+02	-9.144E+02	-2.3E+02	1.85E+03	1.85E+03	1.471E-01	1.714E+01	1.714E+01	9.144E+02	3.2889
70	8.633E-03	3.3E-03	8.633E-03	-4.4E+02	-1.7E+02	-1.049E+03	-3.6E+02	1.83E+03	1.83E+03	1.956E-01	4.032E+01	4.032E+01	1.049E+03	3.3379
71	1.347E-02	3.3E-03	1.347E-02	-4.2E+02	-1.4E+02	-1.128E+03	-3.9E+02	1.80E+03	1.80E+03	2.436E-01	6.797E+01	6.797E+01	1.128E+03	3.3870
72	1.937E-02	4.2E-03	1.937E-02	-3.9E+02	-1.2E+02	-1.170E+03	-3.9E+02	1.77E+03	1.77E+03	2.910E-01	9.917E+01	9.917E+01	1.170E+03	3.4361
73	2.631E-02	5.3E-03	2.631E-02	-3.5E+02	-9.4E+01	-1.187E+03	-3.5E+02	1.74E+03	1.74E+03	3.377E-01	1.331E+02	1.331E+02	1.187E+03	3.4852
74	3.429E-02	6.4E-03	3.429E-02	-3.1E+02	-7.4E+01	-1.189E+03	-3.5E+02	1.70E+03	1.70E+03	3.835E-01	1.689E+02	1.689E+02	1.189E+03	3.5343
75	4.328E-02	7.6E-03	4.328E-02	-2.8E+02	-5.8E+01	-1.179E+03	-3.9E+02	1.66E+03	1.66E+03	4.285E-01	2.062E+02	2.062E+02	1.179E+03	3.5834
76	5.328E-02	8.8E-03	5.328E-02	-2.5E+02	-4.6E+01	-1.161E+03	-3.9E+02	1.63E+03	1.63E+03	4.723E-01	2.443E+02	2.443E+02	1.161E+03	3.6325
77	6.428E-02	1.1E-02	6.428E-02	-2.2E+02	-3.6E+01	-1.139E+03	-3.6E+02	1.59E+03	1.59E+03	5.151E-01	2.828E+02	2.828E+02	1.139E+03	3.6816
78	7.623E-02	1.2E-02	7.623E-02	-1.9E+02	-2.9E+01	-1.114E+03	-3.5E+02	1.55E+03	1.55E+03	5.566E-01	3.215E+02	3.215E+02	1.114E+03	3.7306
79	8.912E-02	1.4E-02	8.912E-02	-1.7E+02	-2.3E+01	-1.087E+03	-3.4E+02	1.51E+03	1.51E+03	5.967E-01	3.601E+02	3.601E+02	1.087E+03	3.7797
80	1.029E-01	1.5E-02	1.029E-01	-1.5E+02	-1.9E+01	-1.060E+03	-3.2E+02	1.47E+03	1.47E+03	6.354E-01	3.984E+02	3.984E+02	1.060E+03	3.8288
81	1.177E-01	1.7E-02	1.177E-01	-1.4E+02	-1.5E+01	-1.032E+03	-3.1E+02	1.43E+03	1.43E+03	6.725E-01	4.362E+02	4.362E+02	1.032E+03	3.8779
82	1.332E-01	1.8E-02	1.332E-01	-1.2E+02	-1.3E+01	-1.004E+03	-2.9E+02	1.40E+03	1.40E+03	7.081E-01	4.736E+02	4.736E+02	1.004E+03	3.9270
83	1.496E-01	2.0E-02	1.496E-01	-1.1E+02	-1.0E+01	-9.776E+02	-2.8E+02	1.36E+03	1.36E+03	7.418E-01	5.103E+02	5.103E+02	9.776E+02	3.9761
84	1.669E-01	2.1E-02	1.669E-01	-9.9E+01	-8.8E+00	-9.518E+02	-2.7E+02	1.32E+03	1.32E+03	7.738E-01	5.463E+02	5.463E+02	9.518E+02	4.0252
85	1.849E-01	2.2E-02	1.849E-01	-9.0E+01	-7.6E+00	-9.272E+02	-2.5E+02	1.29E+03	1.29E+03	8.039E-01	5.818E+02	5.818E+02	9.272E+02	4.0743
86	2.036E-01	2.3E-02	2.036E-01	-8.3E+01	-6.4E+00	-9.037E+02	-2.4E+02	1.25E+03	1.25E+03	8.321E-01	6.165E+02	6.165E+02	9.037E+02	4.1233
87	2.230E-01	2.4E-02	2.230E-01	-7.6E+01	-5.6E+00	-8.815E+02	-2.3E+02	1.22E+03	1.22E+03	8.583E-01	6.506E+02	6.506E+02	8.815E+02	4.1724
88	2.432E-01	2.4E-02	2.432E-01	-7.0E+01	-4.9E+00	-8.607E+02	-2.2E+02	1.19E+03	1.19E+03	8.824E-01	6.840E+02	6.840E+02	8.607E+02	4.2215
89	2.639E-01	2.6E-02	2.639E-01	-6.5E+01	-4.3E+00	-8.411E+02	-2.1E+02	1.15E+03	1.15E+03	9.044E-01	7.169E+02	7.169E+02	8.411E+02	4.2706
90	2.852E-01	2.8E-02	2.852E-01	-6.1E+01	-3.8E+00	-8.229E+02	-2.0E+02	1.12E+03	1.12E+03	9.242E-01	7.491E+02	7.491E+02	8.229E+02	4.3197
91	3.071E-01	3.0E-02	3.071E-01	-5.7E+01	-3.5E+00	-8.061E+02	-1.9E+02	1.09E+03	1.09E+03	9.418E-01	7.807E+02	7.807E+02	8.061E+02	4.3688
92	3.294E-01	3.2E-02	3.294E-01	-5.4E+01	-3.2E+00	-7.905E+02	-1.8E+02	1.06E+03	1.06E+03	9.571E-01	8.119E+02	8.119E+02	7.905E+02	4.4179

93	3.522E-01	2.4E-03	-7.763E+02	-7.763E+02	-5.2E+01	-2.9E+00	-7.763E+02	-3.1E+02	1.03E+03	1.03E+03	9.702E-01	-8.425E+02	-8.425E+02	4.4670
94	3.754E-01	2.5E-03	-7.633E+02	-7.633E+02	-4.9E+01	-2.7E+00	-7.633E+02	-3.0E+02	9.99E+02	9.99E+02	9.809E-01	-8.726E+02	-8.726E+02	4.5160
95	3.989E-01	2.5E-03	-7.516E+02	-7.516E+02	-4.7E+01	-2.5E+00	-7.516E+02	-3.0E+02	9.69E+02	9.69E+02	9.892E-01	-9.023E+02	-9.023E+02	4.5651
96	4.228E-01	2.5E-03	-7.411E+02	-7.411E+02	-4.6E+01	-2.4E+00	-7.411E+02	-3.0E+02	9.40E+02	9.40E+02	9.952E-01	-9.317E+02	-9.317E+02	4.6142
97	4.469E-01	2.5E-03	-7.310E+02	-7.310E+02	-4.4E+01	-2.3E+00	-7.310E+02	-2.9E+02	9.11E+02	9.11E+02	9.988E-01	-9.606E+02	-9.606E+02	4.6633
98	4.712E-01	2.5E-03	-7.238E+02	-7.238E+02	-4.3E+01	-2.2E+00	-7.238E+02	-2.9E+02	8.82E+02	8.82E+02	1.000E+00	-9.893E+02	-9.893E+02	4.7124
99	4.956E-01	2.5E-03	-7.170E+02	-7.170E+02	-4.3E+01	-2.1E+00	-7.170E+02	-2.9E+02	8.53E+02	8.53E+02	9.991E-01	-1.018E+03	-1.018E+03	4.7615
100	5.201E-01	2.5E-03	-7.115E+02	-7.115E+02	-4.2E+01	-2.1E+00	-7.115E+02	-2.9E+02	8.25E+02	8.25E+02	9.955E-01	-1.046E+03	-1.046E+03	4.8106
101	5.446E-01	2.5E-03	-7.071E+02	-7.071E+02	-4.2E+01	-2.1E+00	-7.071E+02	-2.8E+02	7.97E+02	7.97E+02	9.895E-01	-1.074E+03	-1.074E+03	4.8597
102	5.691E-01	2.5E-03	-7.041E+02	-7.041E+02	-4.2E+01	-2.1E+00	-7.041E+02	-2.8E+02	7.70E+02	7.70E+02	9.812E-01	-1.102E+03	-1.102E+03	4.9087
103	5.935E-01	2.4E-03	-7.023E+02	-7.023E+02	-4.3E+01	-2.1E+00	-7.023E+02	-2.8E+02	7.42E+02	7.42E+02	9.705E-01	-1.129E+03	-1.129E+03	4.9578
104	6.177E-01	2.4E-03	-7.018E+02	-7.018E+02	-4.3E+01	-2.3E+00	-7.018E+02	-2.8E+02	7.14E+02	7.14E+02	9.574E-01	-1.157E+03	-1.157E+03	5.0069
105	6.417E-01	2.4E-03	-7.026E+02	-7.026E+02	-4.3E+01	-2.3E+00	-7.026E+02	-2.8E+02	6.87E+02	6.87E+02	9.421E-01	-1.185E+03	-1.185E+03	5.0560
106	6.654E-01	2.3E-03	-7.048E+02	-7.048E+02	-4.5E+01	-2.4E+00	-7.048E+02	-2.8E+02	6.59E+02	6.59E+02	9.245E-01	-1.212E+03	-1.212E+03	5.1051
107	6.888E-01	2.3E-03	-7.085E+02	-7.085E+02	-4.7E+01	-2.6E+00	-7.085E+02	-2.8E+02	6.31E+02	6.31E+02	9.047E-01	-1.240E+03	-1.240E+03	5.1542
108	7.118E-01	2.2E-03	-7.137E+02	-7.137E+02	-4.8E+01	-2.7E+00	-7.137E+02	-2.8E+02	6.03E+02	6.03E+02	8.827E-01	-1.268E+03	-1.268E+03	5.2033
109	7.343E-01	2.1E-03	-7.205E+02	-7.205E+02	-5.0E+01	-3.0E+00	-7.205E+02	-2.9E+02	5.75E+02	5.75E+02	8.586E-01	-1.296E+03	-1.296E+03	5.2524
110	7.563E-01	2.1E-03	-7.289E+02	-7.289E+02	-5.3E+01	-3.3E+00	-7.289E+02	-2.9E+02	5.47E+02	5.47E+02	8.324E-01	-1.324E+03	-1.324E+03	5.3014
111	7.777E-01	2.0E-03	-7.390E+02	-7.390E+02	-5.7E+01	-3.7E+00	-7.390E+02	-2.9E+02	5.19E+02	5.19E+02	8.042E-01	-1.352E+03	-1.352E+03	5.3505
112	7.984E-01	1.9E-03	-7.510E+02	-7.510E+02	-6.1E+01	-4.3E+00	-7.510E+02	-3.0E+02	4.90E+02	4.90E+02	7.740E-01	-1.381E+03	-1.381E+03	5.3996
113	8.184E-01	1.8E-03	-7.649E+02	-7.649E+02	-6.6E+01	-4.9E+00	-7.649E+02	-3.0E+02	4.61E+02	4.61E+02	7.418E-01	-1.410E+03	-1.410E+03	5.4487
114	8.377E-01	1.7E-03	-7.808E+02	-7.808E+02	-7.2E+01	-5.7E+00	-7.808E+02	-3.0E+02	4.31E+02	4.31E+02	7.082E-01	-1.440E+03	-1.440E+03	5.4978
115	8.561E-01	1.6E-03	-7.988E+02	-7.988E+02	-7.9E+01	-6.8E+00	-7.988E+02	-3.1E+02	4.01E+02	4.01E+02	6.727E-01	-1.470E+03	-1.470E+03	5.5469
116	8.736E-01	1.5E-03	-8.190E+02	-8.190E+02	-8.9E+01	-8.4E+00	-8.190E+02	-3.1E+02	3.71E+02	3.71E+02	6.356E-01	-1.500E+03	-1.500E+03	5.5960
117	8.902E-01	1.3E-03	-8.414E+02	-8.414E+02	-1.1E+02	-1.3E+01	-8.414E+02	-3.2E+02	3.40E+02	3.40E+02	5.969E-01	-1.531E+03	-1.531E+03	5.6450
118	9.058E-01	1.2E-03	-8.660E+02	-8.660E+02	-1.3E+02	-1.3E+01	-8.660E+02	-3.2E+02	3.09E+02	3.09E+02	5.562E-01	-1.562E+03	-1.562E+03	5.6941
119	9.203E-01	1.1E-03	-8.926E+02	-8.926E+02	-1.7E+02	-1.7E+01	-8.926E+02	-3.2E+02	2.77E+02	2.77E+02	5.152E-01	-1.594E+03	-1.594E+03	5.7432
120	9.337E-01	9.6E-04	-9.210E+02	-9.210E+02	-2.3E+01	-2.3E+01	-9.210E+02	-3.3E+02	2.45E+02	2.45E+02	4.725E-01	-1.626E+03	-1.626E+03	5.7923
121	9.460E-01	8.3E-04	-9.509E+02	-9.509E+02	-3.3E+01	-3.3E+01	-9.509E+02	-3.3E+02	2.13E+02	2.13E+02	4.286E-01	-1.658E+03	-1.658E+03	5.8414
122	9.571E-01	7.0E-04	-9.816E+02	-9.816E+02	-4.7E+01	-4.7E+01	-9.816E+02	-3.3E+02	1.81E+02	1.81E+02	3.836E-01	-1.690E+03	-1.690E+03	5.8905
123	9.670E-01	5.8E-04	-1.012E+03	-1.012E+03	-6.6E+01	-6.6E+01	-1.012E+03	-3.2E+02	1.49E+02	1.49E+02	3.378E-01	-1.722E+03	-1.722E+03	5.9396
124	9.757E-01	4.5E-04	-1.041E+03	-1.041E+03	-8.6E+01	-8.6E+01	-1.041E+03	-3.1E+02	1.18E+02	1.18E+02	2.911E-01	-1.753E+03	-1.753E+03	5.9887
125	9.831E-01	3.4E-04	-1.067E+03	-1.067E+03	-1.3E+02	-1.3E+02	-1.067E+03	-2.9E+02	8.85E+01	8.85E+01	2.436E-01	-1.783E+03	-1.783E+03	6.0377
126	9.891E-01	2.3E-04	-1.086E+03	-1.086E+03	-1.9E+02	-1.9E+02	-1.086E+03	-2.6E+02	6.12E+01	6.12E+01	1.956E-01	-1.810E+03	-1.810E+03	6.0868
127	9.939E-01	1.4E-04	-1.094E+03	-1.094E+03	-2.8E+02	-2.8E+02	-1.094E+03	-2.2E+02	3.73E+01	3.73E+01	1.471E-01	-1.834E+03	-1.834E+03	6.1359
128	9.973E-01	7.2E-05	-1.076E+03	-1.076E+03	-4.1E+02	-4.1E+02	-1.076E+03	-1.7E+02	1.80E+01	1.80E+01	9.830E-02	-1.853E+03	-1.853E+03	6.1850
129	9.993E-01	2.2E-05	-9.363E+02	-9.363E+02	-5.2E+02	-5.2E+02	-9.363E+02	-9.7E+01	4.85E+00	4.85E+00	4.921E-02	-1.866E+03	-1.866E+03	6.2341
130	1.000E+00	-2.8E-16	Ø.	Ø.	Ø.	Ø.	-1.5E-11	Ø.	Ø.	Ø.	1.110E-15	-1.871E+03	-1.871E+03	6.2832

STAGNATION POINT BETWEEN I= 2 AND I= 3 AT 1.00300

STAGNATION POINT BETWEEN I= 66 AND I= 67 AT 0.00000

TIME = 1.00000E-03   TIME BASED ON CHORD= 2.76315E-04

```
N=   1    DVRTLD(129)=  4.97747E+03
N=   2    DVRTLD(  3)=-3.29525E+03
N=   3    DVRTLD(  3)=-2.28740E+03
N=   4    DVRTLD(  4)=-1.74309E+03
N=   5    DVRTLD(128)= 1.40966E+03
N=   6    DVRTLD(128)= 1.13998E+03
N=   7    DVRTLD(128)= 9.21985E+02
N=   8    DVRTLD(128)= 7.45796E+02
N=   9    DVRTLD(128)= 6.03373E+02
N=  10    DVRTLD(128)= 4.88217E+02
N=  11    DVRTLD(128)= 3.95081E+02
N=  12    DVRTLD(  4)=-3.19732E+02
N=  13    DVRTLD(  4)=-2.58760E+02
N=  14    DVRTLD(  4)=-2.09409E+02
N=  15    DVRTLD(128)= 1.69460E+02
N=  16    DVRTLD(127)= 1.38856E+02
N=  17    DVRTLD(  5)=-1.15002E+02
N=  18    DVRTLD(127)= 9.52469E+01
N=  19    DVRTLD(  5)=-7.88852E+01
N=  20    DVRTLD(127)= 6.53339E+01
N=  21    DVRTLD(  5)=-5.41101E+01
N=  22    DVRTLD(127)= 4.48140E+01
N=  23    DVRTLD(  5)=-3.71144E+01
N=  24    DVRTLD(127)= 3.07371E+01
N=  25    DVRTLD(126)= 2.55617E+01
N=  26    DVRTLD(126)= 2.14397E+01
N=  27    DVRTLD(  6)=-1.79830E+01
N=  28    DVRTLD(126)= 1.50840E+01
N=  29    DVRTLD(  6)=-1.26526E+01
N=  30    DVRTLD(  6)=-1.06135E+01
N=  31    DVRTLD(  6)=-8.90315E+00
N=  32    DVRTLD(126)= 7.46863E+00
N=  33    DVRTLD(126)= 6.26539E+00
N=  34    DVRTLD(126)= 5.25612E+00
N=  35    DVRTLD(  6)=-4.40952E+00
N=  36    DVRTLD(126)= 3.69936E+00
N=  37    DVRTLD(  6)=-3.10363E+00
N=  38    DVRTLD(126)= 2.60389E+00

IS   =66

PRESSURE DERIVATIVE= 1.47781E-01     K.E. DEFECT= 1.17309E-03     UT(1)=-1.65920E-03

1.17309E-03  4.82819E-03  8.67696E-03  1.27202E-02  1.69967E-02  2.14933E-02  2.62353E-02  3.12412E-02  3.65328E-02  4.21348E-02
4.80753E-02  5.43862E-02  6.11030E-02  6.82654E-02  7.59172E-02  8.41065E-02  9.28861E-02  1.02313E-01  1.12451E-01  1.23365E-01
1.35127E-01  1.47816E-01  1.61513E-01  1.76306E-01  1.92286E-01  2.09552E-01  2.28205E-01  2.48353E-01  2.70106E-01  2.93582E-01
3.18897E-01  3.46172E-01  3.75531E-01  4.07096E-01  4.40988E-01  4.77325E-01  5.16219E-01  5.57774E-01  6.02083E-01  6.49225E-01
```

```
6.9926ØE-Ø1  7.52229E-Ø1  8.Ø8143E-Ø1  8.66998E-Ø1  9.28715E-Ø1  9.93238E-Ø1  1.Ø6643E+ØØ  1.13Ø14E+ØØ  1.2Ø214E+ØØ  1.27619E+ØØ
1.35199E+ØØ  1.42922E+ØØ  1.5Ø751E+ØØ  1.58645E+ØØ  1.66553E+ØØ  1.74462E+ØØ  1.82295E+ØØ  1.9ØØ2ØE+ØØ  1.97594E+ØØ  2.Ø4974E+ØØ
2.12124E+ØØ  2.19ØØ8E+ØØ  2.25599E+ØØ  2.31872E+ØØ  2.378Ø3E+ØØ  2.43384E+ØØ  2.486Ø4E+ØØ  2.53462E+ØØ  2.57959E+ØØ  2.621Ø1E+ØØ
2.65899E+ØØ  2.69365E+ØØ  2.72517E+ØØ  2.75371E+ØØ  2.77944E+ØØ  2.8Ø263E+ØØ  2.82339E+ØØ  2.84195E+ØØ  2.85849E+ØØ  2.87323E+ØØ
2.88647E+ØØ  2.89853E+ØØ  2.9Ø974E+ØØ  2.924ØØE+ØØ
```

PRESSURE COEFFICIENT UPSTREAM OF STAGNATION POINT =   2.924ØØE+ØØ

PRESSURE DERIVATIVE= 1.47588E-Ø1   K.E. DEFECT= 1.16993E-Ø3   UT(1)=-1.65713E-Ø3

```
1.16993E-Ø3  4.82Ø25E-Ø3  8.66395E-Ø3  1.271Ø8E-Ø2  1.69727E-Ø2  2.14633E-Ø2  2.61988E-Ø2  3.11979E-Ø2  3.64821E-Ø2  4.2Ø763E-Ø2
4.8ØØ84E-Ø2  5.431Ø2E-Ø2  6.1Ø71E-Ø2  6.81687E-Ø2  7.58ØØ8E-Ø2  8.39853E-Ø2  9.275Ø8E-Ø2  1.Ø2163E-Ø1  1.12282E-Ø1  1.23177E-Ø1
1.34918E-Ø1  1.47583E-Ø1  1.61253E-Ø1  1.76Ø15E-Ø1  1.9196ØE-Ø1  2.Ø9186E-Ø1  2.2779ØE-Ø1  2.4789ØE-Ø1  2.69585E-Ø1  2.92991E-Ø1
3.18227E-Ø1  3.4541ØE-Ø1  3.74663E-Ø1  4.Ø61Ø4E-Ø1  4.39851E-Ø1  4.76Ø19E-Ø1  5.14715E-Ø1  5.56Ø39E-Ø1  6.ØØ877E-Ø1  6.46951E-Ø1
6.96565E-Ø1  7.49Ø98E-Ø1  8.Ø4592E-Ø1  8.62751E-Ø1  9.2375ØE-Ø1  9.87495E-Ø1  1.Ø5375E+ØØ  1.1231E+ØØ  1.1931ØE+ØØ  1.2657ØE+ØØ
1.3398E+ØØ  1.41519E+ØØ  1.491Ø3E+ØØ  1.56786E+ØØ  1.64433E+ØØ  1.72Ø31E+ØØ  1.79532E+ØØ  1.86893E+ØØ  1.94Ø71E+ØØ  2.Ø1Ø24E+ØØ
2.Ø7716E+ØØ  2.14117E+ØØ  2.2Ø197E+ØØ  2.25937E+ØØ  2.31321E+ØØ  2.36339E+ØØ  2.4Ø988E+ØØ  2.45271E+ØØ  2.4919ØE+ØØ  2.5276ØE+ØØ
2.55994E+ØØ  2.5891ØE+ØØ  2.61525E+ØØ  2.63863E+ØØ  2.65945E+ØØ  2.67795E+ØØ  2.69443E+ØØ  2.7Ø915E+ØØ  2.72237E+ØØ  2.73419E+ØØ
2.74457E+ØØ  2.7525ØE+ØØ  2.76255E+ØØ  2.776ØØE+ØØ
```

PRESSURE COEFFICIENT DOWNSTREAM OF STAGNATION POINT =   2.776ØØE+ØØ

STAGNATION POINT AT = Ø.

```
        PRESSURE INTEGRATED FROM GEOMETRICAL TRAILING TO LEADING EDGE
PM  = 1.16947E+Ø3   PM1 = 1.16947E+Ø3   PERCENT ERROR= 2.48863E-12   PERCENT ERROR BASED ON MINMAX= 2.48863E-12
CP  = 9.45467E+Ø1   CF  = 1.897ØØE+ØØ   DRAG = 9.64437E+Ø1   CLP = 1.13367E-11   CLF = 5.95898E-15   LIFT = 1.13427E-11
CMF =-4.28Ø87E-16   CMP =-5.24115E-12   MOMENT=-5.24157E-12

        PRESSURE INTEGRATED FROM GEOMETRICAL LEADING TO TRAILING EDGE
PM  =-1.16947E+Ø3   PM1 =-1.16947E+Ø3   PERCENT ERROR=-1.24432E-12   PERCENT ERROR BASED ON MINMAX= 1.24432E-12
CP  = 9.45467E+Ø1   CF  = 1.897ØØE+ØØ   DRAG = 9.64437E+Ø1   CLP = 3.68848E-11   CLF = 5.95898E-15   LIFT = 3.689Ø8E-11
CMF =-4.28Ø87E-16   CMP =-9.65472E-12   MOMENT=-9.65514E-12
```

AVERAGE DRAG= 9.64437E+01   AVERAGE LIFT= 2.41167E-11   AVERAGE MOMENT=-7.44836E-12

I	XC(I)	PY(JLMN)	VRT(JL)	VRT(JLMN)	VRT(JLM2)	VRTB(JL)	VRT1	PR (L-T)	PR1 (T-L)	V	CPT (L-T)	CPT1 (T-L)	THETA
2	1.000E+00	-2.8E-16	0.762E+02	0.	0.	0.762E+02	-1.6E-12	0.	0.	3.331E-16	-1.167E+03	-1.167E+03	0.0000
3	9.993E-01	-1.2E-05	4.762E+02	4.6E+02	4.1E+02	4.762E+02	8.2E+02	3.84E-01	3.79E-01	-4.921E-02	-1.166E+03	-1.166E+03	.0491
4	9.973E-01	-4.0E-05	5.567E+02	4.8E+02	4.1E+02	5.567E+02	4.1E+01	1.79E+00	1.78E+00	-9.830E-02	-1.165E+03	-1.165E+03	.0982
5	9.939E-01	-8.4E-05	5.805E+02	4.5E+02	3.4E+02	5.805E+02	3.4E+01	4.79E+00	4.79E+00	-1.471E-01	-1.162E+03	-1.162E+03	.1473
6	9.891E-01	-1.4E-04	5.976E+02	4.0E+02	2.6E+02	5.976E+02	6.4E+01	9.94E+00	9.94E+00	-1.956E-01	-1.157E+03	-1.157E+03	.1963
7	9.831E-01	-2.2E-04	6.126E+02	3.6E+02	2.0E+02	6.126E+02	8.9E+01	1.75E+01	1.75E+01	-2.436E-01	-1.149E+03	-1.149E+03	.2454
8	9.757E-01	-3.0E-04	6.257E+02	3.2E+02	1.5E+02	6.257E+02	1.1E+02	2.75E+01	2.75E+01	-2.911E-01	-1.139E+03	-1.139E+03	.2945
9	9.670E-01	-4.0E-04	6.361E+02	2.8E+02	1.1E+02	6.361E+02	1.3E+02	3.97E+01	3.97E+01	-3.378E-01	-1.127E+03	-1.127E+03	.3436
10	9.571E-01	-5.0E-04	6.434E+02	2.5E+02	8.3E+01	6.434E+02	1.5E+02	5.40E+01	5.40E+01	-3.836E-01	-1.113E+03	-1.113E+03	.3927
11	9.460E-01	-6.2E-04	6.476E+02	2.0E+02	6.3E+01	6.476E+02	1.7E+02	7.00E+01	7.00E+01	-4.286E-01	-1.097E+03	-1.097E+03	.4418
12	9.337E-01	-7.3E-04	6.490E+02	1.7E+02	4.9E+01	6.490E+02	1.8E+02	8.74E+01	8.74E+01	-4.725E-01	-1.079E+03	-1.079E+03	.4909
13	9.203E-01	-8.5E-04	6.482E+02	1.6E+02	3.8E+01	6.482E+02	1.9E+02	1.06E+02	1.06E+02	-5.152E-01	-1.061E+03	-1.061E+03	.5399
14	9.057E-01	-9.6E-04	6.455E+02	1.4E+02	3.1E+01	6.455E+02	2.0E+02	1.25E+02	1.25E+02	-5.567E-01	-1.041E+03	-1.041E+03	.5890
15	8.902E-01	-1.1E-03	6.417E+02	1.3E+02	2.5E+01	6.417E+02	2.1E+02	1.45E+02	1.45E+02	-5.969E-01	-1.021E+03	-1.021E+03	.6381
16	8.736E-01	-1.3E-03	6.371E+02	1.1E+02	2.0E+01	6.371E+02	2.1E+02	1.66E+02	1.66E+02	-6.356E-01	-1.001E+03	-1.001E+03	.6872
17	8.561E-01	-1.4E-03	6.321E+02	9.6E+01	1.7E+01	6.321E+02	2.1E+02	1.87E+02	1.87E+02	-6.727E-01	-9.798E+02	-9.798E+02	.7363
18	8.377E-01	-1.5E-03	6.270E+02	8.6E+01	1.4E+01	6.270E+02	2.2E+02	2.08E+02	2.08E+02	-7.082E-01	-9.586E+02	-9.586E+02	.7854
19	8.184E-01	-1.6E-03	6.220E+02	8.3E+01	1.2E+01	6.220E+02	2.2E+02	2.29E+02	2.29E+02	-7.420E-01	-9.372E+02	-9.372E+02	.8345
20	7.984E-01	-1.7E-03	6.175E+02	8.0E+01	1.1E+01	6.175E+02	2.2E+02	2.51E+02	2.51E+02	-7.740E-01	-9.157E+02	-9.157E+02	.8836
21	7.777E-01	-1.8E-03	6.133E+02	7.8E+01	9.5E+00	6.133E+02	2.2E+02	2.73E+02	2.73E+02	-8.042E-01	-8.940E+02	-8.940E+02	.9327
22	7.563E-01	-1.9E-03	6.098E+02	7.6E+01	8.6E+00	6.098E+02	2.2E+02	2.94E+02	2.94E+02	-8.324E-01	-8.723E+02	-8.723E+02	.9817
23	7.343E-01	-1.9E-03	6.069E+02	7.5E+01	8.3E+00	6.069E+02	2.2E+02	3.16E+02	3.16E+02	-8.586E-01	-8.505E+02	-8.505E+02	1.0308
24	7.118E-01	-1.9E-03	6.047E+02	7.4E+01	7.2E+00	6.047E+02	2.2E+02	3.38E+02	3.38E+02	-8.827E-01	-8.287E+02	-8.287E+02	1.0799
25	6.888E-01	-2.1E-03	6.032E+02	7.3E+01	6.7E+00	6.032E+02	2.2E+02	3.60E+02	3.60E+02	-9.047E-01	-8.068E+02	-8.068E+02	1.1290
26	6.654E-01	-2.1E-03	6.025E+02	7.4E+01	6.0E+00	6.025E+02	2.2E+02	3.82E+02	3.82E+02	-9.245E-01	-7.848E+02	-7.848E+02	1.1781
27	6.417E-01	-2.2E-03	6.025E+02	7.6E+01	5.8E+00	6.025E+02	2.3E+02	4.04E+02	4.04E+02	-9.421E-01	-7.628E+02	-7.628E+02	1.2272
28	6.177E-01	-2.2E-03	6.032E+02	7.8E+01	5.7E+00	6.032E+02	2.3E+02	4.26E+02	4.26E+02	-9.574E-01	-7.404E+02	-7.404E+02	1.2763
29	5.935E-01	-2.2E-03	6.047E+02	8.0E+01	5.6E+00	6.047E+02	2.3E+02	4.48E+02	4.48E+02	-9.705E-01	-7.184E+02	-7.184E+02	1.3254
30	5.691E-01	-2.2E-03	6.070E+02	8.2E+01	5.7E+00	6.070E+02	2.3E+02	4.70E+02	4.70E+02	-9.812E-01	-6.961E+02	-6.961E+02	1.3744
31	5.446E-01	-2.3E-03	6.100E+02	8.6E+01	5.8E+00	6.100E+02	2.3E+02	4.93E+02	4.93E+02	-9.895E-01	-6.737E+02	-6.737E+02	1.4235
32	5.201E-01	-2.2E-03	6.137E+02	8.9E+01	6.0E+00	6.137E+02	2.3E+02	5.15E+02	5.15E+02	-9.955E-01	-6.511E+02	-6.511E+02	1.4726
33	4.956E-01	-2.2E-03	6.182E+02	9.3E+01	6.3E+00	6.182E+02	2.4E+02	5.38E+02	5.38E+02	-9.991E-01	-6.284E+02	-6.284E+02	1.5217
34	4.712E-01	-2.2E-03	6.233E+02	1.0E+02	7.0E+00	6.233E+02	2.4E+02	5.61E+02	5.61E+02	-1.000E+00	-6.056E+02	-6.056E+02	1.5708
35	4.469E-01	-2.2E-03	6.291E+02	1.0E+02	7.6E+00	6.291E+02	2.4E+02	5.84E+02	5.84E+02	-9.988E-01	-5.825E+02	-5.825E+02	1.6199
36	4.228E-01	-2.2E-03	6.356E+02	1.1E+02	8.0E+00	6.356E+02	2.4E+02	6.07E+02	6.07E+02	-9.952E-01	-5.593E+02	-5.593E+02	1.6690
37	3.989E-01	-2.2E-03	6.427E+02	1.1E+02	8.2E+00	6.427E+02	2.4E+02	6.31E+02	6.31E+02	-9.892E-01	-5.359E+02	-5.359E+02	1.7181
38	3.754E-01	-2.1E-03	6.504E+02	1.0E+02	9.0E+00	6.504E+02	2.4E+02	6.54E+02	6.54E+02	-9.809E-01	-5.124E+02	-5.124E+02	1.7671
39	3.522E-01	-2.1E-03	6.587E+02	1.1E+02	9.3E+00	6.587E+02	2.5E+02	6.78E+02	6.78E+02	-9.702E-01	-4.886E+02	-4.886E+02	1.8162
40	3.294E-01	-2.1E-03	6.676E+02	1.0E+02	9.9E+00	6.676E+02	2.5E+02	7.02E+02	7.02E+02	-9.571E-01	-4.646E+02	-4.646E+02	1.8653
41	3.071E-01	-2.0E-03	6.769E+02	1.1E+02	1.0E+01	6.769E+02	2.5E+02	7.26E+02	7.26E+02	-9.418E-01	-4.405E+02	-4.405E+02	1.9144
42	2.852E-01	-2.0E-03	6.866E+02	1.1E+02	1.1E+01	6.866E+02	2.5E+02	7.50E+02	7.50E+02	-9.242E-01	-4.161E+02	-4.161E+02	1.9635
43	2.639E-01	-1.9E-03	6.966E+02	1.2E+02	1.1E+01	6.966E+02	2.5E+02	7.75E+02	7.75E+02	-9.044E-01	-3.916E+02	-3.916E+02	2.0126
44	2.432E-01	-1.8E-03	7.069E+02	1.2E+02	1.2E+01	7.069E+02	2.5E+02	8.00E+02	8.00E+02	-8.824E-01	-3.670E+02	-3.670E+02	2.0617

#													
45	2.230E-01	-1.8E-03	7.172E+02	1.3E+02	1.4E+01	7.172E+02	2.5E+02	8.24E+02	8.24E+02	-8.583E-01	-3.422E+02	-3.422E+02	2.1108
46	2.036E-01	-1.7E-03	7.276E+02	1.4E+02	1.6E+01	7.276E+02	2.5E+02	8.49E+02	8.49E+02	-8.321E-01	-3.175E+02	-3.175E+02	2.1598
47	1.849E-01	-1.6E-03	7.377E+02	1.6E+02	1.8E+01	7.377E+02	2.5E+02	8.74E+02	8.74E+02	-8.039E-01	-2.926E+02	-2.926E+02	2.2089
48	1.669E-01	-1.5E-03	7.473E+02	1.6E+02	2.1E+01	7.473E+02	2.5E+02	8.99E+02	8.99E+02	-7.738E-01	-2.679E+02	-2.679E+02	2.2580
49	1.496E-01	-1.4E-03	7.563E+02	1.9E+02	2.5E+01	7.563E+02	2.5E+02	9.23E+02	9.23E+02	-7.418E-01	-2.433E+02	-2.433E+02	2.3071
50	1.332E-01	-1.3E-03	7.643E+02	1.9E+02	3.0E+01	7.643E+02	2.5E+02	9.48E+02	9.48E+02	-7.080E-01	-2.190E+02	-2.190E+02	2.3552
51	1.177E-01	-1.2E-03	7.710E+02	2.1E+02	3.5E+01	7.710E+02	2.4E+02	9.71E+02	9.71E+02	-6.725E-01	-1.951E+02	-1.951E+02	2.4053
52	1.029E-01	-1.1E-03	7.758E+02	2.5E+02	4.1E+01	7.758E+02	2.3E+02	9.95E+02	9.95E+02	-6.354E-01	-1.716E+02	-1.716E+02	2.4544
53	8.912E-02	-9.6E-04	7.785E+02	2.5E+02	5.1E+01	7.785E+02	2.2E+02	1.02E+03	1.02E+03	-5.967E-01	-1.489E+02	-1.489E+02	2.5035
54	7.623E-02	-8.5E-04	7.784E+02	2.7E+02	6.2E+01	7.784E+02	2.2E+02	1.04E+03	1.04E+03	-5.566E-01	-1.271E+02	-1.271E+02	2.5525
55	6.427E-02	-7.4E-04	7.751E+02	3.2E+02	9.2E+01	7.751E+02	1.9E+02	1.06E+03	1.06E+03	-5.151E-01	-1.064E+02	-1.064E+02	2.6016
56	5.328E-02	-6.4E-04	7.678E+02	3.2E+02	1.4E+02	7.678E+02	1.5E+02	1.08E+03	1.08E+03	-4.723E-01	-8.705E+01	-8.705E+01	2.6507
57	4.328E-02	-5.4E-04	7.560E+02	3.8E+02	1.6E+02	7.560E+02	1.5E+02	1.10E+03	1.10E+03	-4.283E-01	-6.924E+01	-6.924E+01	2.6998
58	3.429E-02	-4.4E-04	7.388E+02	3.8E+02	2.2E+02	7.388E+02	1.3E+02	1.11E+03	1.11E+03	-3.835E-01	-5.322E+01	-5.322E+01	2.7489
59	2.631E-02	-3.6E-04	7.152E+02	4.2E+02	2.4E+02	7.152E+02	8.9E+01	1.13E+03	1.13E+03	-3.377E-01	-3.917E+01	-3.917E+01	2.7980
60	1.937E-02	-2.8E-04	6.835E+02	4.3E+02	2.4E+02	6.835E+02	6.0E+01	1.14E+03	1.14E+03	-2.910E-01	-2.723E+01	-2.723E+01	2.8471
61	1.347E-02	-2.1E-04	6.409E+02	4.3E+02	2.2E+02	6.409E+02	4.9E+01	1.15E+03	1.15E+03	-2.436E-01	-1.744E+01	-1.744E+01	2.8962
62	8.633E-03	-1.5E-04	5.825E+02	4.2E+02	2.4E+02	5.825E+02	1.6E+01	1.16E+03	1.16E+03	-1.956E-01	-9.766E+00	-9.766E+00	2.9452
63	4.861E-03	-1.0E-04	4.998E+02	3.9E+02	1.9E+02	4.998E+02	-6.0E-13	1.17E+03	1.17E+03	-1.471E-01	-4.077E+00	-4.077E+00	2.9943
64	2.162E-03	-6.0E-05	3.803E+02	3.1E+02	1.6E+02	3.803E+02	-1.6E+01	1.17E+03	1.17E+03	-9.827E-02	-1.562E+00	-1.562E+00	3.0434
65	5.408E-04	-2.8E-05	2.103E+02	1.7E+02	1.1E+02	2.103E+02	-3.2E+01	1.17E+03	1.17E+03	-4.919E-02	-1.550E-01	-1.550E-01	3.0925
66	0.	3.2E-16	0.	0.	0.	0.	-4.9E+01	1.17E+03	1.17E+03	-3.842E-25	0.	0.	3.1416
67	5.408E-04	2.8E-05	-2.103E+02	-1.7E+02	-1.2E+02	-2.103E+02	-6.8E+01	1.17E+03	1.17E+03	4.919E-02	1.550E-01	1.550E-01	3.1907
68	2.162E-03	6.0E-05	-3.803E+02	-3.1E+02	-2.0E+02	-3.833E+02	-8.9E+01	1.16E+03	1.16E+03	9.827E-02	1.562E+00	1.562E+00	3.2398
69	4.861E-03	1.0E-04	-4.998E+02	-3.9E+02	-2.4E+02	-4.998E+02	-1.1E+02	1.16E+03	1.16E+03	1.471E-01	4.074E+00	4.074E+00	3.2889
70	8.633E-03	1.5E-04	-5.825E+02	-4.2E+02	-2.4E+02	-5.825E+02	-1.3E+02	1.15E+03	1.15E+03	1.956E-01	9.766E+00	9.766E+00	3.3379
71	1.347E-02	2.1E-04	-6.409E+02	-4.3E+02	-2.2E+02	-6.409E+02	-1.4E+02	1.15E+03	1.15E+03	2.436E-01	1.744E+01	1.744E+01	3.3870
72	1.937E-02	2.8E-04	-6.835E+02	-4.3E+02	-2.1E+02	-6.835E+02	-1.7E+02	1.14E+03	1.14E+03	2.910E-01	2.723E+01	2.723E+01	3.4361
73	2.631E-02	3.6E-04	-7.152E+02	-4.2E+02	-1.9E+02	-7.152E+02	-1.9E+02	1.13E+03	1.13E+03	3.377E-01	3.917E+01	3.917E+01	3.4852
74	3.429E-02	4.4E-04	-7.388E+02	-3.8E+02	-1.4E+02	-7.388E+02	-2.2E+02	1.11E+03	1.11E+03	3.835E-01	5.322E+01	5.322E+01	3.5343
75	4.328E-02	5.4E-04	-7.560E+02	-3.5E+02	-1.1E+02	-7.560E+02	-2.3E+02	1.10E+03	1.10E+03	4.285E-01	6.924E+01	6.924E+01	3.5834
76	5.328E-02	6.4E-04	-7.678E+02	-3.0E+02	-9.7E+01	-7.678E+02	-2.4E+02	1.08E+03	1.08E+03	4.723E-01	8.705E+01	8.705E+01	3.6325
77	6.427E-02	7.4E-04	-7.751E+02	-3.0E+02	-7.5E+01	-7.751E+02	-2.4E+02	1.06E+03	1.06E+03	5.151E-01	1.064E+02	1.064E+02	3.6816
78	7.623E-02	8.5E-04	-7.784E+02	-2.5E+02	-5.1E+01	-7.784E+02	-2.5E+02	1.04E+03	1.04E+03	5.566E-01	1.271E+02	1.271E+02	3.7306
79	8.912E-02	9.6E-04	-7.785E+02	-2.5E+02	-4.2E+01	-7.785E+02	-2.5E+02	1.02E+03	1.02E+03	5.967E-01	1.489E+02	1.489E+02	3.7797
80	1.029E-01	1.1E-03	-7.758E+02	-2.2E+02	-2.8E+01	-7.758E+02	-2.5E+02	9.95E+02	9.95E+02	6.354E-01	1.716E+02	1.716E+02	3.8288
81	1.177E-01	1.3E-03	-7.710E+02	-1.9E+02	-2.4E+01	-7.710E+02	-2.5E+02	9.71E+02	9.71E+02	6.725E-01	1.951E+02	1.951E+02	3.8779
82	1.332E-01	1.4E-03	-7.643E+02	-1.6E+02	-1.9E+01	-7.643E+02	-2.5E+02	9.48E+02	9.48E+02	7.080E-01	2.190E+02	2.190E+02	3.9270
83	1.496E-01	1.6E-03	-7.563E+02	-1.6E+02	-1.6E+01	-7.563E+02	-2.5E+02	9.23E+02	9.23E+02	7.418E-01	2.433E+02	2.433E+02	3.9761
84	1.669E-01	1.7E-03	-7.473E+02	-1.5E+02	-1.4E+01	-7.473E+02	-2.5E+02	8.99E+02	8.99E+02	7.738E-01	2.679E+02	2.679E+02	4.0252
85	1.849E-01	1.8E-03	-7.377E+02	-1.3E+02	-1.2E+01	-7.377E+02	-2.5E+02	8.74E+02	8.74E+02	8.039E-01	2.926E+02	2.926E+02	4.0743
86	2.036E-01	2.0E-03	-7.276E+02	-1.3E+02	-1.1E+01	-7.276E+02	-2.5E+02	8.49E+02	8.49E+02	8.321E-01	3.175E+02	3.175E+02	4.1233
87	2.230E-01	2.1E-03	-7.172E+02	-1.2E+02	-1.2E+01	-7.172E+02	-2.5E+02	8.24E+02	8.24E+02	8.583E-01	3.422E+02	3.422E+02	4.1724
88	2.432E-01	2.2E-03	-7.069E+02	-1.1E+02	-1.4E+01	-7.069E+02	-2.5E+02	8.00E+02	8.00E+02	8.824E-01	3.670E+02	3.670E+02	4.2215
89	2.639E-01	2.3E-03	-6.966E+02	-1.1E+02	-1.2E+01	-6.966E+02	-2.5E+02	7.75E+02	7.75E+02	9.042E-01	3.916E+02	3.916E+02	4.2706
90	2.852E-01	2.0E-03	-6.866E+02	-1.2E+02	-9.9E+00	-6.866E+02	-2.5E+02	7.50E+02	7.50E+02	9.242E-01	4.161E+02	4.161E+02	4.3197
91	3.071E-01	2.0E-03	-6.769E+02	-9.3E+01	-9.0E+00	-6.769E+02	-2.5E+02	7.26E+02	7.26E+02	9.418E-01	4.405E+02	4.405E+02	4.3688
92	3.294E-01	2.1E-03	-6.676E+02	-9.3E+01	-8.2E+00	-6.676E+02	-2.5E+02	7.02E+02	7.02E+02	9.571E-01	4.646E+02	4.646E+02	4.4179
93	3.522E-01	2.1E-03	-6.587E+02	-8.9E+01	-7.6E+01	-6.587E+02	-2.4E+02	6.78E+02	6.78E+02	9.702E-01	4.886E+02	4.886E+02	4.4670

STAGNATION POINT BETWEEN I= 2 AND I= 3 AT 1.00000

STAGNATION POINT BETWEEN I= 66 AND I= 67 AT 0.00000

TIME = 2.00000E-03   TIME BASED ON CHORD= 5.52630E-04

PRESSURE INTEGRATED FROM T.E. TO L.E.   PRESSURE INTEGRATED FORM L.E. TO T.E

TIME	FD	FL	FM	PD	PL	PM	DRAG	LIFT	MOMENT	PD	PL	PM	DRAG	LIFT	MOMENT	A.D.	A.L.	A.M.
.0003	2.421	-.000	-.000	137.079	.000	-.000	139.500	.000	-.000	137.079	.000	-.000	139.500	.000	-.000	.000	.000	-.000
.0006	1.897	.000	-.000	94.547	-.000	-.000	96.444	.000	-.000	94.547	.000	-.000	96.444	.000	-.000	.000	-.000	-.000

AIRFOHJ.S001028 LINES PRINTED. (REMOTE RL)

REFERENCES

Bar-Lev, M., and Yang, H. T., 1975. *J. Fluid Mech.*, **48**, 33–55.
Belcher, B. J., Burggraf, O. R., Cooke, J. C., Robins, A. J., and Stewartson, K., 1972. In *Recent Research of Unsteady Boundary Layers*, ed. Eichelbrenner, E. A., 1444–1465.
Blasius, H., 1908. *Z. Math. Phys.*, **56**, 1–37.
Bodonyi, R. J., and Stewartson, K., 1977. *J. Fluid Mech.*, **79**, 669–688.
Bradshaw, P., 1979. *AIAA J.*, **17**, 790–793.
Brown, S. N., and Stewartson, K., 1969. In *Annual Review of Fluid Mechanics*, ed. Sears, W. R., **1**, 45–72.
Buckmaster, J., 1973. *J. Eng. Math.*, **7**, 223–230.
Carter, J. E., 1974. AIAA Paper No. 74-583.
Catherall, D., and Mangler, K. W., 1966. *J. Fluid Mech.*, **26**, 163–182.
Cebeci, T., 1978. *AIAA J.*, **16**, 1305–1306.
Cebeci, T., 1979. *J. Comput. Phys.*, **31**, 153–172.
Cebeci, T., and Wilson, W. B., 1972. *J. Basic Eng.*, **94**, 697–698.
Clements, R. R., and Maull, D. J., 1975. *Prog. in Aero. Sci.*, **16**, 129–146.
Collins, W. M., and Dennis, J. C. R., 1973. *J. Fluid Mech.*, **60**, 105–128.
Cousteix, J., Houdeville, R., and Desopper, A., 1977. In *Unsteady Aerodynamics*, AGARD-CP-227.
Crimi, P., and Reeves, B. L., 1972. "A Method for Analyzing Dynamic Stall of Helicopter Rotor Blades," NASA CR-2009; also AIAA Paper No. 72-0037.
Danberg, J. E., and Fansler, K. S., 1974. "An Investigation of the Moore-Rott-Sears Criterion for Laminar Boundary Layer Separation," Tech. Rep. No. 172, University of Delaware, Newark.
Danberg, J. E., and Fansler, K. S., 1975. *AIAA J.*, **13**, 110–112.
Dean, W. R., 1950. *Proc. Cambridge Philos. Soc.*, **46**, 293–306.
Deffenbaugh, F. D., and Marshall, F. J., 1976. *AIAA J.*, **14**, 908–913.
Despard, R. A., and Miller, J. A., 1971. *J. Fluid Mech.*, **47**, 21–31.
Dwyer, H. A., and Sherman, F. R., 1979. AIAA Paper No. 79-1518.
Fage, A., and Johansen, F. C., 1928, *Phil. Mag.*, **7**, 417–436.
Fansler, K. S., and Danberg, J. E., 1977. *AIAA J.*, **15**, 274–276.
Goldstein, S., 1948. *J. Mech. Appl. Math.*, **1**, 43–69.
Goldstein, S., and Rosenhead, L. 1936. *Proc. Cambridge Philos. Soc.*, **32**, 392–401.
Hayasi, N., 1962. *J. Phys. Soc. Jpn.*, **17**, 194–203.
Howarth, L., 1938. *Proc. R. Soc. London*, A **164**, 547–579.
Inger, G. R., 1974. AIAA Paper No. 74-582.
Inger, G. R., and Swean, R. F., 1973. "Vectored Injection into Non-Adiabatic Boundary Layer Flows Including Incipient Separation," Virginia Polytechnic Institute Eng. Rep. VPI-Aero-003.
Klemp, J. B., and Acrivos, A., 1971. *J. Fluid Mech.*, **53**, 177–191.
Klineberg, J. M., and Steger, J. L., 1974. AIAA Paper No. 74-94.
Koromilas, C. A., and Telionis, D. P., 1980. *J. Fluid Mech.*, **97**, 347–384.
Landau, L. D., and Lifshitz, E. M., 1959. *Fluid Mechanics*, Pergamon, New York.
Lin, C. C., 1956. In *Proc. 9th Int. Congr. Appl. Mech.*, Brussels, **4**, pp. 155–169.
Ludwig, G. R., 1964. "An Experimental Investigation of Separation from a Moving Wall," AIAA Paper No. 64-6.
McCroskey, W. J., 1977. *J. Fluids Eng.*, **99**, 8–38.
Mehta, U. B., 1972. "Starting Vortex, Separation Bubbles and Stall—A Numerical Study of Laminar Unsteady Flow Around an Airfoil," Ph.D. Thesis, Illinois Institute of Technology, Chicago.
Mehta, U. B., 1977. "Dynamic Stall of an Oscillating Airfoil," in *Unsteady Aerodynamics*, AGARD Conference Proceedings CP-227, Paper No. 23.
Mehta, U. B., and Lavan, Z., 1975. *J. Fluid Mech.*, **67**, 227–256.
Mezaris, T. B., and Telionis, D. P., 1980. AIAA Paper No. 80-1420.
Moore, F. K., 1958. In *Boundary Layer Research*, ed. Görtler, H., Springer, New York, pp. 296–311.
Nash, J. F., Carr, L. W., and Singleton, R., 1975. *AIAA J.*, **13**, 167–172.
Nenni, J. P., 1976. "An Asymptotic Approach to the Separation of Two-Dimensional Laminar Boundary Layers," Ph.D. Thesis, Cornell Univ., Ithaca, New York.

O'Brien, V., 1975. *AIAA J.*, **13**, 415–416.

Prandtl, L., 1904. "Über Flüssigkeitsbewegung bei sehr kleiner Reibung," *Proc. III Int. Math. Congr.*, Heidelberg, 484–491.

Proudman, I., and Johnson, K., 1962. *J. Fluid Mech.*, **12**, 161–168.

Riley, N., 1975. *SIAM Rev.*, **17**, 274–297.

Rott, N., 1956. *Q. J. Appl. Math.*, **13**, 444–451.

Sarpkaya, T., 1975. *J. Fluid Mech.*, **68**, 109–128.

Sarpkaya, T., 1979. *J. Appl. Mech.*, **46**, 241–258.

Sarpkaya, T., and Schoaff, R. L., 1979. *AIAA J.*, **17**, 1193–1200.

Sears, W. R., 1956. *J. Aero Sci.*, **23**, 490–499.

Sears, W. R., 1975. *AIAA J.*, **14**, 216–220.

Sears, W. R., and Telionis, D. P., 1972. In *Recent Research of Unsteady Boundary Layers*, ed. Eichelbrenner, E. A., **1**, 404–447.

Sears, W. R., and Telionis, D. P., 1975. *SIAM J. Appl. Math.*, **28**, 215–235.

Shen, S. F., 1968. In *Adv. Appl. Mech.*, ed. Yih, C. S., **18**, 177–220.

Shen, S. F., and Nenni, J. P., 1975. In *Unsteady Aerodynamics*, ed. Kinney, R. B., **1**, 245–259.

Stewartson, K., 1970. *J. Fluid Mech.*, **44**, 247–364.

Tani, I., 1958. *Aero-Res. Inst. Univ. Tokyo*, Rep. No. 331.

Telionis, D. P., 1970. "Boundary Layer Separation," Ph.D. Thesis, Cornell Univ., Ithaca, New York.

Telionis, D. P., 1975. In *Unsteady Aerodynamics*, ed. Kinney, R. B., **1**, 155–190.

Telionis, D. P., 1979. *J. Fluids Eng.*, **101**, 29–43.

Telionis, D. P., 1980. "Analytical Methods for Prediction of Laminar Boundary Layers," in *Special Course on Unsteady Aerodynamics* AGARD Report No. 679. AGARD Lecture Series, Brussels.

Telionis, D. P., and Koromilas, C. P., 1978. In *Nonsteady Fluid Dynamics*, eds. Crow, D. E., and Miller, J. A., 21–32.

Telionis, D. P., and Tsahalis, D. Th., 1974a. *Acta Astron.*, **1**, 1487–1505.

Telionis, D. P., and Tsahalis, D. Th., 1974b. *AIAA J.*, **12**, 614–619.

Telionis, D. P., and Werle, M. J., 1972. "Boundary-Layer Separation from Moving Boundaries," Virginia Polytechnic Institute Eng. Rep., VPI-E-72-13.

Telionis, D. P., and Werle, M. J., 1973. *J. Appl. Mech.*, **40**, 369–374.

Telionis, D. P., Tsahalis, D. Th., and Werle, M. J., 1973. *Phys. Fluids*, **16**, 968–973.

Tollmien, W., 1946. Rep. Aeros. Res. Coun., London, No. 9739.

Tsahalis, D. Th., 1977. *AIAA J.*, **15**, 561–566.

Van Dommelen, L. L., and Shen, S.-F., 1977. "The Laminar Boundary Layer in Lagrangian Description," XIII Biennial Fluid Mechanics Symposium, Olsztyn-Kortowo, Poland.

Vidal, R. J., 1959. "Research on Rotating Stall in Axial Flow Compressors; Part III. Experiments on Laminar Separation from a Moving Wall," Wright Air Dev. Cent. Tech. Rep. 59-75.

Wang, K. C., 1979. "Unsteady Boundary Layer Separation," Martin Marietta Report, MML TR 79-16C.

Werle, M. J., and Davis, R. T., 1972. *J. Appl. Mech.*, **39**, 7–12.

Williams, J. C., 1977. In *Annual Review of Fluid Mechanics*, ed. Van Dyke, M., Wehausen, J. V., Lumley, J. L., **9**, 113–144.

Williams, J. C., 1981, private communication.

Williams, J. C., and Johnson, W. D., 1974a. *AIAA J.*, **12**, 1388–1393.

Williams, J. C., and Johnson, W. D., 1974b. *AIAA J.*, **12**, 1427–1429.

Williams, J. C., and Johnson, W. D., 1975. In *Unsteady Aerodynamics*, ed. Kinney, R. B., **1**, 261–282.

Index